Microbial Genomes

Infectious Disease

SERIES EDITOR: *Vassil St. Georgiev*

National Institute of Allergy and Infectious Diseases
National Institutes of Health

Microbial Genomes, edited by *Claire M. Fraser, PhD, Timothy D. Read, PhD, and Karen E. Nelson, PhD,* 2004

Biological Weapons Defense: *Principles and Mechanisms for Infectious Diseases Counter-Bioterrorism,* edited by *Luther E. Lindler, PhD, Frank J. Lebeda, PhD, and George Korch, PhD,* 2004

Management of Multiple Drug-Resistant Infections, edited by *Stephen H. Gillespie, MD,* 2004

Aging, Immunity, and Infection, by *Joseph F. Albright, PhD and Julia W. Albright, PhD,* 2003

Handbook of Cytokines and Chemokines in Infectious Diseases, edited by *Malak Kotb, PhD and Thierry Calandra, MD, PhD,* 2003

Opportunistic Infections: *Treatment and Prophylaxis,* by *Vassil St. Georgiev, PhD,* 2003

Innate Immunity, edited by *R. Alan B. Ezekowitz, MBChB, DPhil, FAAP, and Jules A. Hoffmann, PhD,* 2003

Pathogen Genomics: *Impact on Human Health,* edited by *Karen Joy Shaw, PhD,* 2002

Immunotherapy for Infectious Diseases, edited by *Jeffrey M. Jacobson, MD,* 2002

Retroviral Immunology: *Immune Response and Restoration,* edited by *Giuseppe Pantaleo, MD and Bruce D. Walker, MD,* 2001

Antimalarial Chemotherapy: *Mechanisms of Action, Resistance, and New Directions in Drug Discovery,* edited by *Philip J. Rosenthal, MD,* 2001

Drug Interactions in Infectious Diseases, edited by *Stephen C. Piscitelli, PharmD and Keith A. Rodvold, PharmD,* 2001

Management of Antimicrobials in Infectious Diseases: *Impact of Antibiotic Resistance,* edited by *Arch G. Mainous III, PhD and Claire Pomeroy, MD,* 2001

Infectious Disease in the Aging: *A Clinical Handbook,* edited by *Thomas T. Yoshikawa, MD and Dean C. Norman, MD,* 2001

Infectious Causes of Cancer: *Targets for Intervention,* edited by *James J. Goedert, MD,* 2000

Infectious Disease

Microbial Genomes

Edited by

Claire M. Fraser, PhD

The Institute for Genomic Research, Rockville, MD

Timothy D. Read, PhD

Biological Defense Research Directorate,
Naval Medical Research Center, Silver Spring, MD

Karen E. Nelson, PhD

The Institute for Genomic Research, Rockville, MD

Foreword by

J. Craig Venter, PhD

The Center for the Advancement of Genomics, Rockville, MD

HUMANA PRESS ✳ TOTOWA, NEW JERSEY

© 2004 Humana Press Inc.
999 Riverview Drive, Suite 208
Totowa, New Jersey 07512

This publication is printed on acid-free paper. ∞

ANSI Z39.48-1984 (American Standards Institute)

Permanence of Paper for Printed Library Materials.

Production Editor: Robin B. Weisberg.

Cover design by Emmanuel Mongodin and Patricia F. Cleary.

Cover illustrations: The cover collage symbolizes the principal steps that occur during a microbial genome sequencing project. From bottom left to upper right : electron micrograph of the bacterium *Thermotoga maritima* (kindly provided by Drs. Reinhard Rachel and Karl O. Stetter), double helix symbolizing the DNA extraction, electropherogram of the sequencing reactions, gene annotation process, metabolic pathway reconstruction based on the genomic data and gene expression assay by microarrays. The circle in the background is the representation of a whole bacterial genome, with its gene annotation and its remarkable features.

Thermotoga maritima electron micrograph is reprinted with permission from Drs. Reinhard Rachel and Karl O. Stetter.

Pseudomonas putida metabolism figure, as appeared in Environ Microbiol. 2002 Dec;4(12):799-808. Complete genome sequence and comparative analysis of the metabolically versatile *Pseudomonas putida* KT2440. Nelson KE et al., is reprinted with permission from Blackwell Publishing.

For additional copies, pricing for bulk purchases, and/or information about other Humana titles, contact Humana at the above address or at any of the following numbers: Tel: 973-256-1699; Fax: 973-256-8341; E-mail: humana@humanapr.com, or visit our Website: www.humanapress.com

Printed in the United States of America. 10 9 8 7 6 5 4 3 2 1

E-ISBN 1-59259-756-4

Library of Congress Cataloging-in-Publication Data

Microbial genomes / edited by Claire M. Fraser, Timothy D. Read, and Karen E. Nelson.
    p. ; cm. -- (Infectious disease)
  Includes bibliographical references and index.
  ISBN 1-58829-189-8 (alk. paper)
   1. Bacterial genomes. I. Fraser, Claire M. II. Read, Timothy D. III. Nelson, Karen E.
IV. Series: Infectious disease
 (Totowa, N.J.)
  [DNLM: 1. Bacteria--genetics. 2. Genomics. 3. Computational Biology.
4. Genetic Techniques. 5. Genetics, Microbial. QW 51 M6267 2004]
  QH434.M527 2004
  572.8'6293--dc22

2003025478

Foreword

Nearly 10 years have passed since the first whole genome shotgun sequencing experiment was undertaken at The Institute for Genomic Research (TIGR) on *Haemophilus influenzae (1)*. The successful completion of that experiment in early 1995 ushered in a new era in science. In 2004, with more than 150 microbial genomes completed and hundreds more underway, the field of microbial genomics is approaching early maturity. Microbial genome projects have been transformed from multiyear to decade-long projects involving entire laboratories or major teams, to projects that can be managed by graduate students or postdoctoral fellows over weeks to months in conjunction with genomic support facilities. Sequencing the *H. influenzae* genome required 4 months in the early TIGR facility, which represented a substantial reduction from the concurrent decade-long microbial genome sequencing efforts. Today in the J. Craig Venter Science Foundation Joint Technology Center (JTC), a dedicated high-throughput DNA-sequencing center that serves TIGR, the Institute for Biological Energy Alternatives (IBEA), and The Center for the Advancement of Genomics (TCAG), the *H. influenzae* genome sequencing would require less than 25% of a single day's sequencing output and could be completed in 4 hours. As DNA-sequencing technology continues its exponential growth, it is likely that within a decade sophisticated DNA-sequencing centers will be able to decode thousands of genomes per day, while small academic laboratories will be able to sequence several microbial genomes daily as part of standard analytical procedures. With the application of genomic technologies to the environment, literally millions of new genes are being discovered from single environmental locations *(2)*. These new approaches, together with the continual expansion of microbial genome sequencing, will move us closer to understanding the more than 10 billion genes that I have estimated to comprise the Earth's gene pool.

Looking ahead, it becomes clear that the principal challenges in microbial genomics are associated with the understanding of genomic structures, evolution, and biology. *Microbial Genomes*, edited by Claire Fraser, Timothy Read, and Karen Nelson, represents not only the most comprehensive attempt to date to bring together in one volume many of the major contributors to modern microbial genomics, but also a broad view of the contributions that sequencing genomes has had on our understanding of microbial metabolism and evolution. This volume also appropriately places considerable emphasis on the bioinformatics tools that are essential for every step in genomic sequencing and analysis. Bioinformatics and computational methods will need to undergo substantial development if the scientific community has any hope of coping with the tsunami of gene and genomic data.

Advances in the post-*H. influenzae* whole genome sequencing era, what I have termed as the genomic era, have led to many practical applications, discoveries, and new approaches to pharmaceutical and vaccine development, microbial forensics, industrial chemistry, and bioremediation. Just a few years ago, ideas and concepts that sounded like science fiction are today being rigorously pursued. For example, combining the heavy metal metabolism pathways from the *Shewanella* genome *(3)* with the genome from the radiation resistant *Deinococcus radiodurans (4)*, the US Department of Energy hopes to have a radiation-resistant species that can metabolize uranium. Other government agencies are using the polymorphic variation that occurs in different microbial isolates to track down the laboratory source of a bioterrorism-associated microbe or even the geographic origin and movement of individuals. Several successful companies are using microbial genomics as a source of new industrial enzymes and chemistries, an area likely to explode with new possibilities as the new emerging field of synthetic genomics grows *(5)*. The challenges are many, but the opportunities are even greater for microbial genomics to impact almost every aspect of our lives. *Microbial Genomes* provides a great reference source that will be useful for years to come in this rapidly moving field.

J. Craig Venter, PhD

REFERENCES

1. Fleischmann RD, Adams MD, White O, et al. Whole-genome random sequencing and assembly of *Haemophilus influenzae* Rd. Science 1995; 269:496–512.
2. Venter JC, Remington K, Heidelberg JF, et al. Environmental genome shotgun sequencing of the Sargasso Sea. Science in press.
3. Heidelberg JF, Paulsen IT, Nelson KE, et al. Genome sequence of the dissimilatory metal ion-reducing bacterium *Shewanella oneidensis*. Nat. Biotechnol 2002; 20:1093–1094.
4. White O, Eisen JA, Heidelberg JF, et al. Genome sequence of the radioresistant bacterium *Deinococcus radiodurans* R1. Science 1999; 286:1571–1577.
5. Smith HO, Hutchison CA III, Pfannkoch C, Venter JC. Generating a synthetic genome by whole genome assembly: phiX174 bacteriophage from synthetic oligonucleotides. Proc Natl Acad Sci USA 2003; 100:15440–15445.

Preface

Over the last 10 years, the field of microbiology has been transformed by the ability to conduct whole genome sequencing with The Institute for Genomic Research (TIGR) located in Rockville, Maryland playing a central role. TIGR scientists published the first three complete microbial genome sequences—*Haemophilus influenzae, Mycoplasma genitalium,* and *Methanococcus jannaschii*—and have subsequently contributed more than 40 of the more than 150 complete and published genomes that are available as of January 2004. It was therefore an exciting opportunity when Humana Press proposed a book that surveyed the current state of the field. In choosing authors for the chapters, we turned in many cases to the expertise of our colleagues at TIGR. The external contributors represent a diverse range of experience spread over the globe.

The field of microbial genomics is vast. In selecting themes for chapters, however, we quickly realized that we could either opt for broad coverage at a relatively introductory level or instead focus in depth on topics we considered of critical importance. We chose the former approach. We hope *Microbial Genomes* will convey the great expanse of the subject and that it will be of interest both to readers new to the field, as well as to those with specialized knowledge. As Editors, it was interesting to find that despite very disparate titles, certain recurring themes—horizontal gene transfer, the importance of comparative genomic analysis, and microarray-based gene expression analysis—were threaded throughout the volume.

Microbial Genomes is divided into six major parts. Believing it was important to have a section devoted to the history of microbial genomics so as to allow readers to understand how the field developed, the editors commissioned Hamilton Smith to prepare the Introduction. Bioinformatics As a Tool in Genomics describes some of the most common computational tools for genomics and their application. Core Functions deals with metabolism, transporters and cell cycle processes that are found in every microbial genome. The Evolution of Microbial Genomes is a series of chapters that aim to show how genomics can be used to reconstruct the history and dynamism of the microbial world. In A Survey of Microbial Genomes we have organized a series of chapters that deal with selected groups of organisms. Although not every microbial genome is covered (and several are described in multiple chapters), the aim is to provide an indication of the biological information that can be extracted by genomic studies. Finally, Applications of Genomic Data describes how the genome sequences are being used to tackle the most important issues in microbiology. It should also be noted that many genomes listed as unfinished when the chapters were written in late 2002/early 2003 have now been completed. Interested readers should consult a website such as the TIGR microbial database (www.tigr.org/tdb/mdb/mdbcomplete.html) for the latest details.

The successful publication of this book would not have been possible without the help of Trina Eacho. Trina devoted many weekends to the organization of chapters and to the formatting of manuscripts. The editors gratefully recognize her significant contribution.

We realize of course, that this book has some omissions—not every important story could be covered. We hope readers will find inspiration from the diverse chapters and authors that have been assembled. In its truest sense, microbial genomics, consisting of evolutionary and population biology, gene expression analysis, proteomics and all the other studies that begin from the knowledge of DNA sequence, encompasses the very breadth and scope of the science of microbiology itself.

Claire M. Fraser, PhD
Timothy D. Read, PhD
Karen E. Nelson, PhD

Contents

Contributors

SIV G. E. ANDERSSON, PhD • *Department of Molecular Evolution, Uppsala University, Uppsala, Sweden*

STEPHEN D. BENTLEY, PhD • *The Pathogen Group, The Sanger Centre, Wellcome Trust Genome Campus, Hinxton, Cambridge, UK*

SVEND BIRKELUND, MD, PhD, DMSc • *Department of Medical Microbiology and Immunology, University of Aarhus, Aarhus C, Denmark*

FIONA S. L. BRINKMAN, PhD • *Department of Molecular Biology and Biochemistry, Simon Fraser University, Burnaby, British Columbia, Canada*

ROLAND BROSCH, PhD • *Unite de Genetique Moleculaire Bacterienne, Institut Pasteur, Paris, France*

C. ROBIN BUELL, PhD • *Department of Plant Genomics, The Institute for Genomic Research, Rockville, MD*

GUNNA CHRISTIANSEN, MD, DMSc • *Department of Medical Microbiology and Immunology, University of Aarhus, Aarhus C, Denmark*

FREDERICK M. COHAN, PhD • *Biology Department, Wesleyan University, Middletown, CT*

STEWART T. COLE, PhD • *Unite de Genetique Moleculaire Bacterienne, Institut Pasteur, Paris, France*

SHILADITYA DASSARMA, PhD • *Center of Marine Biotechnology, University of Maryland Biotechnology Institute, Baltimore, MD*

ARTHUR L. DELCHER, PhD • *Department of Bioinformatics, The Institute for Genomic Research, Rockville, MD*

EDWARD F. DELONG, PhD • *Monterey Bay Aquarium Research Institute, Moss Landing, CA*

ROBERT A. FELDMAN, PhD • *SymBio Corporation, Menlo Park, CA*

DERRICK E. FOUTS, PhD • *Department of Microbial Genomics, The Institute for Genomic Research, Rockville, MD*

JOANNA L. FUEYO, PhD • *IBM Life Science Solutions Consulting, Cambridge, MA*

MALCOLM J. GARDNER, PhD • *Department of Parasite Genomics, The Institute for Genomic Research, Rockville, MD*

STEVEN R. GILL, PhD • *Department of Microbial Genomics, The Institute for Genomic Research, Rockville, MD*

J. PETER GOGARTEN, PhD • *Department of Molecular and Cell Biology, University of Connecticut, Storrs, CT*

STEPHEN V. GORDON, PhD • *Veterinary Laboratories Agency, Addlestone, Surrey, UK*

GUIDO GRANDI, PhD • *Chiron Corporation, Siena, Italy*

MARK E. HANCE • *Department of Microbial Genomics, The Institute for Genomic Research, Rockville, MD*

THOMAS E. HANSON, PhD • *Department of Microbiology and the Plant Molecular Biology/Biotechnology Program, The Ohio State University, Columbus, OH*

DAVID A. HOPWOOD, PhD • *Department of Molecular Microbiology, John Innes Centre, Norwich Research Park, Colney, Norwich, UK*

KATHERINE H. KANG • *Department of Microbial Genomics, The Institute for Genomic Research, Rockville, MD*

ROBERT M. KELLY, PhD • *Department of Chemical Engineering, North Carolina State University, Raleigh, NC*

MICHAEL T. LAUB, PhD • *Bauer Center for Genomics Research, Harvard University, Cambridge, MA*

VEGA MASIGNANI, PhD • *Chiron Corporation, Siena, Italy*

HARLEY H. MCADAMS, PhD • *Department of Developmental Biology, Stanford University, Palo Alto, CA*

GARRY S. A. MYERS, PhD • *Department of Microbial Genomics, The Institute for Genomic Research, Rockville, MD*

KAREN E. NELSON, PhD • *Department of Microbial Genomics, The Institute for Genomic Research, Rockville, MD*

LORRAINE OLENDZENSKI • *Department of Molecular and Cell Biology, University of Connecticut, Storrs, CT*

JULIAN PARKHILL, PhD • *The Wellcome Trust Sanger Institute, Wellcome Trust Genome Campus, Hinxton, Cambridge, UK*

IAN T. PAULSEN, PhD • *Department of Microbial Genomics, The Institute for Genomic Research, Rockville, MD*

MARIAGRAZIA PIZZA, PhD • *Chiron Corporation, Siena, Italy*

RINO RAPPUOLI, PhD • *Chiron Corporation, Siena, Italy*

JACQUES RAVEL, PhD • *Department of Microbial Genomics, The Institute for Genomic Research, Rockville, MD*

TIMOTHY D. READ, PhD • *Biological Defense Research Directorate, Naval Medical Research Center, Silver Spring, MD*

QINGHU REN, PhD • *Department of Microbial Genomics, The Institute for Genomic Research, Rockville, MD*

FRANK T. ROBB, PhD • *Center of Marine Biotechnology, University of Maryland Biotechnology Institute, Baltimore, MD*

CARSTEN ROSENOW, PhD • *Affymetrix, Santa Clara, CA*

STEVEN L. SALZBERG, PhD • *The Institute for Genomic Research, Rockville, MD and Departments of Computer Science and Biology, Johns Hopkins University, Baltimore, MD*

LUCY SHAPIRO, PhD • *Department of Developmental Biology, Stanford University, Palo Alto, CA*

ALLAN C. SHAW, PhD • *Department of Medical Microbiology and Immunology, University of Aarhus, Aarhus C, Denmark and Novo Nordisk Research and Development Center China, Beijing, China*

KEITH R. SHOCKLEY, PhD • *Department of Chemical Engineering, North Carolina State University, Raleigh, NC*

HAMILTON O. SMITH, PhD • *Institute for Biological Energy Alternatives, Rockville, MD*

F. ROBERT TABITA, PhD • *Department of Microbiology and the Plant Molecular Biology/Biotechnology Program, The Ohio State University, Columbus, OH*

JOHN L. TELFORD, PhD • *Chiron Corporation, Siena, Italy*

NICHOLAS R. THOMSON, PhD • *The Wellcome Trust Sanger Institute, Wellcome Trust Genome Campus, Hinxton, Cambridge, UK*

BRIAN TJADEN, PhD • *Department of Computer Science, Wellesley College, Wellesley, MA*

BRIAN B. VANDAHL, MSc • *Department of Medical Microbiology and Immunology, University of Aarhus, and Loke Diagnostics ApS, Aarhus C, Denmark*

J. CRAIG VENTER, PhD • *The Center for the Advancement of Genomics, Rockville, MD*

LAWRENCE P. WACKETT, PhD • *Department of Biochemistry, Molecular Biology an Biophysics, University of Minnesota, St. Paul, MN*

OWEN WHITE, PhD • *The Institute for Genomic Research, Rockville, MD*

OLGA ZHAXYBAYEVA • *Department of Molecular and Cell Biology, University of Connecticut, Storrs, CT*

I INTRODUCTION

History of Microbial Genomics

Hamilton O. Smith

INTRODUCTION

Genomics[1] is a new field of science that analyzes and compares the complete genome sequences of organisms. Sequence provides the most fundamental information about an organism. The genes and regulatory sites encoded in the sequence specify the "parts list" and "operating instructions" for the organism and yields clues to its evolution. Sequence is now considered the natural starting point for the study of many new organisms. The development of genomics is one of the most dramatic results of the enormous advances in medicine and biology in the 20th century; emerging in the final decade, genomics is providing a solid foundation for medicine and biology in the 21st century.

My focus is on microbial genomics. Its emergence as a new field of study is closely linked to the Human Genome Project (HGP). There is little doubt that without the HGP, the sequencing of the first microbes would have been significantly delayed. Two of the pilot projects proposed in the initial 5-year plan of the HGP were to sequence the model laboratory microbes *Escherichia coli* and *Saccharomyces cerevisiae*. These were the earliest microbial projects, beginning around 1990.

It is thus necessary to review briefly the history of the HGP. In addition, the state of knowledge before the HGP is examined. Genomics and the HGP did not arise from a vacuum. A huge amount of biological research and discovery took place throughout the 20th century that provided the fundamental knowledge and technology to enable the HGP. Without knowledge of the chemical nature of deoxyribonucleic acid (DNA) and the invention of DNA sequencing, there could be no modern genomics. This chapter therefore includes a description of the events and discoveries that set the stage for modern genomics. Also included are descriptions of some of the landmark sequencing projects that established microbial genomics as a prominent new field of scientific investigation.

This brief historical account concentrates mainly on the sequencing aspects of microbial genomics. A postsequencing phase that focuses on the analysis of sequence information is rapidly evolving, but is perhaps too current to be a suitable subject for historical treatment. In addition, much of this book focuses on genomic analysis and the various methods for extraction of biological information from genomic sequences.

---

[1]The word, *genomics* is of recent origin. Tom Roderick of the Jackson Laboratories proposed it in 1986 as the name for the new field of science that deals with whole genome sequences and related high-throughput technologies *(10)*. In 1987, *Genomics* became the name of a new journal founded by Victor McKusick.

From: *Microbial Genomes*
Edited by: C. M. Fraser, T. D. Read, and K. E. Nelson © Humana Press Inc., Totowa, NJ

In relating historical events, I cannot help but be biased to a significant degree by my experience. It is natural to provide greater detail and emphasis to those events person-ally witnessed and that can be described from memory. Finally, I apologize for the fact that this brief history is not more comprehensive, and that it does not treat the historical contributions of all participants in a particularly fair or balanced way.

GENES, CHROMOSOMES, GENOMES, BACTERIOPHAGE, BACTERIA, AND DNA

The word *genome* was introduced into the scientific vocabulary by Winkler in 1920 as a conjunction between GENe and chromosOME *(2)*, and it stood for the complete hap-loid set of chromosomes and genes. However, genes were only superficially understood. They were the units of inheritance that determined the visible or measurable characteris-tics of plants and animals. Their chemical nature was totally unknown. Classical genetic studies of plants, flies, and humans could not yield the answer. Instead, it was the study of very simple organisms, bacteria and bacteriophages, from the 1930s to the 1950s, that led to fundamental discoveries of the true nature of chromosomes, genes, and DNA.

It is interesting that bacteria initially were thought to be different from other organisms because they did not have nuclei or chromosomes of the sort seen in higher organisms. They were able to adapt rapidly to new environments by what appeared to be nongene-tic mechanisms. In 1943, Luria and Delbruck used fluctuation analysis to show that adaptation was simply genetic selection of previously existing mutations in bacterial populations *(3)*. They concluded that bacteria possessed inherited characteristics and genes like other organisms. In 1944, Avery and colleagues *(4)* at the Rockefeller Institute proved that DNA was the carrier of heredity in the genetic transformation of bacteria.

In 1946, Lederberg and Tatum *(5)* discovered that mixing two strains of *E. coli* carry-ing different mutations resulted in new recombinant strains. Subsequently, in experiments conducted in several laboratories over a decade or so, it was found that only certain strains were fertile, and that male strains sequentially transferred part of their chromo-some to female strains during a mating event *(6)*. The time of entry of genetic markers allowed the construction of genetic maps of *E. coli*, and as more and more markers were mapped, the maps were eventually shown to be circular.

Meanwhile, in 1953, Watson and Crick deduced from X-ray diffraction photographs that DNA was a double helix of two antiparallel DNA strands *(7)*. They suggested that the sequence of the bases along the helix carried the genetic information. They further suggested that DNA replicated semiconservatively by copying new daughter strands from the complementary parental strands.

In 1963, Cairns labeled *E. coli* DNA with tritiated thymidine *(8)*. Cairns showed by autoradiography that the *E. coli* genome was a single circular DNA molecule that repli-cated semiconservatively.

These experiments firmly established that bacteria are genetic organisms with DNA chromosomes.

By 1990, at the dawn of the genomics era, genetic maps of *E. coli (9)*, *Salmonella typhimurium (10)*, and *Bacillus subtilis (11)* were quite detailed and contained hundreds of mapped genes. A low-resolution comparative genomics was even possible with these maps. It was scarcely evident at that time that the era of recombination-based maps would soon be finished. In the mid-1990s, the determination of the complete sequences of these bacteria largely supplanted the achievements of the prior decades of laborious genetic mapping.

INVENTION OF DNA SEQUENCING

There could be no real microbial genomics without DNA sequencing. It might be argued that genomics, the study of an organism's genome as a whole, arose with genetic mapping and restriction mapping. But, the ability to look at the complete genetic information contained in a haploid set of chromosomes (i.e., to examine all of the genes of the organism) and to compare them with other organisms—the essence of genomics—is not possible with the relatively restricted information gained from such "crude" methods. Only by knowing the complete assembled sequence of the genome can these qualities be examined. Therefore, genomics as a full-fledged science could not have pre-dated the invention of DNA sequencing.

In June 1975, at a Gordon Conference in New England, an important event took place. Two separate groups dramatically announced the invention of DNA sequencing, and each had invented a different method. Even though neither method was perfected at the time of the meeting, many realized that a new era of DNA sequencing was about to begin.

Both methods started with DNA molecules with a fixed 5' end. The methods differed in the way base-specific cleavage or termination was achieved. DNA species terminating in each of the four different bases were separated in stepladder fashion in separate lanes by polyacrylamide gel electrophoresis. Initially, Maxam *(12)* presented a base-specific chemical cleavage procedure such that A and G could be read in one lane of an electrophoretic gel and C and T in the other lane. The differing intensities of the A and G bands and of the C and T bands allowed discrimination of the four bases in a stepladder fashion. In early 1976, Maxam and Gilbert's method was perfected into a four-lane procedure such that positions of each of the bases in the sequence were displayed in separate lanes, but they did not publish the detailed procedure until 1980. Sanger's group presented an eight-lane, so-called ±, enzymatic method utilizing DNA polymerase. In 1977, Sanger and coworkers improved this enzymatic method using dideoxy chain-terminating nucleotides, one for each base *(13)*. With the introduction of ultrathin gels by Sanger and Coulson in 1978 *(14)*, it became possible to read sequences of several hundred bases on a single sequencing gel.

INVENTION OF AUTOMATED SEQUENCING MACHINES

In the late 1970s and early 1980s, academic laboratories began routinely sequencing several kilobases of DNA using either method. However, Maxam and Gilbert's chemical sequencing method proved to be more difficult and soon gave way to the enzymatic method of Sanger and coworkers. The φX174 bacteriophage genome (5386 bp) was the first complete genome to be sequenced in 1977 by Sanger's group *(15)*. The bacteriophage lambda genome (48,502 bp) was sequenced in 1982 by Sanger's laboratory and was the largest sequencing project up to that time *(16)*. Sequencing was labor intensive; it required labeling with radioactive nucleotides, autoradiography, and laborious reading of films, an inherently error-prone process. All of this would change in the next half dozen years, and sequencing would become faster and more automatic.

Around 1985, Hood and Smith at the California Institute of Technology in Pasadena showed that four different fluorescent nucleotide labels could be incorporated into DNA, enabling an automated laser readout of bases on a sequencing gel apparatus. They announced the first automated DNA sequencing machine in June 1986 *(17)*. In late 1987, Applied Biosystems Inc., using Hood's technology, marketed the first automated machine. Each machine was able to produce 10,000 to 20,000 bp of raw sequence data per day. By comparison, today's capillary sequencers, such as the ABI Prism 3700 (Applied Biosystems, Foster City, CA), produce greater than 0.5 million bp of raw sequence data per day per machine.

THE HUMAN GENOME PROJECT

The history of microbial genomics is inseparably linked with the US Department of Energy (DOE) and the HGP. It is hard to imagine that microbial genomics would be as advanced if there had been no HGP. The HGP seized people's imaginations, provided the funds, and stimulated and motivated scientists as never before to develop tools and strategies for sequencing.

In March 1986, Charles DeLisi and David Smith of the Office of Health and Environmental Research at DOE, hosted a meeting in Sante Fe, New Mexico, of about 30 scientists to discuss the feasibility of sequencing the human genome. I had the good fortune to be present and recall being surprised that there was almost unanimous enthusiasm for the project even though sequencing was not advanced enough at the time—automated sequencers were still a year away—to ensure a rational undertaking of such a monumental task. Discussion ranged primarily around tactics and cost. Various strategies were discussed, including yeast artificial chromosome, phage, and cosmid maps; random shotgun sequencing, and cDNAs. The map-based approach seemed to win the most support. A large number of yeast artificial chromosome and cosmid clones would be mapped to form overlapping coverage of the human genome. Then, the individual clones would be sequenced. A realistic cost was considered to be $1 per finished base or $3 billion for the project.

Almost simultaneous with the meeting, Dulbecco published a commentary in *Science* *(18)* strongly supporting and promoting a human genome project. In September 1986, to get the program going, DeLisi relocated $5.3 million at the DOE for pilot studies at the DOE national laboratories. A DOE advisory committee was set up in 1987; it recommended $1 billion for mapping and sequencing over 7 years and suggested that the DOE should lead the US effort. With momentum building, the National Research Council endorsed the HGP in 1988 and called for a phased approach with a scale-up to $200 million a year.

That same year, James Wyngaarden, then director of the National Institutes of Health (NIH), decided that the HGP was health related, and that the NIH should be the major player. In effect, the NIH seized the lead, somewhat belatedly, from the DOE. James Watson was selected to head the newly formed Office of Human Genome Research, and funds were appropriated to initiate the project. A Memorandum of Understanding for cooperation between the DOE and NIH was drawn up, with the NIH playing the leadership role.

In 1990, the DOE and NIH presented a joint 5-year plan to Congress and outlined a 15-year HGP. October 1, 1990, marked the official beginning of the HGP. The importance of these events to the history of microbial genomics became apparent in the initial 5-year plan, which outlined plans to sequence several model organisms, including the two most widely studied microorganisms, *E. coli* and yeast.

The E. coli *Genome: The Story of One Man's Dedication*

In 1983, Blattner was the first to suggest sequencing the *E. coli* genome *(19)*. *Escherichia coli* was the single most important bacterium and was used by thousands of laboratories for research into the fundamental processes of living systems and as a host for recombinant DNA work. Blattner, a faculty member of the Laboratory of Genetics at the University of Wisconsin–Madison, has devoted his career to studies of *E. coli* and phage lambda. He developed the widely used Charon phages in 1977 as safe vectors for cloning *(20)*.

As early as 1988, Blattner began constructing a set of minimally overlapping 15- to 20-kb phage lambda clones to be used for sequencing the genome of *E. coli*. He obtained an HGP center grant to begin sequencing in 1990. His initial strategy was to sequence overlapping clones spanning sections of a few hundred kilobases. From 1992 to 1995, a total of 1.92 Mb (positions 2,686,777 to 4,639,221 on the genome) was completed using radioactive manual sequencing.

In 1995, Blattner came under increasing pressure to speed up the sequencing, particularly after whole genome shotgun sequencing proved to be far more efficient (*see* below). He was forced to defend himself in a competitive renewal of his grant. In his successful renewal application, he adopted a new strategy that guaranteed completion in about a year.

The new strategy abandoned the phage clones and used I-SceI, a rare-cutting restriction enzyme, fragments of about 250 kb each; these covered most of the remaining genome. These fragments were shotgun sequenced using automated sequencers. Blattner finished on schedule and published the *E. coli* genome sequence in September 1997 *(21)*, an event long awaited and of great importance to both the *E. coli* research community and the countless scientists that use the bacterium as a molecular genetics workhorse in their laboratories.

The Yeast Genome: A Successful International Collaboration

The *S. cerevisiae* sequence, the other model system pilot project for the HGP started in the late 1980s and was completed in 1996 *(22)*. It was the largest microbial project of its time at 12 Mb and 6000 genes distributed on 16 chromosomes. The sequencing of yeast has been called the largest decentralized experiment in modern molecular biology *(23)*. More than 600 scientists contributed from over 100 US, Canadian, European, and Japanese laboratories. The yeast community, under the leadership of Andre Goffeau, Professor Extraordinaire at the Universite Catholique de Louvain in Belgium, served as a model for efficiency, cooperation, and organization. That cooperation has extended into the postsequencing phase as well, with many laboratories working together to determine the function and expression patterns of the genes.

Haemophilus influenzae Rd: *The First Completed Microbial Genome*

Haemophilus influenzae was never selected or even considered as a model organism, and its examination started years later than *E. coli* and yeast. How it became the first microbial genome completed is illuminating.

Sol Goodgal at the University of Pennsylvania in Philadelphia was the first to propose sequencing *H. influenzae* Rd. He submitted a grant to the HGP in 1988 in which he described a random transposon insertion strategy, but the grant was not funded because of concerns regarding the feasibility of the method. In the summer of 1993 at Radcliffe Hospital in Oxford, England, Richard Moxon, a long-time investigator of virulence determinants in serotype b strains of *H. influenzae*, visited me at Johns Hopkins in Baltimore, Maryland, with a proposal to sequence *H. influenzae*. We had worked together for several years while he was on the Johns Hopkins medical faculty. He proposed that our laboratories jointly sequence *H. influenzae*. As I recall, I had little enthusiasm for the idea. I argued that the project was too big for our academic laboratories and predicted that funding would be virtually impossible to obtain. In retrospect, that judgment was probably correct.

That same summer, I had the good fortune to meet J. Craig Venter (Fig. 1) at a Genome Ethics meeting in Bilbao, Spain. He invited me to join the Scientific Advisory Council of The Institute for Genomic Research (TIGR). Venter founded TIGR in June 1992 to generate expressed sequence tags (ESTs) as a means of rapidly discovering human genes. He had pioneered the method, which involved random sequencing of cDNA clones derived from many different tissues.

The first TIGR Scientific Advisory Council meeting was held in September 1993. In discussions at that meeting, I learned that the EST work would be nearing completion in a few months. TIGR had about 30 model 373 ABI sequencing machines and was producing about 400,000 bp of sequence reads per day (Fig. 1). Out of the blue, I asked Venter if he would be interested in sequencing the *H. influenzae* Rd bacterial genome. My laboratory at Hopkins had determined the genome size was approx 1.9 Mbp by pulse field gel electrophoresis, and the 40% G+C base composition was favorable for sequencing. Venter appeared very interested. I volunteered to construct and map library clones to serve as substrate for sequencing.

On returning to Hopkins, I informed my laboratory group of the opportunity. To my surprise, I was met with skepticism. Everyone was busy with their own projects, it would take at least a year to generate the clone map, there was not enough money in the laboratory budget to do it, and it would take months to obtain grant funding. A few weeks later, Venter asked me how I was progressing with the libraries. I admitted that I could not construct and map clones as originally proposed and suggested that we should discuss another strategy.

It was apparent to me that TIGR was best at random high-throughput sequencing. Why not make a single library of small insert clones and simply generate random sequence reads (or tags) and assemble them by computer? A program had already been developed at TIGR for assembling the tens of thousands of ESTs into consensus cDNAs. The EST strategy could be readily adapted to the assembly of a bacterial genome.

The random whole genome shotgun strategy was presented to the entire TIGR staff in November 1993 and generated overwhelming enthusiasm. An NIH grant was submitted, but was turned down because of questions regarding feasibility of the random shotgun-and-assembly approach. Venter, however, was totally confident that the strategy would work. He decided to use TIGR endowment funds for the project and gave the go-ahead.

A library of 1.8 kb *H. influenzae* Rd DNA fragments was constructed at Hopkins and tested at TIGR in early 1994. Robert Fleischman, a staff scientist at TIGR (Fig. 1), was

Fig. 1. Top, left: Claire M. Fraser, president of the Institute for Genomic Research (TIGR), Rockville, MD. Top, right: J. Craig Venter, founder and chairman of the board of TIGR. Middle, left: some members of the team that sequenced *Haemophilus influenzae* at TIGR in 1994; right, foreground: Hamilton Smith standing next to J. Craig Venter. Robert Fleischman is pointing to the computer screen. Middle, right: Owen White, director of informatics at TIGR. Bottom: the sequencing laboratory at TIGR as it existed in 1995.

appointed as the team leader. Sequencing of the random clone library was initiated at TIGR in April 1994 and completed 4 months later. This "draft" sequence consisted of 24,000 sequence reads, averaging 500 bp in length, plus the sequence from both ends of about 300 phage lambda clones with 16-kb to 20-kb inserts. The assembler software was developed by TIGR's Granger Sutton and was tested and perfected during the sequencing phase. This amounted to about an average of 6 sequencing reads per base of the 1.8-Mb genome (6× coverage in genomics jargon). The few dozen short gaps in the original assembly were bridged primarily by polymerase chain reaction using flanking primers in known sequence to either side of the gaps. The sequence was completed and annotated in just 1 year *(24)*.

THE DORMY HOUSE WORKSHOP AND *MYCOPLASMA GENITALIUM*

In early 1995, with Venter's encouragement and approval, Moxon and I organized a small workshop for the purpose of presenting the first bacterial genome sequence. It was considered a potentially historic event and was to be held in Dormy House, Worcestershire, England, in April 1995, with the Wellcome Trust as host. As the date drew near, there was speculation at the Wellcome Trust that the *H. influenzae* sequence was not complete, that Venter would not deliver, or that Venter would not allow public access to the data.

There were, in fact, legitimate reasons to doubt Venter's ability to release the data. TIGR was a not-for-profit research institute, but it was affiliated with Human Genome Sciences (Gaithersburg, MD), a biotech company that owned all of TIGR's data. In the weeks following the completion of the *H. influenzae* project, Venter went to great lengths to bargain with William Haseltine, president of Human Genome Sciences, Rockville, MD, for the right to publish the *H. influenzae* genome. His bargaining paid off, and he was able to deliver the *H. influenzae* sequence for the meeting. However, the clash with Haseltine eventually meant giving up $35 million of TIGR's endowment. Venter wanted independence and the freedom to publish research data produced at TIGR.

It was characteristic of Venter that, even as *H. influenzae* was in the final stage of closure and annotation, he began to think about doing a second genome. He was anxious to demonstrate the power of the shotgun method and to show that *H. influenzae* was not somehow a fluke. At lunch in February 1995, discussion revolved around the best choice for a second genome. *Mycoplasma genitalium*, the smallest known bacterium, with a genome estimated at no more than 600 kb was the obvious choice.

I was delegated to call Clyde Hutchison at the University of North Carolina to enlist his collaboration. Hutchison had been studying *M. genitalium* for several years and thought it probably possessed close to a minimal genome for free-living life under laboratory conditions. He had already carried out a limited random shotgun sequencing of about 350 clones *(25)* and had been invited to the Dormy House workshop to present his results. I convinced him that the whole sequence could be accomplished before the meeting, merely 2 months away. Hutchison quickly checked with his coworkers and then agreed to the collaboration; he sent about 10 µg of DNA for the library construction.

The sequencing, led by Claire Fraser (Fig. 1), Venter's wife, proceeded smoothly and was completed in less than 4 weeks. Annotation revealed about 470 genes. The sequence was presented at the workshop as the second genome, and was of extraordinary interest because of its minimal size.

AN EXPLOSION OF MICROBIAL SEQUENCING
BECAUSE OF THE WHOLE-GENOME SHOTGUN METHOD

Microbial genomics was well on the way to being established as a full and legitimate field of study with the start of the *E. coli* and yeast projects, but it did not gain a sure footing until the first cellular genome sequence was solved, that of the bacterium *H. influenzae* Rd in 1995 *(24)*. The importance of that work was twofold. It demonstrated the value of having a complete catalog of an organism's genes. Second, it clearly established the potential of the whole genome shotgun method to solve other microbial genomes, whether previously studied or not.

The importance of the whole-genome shotgun-and-assembly strategy to the sequencing of microorganisms should not be underestimated. The great explosion of microbial sequencing projects carried out in the half dozen or so years since publication of the *H. influenzae* genome in 1995 has been made possible largely by application of the whole genome shotgun method *(24)*. Previously unknown and unstudied organisms from the environment or virtually all sources are amenable to sequencing. All that is needed are a few micrograms of genomic DNA.

Generally, a small insert (2-kb) library and at least one larger insert (usually 10-kb) library are constructed. Assembly is greatly facilitated by sequencing both ends of each insert from a sufficient number of randomly selected clones to yield 6× to 10× overall sequence coverage of the genome. The sequences obtained from the two ends of each insert are a known distance apart. This distance information greatly facilitates the assembly process.

TIGR has continued to exploit the whole genome shotgun method. As of September 2002 they had completed 21 microbial genomes.

THE DOE AND THE MICROBIAL GENOME INITIATIVE

The DOE, again demonstrating their leadership role in genomics, was the first to start a microbial genome project (MGP) in 1994, with a focus on nonpathogenic microbes, particularly those with environmental, phylogenetic, commercial, or energy relevance. In early discussions with TIGR, they agreed to provide funding for the *M. genitalium* project because of DOE interest in defining a minimalist genome. In 1995, with the great success of the *M. genitalium* project, DOE signed a 3-year cooperative agreement with TIGR and the University of Illinois to sequence *Methanococcus jannaschii* and other organisms relevant to their interest. They also signed agreements with Genome Therapeutics Corporation (Waltham, MA), the University of Utah in Salt Lake, and Recombinant BioCatalysis Inc. (now Diversa Corp., San Diego, CA) to sequence *Methanobacterium thermoautotrophicum, Pyrococcus furiosus*, and *Aquifex aeolicus* VF5, respectively. *Aquifex* is a bacterium, but the other three are members of the Archaea. Thus, with a single stroke they hoped to obtain the first complete genome information on three different archaea belonging to the so-called third domain of life.

The archaea were originally classified as bacteria, but in 1977, Carl Woese of the University of Illinois in Urbana carried out a comparative study of 16S ribosomal DNA sequences in two methanogenic bacteria and discovered that, although the two were closely related, they bore little resemblance to typical prokaryotic bacteria *(26)*. He proposed the name archaebacteria for these primitive organisms *(27)*. By 1987, sufficient biochemical and 16S ribonucleic acid (RNA) taxonomic evidence had accumulated for Woese and coworkers *(28)* to propose an entirely new taxon, above the level of kingdom, called a *domain*. They divided life on earth into three domains: the Archaea, the Bacteria, and the Eukarya.

Methanococcus jannaschii was the first of the archaea to be completely sequenced *(29)*. When the genes were examined, nearly two-thirds were unlike any previously found; in general, there was confirmation that genes for transcription, translation, and replication were more similar to eukaryotic genes, whereas biosynthetic and metabolic genes were more similar to bacterial genes. This placed *M. jannaschii* clearly into a third domain differing evolutionarily from both the Bacteria and the Eukarya. It appeared that Woese's hypothesis was substantiated.

The DOE MGP has continued to support many microbial sequencing projects. It has been extraordinarily successful in its 7-year history. As of January 2002, DOE MGP grantees had sequenced and published concerning 6 archaea and 8 bacteria; completed but not yet published information on 10 additional bacteria; and draft sequenced 14 bacteria, 2 archaea, and 1 eukarya.

JAPAN AND *SYNECHOCYSTIS* SPECIES PCC 6803: A PHOTOSYNTHETIC CYANOBACTERIUM

The Kazusa DNA Research Institute of Japan (Kisarazu) could easily have been the first to complete a microbial genome. Few were aware of their remarkable progress on *Synechocystis* sp PCC6803. In 1994, they published a physical map of the genome *(30)*, and in 1996, they completed, annotated, and published the entire 3.57-Mbp genome sequence *(31)*, a truly remarkable achievement. It was the largest genome of its time and preceded both yeast and *E. coli* by about a year.

NATIONAL INSTITUTES OF ALLERGY AND INFECTIOUS DISEASE AND HUMAN PATHOGENS

The National Institutes of Allergy and Infectious Disease (NIAID) has played a role complementary to that of the DOE. It funded the sequencing of 15 bacterial pathogens important in human disease as of September 2002. It is also currently funding or partially funding 44 bacterial, fungal, and parasite genome sequencing projects relevant to human health.

THE SANGER INSTITUTE AND THE WELLCOME TRUST

The largest genome center outside of the United States is the Sanger Institute (Hinxton, UK). It is a research institute funded by the Wellcome Trust to provide a major focus in the United Kingdom for mapping and sequencing the human genome and the genomes of other organisms. The Wellcome Trust is the world's largest charity, with approximately £15 billion of assets. Its goal is to be a leader in promotion of research into animal and human health. Under the direction of John Sulston, the Sanger Institute, with nearly 600 staff, contributed almost one-third of the public human genome effort.

Sanger Institute has also become a powerhouse in microbial genomics, focusing on both pathogens and model organisms. As of September 2002, Sanger Institute had published 7 bacterial genome sequences, finished an additional 7, and had some 24 bacterial sequencing projects in various stages of completion. Sanger sequenced or is sequencing part or all of 5 fungi: *Schizosaccharomyces pombe, Aspergillus fumigatus, Pneumocystis carinii, Candida albicans,* and *S. cerevisiae.* In addition, they are sequencing or participating in the sequencing of 11 protozoal pathogens, including the recently completed *Plasmodium falciparum,* the malarial parasite.

DOE JOINT GENOME INSTITUTE

The remarkable growth in microbial sequencing continues unabated. The DOE Joint Genome Institute (JGI), in Walnut Creek, California, has become a very significant contributor of microbial genome sequences in the past several years. JGI has completed the sequencing of the autotrophic nitrifying bacterium *Nitrosomonas europaea,* as well as the four photosynthetic bacteria: *Prochlorococcus marinus* strains MED4 and MIT9313, members of the polyphyletic group of cyanobacteria; *Rhodopseudomonas palustri,* a purple nonsulfur phototrophic bacterium; and *Synechococcus* sp strain WH8102, a marine unicellular cyanobacterium. In addition, the JGI is surveying many environmentally interesting microbes relevant to the DOE mission. It carried out draft sequencing of 17 microbial organisms in 2001; in 2002, they draft sequenced another 6.

OTHER NOTEWORTHY MICROBIAL SEQUENCING PROJECTS

Bacillus subtilis, a Gram-positive bacterium, was one of the important early laboratory organisms to be completely sequenced in November 1997 by a collaboration of European and Japanese laboratories led by the Institut Pasteur, Paris, France *(32). Rickettsia prowazekii,* the causative agent of epidemic typhus and an obligate intracellular parasite thought to have been the evolutionary source of mitochondria, was sequenced, and the information was published in November 1998 by a team at the University of Uppsala, Sweden *(33). Pseudomonas aeruginosa* PAO1 (6.3 Mb), an important environmental bacterium and a pathogen in cystic fibrosis and immunologically compromised patients, was sequenced in a collaborative effort among the Cystic Fibrosis Foundation, the University of Washington Genome Center in Seattle, WA, and the Pathogenesis Corporation, Seattle, WA; this information was published in August 2000 *(34).*

The list of determined microbial sequences continues to grow at a rapid rate (Fig. 2) (www.tigr.org). Microbial species outnumber all others on earth and contribute significantly to the earth's biomass. It seems likely that the number of completed microbial genomes will continue to increase.

BEYOND SEQUENCING

As microbial genome sequences have accumulated, methods for their analysis are being actively developed. The sequence of an organism contains not only the gene sequences and the regulatory information, but also a written, though somewhat blurred, record of the organism's evolution. Extracting this information from a sequence is both a computational and an experimental task.

All of the published genome papers include preliminary attempts at gene annotation and the recording of unusual sequence features, such as repeats and duplications. Gene-finding programs have evolved significantly over the past 7 to 8 years. Glimmer (Gene Locator and Interpolated Markov Modeler) *(35)* is currently the standard for bacteria and

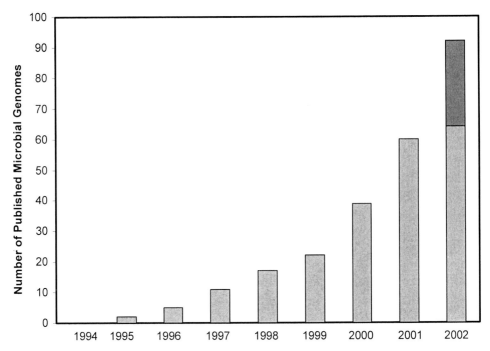

Fig. 2. Cumulative number of published microbial genomes. The data are from the TIGR Microbial Genome Database as of May 30, 2002. There are 28 genomes that are finished but not yet published included for 2002 because most of these will be published before the end of 2002.

archaea. Gene-functions are generally predicted by database comparisons with other similar genes of known function. It is remarkable, however, that with each new genome the number of new proteins with no database matches continues to grow. Generally, about a third of the genes in each new organism do not have a database match.

Genomes are being compared at the level of individual genes, clusters of genes, and whole genomes with the aim of discovering examples of vertically or horizontally transmitted genes that give clues to evolutionary mechanisms and relationships. A computational system for comparative genomics, called MUMmer *(36)*, that makes it possible to align the whole genome sequences of two different organisms has been in use for about 3 years.

Progress in finding regulatory sites has been slow. No entirely satisfactory software exists, for example, to find promoters. However, there has been more success with transcriptional terminators. TransTerm, for example, is a program that finds rho-independent transcription terminators in bacterial genomes *(37)*. Software for finding more complex regulatory structures is yet to be developed.

Experimental methods are focusing on determining the functions of all the genes and analyzing their patterns of expression in different physiological states of growth. Transposon mutagenesis is one of the most widely used methods to inactivate or "knock out" genes *(38)*. This enables determining whether a gene is essential in a particular growth environment. Determination of function usually requires more detailed biochemical or structural analysis. The yeast two-hybrid system *(39)* is being used extensively to examine interaction of the protein products of genes as additional clues to function.

Gene expression and regulation are being studied by global microarray methods *(40)*. DNA from each gene (usually obtained by polymerase chain reaction using primers that flank the gene) is spotted in very dense microarrays on membranes or glass slides. Labeled RNA is made from control cells grown under standard conditions or cells grown in a particular environment. The RNA is hybridized to the arrays, and the amount bound is measured. Thus, the level of expression of each gene can be simultaneously determined.

These various methods are just beginning to be exploited fully. The microbial genomics of the next few decades will give unparalleled knowledge of how cells work.

REFERENCES

1. Jenkins NA, Kucherlapati RS, McKusick VA. Genomics as it enters the second decade. Genomics 1977; 45:243.
2. Ruddle F. Hundred-year search for the human genome. Annu Rev Genomics Hum Genet 2001; 2:1–8.
3. Luria S, Delbruck M. Mutations of bacteria from virus sensitivity to virus resistance. Genetics 1943; 28:491–511.
4. Avery O, Macleod C, McCarty M. Studies on the chemical nature of the substance inducing transformation of pneumococcal types. J Exp Med 1944; 79:137–157.
5. Lederberg J, Tatum EL. Novel genotypes in mixed cultures of biochemical mutants of bacteria. Cold Spring Harb Symp Quant Biol 1946; 11:113–114.
6. Hayes W. Recombination in *E. coli* K12: Unidirectional transfer of genetic material. Nature 1952; 169:118–120.
7. Watson JD, Crick FHC. Molecular structure of nucleic acids: a structure for deoxyribose nucleic acid. Nature 1953; 171:737–738.
8. Cairns J. The bacterial chromosome and its manner of replication as seen by autoradiography. J Mol Biol 1963; 6:208, 213.
9. Bachmann BJ. Linkage map of *Escherichia coli* K-12, edition 8. Microbiol Rev 1990; 54:130–197.
10. Sanderson KE, Roth JR. Linkage map of *Salmonella typhimurium*, edition 7. Microbiol Rev 1988; 52:485–532.
11. Piggot PJ, Hoch JA. Revised genetic linkage map of *Bacillus subtilis*. Microbiol Rev 1985; 49: 158–179.
12. Maxam AM, Gilbert W. Sequencing end-labeled DNA with base-specific chemical cleavages. Methods Enzymol 1980; 65:499–560.
13. Sanger F, Nicklen S, Coulson AR. DNA sequencing with chain-terminating inhibitors. Proc Natl Acad Sci USA 1977; 74:5463–5467.
14. Sanger F, Coulson AR. The use of thin acrylamide gels for DNA sequencing. FEBS Lett 1978; 87:107–110.
15. Sanger F, Air GM, Barrell BG, et al. Nucleotide sequence of bacteriophage φX174 DNA. Nature 1977; 265:687–695.
16. Sanger F, Coulson AR, Hong GF, Hill DF, Petersen GB. Nucleotide sequence of bacteriophage lambda DNA. J Mol Biol 1982; 162:729–773.
17. Smith LM, Sanders JZ, Kaiser RJ, et al. Fluorescence detection in automated DNA sequence analysis. Nature 1986; 321:674–679.
18. Dulbecco R. A turning point in cancer research: sequencing the human genome. Science 1986; 231:1055–1056.

19. Blattner F. Biological frontiers. Science 1983; 222:719–720.
20. Blattner FR, Williams BG, Blechl AE, et al. Charon phages: safer derivatives of bacteriophage lambda for DNA cloning. Science 1977; 196:161–169.
21. Blattner FR, Plunkett G 3rd, Bloch CA, et al. The complete genome sequence of *Escherichia coli* K-12. Science 1997; 277:1453–1474.
22. Goffeau A, Barrell BG, Bussey H, et al. Life with 6000 genes. Science 1996; 274:546–567.
23. Mewes HW, Albermann K, Bahr M, et al. Overview of the yeast genome. Nature 1997; 387:7–65.
24. Fleischmann RD, Adams MD, White O, et al. Whole-genome random sequencing and assembly of *Haemophilus influenzae* Rd. Science 1995; 269:496–512.
25. Peterson SN, Hu PC, Bott KF, Hutchison CA 3rd. A survey of the *Mycoplasma genitalium* genome by using random sequencing. J Bacteriol 1993; 175:7918–7930.
26. Balch WE, Magrum LJ, Fox GE, Wolfe RS, Woese CR. An ancient divergence among the bacteria. J Mol Evol 1977; 9:305–311.
27. Woese CR, Fox GE. Phylogenetic structure of the prokaryotic domain: the primary kingdoms. Proc Natl Acad Sci USA 1977; 74:5088–5090.
28. Woese CR, Kandler O, Wheelis ML. Towards a natural system of organisms: proposal for the domains Archaea, Bacteria, and Eucarya. Proc Natl Acad Sci USA 1990; 87:4576–4579.
29. Bult CJ, White O, Olsen GJ, et al. Complete genome sequence of the methanogenic archaeon, *Methanococcus jannaschii*. Science 1996; 273:1058–1073.
30. Kotani H, Kaneko T, Matsubayashi T, Sato S, Sugiura M, Tabata SA. Physical map of the genome of a unicellular Cyanobacterium *Synechocystis* sp strain PCC6803. *DNA Res* 1994; 1: 303–307.
31. Kaneko T, Sato S, Kotani H, et al. Sequence analysis of the genome of the unicellular Cyanobacterium *Synechocystis* sp strain PCC6803. II. Sequence determination of the entire genome and assignment of potential protein-coding regions. *DNA Res* 1996; 3:109–136.
32. Kunst F, Ogasawara N, Moszer I, et al. The complete genome sequence of the Gram-positive bacterium *Bacillus subtilis*. Nature 1997; 390:249–256.
33. Andersson SG, Zomorodipour A, Andersson JO, et al. The genome sequence of *Rickettsia prowazekii* and the origin of mitochondria. Nature 1998; 396:133–140.
34. Stover CK, Pham XQ, Erwin AL, et al. Complete genome sequence of *Pseudomonas aeruginosa* PA01, an opportunistic pathogen. Nature 2000; 406:959–964.
35. Delcher AL, Harmon D, Kasif S, White O, Salzberg SL. Improved microbial gene identification with Glimmer. Nucleic Acids Res 1999; 27:4636–4641.
36. Delcher AL, Kasif S, Fleischmann RD, Peterson J, White O, Salzberg SL. Alignment of whole genomes. Nucleic Acids Res 1999; 27:2369–2376.
37. Ermolaeva MD, Khalak HG, White O, Smith HO, Salzberg SLJ. Prediction of transcription terminators in bacterial genomes. J Mol Biol 2000; 301:27–33.
38. Hutchison CA, Peterson SN, Gill SR, et al. Global transposon mutagenesis and a minimal Mycoplasma genome. Science 1999; 286:2165–2169.
39. Bartel PL, Fields S, eds. The Yeast Two-Hybrid System (Advances in Molecular Biology). Oxford, UK: Oxford University Press; 1977.
40. Khodursky AB, Peter BJ, Cozzarelli NR, Botstein D, Brown PO, Yanofsky C. DNA microarray analysis of gene expression in response to physiological and genetic changes that affect tryptophan metabolism in *Escherichia coli*. Proc Natl Acad Sci USA 2000; 97:12,170–12,175.

II   BIOINFORMATICS AS A TOOL IN GENOMICS

2

Tools for Gene Finding
and Whole Genome Comparison

Steven L. Salzberg and Arthur L. Delcher

OVERVIEW

This chapter describes two computational methods for genome analysis: gene finding and whole genome comparison. Finding the genes in the genome of a prokaryotic organism can be accomplished rapidly and accurately with a computational method that scans the genome and analyzes statistical features of the sequence. Following this step, which identifies more than 99% of the genes, the predictions can be refined by searching for nearby regulatory sites and by aligning protein sequences to other species. These steps can be automated using freely available software and databases.

Whole genome comparison refers to the problem of aligning the entire deoxyribonucleic acid (DNA) sequence of one organism to that of another, with the goal of detecting all similarities as well as any rearrangements, insertions, deletions, and polymorphisms. With the increasing availability of complete genome sequences from multiple, closely related species, such comparisons are becoming a powerful tool in genome analysis. This computational task can be accomplished in minimal time and space using suffix trees, a data structure that can be built and searched in linear time.

INTRODUCTION

Gene finding in prokaryotic genomes (bacteria, archaea, viruses) is remarkably accurate, in contrast to the still-difficult problem of finding genes in higher eukaryotes. The absence of introns removes one of the major barriers to computational analysis of the genome sequence, with the result that the gene finder described in this chapter finds over 99% of the genes in most genomes without any human intervention. Gene finding in single-cell eukaryotes is of intermediate difficulty, with some organisms (e.g., *Trypanosoma brucei*) having so few introns that a bacterial gene finder suffices to find their genes. Others, such as *Plasmodium falciparum*, have numerous introns, requiring a special-purpose gene finder such as GlimmerM *(1,2)*. Gene finding in these simple eukaryotes is much easier, and therefore more accurate, than in the human or mouse, but it is still more difficult than for bacteria.

Using modern bioinformatics software, finding the genes in a bacterial genome results in a highly accurate, rich set of annotations that provide the basis for further research into the functions of those genes. In this chapter, we review in detail the computational

From: *Microbial Genomes*
Edited by: C. M. Fraser, T. D. Read, and K. E. Nelson © Humana Press Inc., Totowa, NJ

methods involved in automated gene finding, focusing in particular on the Glimmer system *(3,4)*. We briefly touch on how these initial gene predictions can be further refined using nearby sequence signals, such as ribosome-binding sites (RBSs).

As the number of complete genomes has grown rapidly in recent years, an increasing number of closely related species have been sequenced. This has created a need for software that can take a pair of complete genomes and align one to the other. In addition to revealing large-scale similarities between organisms, these alignments have led to new discoveries about genome evolution. The most efficient systems for whole genome alignment can align two bacterial genomes in less than 1 minute on a conventional desktop computer. Following the discussion of gene finding, this chapter describes the key technology behind one of these systems, MUMmer *(5,6)*, and provides examples of the kind of alignments the system can perform.

GLIMMER: AN INTERPOLATED
MARKOV MODEL FOR GENE FINDING

Bacterial genomes are packed full of genes. In a typical bacterium, 90% of the DNA sequence consists of protein-coding regions, separated by relatively short spacer elements that themselves often contain other regulatory sequences. (Archaea and viruses share these features, but for the sake of discussion, we discuss only bacteria.) In addition, the protein coding sequences are rarely interrupted by introns. Therefore, a computational gene finder can employ a very simple strategy to identify genes: Just find every open reading frame (ORF) longer than some fixed length (for example, 500 bp) and call the region from its first start codon (ATG or GTG) to its end a gene. We call this the simple gene finder, or SGF. The SGF strategy often fails, but with some refinement, it can turn into a useful subroutine for our gene finder.

First, we need to define an ORF: an ORF is a sequence of nucleotide triplets without any in-frame stop codons (TAA, TAG, or TGA). We are interested in the longest possible ORFs, so we assume that the ORF is bounded on each side by a stop codon. The reading frame is defined by the bounding stop codons, so the number of bases between them is a multiple of 3. Note that ORFs in different reading frames overlap. Because 3 of 64 DNA triplets are stop codons, we expect to find stop codons relatively frequently in noncoding DNA. The precise frequency depends on the GC content of the genome; for example, *Mycobacterium tuberculosis*, with a GC content of 66%, contains relatively few stop codons compared to organisms with a GC content of less than 50%. With this constraint in mind, we can decide on a reasonable minimum ORF length to use for the SGF algorithm. Glimmer (described below) computes a default ORF length using a statistical formula that estimates the length above which only 1 ORF per million basepairs (Mbp) will occur by chance in noncoding DNA. For simplicity, we assume that this length is 600 bp.

The SGF algorithm is almost set: We can quickly scan a genome in all six reading frames, identify all ORFs longer than 600 bp in length, and call these ORFs genes. Of course, we will miss all the short genes, but that can be addressed later. First, SGF needs to be modified to consider one more constraint: genes should not overlap. Gene overlap is not absolutely forbidden, but only a very small number of genes have been conclusively shown to have overlapping coding regions, and those overlaps generally are very small. (One example is a pair of genes overlapping in the sequence TGATG, for which the TGA is a stop codon and the ATG is a start codon.) To be safe, SGF should discard all of the ORFs from its initial scan that overlap another ORF. Note that if we use 600 bp as the definition of a "long ORF," then ORFs less than 600 bp are ignored. Therefore, if a 600-bp ORF overlaps a shorter ORF, the algorithm will call the longer ORF a gene.

SGF has two serious shortcomings. First, it does not identify any genes shorter than the predefined length of 600 bp. We know that many such genes exist, and we obviously would like to include them in our genome annotation. Second, it is unable to make a choice when two or more ORFs overlap, so we obtain no genes in those regions. For high-GC organisms, this can represent a high percentage of the genome. Moreover, the absence of stop codons in an ORF on one DNA strand decreases the likelihood of stop codons on the opposite strand, which tends to produce overlapping ORFs on opposite strands.

The Glimmer system attempts to find all the genes in a genome, not just those above a fixed length. The core of the Glimmer algorithm is an interpolated Markov model (IMM), a statistical method that estimates the probability that any sequence is a coding region. To build this IMM, Glimmer needs a training set, a representative sample of genes from the genome. This chicken-and-egg problem—how to find the genes for training before running the gene finder—can be solved rather elegantly using SGF. The SGF algorithm is a very conservative gene finder that finds only a limited number of genes, but it has a very low false-positive rate. Therefore, we can use SGF to find a training set for Glimmer. The Glimmer package includes the SGF algorithm in a program called "long-orfs," which can be used on any genome to identify a training set for the larger system. For a typical 2-Mbp bacterial genome, SGF will yield as much as 1 Mbp of training data. Alternatively, BLAST *(12)* last can be used to search the databases for all proteins from other organisms with sequence similarity to the organism being analyzed. These proteins will also serve as a good training set for Glimmer. The automated annotation system at The Institute for Genome Research does both: The initial run uses long-orfs, but the system is rerun after performing BLAST searches against all predicted proteins against the database and collecting a new training set.

BACKGROUND: MARKOV MODELS

An IMM is a probabilistic modeling technique based on Markov chains. First, it will help to explain Markov chains and then will show how they can be extended to create IMMs. A Markov chain can be conceptualized as a series of states, each emitting a character from the alphabet {A,C,G,T}. To decide which character to emit, the state uses a probability distribution for those four values. A letter is chosen at random according to the probability distribution; thus, if $p(A) = 0.2$, then the letter A is output 20% of the time. The model then makes a transition to the next state.

An extremely simple Markov model is shown in Fig. 1A; it uses probabilities 0.2, 0.3, 0.4, and 0.1 for the four nucleotides. This model can output a DNA sequence by simply "running" it, generating one character in each execution cycle. In the trivial model of Fig. 1, every position of the sequence will be generated by state S_0; therefore, all positions will have the same probability distribution. Clearly, this is not a good model of

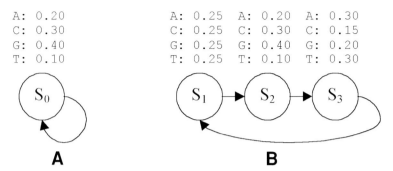

Fig. 1. (A) A simple one-state Markov chain that outputs nucleotides according to a single probability distribution regardless of sequence position. **(B)** A three-state Markov chain that outputs nucleotides using different probabilities in codon positions 1, 2, and 3, cycling indefinitely through three states. All edges (arrows) have probability 1.0 in both models.

DNA. A better model for protein-coding sequence might be to use three states, one for each codon position. This allows use of different probabilities for each position. Such a model is illustrated in Fig. 1B.

Clearly, we do not want to generate a made-up sequence; rather, we want to read a sequence and decide whether or not it is a gene. To do this, we compute the probability that the model would generate the given sequence. "Inverting" the model like this is surprisingly simple and works as follows. Starting at the beginning of the sequence, each state "eats" one character and determines its probability. If the Markov model reads a G, then it looks up the probability of G in that state. This probability is then multiplied by the probabilities of all preceding characters. The score of the sequence is just the product of all the probabilities from all the states. In practice, one cannot multiply thousands of numbers smaller than 1: The product will be a tiny number, such as 10^{-3000}, and computers simply cannot store such numbers without special data structures. To avoid this problem, the probabilities are all stored as logarithms, and then these logs are simply added, which is mathematically equivalent to multiplying the probabilities.

At this point, there is a single score for the sequence, which does not tell whether or not it is a gene. Obviously, the score will be lower as the sequence gets longer because all the probabilities are less than 1. However, the probability that the sequence is a gene does not get lower with increasing length; on the contrary, it usually gets higher. To make a decision about a gene, we need to compare our number to something else. To do this, we need to create a model of a "nongene."

One possible model is just random DNA: We can create a trivial model (see Fig. 1A) using the probabilities of A, C, G, and T computed from our entire genome. To understand what we do next, we first must be very clear about the interpretation of these scores. The score generated by a sequence S and a model M is a probability; more accurately, it is the probability that the model generated that exact sequence S: $p(S|M)$.

We have now described a simple model of random (or noncoding) DNA, shown in Fig. 1A, which we call M_N. Our three-state model of coding DNA shown in Fig. 1B is M_C. We now know how to compute $p(S|M_C)$ and $p(S|M_N)$, but neither of these is precisely what we want. If we assume that our models are good ones (and we refine them further below), then we need to compare the probability that the model of coding regions generated the sequence $p(M_C | S)$ to the probability that it came from a model of noncoding regions. If we can compute that it is much more likely that the sequence was generated by our coding model, then we can confidently assert that the sequence is a gene.

To allow us to compute what we need, we use Bayes' Law, which tells us that

$$p(M | S) = \frac{p(S | M)\, p(M)}{p(S)}$$

We would like to compute the probability of the model given the sequence and compare this value for the two models, $p(M_C | S)$ vs $p(M_N | S)$. Note that we do not know the prior probabilities of either the sequence, $p(S)$, or of the models, $p(M_C)$ and $p(M_N)$. The usual mathematical trick Glimmer uses to get around this difficulty is to assume that all models are equally probable. Alternatively, it can be assumed that one model is always a constant factor more likely than the other and use that. Once this is done, the fact that $p(M_C | S)$ and $p(M_N | S)$ sum to 1.0 can be used to determine the value of $p(S)$.

INTERPOLATED MARKOV MODELS

The Markov models described in the preceding section are zero-order Markov chains, a very simple type of model in which the probability of any nucleotide depends only on the state of the model. Glimmer uses a more sophisticated model in which up to 8 bases are used to compute the probability of each base. These 8 bases are chosen from among the 12 bases preceding the base in question.

A full explanation of interpolated Markov models can be found in articles describing Glimmer *(3,4)*, but a brief summary is given here. The simplest extensions of zero-order Markov chains are first-order Markov chains, in which each probability is computed based on one other position, usually the position just prior to it. Thus, rather than computing $p(A)$ in our data, we would compute $p(A|b)$, where b takes on the values A, C, G, T. We could use these probabilities in our codon model from Fig. 1B. Both empirical evidence and mathematical proof show that a first-order model is never inferior to a zero-order model. This could be extended quickly to a second-order or third-order model by computing probabilities based on the two or three previous bases. In fact, it can easily be shown mathematically that higher-order Markov models are always superior to lower-order ones.

The only caveat to this mathematical fact is empirical: If the probabilities are not correct, then a higher-order model is not superior and may in fact be inferior. A quick thought experiment shows that, for a typical bacterial genome, the amount of training data will only be sufficient for a fifth- or six-order model. For example, a fifth-order model requires $4^6 = 4096$ probabilities because the probability of all four bases must be estimated following every combination of 5-mers. If the training algorithm (long-orfs, see the section on Glimmer) yields 1 Mbp of data, then each of the 1024 5-mers will be sampled an average of 1000 times, which should be adequate to estimate 4 probabilities. For each increase in the length of the model, the number of samples drops by a factor of 4. Furthermore, many hexamers are underrepresented in bacterial genomes, and the estimates of these probabilities will therefore be inaccurate.

The solution to this problem is to use higher-order Markov models when the data are plentiful and lower-order models otherwise. IMMs are designed precisely with this solution in mind. Internally, Glimmer constructs nine different Markov models, ranging from zero-order through eighth-order. It then computes statistics, using the training data, on which Markov models have sufficient data for every oligomer up to size 8. These statistics in essence tell the IMM whether or not it has enough data to accurately estimate four probabilities (for A, C, G, T) for a given oligomer. The system then tries to use the longest oligomer possible to score each base in a sequence. The scoring function is a product of these probabilities across an ORF.

SCORING

For any region of a genome containing an ORF, the primary decisions that a bacterial gene finder needs to make are whether that region contains a gene, and, if so, what reading frame contains it. There are seven choices: Each of the three reading frames on either DNA strand might be a gene, or there might be no gene. Our default assumption is that only one of these choices is correct; although overlapping genes do exist, they are very rare, and in general the overlap is only a few nucleotides. Glimmer can be set to find overlapping genes; by default, it permits a small amount of overlap.

Glimmer builds several IMMs and uses them all when scoring ORFs and deciding which are true genes. It constructs IMMs for the first, second, and third codon positions; to score an ORF, it uses a different IMM in each position. This is known as a "three-periodic" Markov model because the system cycles through the three models with every three positions in an ORF. To score a particular ORF, the system just multiplies the probabilities given by the three-periodic IMM in that ORF's reading frame. To score a region of a genome, then, the system can simply construct the six alternative reading frames (some containing stop codons, of course) and score them all. For the seventh choice (no gene), the system uses a simple model of random DNA similar to that shown in Fig. 1A. All these scores are computed, and the highest-scoring model "wins" if its score is greater than 90% of the total (this threshold can be changed easily by the user). Thus, Glimmer identifies a gene only when an in-frame ORF wins among these competing models.

The Glimmer 2.0 system includes several additional steps that attempt to resolve overlaps between genes. The final output is a detailed gene table showing all the IMM scores, including scores of those ORFs that did not make the final cut, and a shorter list showing all predicted genes and their locations in the genome. One nice feature of Glimmer (1.0 and 2.0) is that the IMM model itself is output as a stand-alone file by a separate module, which permits the user to train on one species and find genes in another. This turns out to be surprisingly useful; there have been numerous instances when a scientist wanted to find genes in a relatively small fragment of DNA for which no training data were available. In this case, it is easy to train on a related organism, using the closest species from among the growing list of complete genomes, and then to run the gene-finding module on the sequence fragment.

Our results *(4)* showed that Glimmer 2.0 finds, without any human intervention, more than 99% of the genes in most bacterial genomes. There is also a small false-positive rate associated with this, but this rate is difficult to quantify precisely. The number of additional gene predictions—those not matching published annotation—ranges from 5 to 15% for most genomes, but without further evidence, it is hard to know how many of these predictions might represent genuine, previously undetected genes. Even today, new protein-coding genes are being discovered in *Escherichia coli*, the most extensively studied of all bacterial genomes, and other genomes will require much more analysis before it can be certain that all their proteins have been identified. For example, a study by Wassarman and colleagues *(7)* identified six new protein sequences in what were previously annotated as intergenic regions of *E. coli*. All had been predicted correctly as genes by Glimmer.

USING RIBOSOME-BINDING
SITES TO IDENTIFY START CODONS

One last piece of evidence that can improve bacterial gene finding is the RBS, a short sequence just upstream of the start codon that is present for most genes in most species. This sequence, known as the Shine–Delgarno sequence, is complementary to the 3' end of the 16S ribosomal RNA and is usually a variation on AGGAG (in *E. coli* and many other bacteria). We have developed an algorithm to use the RBS to improve the predictions of the start sites as output by Glimmer, GeneMark *(8)*, or any other bacterial gene finder. The RBSfinder program, which is freely available from the Glimmer home page (www.tigr.org/software/glimmer), runs as a postprocessor to the Glimmer program *(9)*. The user simply feeds it the genome along with Glimmer's predictions, and using the Shine–Delgarno sequence as a guide, it attempts to find RBSs for every gene.

If the 16S ribosomal RNA is unknown for a given genome, then the program can be allowed to figure it out. For this task, it uses a variation on a technique known as Gibbs sampling. The basic idea is as follows. First, it is assumed that many of the genes in the Glimmer output are likely to have the correct start site. Therefore, the RBS is already present within a window about 10–15 bases upstream from the predicted start codon. Using this as a guide, the algorithm extracts a short window upstream from all gene predictions and then iteratively searches through these windows for the most common motif. The user can adjust the length of the window and the motif. The sampling algorithm refines this motif until it reaches stability (usually very quickly). Finally, using either the RBS provided by the user or the motif discovered by the Gibbs sampling routine, it scans all predicted genes and defines a position weight matrix *(10)* that captures the probabilities of each base in the RBS. This matrix functions as a zero-order Markov model that can be used to score any putative sequence.

Using the position weight matrix, the algorithm then scans the list of genes and attempts to modify any start sites that do not have a high-scoring RBS. For each start codon, it looks both upstream and downstream in an effort to find alternative starts that have good RBSs. If it finds a good site at the proper distance from another start codon, then it shifts the start codon. Of course, it only shifts the start codon if the reading frame of the predicted gene is preserved. The program then prints out the complete list of genes, including both the old and new start codon positions for those genes that it adjusted, as well as the location and sequence of the predicted RBS.

All of the predictions from Glimmer can be confirmed and modified (if necessary) by aligning the predicted proteins to a protein sequence database. The complete process of annotation, which includes this step, is the subject of Chapter 3, this volume.

WHOLE GENOME COMPARISON

When The Institute for Genome Research first completed the genome of *M. tuber-culosis* strain CDC1551 in 1999, one of the first analysis problems was to compare it to the genome of the closely related laboratory strain H37Rv *(11)*, whose genome had been published a year earlier. Each of these genomes is approximately 4.4 Mbp in length, and at that time, no software existed to align such long sequences. On average, these two genomes are more than 99% identical, differing by approximately 1000 single-nucleotide differences plus a small number of insertion sequences. The computational problem faced at the time was to align the sequences so that all differences could be discovered.

The problem of aligning two genes, or aligning a gene to a large database of genes, has been solved for quite some time, and efficient implementations are available in the latest BLAST (Basic Local Alignment Search Tool) *(12)* and FASTA *(13)* systems. Until the completed identification of the second *M. tuberculosis* strain, however, no one had needed a program for aligning megabase-scale sequences. The algorithms behind BLAST and FASTA require quadratic amounts of computing time (i.e., the time needed is a function of the square of the size of the input), which is acceptable for short inputs such as protein sequences, but is prohibitively expensive for whole genomes. Therefore, our group created a new algorithm based on maximal unique matches (MUMs) and suffix trees, which we called MUMmer.

If two genomes are closely related, then it might reasonably be expected that they would contain many DNA sequences that match exactly; further, these sequences would be found in the same order and orientation in both genomes. MUMmer is designed to find all such sequences and then to cluster them in larger blocks. The matching sequences are MUMs, defined as subsequences that match both genomes exactly, that occur just once in each genome, and that are maximal in length. This last requirement means that if the sequences are extended on either end by 1 bp, they will no longer match each other, as illustrated in Fig. 2.

The computationally intensive step of the MUMmer algorithm involves exhaustively finding all MUMs in a pair of genomes. These are then grouped into clusters that represent longer regions of similarity. The clustering step can be controlled by the user, who can specify the maximum separation between MUMs within a cluster. MUMmer also includes two additional packages that permit alignment of partially assembled genomes as well as alignments based on protein sequence similarity, explained below. First, we summarize the algorithm for building suffix trees and searching them for matches. This is only meant as a summary; for the complete details, please see the original publication *(5)*.

A suffix tree is a data structure that contains all of the subsequences from a particular sequence. An example of a tree for the sequence atgtgtgtc$ is shown in Fig. 3. (The $ is a special character used for convenience to mark the end of the sequence.) Nodes in the tree represent positions in the sequence, and each leaf node (the bottom-most nodes) represents a *suffix*, defined as a sequence starting at any position and extending all the way to the end of the sequence. In the figure, the leaves (square boxes) are labeled with the position from the reference sequence where that suffix begins. Edges are labeled with subsequences from the original input sequence. Labels of edges under the same node must begin with different letters, and each nonleaf node must have at least two edges under it. To reconstruct the suffix for any leaf node, you simply trace the path

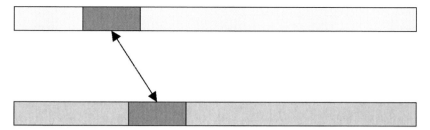

Fig. 2. Two genomes are represented by the upper and lower rectangles, and a maximal unique match (MUM) is shown schematically in gray. The sequence of the MUM occurs exactly once in each genome, and it cannot be extended on either side without creating a mismatch.

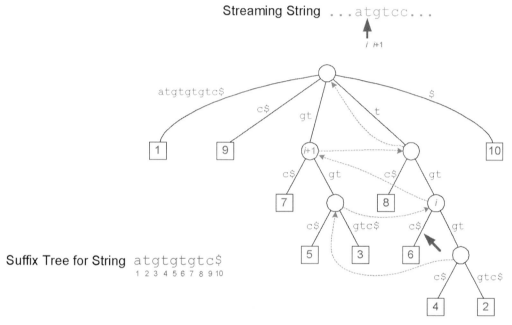

Fig. 3. A sample suffix tree showing the streaming behavior for finding matches between a query and a reference. The arrows indicate that the streaming string (the query) matches at the position shown by the arrow to the position shown in the tree by a second arrow. Node 6 in the tree, just below the arrow, corresponds to position 6 in the reference string, indicating that both the query and the reference contain the sequence tgtc.

from the root down to that leaf, concatenating all of the labels encountered on edges along the way. Although a sequence of length n contains $(n^2 + n)/2$ total subsequences, the suffix tree groups together all common subsequences in a data structure with exactly n leaves and at most $n - 1$ nonleaf nodes.

The original MUMmer algorithm *(5)* built a suffix tree containing two input sequences and then found all MUMs between them (see Fig. 3). The latest version of the system stores only one sequence (one genome) in the suffix tree, which we call the *reference sequence*. The second sequence is then "streamed" against the suffix tree, meaning that we pass the sequence through the tree, marking all locations where it matches. This

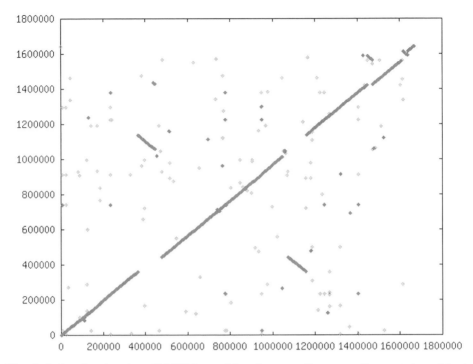

Fig. 4. A dot plot showing all MUMs resulting from an alignment of two strains of the bacterium *Helicobacter pylori*, 26695 (horizontal axis) and J99 (vertical axis). Long diagonals indicate aligned regions; diagonals with a slope of −1 indicate large-scale chromosomal reversals.

technique was introduced by Chang and Lawler *(14)*; for a fuller description, see the book by Gusfield *(15)*. Because the tree contains all the sequences in the first genome, every place that the second genome matches can be found using the streaming method. The main difference from the original algorithm is that the matches are unique to the first genome, but not necessarily to both. For those who want the uniqueness guaranteed in both genomes, the original algorithm is included with the package.

The result of this streaming algorithm is that, in time proportional to the length of the query sequence, all maximal matches between it and a unique string in the reference sequence can be identified. One advantage of this streaming algorithm is that, once the suffix tree has been built for the reference sequence, arbitrarily long, multiple queries can be streamed against it. In fact, we used these programs to compare two assemblies of the entire human genome (each approximately 2.7 billion characters), using each chromosome as a reference and then streaming the other entire genome past it (A. Halpern, personal communication, 2002). The streaming algorithm also greatly speeds up a comparison of a small genome—or a single chromosome—against a large, multichromosome genome: Simply provide the smaller genome as the reference and let the larger genome be the query, and the amount of space used will be proportional to the smaller sequence.

The output of the main alignment routines in MUMmer is a comprehensive list of exact matches. The simplest way to view the alignment is a simple dot plot in which diagonals indicate regions of alignment. An example is shown in Fig. 4, which shows

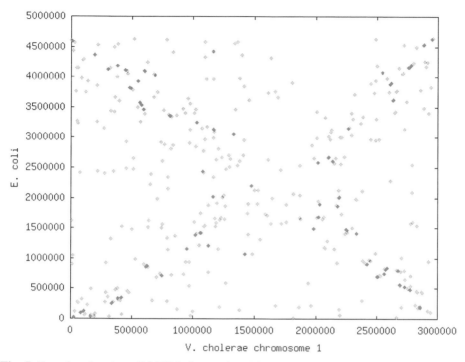

Fig. 5. Dot plot showing all MUMs larger than 20 bp between *E. coli* and *V. cholerae*, demonstrating the X alignment that results from numerous large-scale chromosome inversions, all symmetric about the origin of replication.

the alignment between two *Helicobacter pylori* strains. (This alignment takes less than 20 seconds to generate on a standard desktop computer running Linux.) The figure makes it instantly clear that the two strains are largely aligned with one another, and it also highlights four large chromosomal segments that have been inverted. The origin of replication is located between the third and fourth inversions *(16)*; therefore, these inversions are all symmetrically positioned around the origin. As we showed in a larger-scale study, bacterial chromosomes undergo frequent inversions around the origin of replication, and numerous species display a typical "X alignment" when they are aligned to one another and displayed in this fashion *(17)*. The *H. pylori* strains are just beginning to form an X, but after many more inversions, the pattern will become more pronounced as the conserved segments (diagonals in the dot plot) are broken into smaller and smaller pieces. An example of a more distant, but still detectable, alignment is that between *E. coli* and *Vibrio cholera*, shown in Fig. 5.

The latest version of MUMmer is available from our Web site at www.tigr.org/software/ mummer. It includes a viewer (DisplayMUMs) that will help the user navigate through the output.

COMPARING ASSEMBLIES OF PARTIAL SHOTGUN DATA

Because the suffix tree algorithm is both time and space efficient, it is able to align large eukaryotic chromosomes with only slightly greater requirements than bacterial genomes. It was used to align all the chromosomes of the model plant *Arabidopsis thaliana*, which was a key step in the analysis that led to the discovery of a recent whole genome duplication in that species *(18)*. It was used on an even larger scale to align the chromosomes of the human genome to one another, uncovering numerous large-scale ancient duplications involving large chunks of chromosomes *(19)*.

In addition to its use for comparing whole genomes and chromosomes, MUMmer is especially helpful in comparing incomplete genomes. Just as completed genomes are growing in number, so are incomplete genomes, some of which will not be completed for many years, if ever. One reason is that closing all the gaps and completing a genome is more difficult and time consuming than the shotgun sequencing phase, which can be accomplished in a rapid, high-throughput manner. Many genomes have already been sequenced to some level of coverage, ranging from 1× to 8×, and then the information is published (or released), with no ongoing effort to complete them.

MUMmer enters the picture because it can align two incomplete genomes to one another as easily as it aligns complete genomes. The inputs to the system can be a pair of multi-FASTA files representing genomes at different stages of completeness; either or both can be incomplete. It will then align all the contigs (contiguous segments of DNA, as produced by a genome-assembly program) to one another just as quickly as it would if they were completed. The system includes a separate program called NUCmer, which can then cluster the matches together and identify the order and orientation of the contigs with respect to one another. For example, if three small contigs in genome A map to one large contig in genome B, then NUCmer will show where they match and will show their orientation.

For more distantly related species, MUMmer alignments may be unable to detect DNA sequence similarity due to evolutionary divergence of the sequences. Protein sequences remain similar over much longer periods of time, so we designed another program, PROmer, which uses MUMmer to align genomes (or partial genome sequences) based on protein similarity. PROmer translates all its input sequences in all six reading frames, generating all possible protein sequence fragments longer than a preset minimum (easily changed by the user). It then concatenates these sequences, finds all MUMs (using the amino acid alphabet and shorter "words"), and then maps the matches back to the original DNA sequence coordinates. The resulting maps are presented in a form that allows the user to determine quickly the order and orientation of any contigs that align to one another.

We have used this in our own recent work to align the genomes of *P. falciparum* and *Plasmodium yoelii*, parasites that cause human and mouse malaria, respectively. Although work on *P. falciparum* was nearly completed at the time of analysis, *P. yoelii* was sequenced only to 5× coverage because of funding constraints, and the resulting assembly yielded thousands of contigs. The PROmer system was able to map large numbers of these contigs to the *P. falciparum* genome, creating "pseudo-contigs" based on the syntenic map. These pseudo-contigs consisted of ordered and oriented assemblies with no other information linking them; with the putative order suggested by the map, we were able to go back and run polymerase chain reactions to link many of them, confirming the original alignment-based map. The speed of PROmer and MUMmer was essential for this project because the genome sequences of both species were reassemblied many times, each time requiring that all the alignments be rerun.

ACKNOWLEDGMENTS

This work was supported in part by grants IIS-9902923 to S. L. S and IIS-9820497 to A. L. D from the National Science Foundation and by grant R01-LM06845 to S. L. S from the National Institutes of Health.

REFERENCES

1. Salzberg SL, Pertea M, Delcher AL, Gardner MJ, Tettelin H. Interpolated Markov models for eukaryotic gene finding. Genomics 1999; 59:24–31.
2. Pertea M, Salzberg SL. Computational gene finding in plants. Plant Mol Biol 2002; 48:39–48.
3. Salzberg SL, Delcher AL, Kasif S, White O. Microbial gene identification using interpolated Markov models. Nucleic Acids Res 1998; 26:544–548.
4. Delcher AL, Harmon D, Kasif S, White O, Salzberg SL. Improved microbial gene identification with GLIMMER. Nucleic Acids Res 1999; 27:4636–4641.
5. Delcher AL, Kasif S, Fleischmann RD, Peterson J, White O, Salzberg SL. Alignment of whole genomes. Nucleic Acids Res 1999; 27:2369–2376.
6. Delcher AL, Phillippy A, Carlton J, Salzberg SL. Fast algorithms for large-scale genome alignment and comparison. Nucleic Acids Res 2002; 30:2478–2483.
7. Wassarman KM, Repoila F, Rosenow C, Storz G, Gottesman S. Identification of novel small RNAs using comparative genomics and microarrays. Genes Dev 2001; 15:1637–1651.
8. Lukashin AV, Borodovsky M. GeneMark.hmm: new solutions for gene finding. Nucleic Acids Res 1998; 26:1107–1115.
9. Suzek BE, Ermolaeva MD, Schreiber M, Salzberg SL. A probabilistic method for identifying start codons in bacterial genomes. Bioinformatics 2001; 17:1123–1130.
10. Claverie J-M, Audic S. The statistical significance of nucleotide position-weight matrices. Comput Appl Biosci 1996; 12:431–440.
11. Cole ST, Barrell BG. Analysis of the genome of *Mycobacterium* H37Rv. Novartis Found Symp 1998; 217:160–172, discussion 172–167.
12. Altschul SF, Gish W, Miller W, Myers EW, Lipman DJ. Basic Local Alignment Search Tool. J Mol Biol 1990; 215:403–410.
13. Pearson WR. Flexible sequence similarity searching with the FASTA3 program package. Methods Mol Biol 2000; 132:185–219.
14. Chang WI, Lawler EL. Sublinear expected time approximate string matching and biological applications. Algorithmica 1994; 12:327–344.
15. Gusfield D. Algorithms on Strings, Trees, and Sequences: Computer Science and Computational Biology. New York: Cambridge University Press; 1997.
16. Salzberg SL, Salzberg AJ, Kerlavage AR, Tomb JF. Skewed oligomers and origins of replication. Gene 1998; 217:57–67.
17. Eisen JA, Heidelberg JF, White O, Salzberg SL. Evidence for symmetric chromosomal inversions around the replication origin in bacteria. Genome Biol 2000; 1: research11.01–09.
18. The Arabidopsis Genome Initiative. Analysis of the genome sequence of the flowering plant *Arabidopsis thaliana*. Nature 2000; 408:796–815.
19. Venter JC, Adams MD, Myers EW, et al. The sequence of the human genome. Science 2001; 291: 1304–1351.

3

Bacterial Genome Annotation at TIGR

Owen White

INTRODUCTION

The overarching goal of annotation is the identification of the functions of all the genes in a genome as accurately as possible. During the course of annotating a large number of microbial genomes, The Institute for Genomic Research (TIGR) has developed and implemented a number of automated annotation methods and tools that, together with manual curation by a team of trained doctoral scientists, are used for the analysis of a completed genome sequence. The annotation tools applied to genomes have evolved over the past 9 years to include a combination of automated gene identification, identification of noncoding features such as regulatory sites and repetitive sequences, and assignment of database matches and role categories to genes. An outline of the approaches routinely used for prokaryotic annotation is summarized below. First, I briefly discuss another crucial issue: the challenge of consistent bacterial annotation.

BACTERIAL DATA CONSISTENCY

As many scientists utilizing the data from bacterial genome projects know, use of that information is problematic; the functional assignments derived from computational analysis are not as reliable as when individual genes are characterized experimentally at the bench. Is *in silico* assignment to blame? Perhaps it is, given that unless biochemical or cellular activity has been proven, functional assignments are only a hypothesis, which in some cases are bad. However fallible, experimental assignment is also problematic because the way itself annotation is described. Annotation data types are embodied in concepts like common name (e.g., alcohol dehydrogenase), genetic names (*adh1*), or Enzyme Commission (EC) numbers (e.g., 1.1.1.1). Gene function is also described by membership in metabolic pathways, translation start sites, and transporter classifications.

One problem arises from the heterogeneity used in describing data types. One example is the data type role used to describe activity of a gene in the cell. At TIGR, biological roles have been categorized based on a system originally adapted from ref. *1*. Other categories for representing roles have also been developed, such as the Gene Ontology developed by Ashburner and coworkers (*2*). The choice of category to represent the biological role data type has varied between laboratories and poses an obstacle to meaningful comparison between genomes.

From: *Microbial Genomes*
Edited by: C. M. Fraser, T. D. Read, and K. E. Nelson © Humana Press Inc., Totowa, NJ

Another issue is that heterogeneous criteria used to assign annotation data types. For example, assigning the function of a gene is one of the most central tasks in annotation and accompanies nearly all published description of a prokaryotic genome project. Unfortunately, assignments have been made in the absence of a common standard. Examples in Fig. 1 show the many synonymous ways of assigning common names as they appear in various GenBank records. Variations in the annotation tags of GenBank records such as "/gene" may appear trivial, but such problems make comprehensive retrieval of genes of the same function nearly impossible. Confounding this problem, genes with inconsistent annotation have themselves been used for assignments of genes from new genomes, resulting in transitive error propagation.

It is important to recognize that heterogeneous annotation is not necessarily incorrect. An analogy comparing annotation to books in a library might be useful in this context. Most libraries have fixed operational procedures for placing books on their shelves. The procedures form a framework for placing books that can be effectively retrieved at any given library. The books assigned to the category art history at two different libraries can be compared, and it is common to find different sets of books assigned to that category even if exactly the same sets of books were kept somewhere at both libraries. It may be that one library used the Dewey Decimal system and the other used the Library of Congress assignment categories, an example of heterogeneous data types. Alternatively, differences can also arise when both libraries used the Dewey Decimal system, but the same book was assigned to different Dewey Decimal numbers, an example of heterogeneous assignment criteria. The placement of books on the shelves of each library is not necessarily wrong in each library. Book storage simply lacks consistency, and queries for books across institutions are less meaningful.

Similarly, the annotations for completed prokaryotic genomes are not wrong. The issue is that successful comparisons across all prokaryotes is only possible when annotation is made using identical data types and identical criteria for assignment. These are currently absent in the annotation for complete genomes in the public archives.

FUNCTIONAL ASSIGNMENT OF BACTERIAL PROTEINS

To reduce the variation in data type assignment at TIGR, a series of standard operational procedures (SOPs) for the annotation of microbial genomes has been developed. These SOPs are described in Table 1 and include protocols for the assignment of a number of annotation data types, such as coding regions, function, common names, biological roles, creation of protein families, hydrophobicity analysis, identification of candidate transporters, repeat regions, and structural ribonucleic acids (RNAs).

Fig. 1. Heterogeneity in data types used in prokaryotic GenBank annotation. (**A**) Displayed in the text of this figure are the fields typically used in GenBank records to display annotation information associated with individual genes. The tags (e.g., "/gene") are used as the computer parsable field to get annotation information. The information here displays the heterogeneous usage of the tags associated with different organisms. In some cases the /gene tag is used to store the genetic name purU, but varies for some organisms. The /product field typically stores a common name (e.g., "formyltetrahydrofolate deformylase"), but is also used to store a genetic

A

Escherichia coli
> /gene="purU"
> /EC_number="3.5.1.10"
> /function="enzyme; Purine ribonucleotide biosynthesis"
> /product="formyltetrahydrofolate deformylase; for
>   purT-dependent FGAR synthesis"

Campylobacter jejuni
> /gene="purU"
> /EC_number="3.5.1.10"
> /product="formyltetrahydrofolate deformylase"

Methylobacterium sp. CM4
> /gene="PurU"
> /product="purU protein"

Synechocystis sp. PCC6803
> /gene="purU"
> /product="phosphoribosylglycinamide formyltransferase"

Rhodospirillum rubrum
> /gene="purU"
> /product="formyltetrahydrofolate deformylase"

Halobacterium sp. NRC-1
> /gene="purU"
> /product="formyltetrahydrofolate deformylase"

Helicobacter pylori, strain J99
> /gene="purU"
> /product="FORMYLTETRAHYDROFOLATE HYDROLASE"

Helicobacter pylori
> /gene="HP1434"
> /product="formyltetrahydrofolate hydrolase (purU)"

Streptomyces coelicolor
> /gene="SCD10.35"
> /product="putative formyltetrahydrofolate deformylase (fragment)"

Mycobacterium tuberculosis H37Rv
> /gene="purU"
> /product="purU"

Corynebacterium sp.
> /gene="purU"
> /product="10-formyltetrahydrofolate hydrolase"

name in some cases. The EC number, although explicitly annotated in *E. coli*, is frequently found in other portions of the GenBank record, such as the common name of the genes as shown in part B. Heterogeneity is also observed in the annotation in specific datatypes, as observed in the variation of common names placed in the /product tag.

B

32 genes for DNA polymerase III, alpha subunit (dnaE).

3 DNA polymerase III
1 DNA-directed DNA polymerase (EC 2.7.7.7) III alpha chain
2 probable DNA-directed DNA polymerase (EC 2.7.7.7) III alpha chain 3
1 DNA-directed DNA polymerase (EC 2.7.7.7) III alpha chain, spliced form 1
1 probable DNA polymerase III, alpha chain
18 DNA polymerase III, alpha chain
1 putative DNA polymerase III alpha chain
3 DNA polymerase III alpha subunit
1 DNA polymerase III, alpha subunit
1 DNA pol III alpha

EC number: 18
genetic names: 6
functional info: 13

----

18 genes for homoserine acetyltransferase

2 probable homoserine O-acetyltransferase
2 putative homoserine O-acetyltransferase
9 homoserine O-acetyltransferase
2 homoserine-o-acetyltransferase
1 probable metA protein
2 homoserine O-trans-acetylase (yeast)

EC number: 3
genetic names: 6 (metA, met2)
functional information: 2
----
20 genes for glycogen operon protein (glgX).

5 glycogen operon protein GlgX
1 glycogen operon protein GlgX (EC 3.2.1.-)
1 longer ORF due to differences from ECOGLG
4 glycogen debranching enzyme
2 putative glycogen debranching enzyme
1 probable glycosyl hydrolase
1 putative glycosyl hydrolase
2 glycogen operon protein (EC 3.2.1.-) glgX
1 probable glycogen hydrolase (debranching)
2 glycosyl hydrolase family protein

genetic names: 8 (treX, glgX, glgX2)
EC number: 6
functional information: 3

Fig. 1. (B) A listing of the common names associated with DNA polymerase III, alpha sub-unit (*dnaE*), homoserine acetyltransferase (*metA* or *met2*), and glycogen operon protein (*glgX*) extracted from GenBank using a keyword search in the Entrez literature search server. The counts of each name for these genes in the GenBank annotation are shown. For the number of GenBank entries for these genes that had EC numbers, genetic names are listed. The counts for genes that had functional information, such as the biological role of these genes, are also listed.

Table 1
Standard Operational Procedures Used in Current TIGR Annotation

Curated annotation. *See* text.

Gene finding. Anonymous DNA sequence is initially searched using Glimmer *(8)*, a system that assigns probabilities to potential coding regions. Glimmer has about a 99% sensitivity for identification of known genes. (See Chapter 2.)

HMM searches. Predicted coding regions are searched against the TIGRFAM HMMs and Pfam. Graphical user interfaces have been developed for rapid evaluation and assignment of function for anonymous genes to an HMM.

Hydrophobicity, membrane-spanning regions, and signal peptides. The identification of signal peptides and membrane-spanning domains is examined in the context of database matches to identify biologically relevant genes.

Intergenic analysis. Glimmer identifies predicted coding regions, and these proteins are searched for similarity to other known proteins. In some cases (such as regions of the genome that have been laterally transferred), genes that have a sufficiently unusual composition are not detected by Glimmer. To correct for this, the genome is scanned either for regions that contain open reading frames (ORFs) without any similarity matches or for regions that do not contain ORFs. All six reading frames from these "intergenic" regions are examined for sequence matches; if any are found in a translation, the end points of an ORF are determined from the position of the pairwise alignment in the region. These candidate genes are then evaluated by annotators prior to placement into final annotation.

Insertion sequence elements. Insertion sequence (IS) elements are relatively small and simple in structure, containing only one or a few ORFs, of which one or two encode the transposase enzyme. The boundaries of an IS element are identified manually by examination for left and right end sequences, which are the DNA substrate for the transposase enzyme. These left and right end sequences often contain imperfect inverted repeats, and they are sometimes flanked by a characteristic target site duplication (direct repeat) created by the transposition process. Newly discovered IS elements are named based on a simple nomenclature system that uses the first one or two letters of the genus followed by the first two letters of the species and a unique identifying number. As a guideline, IS elements that are more than 90% identical at the nucleotide level get the same name. These names can be reserved at the ISFinder database (http://www-is.biotoul.fr/), which is a specialized Web site dedicated to the recording, classification, and description of IS elements.

Origin of replication. Potential replication origins in microbial genomes are located by a method that examines short oligomers whose orientation is preferentially skewed around the origin *(9)*. These regions are also examined in the context of genes that are frequently observed near origins, and potential replication origins are assigned.

Paralogous gene families. Paralogous genes represent gene duplications within an organism. Identification of such genes is important because increased duplication of genes is associated with biological activity that is specific to that organism's environmental niche *(10,11)*. Collection of genes into paralogous families increases the confidence of each individual gene's assignments. Methods for the identification and annotation of paralogous genes are simple and involve searching against all proteins from the candidate organism using fairly stringent search parameters and inspecting the results. However, no single match criterion has been used to collect proteins into paralogous families. The degree of similarity between paralogous genes is the result of duplication that occurred over many different evolutionary time periods, is still unavoidably the subject of interpretation, and varies for each gene family and for each organism.

Pairwise protein searches. Predicted coding regions are searched against the nonredudundant database of publicly available proteins using the BLAST algorithm. BLAST matches are collected in a minidatabase. An extended portion of the predicted coding region is then aligned at the DNA-level hits from the minidatabase using PRAZE, a pattern-matching program that employs a modified Smith–Waterman algorithm. PRAZE has the capability of generating alignments across gapped regions and into other frames; therefore, it is particularly useful to identify frameshifts.

Start/stop definition and ORF management. Translational start site accuracy is currently about 75%. Annotators using a graphical user interface inspect the Glimmer results, compare the match against lengths of orthologous proteins, and examine upstream genes to best identify potential starts of translation. Regions containing potential frameshifts are identified and typically are resequenced using alternative sequencing chemistries. Electropherograms are examined in the context of the overall assembly, and authentic frameshifts are repaired. Approximately 200 frameshifts are found and resolved in a typical bacterial shotgun sequencing project.

Structural RNAs. Transfer RNAs are identified by tRNAscan. Ribosomal RNA genes and other structural RNAs are identified manually.

Our annotation process may be thought of as a two-step process in which an initial automatic annotation is used for preliminary assignments of function to genes, followed by a second stage of manual curation. For automated assignment, the initial searches are performed against proteins using a nonredundant database of all proteins available from the public archives. The search algorithm employed for these searches is Basic Local Alignment Search Tool (BLAST)-Extend-Repraze (BER). This program first executes a BLAST search *(3)* of each protein against the nonredundant database and stores all significant matches. Then, a modified Smith–Waterman alignment *(4)* is performed on the protein against the previous subset of matches of BLAST hits. To identify potential frameshifts or point mutations in the sequence, the gene is extended 300 nucleotides upstream and downstream of the predicted coding region.

During automated functional assignment, the best matches to BER searches are identified and used to assign a common name, gene symbol, and an EC number. The software that performs these assignments tests for a full-length match (at least 80% of the length of the subject) with a high-percentage identity (at least 35%). If more than one match is found, the program attempts to choose a match with a name that follows the TIGR naming conventions and assigns a role.

The program also tests for matches to the TIGRFAM data set (*5* and described in the next section). If the chosen BER match is a hypothetical protein from another species or if no pairwise matches meet the match criteria, the algorithm will go back to the TIGRFAM matches and look for any weak homologous hits. If any hits exist, the protein will be assigned a family name based on the TIGRFAM name, for example, "transcriptional regulator, TetR family." Proteins with a pairwise match to a hypothetical protein from another species, but no TIGRFAM hit, are named "conserved hypothetical protein." Protein with no TIGRFAM or BER matches remains a "hypothetical protein."

Data reflecting the success of the automated annotation are shown in Table 2. In the first part of the table, the quality of automated assignment of EC numbers and genetic names is assessed. The computationally derived assignments were compared to those made by a human annotator. One measure considered was the specificity of the automated method, which is the percentage of automated assignments that were correct. Considering all genes from five genomes for which the automated procedure attempted to assign a genetic name or EC number, the average specificities of automated assignment were 50.4 and 49.9%, respectively. *Sensitivity* measures the percentage of correct assignments made in comparison to the total number of possible correct assignments. For all genes that should have received assignments, the sensitivities for automated assignment were 75.3 and 51.9% for genetic names and EC numbers, respectively.

Table 2 also presents the results of automated vs manual assignment of common names to genes and shows that an average of 47.5% of the automatically assigned common names were altered (improved) by manual curation. Unlike the accuracy evaluation of automatic assignments of genetic names and EC numbers, it was not possible to determine if the automatically assigned common names were correct based on simple string comparison. This is because many cases for which manual and automated assignments differed were due to minor syntactic changes between assignments that did not alter the meaning of the gene name. In other words, the automated assignments were not strictly speaking erroneous, but they were not as precise as they could have been. Such

Table 2
Automated and Manual Annotation Comparison

Part A

	B. anthracis		B. suis		S. agalactiae		P. putida		S. oneidensis		Total	
	Filled	Null	Filled	Null	Filled	Null	Filled	Null	Filled	Null	Filled	Null
Genetic name												
Correct	580	1,964	601	1,109	440	749	668	1,912	815	1,321	3,104	7,055
False positives	817	200	421	147	328	67	883	410	611	195	3,060	1,019
False negatives	200	529	147	333	67	263	410	588	195	388	1,019	2,101
Sensitivity/specificity	74.4/41.5	78.8/90.8	80.3/58.8	76.9/88.3	86.8/57.3	74/91.8	62/43.1	76.5/82.3	80.7/57.2	77.3/87.1	75.3/50.4	77.1/87.4
EC number												
Correct	331	2,488	251	1,516	140	1,086	382	2,798	358	2,065	1,462	9,953
False positives	406	336	254	257	171	187	374	319	262	257	1,467	1,356
False negatives	336	386	257	240	187	165	319	343	257	233	1,356	1,367
Sensitivity/specificity	49.6/44.9	86.6/88.1	49.4/49.7	86.3/85.5	42.8/45	86.8/85.3	54.5/50.5	89.1/89.8	58.2/57.7	89.9/88.9	51.9/49.?	87.9/88

Part B

	B. anthracis	B. suis	S. agalactiae	P. putida	S. oneidensis	Total
Common names						
Predictions	3,561	2,278	1,584	3,873	2,942	1.4238
Altered	1,678	970	521	2,492	1,097	6,758
	47.1%	42.6%	32.9%	64.3%	37.3%	47.5%

The results of annotation processes for *Bacillus anthracis*, *Brucella suis*, *Pseudomonas putida*, *Shewanella oneidensis*, and *Streptococcus agalactiae* are shown. Part A describes the outcome of automated assignments of genetic names and EC numbers. An automated prediction either assigns could a genetic name or an EC number (filled) or could leave those fields empty (null). Once those computational assignments were made, human annotators inspected all information associated with each gene (such as BLAST and HMM search results) and verified whether those assignments were correct or required improvement. Results for correct assignments, false positives, and false negatives are shown as counts. Values for sensitivity were calculated as correct /(correct + false negatives); specificity was calculated as correct /(correct + false positives). Part B shows the outcome of automatically assigned common names in comparison to the final assignment of common name determined by manual annotation. Genes listed as altered were determined by a case-insensitive string-matching algorithm.

changes to names may involve disambiguation (e.g., conversion of "probable spore germination protein C" to "spore germination protein GerPC") or segregation of information found in the common name to other database fields (e.g., removal of EC numbers such as "peroxide dismutase [EC 1.15.1.1]" to yield "superoxide dismutase"). Some proteins had matches to proteins in the public archives, with names containing vestigial information from experimental isolation of those genes (e.g., "RXA00030, putative"), and were converted to "conserved hypothetical." Other cases simply reflect adherence to nomenclatural standards (e.g., conversion of "peptide ABC transporter" to "oligopeptide ABC transporter, ATP-binding protein"). Overall, relatively low success for automated assignment is exhibited in the data shown in Table 2; the values for sensitivity and specificity of automatic assignments and the number of common names requiring manual correction serve to underscore the importance of subsequent manual curation to produce quality bacterial genome annotation.

The SOPs developed at TIGR to assign gene function include methods to determine common name, EC number, genetic name, and role category. To make these assignments, curators view the data using a graphical user interface, Manatee (Fig. 2) that allows annotators to view all the data accumulated for each predicted coding region easily. A main information page displays identifying information and summarizes results from the various search programs. TIGRFAM scores can be viewed, and the user can follow links to internal and external pages that fully describe any particular model or to the multiple alignment of the predicted protein to the proteins that seeded the model. InterPro *(6)* motifs found in the predicted protein are listed along with the subsequence from the protein that matches the motif. A link can also be made to the Interpro documentation.

Manatee also displays a table summarizing the BER search results links to a display of the pairwise alignments. Links within the alignment display take the user to the source database entry of the specific match. At these databases, annotators can view information about the match proteins, such as active sites, membrane-spanning regions, and sites for deoxyribonucleic acid (DNA) binding. Annotators also check to see whether this information was experimentally determined or inferred from sequence similarity. Using this information, annotators can assess whether the protein from the predicted coding region has the same motifs, domains, and functionality as the match protein.

The information page also displays physical characteristics of the predicted coding region. The coordinates of the gene, the gene and protein lengths, molecular weight, and isoelectric point are listed. The annotator can link to views of the DNA and protein sequences, membrane-spanning regions, and a graphical view of the region surrounding the gene on the genome.

Distilling all of this information from a wide variety of sources to an accurate gene assignment is a complex task. All effort is made to attach reliable information to each gene, but annotators are wary of inferring too much from sequence similarity. This has led to a conservative approach to gene naming and a system of nomenclature in which the specificity of the name reflects confidence in the assignment. If there are multiple lines of evidence indicating that a protein has a specific function, including hidden Markov model (HMM; *7*) matches, multiple full-length pairwise matches with percentage identity greater than 30%, and conserved Interpro motifs (when applicable), then the fully descriptive name of the protein is used, and a gene name is assigned. For example, a name reflecting high confidence is "ribose ABC transporter, permease protein

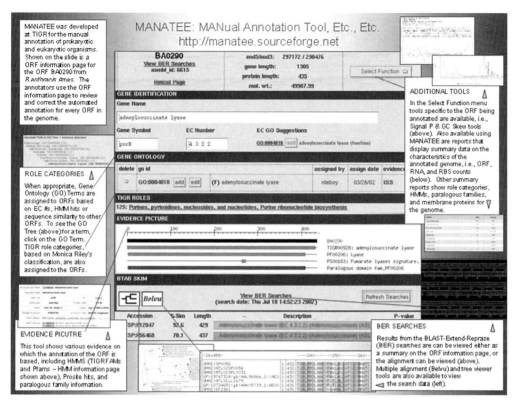

Fig. 2. The Manatee annotation system. This system is based on an intranet browser Web form and clickable interfaces. The figure depicts several sections of the Manatee interface that allow biologists to identify genes quickly and make high-quality functional assignments. Annotation projects may be evaluated from several different views. Users may view data by chromosome, gene ontology, paralogous protein, or InterPro domain. The interface also supports selection of genes based on attributes such as Enzyme Commission numbers, gene symbols, and matches to HMM or COG accessions. Users may review summaries such as frameshift reports or repeat analysis. Manatee also provides summary reports on annotation workflow, the overall genetic content of a genome, contact information for annotators, and graphical views of the complete genome. Requirements: Manatee is written in Perl and is run on a Web server such as Apache. Manatee also requires at least one MySQL (or Sybase) project database and associated search files. The project databases use a data model and schema developed at TIGR for representing eukaryotic or prokaryotic data. Example databases, documentation, and source code are available at manatee.sourceforge.net.

(*rbsC*)." However, if the search indicates multiple sugars could be the substrate for this transporter, a more general name would be given to the protein, such as "sugar ABC transporter, permease protein," without a gene name assigned. Further, if the type of substrate that the transporter carried was not clear, it would be named "ABC transporter, permease protein." Finally, if even the type of transporter could not be confidently assessed, then it would be named "transporter, putative." In some cases, only membership in a defined protein family is confidently known; in these cases, only the family name is assigned to the protein. In this way only, as much information is included in the common name of the protein as can be confidently determined.

THE TIGRFAM GENE FAMILY DATA SET

Sequence similarity is the most commonly used method for assignment of putative function to a newly discovered gene. Other sequence-based strategies for functional prediction, such as protein motif searching, and specialized composition algorithms (e.g., those that measure signal peptide or membrane-spanning domains) supplement similarity. At present, however, most functional assignments are the result of "transitive" assignment by pairwise comparisons of anonymous genes against the public protein archives. The name assignment usually reflects cautious interpretation for every gene; however, many false assignments may enter the annotation process. For example, functional assignments may reflect the best match of a protein to a query sequence, but still not necessarily match its true function. Annotators may have extensive knowledge for some, but not all, genes in a genome and only make correct assignments based on their area of expertise. For these and other reasons, the data from complete microbial sequencing have become problematic; accurate assignment by transitive annotation for subsequence is nearly impossible when the query sequence already contains ambiguously assigned function.

Structuring genes into ortholog families (groups of genes related by function) helps overcome problems associated with transitive assignment. Transitive assignment fails because no single assignment criteria such as a BLAST probability can be used to make correct pairwise assignments for all bacterial proteins. However, manual collection of genes into an orthologous family serves to define the criteria that place individual genes into a family and thereby establishes the criteria that can be used to add new members to that group. The multiple alignments of the family will contain the range of permitted similarities between genes, and the domains within the group, that are critical for correct identification. In favorable contrast to pairwise sequence similarity searches, the redundancy produced from multiple bacterial genome projects aids, rather than confounds, functional prediction because biologically relevant positions along the gene alignment will have greater statistical significance. Curation of family assignment requires that the function for each gene is reassessed as the genes are placed into ortholog groups. Misassigned functions typically associated with transitive annotation are corrected during this process.

Another benefit of ortholog families is that they are used for construction of HMMs and are searched against newly sequenced genomes. HMMs can then be used in searching, scoring, and classifying anonymous genes as members of an orthologous family. The TIGR data set of orthologous proteins (i.e., TIGRFAMS; 5) has been constructed into HMMs and is significantly more sensitive than classical sequence similarity. Each TIGRFAM object also has value in annotation not only because of search sensitivity, but also because the entry, maintained in a database, can itself be annotated extensively. Cut-off scores assigned to each family have been generated for each TIGRFAM entry that, when used for gene assignment, reliably assign gene function to anonymous proteins. The comment and literature citations in a TIGRFAM entry can explain how the presence of a matching region should be interpreted. Confirmed identifications and exceptions can be listed, and pitfalls for annotation may be described. TIGRFAM entries also may contain summaries of alternate name clarifications of nomenclature.

Both the TIGRFAM and protein families (Pfam) *(12)* HMMs are used in annotation; however, they differ greatly in focus. Pfam is a collection of HMMs designed to provide broad coverage of both prokaryotic and eukaryotic protein sequences. Pfam models generally are designed to encompass as many homologs as possible with a single, broad model. Pfam models are strictly nonoverlapping; no residue of any sequence should be part of two different regions scoring above the trusted cutoff to two different Pfam HMMs. The thoroughness and care in adding diverse sequences to curated alignments affords these HMMs excellent sensitivity. These HMMs work quite well and identify regions not easily marked in compilations of BLAST search results.

However, the breadth of Pfam models is a liability as well as a feature; a single model may hit a family so diverse in function that naming the protein based on its Pfam model name may be general to the point of being uninformative. Most TIGRFAM models overlap with one or more Pfam models, but cover the same proteins in a different way. In general, the hit region is longer, and the family typically has fewer members. A single Pfam family may be split into five or more TIGRFAM models. On the other hand, a single TIGRFAM model may represent a long protein with a number of different domains, each described by a separate Pfam HMM.

The motivation for orienting the scope of TIGRFAM models toward narrowly drawn families of proteins with the same or similar function and full-length homology is to match the goal of microbial annotation. This goal is to assign functionally descriptive names to whole proteins rather than produce an explicit feature table of domain boundaries and active sites. Approximately 25% of the proteins in a bacterial genome that can receive a functional assignment have membership in a TIGRFAM gene family.

THE COMPREHENSIVE MICROBIAL RESOURCE

Many Web sites have been constructed that contain data from complete bacterial genomes. A subset of these sites is listed in Table 3. In addition, TIGR has developed the Comprehensive Microbial Resource (CMR), a data warehouse of all completed microbial genomes. Users can access the CMR at www.tigr.org/CMR. To query biological elements such as function, role category, protein similarity, hydrophobicity, or genetic symbol across all bacteria, the CMR database was populated by carefully parsing relevant data types from each individual genome and storing those data as separate objects. This makes cross-genome analyses both easier and more meaningful. For example, it is possible to retrieve genes from all completed bacterial genomes that have been assigned the same biological role (i.e., "Display all genes involved in amino acid biosynthesis"). Complex queries that use many of the attributes, such as taxonomy, similarity to other proteins, Gram staining, or chromosome topology, allow retrievals like "Display all transporters with >5 membrane-spanning domains, and have a MW of 36–51 kilodalton."

There are several levels at which data can be reported in the CMR. For example, at the gene level, users are able to view an individual gene's primary common name assignment, the common name assigned by the automated annotation process, the genetic symbol, the gene's EC number, the isoelectric point, the molecular weight, the DNA sequence, and the protein. Users may also view the hydrophobicity profile; links to

Table 3
Microbial Genome Resources

Clusters of Orthologous Groups (COGs) of proteins. Phylogenetic classifications of proteins encoded in complete genomes. Each COG consists of individual proteins or groups of paralogs and corresponds to an ancient conserved domain. http://www.ncbi.nlm.nih.gov/COG/

DNA Structural Analysis of Sequenced Microbial Genomes. A method of visualizing structural features within large regions of DNA. http://www.cbs.dtu.dk/services/GenomeAtlas/

European Bioinformatics Institute (EBI). Manages databases of biological data, including nucleic acid, protein sequences, and macromolecular structures. http://www.ebi.ac.uk/

Genome Information Broker (GIB) for microbial genomes. Completed microbial genomes compiled from the International Nucleotide Sequence Database (DDBJ/EMBL/GenBank). http://gib.genes.nig.ac.jp/

High-quality Automated and Manual Annotation of Microbial Proteomes (HAMAP). A project that aims to annotate automatically a significant percentage of proteins originating from microbial genome-sequencing projects. http://us.expasy.org/sprot/hamap/

Kyoto Encyclopedia of Genes and Genomes (KEGG). An effort to computerize current knowledge of molecular and cellular biology in terms of the information pathways that consist of interacting molecules or genes and to provide links from the gene catalogs produced by genome-sequencing projects. http://www.genome.ad.jp/kegg/kegg2.html

Microbial Genome Database (MBGD). The aim is to facilitate comparative genomics from various points of view, such as ortholog identification, paralog clustering, and motif analysis. http://mbgd.genome.ad.jp/

National Center for Biotechnology Information (NCBI). Serves as a national resource for molecular biology information. NCBI creates public databases, conducts research in computational biology, develops software tools for analyzing genome data, and disseminates biomedical information. http://www.ncbi.nlm.nih.gov/

Paulsen transporter page. Provides the genomic comparison of membrane transport systems. www.membranetransport.org

Protein Extraction, Description, and ANalysis Tool (PEDANT). Provides genome analysis and annotation for all the genomes that have been sequenced. http://pedant.gsf.de/

The Genome Channel. Prediction of prospective gene and protein models for analysis, including computer-annotated genomes. http://compbio.ornl.gov/channel/

What Is There? (WIT). Supports the curation of functional assignments made to genes and the development of metabolic models. http://wit.mcs.anl.gov/WIT2/

other resources, such as Swiss-Prot; secondary structure; and third-position GC skew. Users examining an individual gene may also wish to link to genes from other organisms with a similar function. The service that allows users to do this explicitly displays the evidence that links genes of similar function. Types of evidence may be membership in the TIGRFAM or Pfam protein families, Clusters of Orthologous Groups (COGs) of proteins, sequence similarity, common EC numbers, or common role categories.

Another level of data display on the CMR is for an individual microbial genome. Information of this kind includes graphical displays that show genes placed linearly on a region of the chromosome or as a complete circle for an entire chromosome. Several CMR pages give overviews, such as of codon usage, GC plots, computer generated two-dimensional (2D) gels, and tables summarizing information such as average gene size, number of coding regions, and the like. The CMR also presents comparative information between microbial genomes. The pages make use of many of the data types associated with individual genes, such as hydrophobicity, matches to protein motifs, role categories, and EC numbers, and can be used for large-scale comparisons between organisms. Other genome-to-genome services compare protein similarities and whole genome alignments between different species.

ADDITIONAL SERVICES

Other activities are under way at TIGR that may be interesting to scientists involved in microbial annotation. My group provides the Annotation Engine, an automated annotation of prokaryotic sequence offered at no cost to the sequencing center. The service runs a bacterial genome sequence through the annotation pipeline. The resulting automated annotation from the sequence is returned to the sequencing center. This includes coordinates of open reading frames (ORFs) and RNAs; common name, gene symbol, EC numbers, TIGR roles, and Genome Oncology terms for proteins; underlying BER search results; HMM; InterPro; signal peptide analysis; and membrane spans. These data are available as a MySQL database, which can then be used with TIGR's manual annotation tool Manatee.

The Annotation Engine serves several functions. It initializes the annotation in a format to promote consistency of data types. It also encourages genome centers to use SOPs that embody uniform annotation standards and allow straightforward reincorporation of annotation back into the CMR data management system for display at the CMR (when the sequencing center agrees). It allows the genome center to display their preliminary annotation locally at their own Web site, promoting recognition of their sequencing effort, with a relatively small bioinformatics resource investment. Because some of the annotation effort is decentralized, it distributes the labor involved for annotation and scale with increased genome sequencing.

Other services offered are training classes on microbial annotation and numerous software packages that may be of value to genome annotation. Check TIGR's Web site www.tigr.org for more information.

ACKNOWLEDGMENTS

This work was supported by the US Department of Energy, Office of Biological and Environmental Research, Cooperative Agreement DE-FC02-95ER61962, amendment 8.

REFERENCES

1. Riley M. Functions of the gene products of *Escherichia coli*. Microbiol Rev 1993; 57:862–952.
2. Ashburner M, Ball CA, Blake JA, et al. Gene Ontology: tool for the unification of biology. The Gene Ontology Consortium. Nat Genet 2000; 25:25–29.
3. Altschul SF, Gish W, Miller W, Myers EW, Lipman DJ. Basic Local Alignment Search Tool. J Mol Biol 1990; 215:403–410.
4. Waterman M. General methods of sequence comparison. Bull Math Biol 1984; 46:473–500.
5. Haft DH, Selengut JD, White O. The TIGRFAMs database of protein families. Nucleic Acids Res 2003; 31:371–373.

6. Mulder NJ, Apweiler R, Attwood TK, et al. The InterPro Database, 2003 brings increased coverage and new features. Nucleic Acids Res 2003; 31:315–318.

7. Eddy SR. Hidden Markov models. Curr Opin Struct Biol 1996; 6:361–365.

8. Delcher AL, Harmon D, Kasif S, White O, Salzberg SL. Improved microbial gene identification with Glimmer. Nucleic Acids Res 1999; 27:4636–4641.

9. Salzberg SL, Salzberg AJ, Kerlavage AR, Tomb JF. Skewed oligomers and origins of replication. Gene 1998; 217:57–67.

10. Fraser CM, Gocayne JD, White O, et al. The minimal gene complement of *Mycoplasma genitalium*. Science 1995; 270:397–403.

11. Tomb JF, White O, Kerlavage AR, et al. The complete genome sequence of the gastric pathogen *Helicobacter pylori*. Nature 1997; 388:539–547.

12. Sonnhammer ELL, Eddy SR, Birney E, et al. Pfam: multiple sequence alignments and HMM-profiles of protein domains. Nucleic Acids Res 1998; 26:320–322.

4

Bioinformatics and Microbial Pathogenesis

Fiona S. L. Brinkman and Joanna L. Fueyo

Infectious diseases are the leading cause of mortality for people younger than 40 years of age, and may be the leading cause of death for all persons worldwide if the infectious origins of some diseases (respiratory, digestive, etc.) are taken into consideration *(1)*. The increased prevalence of antibiotic resistant bacteria continue to threaten, as do new threats from infectious biological weapons and emerging and reemerging infections. To combat this problem, the genomes of an increasing number of infectious agents have been sequenced to gain fundamental knowledge of their genetic makeup and insight into how they may be controlled. There has also been a corresponding increase in the popularity of whole genome approaches to the study of microbial pathogens (i.e., microarray analysis, in vivo expression technology, proteomic approaches, etc). The fundamental premise is that, by studying microbial genes in the highly parallelized manner typical of genome-based experimentation, discoveries of mechanisms of microbial pathogenesis may be accelerated and the process leading to the identification of antiinfective drug and vaccine targets shortened. However, the resulting exponential growth of genome-based data has created a critical need for new bioinformatic methods and tools to organize and analyze this information. Knowledge of such methods and tools is now increasingly necessary for any interdisciplinary microbiology laboratory studying infectious disease.

We therefore present an overview of computational approaches that may be used to aid analysis of microbial pathogen genomes and investigations of microbial pathogenicity. It should be noted that computational approaches specific for the analysis of pathogen genomes and associated data are still very limited, particularly for *de novo* virulence gene discovery. Also, controversies surrounding the definition of virulence, and what comprises a virulence factor, complicate analysis. Such controversies also make it difficult to determine the accuracy of computational approaches in identifying virulence genes or pathways. However, as illustrated in this chapter, there have been some successes in this area, and there are many obvious analyses and tools that remain to be developed. The future looks bright for bioinformatics research and its potential to aid the study of important infectious diseases.

GENERAL CONSIDERATIONS

Many general bioinformatics approaches described in other chapters of this book are applicable to the study of microbial pathogens; for brevity, a selection of such methods is

From: *Microbial Genomes*
Edited by: C. M. Fraser, T. D. Read, and K. E. Nelson © Humana Press Inc., Totowa, NJ

briefly described here. This chapter focuses on methods developed that are more specific to the study of pathogens and pathogenicity and describes selected bioinformatics methods in the context of pathogen studies. Other informatic approaches, such as epidemiological methods, are not covered, although such methods will likely increasingly be integrated with the methods described here as the field moves more toward population-based genomics. Phylogenetic approaches are also not covered, although there are numerous examples for which phylogenetic and related evolutionary analyses have provided significant insight into the evolution of virulence and mechanisms of pathogenesis (i.e., refs. *2–5*). The chapter has a slight emphasis on approaches applicable for the study of bacterial pathogens, rather than viral, protozoal, or fungal pathogens, and there is a focus on select recently developed approaches. Readers are therefore encouraged to explore the Web sites mentioned and referenced literature for further information. Such referenced material also frequently provides more information on the assumptions made for each method and the limitations and appropriate use of each analysis, as well as advice on interpretation of the results. Armed with such knowledge, computational analyses can avoid overinterpretation by those too enthusiastic about their use, and conversely underutilization by those distrustful of them.

Pathogen bioinformatics, and the use of computational approaches to specifically aid microbial pathogenesis, is a relatively new field, but is growing rapidly *(6–8)*. The approaches mentioned in this chapter represent the framework for what will likely become increasingly sophisticated approaches for the bioinformatic analysis of pathogens and mechanisms of microbial pathogenesis.

IDENTIFICATION OF VIRULENCE FACTORS THROUGH HOMOLOGY OR MOTIF-BASED ANALYSES: THE NEED FOR VIRULENCE GENE DATABASES AND THE DEVELOPMENT OF DISEASE-SPECIFIC ONTOLOGIES

One of the most frequently used bioinformatics approaches for the analysis of pathogen genomes involves the identification of genes homologous to known virulence factor genes. Ironically, no comprehensive tool has yet been developed to facilitate this common analysis. General methods for sequence similarity searches against a database are instead frequently used in a customized manner and include BLAST (Basic Local Alignment Search Tool), Position-Specific Iterated BLAST (PSI-BLAST), and FASTA *(9–11)*; less frequently used is the more sensitive, but slower, Smith–Waterman approach *(12)*. Other, protein structure based, methods such as the Vector Alignment Search Tool (VAST; http://www.ncbi.nlm.nih.gov/Structure/VAST/vast.shtml) will likely increasingly be used as more protein structures of virulence factors are deduced. Information about BLAST and related methods is best obtained from a bioinformatics textbook *(13)* or articles that describe issues to watch out for when performing such analyses *(14)*. Of course, all such database search methods depend on researcher knowledge of which genes in a database are virulence genes or development of a specific database of virulence genes for the analysis. However, there are currently few publicly available virulence factor databases; most commonly, a manual or semimanual review of sequence similarity results is required, with examination for similarity to selected virulence factors, or an in-house list of virulence factor genes is constructed.

One of the reasons for the lack of virulence factor databases is that defining a virulence factor is not trivial (the "what is virulence" debate) and can be quite controversial. The traditional Koch's postulates that define a disease-causing organism have had to be revised to the "molecular Koch's postulates" for the definition of virulence factors. Now, the definition of a virulence factor is being further refined, as the complex interplay among a microbe, its genes, its environment, and the genetic susceptibility of the host is increasingly appreciated *(6)*. However, some databases of limited scope have been constructed, such as the BacBix database of virulence factors (http://www.jenner.ac.uk/BacBix3/Welcomehomepage.htm), which is a collection of entries from the PRINTS database *(15)*. Swiss-Prot and other manually or semimanually curated gene data sets can also be sources of data sets of known virulence genes if appropriate queries are made and they are subject to further curation. Major genome centers or molecular pathogenesis laboratories have collected, or are in the process of collecting, lists of "virulence" genes of interest; some of these lists will be made publicly available.

One of the features that would be most useful for future development of such databases is the development of appropriate *ontologies* (a set of defined vocabularies) to define the different kinds of virulence-associated attributes and their relationships. With a complex attribute such as virulence, which is dependent on so many different factors, different levels of gene function and the interplay between the genes and their environment must be taken into consideration. Virulence gene databases will only become useful resources if each gene in the database is classified with such ontologies and is linked to detailed, organized information about the conditions in which a given virulence factor has been found to play a role in disease. Only then will contextual evaluation of a homolog of a virulence factor lead to more accurate predictions of their true role in virulence or the conditions under which they may play a role in virulence.

Related to the sequence or structural similarity search approach, another method for identification of putative virulence factors in genomes involves examination of the presence of motifs or protein domains associated with virulence factors. Again, the approaches and resources used, such as InterPro (which provides integrated access to PROSITE, Pfam, PRINTS, ProDom, SMART, and TIGRFAMs databases) are better described elsewhere *(16)*. Many of these resources have not been specifically designed for the identification of genes involved in pathogenesis, so their use frequently involves certain targeted analyses relevant to a researcher's particular interest.

Note that all of these methods can also be easily translated into a search for drug or vaccine targets if an appropriate database or knowledge of known targets is developed. The philosophy for such an approach is that if a given antimicrobial drug, for example, binds a particular protein in one microorganism, then homologs of that protein in another, similar microorganism may also be a suitable drug target. Again, appropriate ontologies need to be developed for such an approach to have a reasonable level of accuracy in correctly predicting targets.

Regardless of whether virulence factors or therapeutic targets are being identified, caution must be utilized in such sequence similarity and motif-based analyses, since there has been little evaluation of the accuracy of such approaches in new virulence gene detection in a genome. Again, what may be a virulence factor gene in one pathogen's genome may not play a role in virulence in another pathogen or under a different condition. However, there are a number of studies, discussed below, that illustrate the utility of such approaches in helping prioritize which genes in a genome could constitute suitable targets for further laboratory investigation. Such methods will therefore remain a commonly used tool; it is hoped in the future there will be more formalization of the process, with associated analyses of accuracy. As more genomes are analyzed in this fashion, trends in the success rate of such methods for different gene families may become more apparent, for example, thereby increasing the effectiveness of these approaches.

IDENTIFYING PATHOGENICITY ISLANDS
AND RELATED SEQUENCES: GENOME COMPOSITION APPROACHES

Pathogenicity islands (PAIs), first discovered and named in uropathogenic *Escherichia coli* in the late 1980s *(17)*, have been studied intensively in microbial pathogenesis laboratories because they represent the intersection of two interesting phenomena: bacterial pathogenesis and horizontal gene transfer (HGT). As the name implies, genes associated with bacterial pathogenesis were found clustered on such islands (i.e., *18, 19*), and accumulating evidence suggests that such islands have horizontal origins *(20, 21)*. As more prokaryotic genomes become available, it is evident that the concept of PAIs can be extended to other genetic elements that share the general structure of PAIs but encode functions other than virulence *(22)*. These genetic elements, collectively called *genomic islands*, encode genes involved in such diverse cellular functions as secondary metabolism (metabolism islands), antibiotic resistance (resistance islands), and secretion (secretion islands). All appear to confer some sort of functional advantage to the organism, hence the prevalence of genes involved in pathogenesis in such regions for microorganisms that can successfully infect humans.

Features commonly associated with genomic islands include the presence of flanking repeats, mobility genes (e.g., integrases, transposases), proximal transfer ribonucleic acids (tRNAs), and atypical guanine and cytosine content (%G+C; *23*). tRNAs are known phage integration sites *(24,25)*, and they may serve as integration sites for mobile genetic elements that become PAIs. %G+C and additional species-specific "deoxyribonucleic acid (DNA) signatures" (e.g., dinucleotide bias and codon usage profile) have also been proposed as useful in identifying islands *(26,27)*, which frequently exhibit DNA signatures distinct from the rest of the genome.

Although many of these features have been adopted separately in applications for analyzing genomic islands, there is currently only one computational tool that integrates multiple features for island detection: IslandPath, a Web-accessible service, graphically displays island-associated features in full-genome context *(28)*. It displays %G+C of predicted open reading frames, dinucleotide bias for gene clusters (an independent genome composition measure), known or probable mobility genes, and tRNAs. These features are represented by different symbols in a compact graphical display at gene resolution (Fig. 1; http://www.pathogenomics.sfu.ca/islandpath/); users can further interact by setting different %G+C cutoffs and linking to gene annotations and analyses at the National Center for Biotechnology Information. IslandPath is proposed to provide a quick way to browse a genome for putative islands that may be of interest for further computational or "wet-lab" characterization. The first version of this application permits analysis of all currently available fully sequenced bacterial and archaeal genomes (http://www.ncbi.nlm.nih.gov/PMGifs/Genomes/micr.html), and the data set is updated regularly.

There are some cautionary notes regarding such composition-based approaches: First, certain DNA signal-based approaches by themselves have been shown to be poor indicators of HGT *(29)*. However, IslandPath complements multiple DNA signal analyses with additional annotation features that correlate with known pathogenicity islands as a way to overcome this and is in the process of adding information about homology to known virulence genes. Of course, IslandPath and other genome composition-based approaches cannot be effective in the identification of islands or virulence genes obtained from organisms with similar DNA signals or virulence genes that have not been subject to HGT, so such approaches only identify a subset of possible virulence factor candidates. However, interestingly, the majority of the classic virulence genes are in genomic regions with unusual DNA signals (F. S. L. Brinkman, 2003, unpublished; see IslandPath's on-line help file for examples).

The approach of identifying virulence genes by first identifying PAIs has clearly been successful in microbial pathogenesis research, as illustrated by the number of publications on the subject *(23)*. It is hoped that such approaches may also aid detection of other genes of interest in pathogens, such as antibiotic resistance genes, because such genes are noted for also being present on genomic islands, which share similar features that may be computationally identified *(22)*.

IDENTIFYING PATHOGENICITY ISLANDS AND VIRULENT SPECIES-SPECIFIC SEQUENCES: COMPARATIVE GENOMIC APPROACHES

Comparative genomic approaches for the bioinformatic analysis of pathogens are based on a simple fundamental premise: Differences or similarities in the ability of two microbes to cause disease will be reflected in their genomic sequence data. A number of comparative genomics tools have been developed that may be used to compare whole microbial genomes, identifying how the presence or absence of genes correlates with a particular disease phenotype. For such comparisons, there are really two main approaches: The most popular approach involves the comparison of closely related genomes, identifying differences that may correlate with pathogenicity. The second approach is the converse: It compares very different genomes of species that cause similar infections and identifies similarities in their genomes that may correlate with a particular virulence phenotype.

For the first approach, the Artemis Comparison Tool (ACT; http://www.sanger.ac.uk/Software/Artemis/) is an example of an application widely used for whole genome comparisons of relatively closely related microbes. ACT's sequence comparison is usually the result of a BLASTN or TBLASTX search that has been further processed by MSPcrunch *(30)* to refine BLAST high-scoring segment pairs. This BLAST analysis must be processed separately first because ACT is essentially a viewing tool only. An example of the resulting output for this intuitive Java-based application is illustrated in Fig. 2.

Another example of such comparative genomic tools that is more specific for the analysis of pathogens is the Enteric, Menteric, and Maj suite of applications, which is centered around the analysis of enteric microbial genomes *(31)* (http://bio.cse.psu.edu/). The Web-based Enteric tool produces a graphical view of pairwise alignments between a reference *E. coli* genome and sequences from each of several related organisms, covering 20 kb around a specified position. Menteric produces nucleotide-level multiple align-

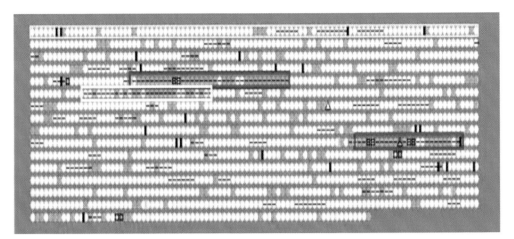

Fig. 1. An example of IslandPath's graphical display for a complete genome (*Helicobacter pylori* 26695) that facilitates the viewing of known pathogenicity islands and the identification of putative new islands. Each circle in the graph corresponds to a predicted protein-coding open reading frame in the genome. Circle colors indicate %G+C (yellow, above a chosen high cutoff; pink, below a chosen low cutoff; and green, between the cutoff values). Strike lines across the circles represent regions with dinucleotide bias according to set criteria (see ref. *28*). Vertical bars denote transfer RNA and ribosomal RNA (rRNA) locations (tRNA, black; rRNA, purple; both, dark blue). Black squares mark known or probable transposase genes, and black triangles indicate the location of known or probable integrase genes. Regions containing several of these features may represent putative genomic islands. For this example, three known or putative islands in *H. pylori* 26695 are marked with colored boxes: yellow box, CAG pathogenicity island; blue box, contains *virB* homologues and is not present in *H. pylori* J99; red box, plasticity zone that contains different genes for J99 and 26695. Note how well long regions of dinucleotide bias (marked with long strike lines) correlate with the probable or known genomic islands in this genome. The utility of IslandPath is further illustrated on-line (http://www.pathogenomics.sfu.ca/islandpath/) by an analysis of *Salmonella* genomes, revealing both known islands and putative islands newly identified by IslandPath.

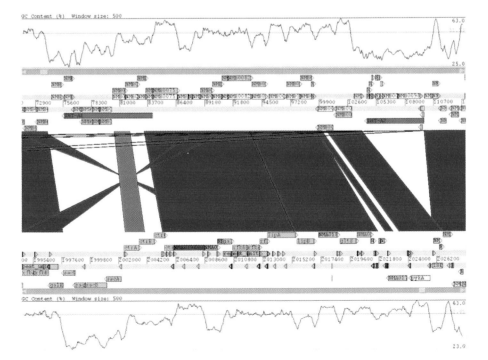

ments of the same sequences, together with open reading frame and regulatory site annotations, in a 1-kb region surrounding a given position. Both use a BLASTZ algorithm developed for such genome comparisons *(32)*. Maj combines features of Enteric and Menteric in a Java based viewer that has additional functionality, such as a zooming function.

Such comparative genomic tools, as well as custom-made analyses, may be used to perform comparisons between pathogens that cause disease and close relatives that are avirulent or comparisons between related pathogen species or between strains with differing virulence phenotypes. There are numerous recent examples in the literature for which such approaches have provided new directions for microbial pathogenesis research *(31,33–38)*. However, it should be noted that there is a growing appreciation for the complex variability between microbial species and strains, as illustrated by the comparison of *E. coli* K12 and *E. coli* O157 genomes *(38)*. Clearly, such tools will increasingly need to handle more complex views of an increasing number of genome sequences.

For the second comparative genomics approach, the reverse approach is taken: The researcher compares the genomes of evolutionarily distant species that cause similar types of infections or that inhabit the same pathogenic niche to determine which sequences are conserved between the genomes.

Target-BLAST *(39)* is an example of such an approach; it is based on the premise that different bacterial species that share common features, such as the ability to cause a similar spectrum of human diseases, may have sequences in common that facilitate colonization of the host or establishment of infection. The initial Target-BLAST algorithm *(39)* focused on the identification of DNA and protein sequences unique to organisms that colonize and cause disease in the human respiratory tract, namely, *Streptococcus pneumoniae* and *Haemophilus influenzae*. The method relies on iterative applications of BLAST and PSI-BLAST, followed by secondary feature analysis, for the analysis of whole and partial genomic and proteomic sequence data to identify sequences conserved between the above-named human respiratory tract pathogens and other known bacterial pathogens of the human respiratory tract.

Using Target-BLAST, protein sequences were identified that were uniquely conserved among infection-causing pathogens in the same niche: the oronasopharynx and the lower respiratory tract in humans *(39)*. Twelve sequences were uniquely conserved among *H. influenzae, S. pneumoniae,* and in most cases at least one other bacteria that infects the human respiratory tract, to the exclusion of all other nonrespiratory tract genomes analyzed. Of these 12, 4 were known virulence factors, and the remaining 8

Fig. 2. (*Opposite page*) Screen shot of the main window of the Artemis Comparison Tool (ACT), a tool that can aid comparative genomic analysis of pathogen genomes or pathogen vs nonpathogen genome comparisons. Each red/pink box corresponds to single BLAST match, with red representing a good match and white/pale pink a lower-scoring match. The two sequence views in ACT above and below the red/pink boxes show the forward and reverse strands of the sequences (subject sequence at top, query sequence at bottom) and a representation of the three translation frames in each direction. The %G+C content, using a sliding window approach, is also shown. In this example, a comparison of the genomes of *Neisseria meningitidis* MC58 (serogroup B) and *N. meningitidis* Z2491 (serogroup A) is shown: two sero-groups of the causative agent of meningococcal meningitis that have differing virulence phenotypes. Using this tool, genomic regions specific to a microbe that has a particular disease phenotype can be identified, facilitating the discovery of genomic regions with potential roles in microbial pathogenesis. (Image supplied by the Pathogen Sequencing Unit, Sanger Institute, UK.)

were then selected for molecular biological, genetic, and animal model of infection studies. According to these further laboratory studies, this method of screening correctly predicted virulence factors the majority (87.5%) of the time for human respiratory tract pathogens, therefore supporting the hypothesis that pathogens inhabiting the same pathogenic niche share common sequences that play a role in infection. Therefore, thus far, Target-BLAST has been a useful approach for *de novo* prediction of virulence-associated sequences from whole and partial genomic sequence data and could easily be adapted to identify sequences common to microbes that share other features. One benefit of the Target-BLAST approach is that it can easily handle both partial, unannotated DNA sequences and complete, annotated genomes, which enables the simultaneous discovery of genes conserved in both complete and partially sequenced genomes.

Another related approach is illustrated by SEEBUGS *(40)*. SEEBUGS can be used to search complete, annotated genome data using the FASTA algorithm *(41)* for concordant protein sequences of known antibiotic targets. That is, it can be used to perform subtractive genome analysis to search for genes from one organism that are found in another set of organisms, but not found in a third set of organisms. This has been used for the identification of genes shared between bacterial pathogens, but not apparently present in the host genome, which may represent suitable initial candidates for investigation as broad spectrum antibiotic targets. It should be noted that this method is limited to complete, annotated genomes, which means that genes or proteins conserved in partial genomic sequence data would be missed. The approach based on sequence similarity used by this and other similar methods is also limiting as it does not take into consideration differences in rates of gene evolution. Although such methods appear to be a good first screen for target identification *(42)*, it is hoped that future methods of concordance analysis may use more sophisticated evolutionary approaches, further increasing their ability to detect suitable targets.

It should be noted here that microarray-based comparative genomics approaches also result in the development of new bioinformatic tools for analysis of such data. Microarray-based analysis is not discussed further here as it is described elsewhere in this book, but such analyses have been successful in furthering microbial pathogenicity research. For example, a comparison of endemic and pandemic strains of *Vibrio cholerae* identified genes that correlated with pandemic vs endemic strains *(43)*.

Comparative metabolic pathway analysis, such as performed using MetaCyc and Pathway Tools or using KEGG (Kyoto Encyclopedia of Genes and Genomes), is also a powerful comparative method that moves away from the genome-centric view and toward the cellular view (see description of MetaCyc, Pathway Tools, and KEGG in the following section, "Predicting Surface-Exposed and Secreted Proteins"). It should also be noted that whole genome gene function comparisons and metabolic pathway comparisons can greatly aid the identification of suitable drug targets and vaccine candidates through the identification of the specificity of a gene or pathway to a given microbial species *(44,45)*. Cluster of Orthologous Groups (COG) analysis *(46)* is another common gene function-oriented approach used. COGs are identified by comparing protein sequences encoded in over 40 complete genomes and identifying individual proteins or families of proteins present in at least 3 phylogenetic lineages and thought to reflect an orthologous group (*orthologous genes* are genes related to each other and diverged due to speciation of the species compared). Genes within a given COG are thought more likely to have a similar function (vs *paralogs*, which are genes that diverged after gene duplication), so comparative analysis of genomes through classification of genes into COGs can aid the identification of genes unique to a given organism or common to a set of organisms.

As more genomes are sequenced and the move is toward genomics that are more population based, all such comparative genomics tools will play an increasingly important role in pathogen bioinformatics research.

PREDICTING SURFACE-EXPOSED AND SECRETED PROTEINS

Computational prediction of the subcellular localization of proteins is a valuable tool for genome analysis and annotation because a protein's subcellular localization can provide clues regarding its function in an organism. For microbial pathogens, the prediction of proteins on the cell surface is of particular interest because of the potential of such proteins to be primary drug or vaccine targets (it is generally thought to be easier to develop drugs that do not have to cross lipid membranes, and vaccine candidates must be accessible to the immune system). In addition, prediction of proteins that are secreted is of interest because of their potential role in interaction with the host on infection and therefore their potential role in pathogenesis. There are a number of classic secreted toxins (exotoxins), proteases, and the like that are well-known virulence factors, such as cholera toxin, hemolysins, hyaluronidase, and immunoglobulin A protease. A number of secreted proteins have also been found to be excellent drug or vaccine targets, for example, the pertussis toxin, found in the second-generation diphtheria, tetanus, and pertussis vaccines used today.

A protein's subcellular localization is influenced by several features in the protein's primary structure, such as the presence of a signal peptide or membrane-spanning α-helices. Several algorithms have been developed to identify such single features for both prokaryotic and eukaryotic microbes (for a review, see ref. *47*). In addition, comprehensive tools have been developed that analyze several features at once. A list of some of the currently available methods that can aid prediction of potential surface-exposed or secreted proteins is summarized in Table 1. This table also lists methods that may aid topological prediction of membrane proteins, which can aid prediction of surface-exposed sequences.

Prediction of protein subcellular localization is also a powerful approach for the identification of pathogen proteins that may play a role in interaction with the host or may be suitable therapeutic/prophylactic targets; examples of successful predictions are mentioned below. However, it should be emphasized that the accuracy of each predictive method differs greatly *(47,48)*. Some methods also have significant pitfalls that should be considered. For example, the original version of PSORT does not predict secreted proteins, and it forces classification of a given protein into one of the remaining possible localization sites. Therefore, it tends to overpredict certain localizations of proteins *(48)*. Most predictive methods for eukaryotes are also based on animal, yeast, and plant sequences, so predictions for protozoan pathogens may be not as accurate. Likewise, methods for the prediction of subcellular localization of bacterial proteins are primarily based on Gram-negative and Gram-positive bacteria, so predictions

Table 1
Tools That Aid the Prediction of Putative Secreted Proteins, Surface-Exposed Proteins, and Surface-Exposed Sequences From Pathogen Genome Sequences[a]

Prediction of possible surface-exposed membrane proteins, or their topology, for eukaryotes and Gram-positive bacteria
• HMMTOP *(73)*
• TMHMM *(74)*
• DAS *(75)*
• TMpred *(75a)*
• SOSUI (Tokyo University of Agriculture and Technology)
• TMAP (Karolinska Institut, Sweden)
• TopPred 2 (Pasteur Institute)
• Janulczyk and Rasmussen *(76)* for cell wall-attached proteins in Gram-positive bacteria

Prediction of possible surface-exposed outer membrane proteins (OMPs), or their topology, for Gram-negative bacteria
• PSORT-B—Support Vector Machine, BLAST, and pattern-based OMP prediction *(48)*
• Neural network-based prediction of OMP β-strand topology *(77)*
• Neural network-based prediction of OMP transmembrane region topology *(78)*
• Hidden Markov model-based prediction of OMPs *(79)*
• BBF (Beta-Barrel Finder) *(80)*
• Hydrophobicity-based prediction of OMPs *(81)*
• Sequence composition and pattern-based OMP predictor *(82)*

Prediction of secreted proteins
• PSORT-B *(48)* for Gram-negative bacterial sequences
• PSORT II *(84)* for eukaryotic sequences
• iPSORT *(84)* for classification of N-terminal sorting signals for eukaryotes
• TargetP *(47)* for classification of N-terminal sorting signals for eukaryotes
• SignalP *(85)* for identification of signal peptides in eukaryotes and bacteria
• SubLoc *(86)* uses Support Vector Machine to assign a prokaryotic or eukaryotic protein to some subcellular localization sites, including the extracellular location
• NNPSL *(87)* uses amino acid composition to assign a prokaryotic or eukaryotic protein to some subcellular localization sites, including the extracellular location
• PSORT *(88)* for lipoprotein motif analysis
• ExProt *(89)* for prokaryotic sequences (not to be confused with EXProt, a database of proteins of experimentally verified function)
• Domain projection method *(90)* for eukaryotic sequences
• Gomez et al. *(91)* for Gram-positive bacterial sequences
• Guttman et al. *(52)* for prediction of type III secretion effector proteins for Gram-negative bacteria

---

[a]Such proteins may play a role in interaction of the pathogen with its host, and their sequences are of interest as possible drug or vaccine targets. Note that the currently available tools still predict only a subset of possible secreted or surface-exposed proteins. For example, type IV secretion signals are currently not predicted. General approaches applicable to all localizations (see text for comments) include BLAST and related sequence similarity approaches and InterPro and related motif- and domain-based approaches. DAS, dense alignment surface.

may be less accurate for bacteria that fall outside these groups, such as the Spirochetes and Chlamydiaceae. The data sets used to train some of these programs are quite small, so users are encouraged to examine the primary literature associated with each method.

A selection of training data sets, which may be used to develop individual customized analysis, is available from the psort.org Web site, and psort.org also provides links to methods available as Web-based applications.

Table 1 also lists some more general approaches for analysis of protein subcellular localization; these approaches are useful if appropriate criteria are used. For example, it has been shown that subcellular localization is a fairly conserved trait *(49)*, so simple approaches based on sequence similarity have been developed, such as SCL-BLAST (SubCellular Localization–BLAST *[48]*). However, because of the domain nature of proteins, accuracy is increased if both the query and the subject proteins of a sequence similarity approach are of similar length. This reduces the potential for incorrect localization predictions based on similarity to a single domain of a protein in a database, a protein with a domain that may reside in different localization sites. SCL-BLAST has a calculated precision of 97% based on the use of a data set of 1443 proteins of known localization as its database *(48)*; its accuracy will likely only increase as more proteins of known localization are added to the data set. Another general approach, the use of motifs associated with proteins of particular subcellular localizations, also works well as long as motifs with high precision/specificity are utilized *(48)*. For example, the PROSITE motif PS00695 (ENT_VIR_OMP_2) is highly specific for identifying a particular family of outer membrane proteins of Gram-negative bacteria.

There are a number of success stories in which analysis of protein localization has aided in the study of microbial pathogenesis or therapeutic/prophylactic target research (see Box 1). Vaccine candidates against the causative agents of meningococcal meningitis and periodontal disease have been identified through such bioinformatic analyses of genomic sequences for surface proteins *(36,50,51)*. Pizza et al. *(51)* reported what has been described as the first definitive demonstration of the potential of genomic information to expand and accelerate the development of vaccines against pathogenic organisms. In their study, they computationally predicted surface-exposed proteins from a *Neisseria meningitidis* genome sequence and then used this cohort for a large-scale screen of vaccine candidates in the laboratory. They successfully identified two promising candidates. In their study, they reported using a wide range of bioinformatic tools, including Pattern-Hit Initiated BLAST (PHI-BLAST), FASTA, MOTIFS, FIND-PATTERNS, PSORT, ProDom, Pfam, and Blocks to predict features typical of surface-associated proteins, such as transmembrane domains, signal peptides, homology to known surface proteins, lipoprotein signatures, outer membrane anchoring motives, and host cell binding domains such as the sequence RGD (Arg-Gly-Asp) *(51)*. This customized approach is typical of what has been used by a number of research groups and reflects the lack of development of comprehensive, publicly available, bioinformatics tools specifically designed for the identification of vaccine candidates.

New patterns associated with proteins of certain localizations have also recently been uncovered through bioinformatic analyses; these patterns may aid computational identification of proteins of importance in microbial pathogenesis. For example, type III secreted effector proteins play central roles in virulence in a number of bacterial pathogens, yet are notoriously difficult to identify using common bioinformatics approaches because they share no readily detectable similarities at the sequence level. Guttman et al. *(52)* examined 13 experimentally determined effector proteins in the plant pathogen *Pseudomonas syringae* and noted that the amino-terminal regions of the effectors had

---

Box 1
Bioinformatics and Laboratory Research Working Together to Identify Novel Virulence Factors and Antiinfective Targets in Microbial Pathogen Genomes: The *Streptococcus pneumoniae* Example[a]

Streptococcus pneumoniae is the leading bacterial cause of acute respiratory infection and otitis media (ear infection) and causes more than 1 million deaths every year worldwide. The emergence of drug-resistant pneumococcal strains and the inadequacy of available vaccines have led to new calls for a search for novel antibiotic and vaccine targets. A complete genome sequence made it possible to investigate thousands of genes in this pathogen for their role in infection and to identify many essential genes necessary for pathogen survival which may serve as novel therapeutic targets. Both during and after the complete genomic sequence of a virulent strain of *S. pneumoniae* (TIGR4/N4) was determined, open reading frames were computationally predicted and analyzed by Wizemann et al. *(93)* and Tettelin et al. *(94)*. Over 2000 putative genes identified were computationally examined for features that indicate they could encode cell surface proteins or virulence factors. In the Wizemann et al. study, the search was for protein motifs associated with transport (signal peptidase I or II or type IV prepilin signal sequences), cell wall anchorage, choline binding, or integrin binding, as well as putative genes with similarity to known surface-exposed virulence factors in other bacteria. There were 130 putative genes or gene fragments identified, and the products of 108 of these were successfully expressed and purified for further laboratory study. Four of these were found both to confer protection against disseminated *S. pneumoniae* in a mouse model of infection and to show immunogenicity in the human host. After one candidate was found to have a notable repeated histidine triad motif in its sequence, the genome sequence was reexamined to look for other genes with this motif *(95)*. This analysis identified three additional genes, each with five or six copies of this motif, and two of these proteins were protective immunogens. In other words, this approach involved bioinformatic analysis of the whole genome sequence for gene candidates, followed by the expression and testing of candidate proteins, followed by further computational analysis using features detected in promising candidates identified in the laboratory, followed by further laboratory examination of the newly identified candidates.

While the above study focused on the identification of vaccine candidates, of which some were found to play a role in virulence, a study by Gosink et al. *(96)* focused on the identification of novel *S. pneumoniae* virulence factors. They performed genome wide computational searches with the C-terminal choline-binding region of *cbpA*, identifying six new members of this family. Constructed strains comprising knockouts of these family members were then examined for their adherence to eukaryotic cells, colonization of the rat nasopharynx, and ability to cause sepsis; a clear majority played a role in virulence. One newly identified factor, CbpG, shares structural similarity to proteases and may be an excellent candidate for target-based drug development.

[a]See the review by Di Guilmi and Dessen *(92)* for more examples involving *S. pneumoniae*. See the text for examples involving the study of other pathogens.

---

notably similar amino acid compositions. They therefore developed a sequence composition–based analysis that identified 15 previously unknown proteins as potential effectors. They subsequently showed that the secretion of two of these putative effectors was type III dependent.

Bannantine et al. *(53)* also noted a particular secondary structure motif associated with Chlamydial proteins that locates to the inclusion membrane in which the Chlamydial intracellular pathogens reside. Using this motif, they were able to identify additional novel proteins that end up in this inclusion membrane, demonstrating the localization experimentally through antisera studies.

PREDICTING SURFACE-EXPOSED/SECRETED NONPROTEIN COMPOUNDS AND METABOLIC PATHWAY ANALYSIS

In addition to proteins, a number of other nonproteinaceous compounds are found on microbial cell surfaces or secreted. Some of these compounds are well-known virulence factors, such as the fever-inducing lipopolysaccharide (LPS) of Gram-negative bacteria (endotoxin), polysaccharide capsules, and nonprotein toxins such as hydrogen sulfide. For bioinformatic characterization of such factors, frequently a metabolic pathway-based approach is used in combination with analyses of homology to known genes in the relevant pathways. For example, identification of LPS biosynthesis genes in newly sequenced pathogen genomes frequently involves a BLAST analysis of homologs to known LPS biosynthesis genes. However, such analyses of newly sequenced pathogen genomes may be further enhanced by pathway-specific analyses that can investigate, using a whole genome–based approach, the presence or absence of particular pathways in a genome (see Chapter 6).

The use of MetaCyc and the Pathway Tools software is an example of such an approach *(54,55)*. Pathway Tools is a software environment that constructs a database of pathways, genes, proteins, and related data to facilitate analysis, annotation, and visualization of both metabolic and genetic networks. There are four main components of the Pathway Tools: The PathoLogic component supports creation of new databases from the annotated genome of an organism. The Pathway/Genome Navigator provides query, visualization, and Web publishing services for the created databases. The Pathway/Genome Editors support interactive updating of the databases. The fourth component is the Pathway Tools ontology, which defines the schema of the databases. The most well-known example of its use is the EcoCyc database (http://ecocyc.org), which was based on the first *E. coli* K12 genome.

Subsequent to EcoCyc's development, MetaCyc (BioCyc) was developed; it is a more general metabolic pathway database that, when first published in 2002, describes a combined 445 pathways and 1115 enzymes occurring in 158 organisms *(55)*. The pathways in MetaCyc were determined experimentally, based on integrated information from multiple literature sources, and are labeled with the species in which they are known to occur, based on literature references examined to date. Note that microarray data may be overlaid on such pathways, permitting a more pathway-based analysis of gene expression changes. Such resources, including the KEGG database *(56)*, are useful for analysis of metabolic pathways not only relevant to virulence, but also for identification of pathogen growth requirements. Investigations of growth requirements of an organism, and associated pathways, can provide insight into which pathways may be disrupted to attenuate or kill the organism; therefore, such analyses can aid the identification of new antimicrobial drug targets.

SURFACE-EXPOSED AND SECRETED COMPOUNDS

Identifying Phase-Variable Genes

Phase-variable genes encode proteins that undergo rapid phenotypic switching. Such genes are of interest because they often encode surface-exposed proteins, secreted proteins, or proteins that create surface-exposed compounds (i.e., LPS) that play a role in pathogenesis. The phenotypic switching occurs through an on/off switching of expression of a given protein, or family of proteins, such that the bacterium is continually changing the protein repertoire expressed (either expression is on or off or different variants of the same gene family are expressed). This process, which permits rapid cell surface change, is thought to aid the bacterium in evading the host's immune system and in adapting quickly to a new environment (such as infection of a different tissue type). One elegant mechanism for this switching involves changes in the number of repeated nucleotides in homopolymeric (mononucleotide) tracts or short tandem repeats, which are located in a gene or in a regulatory region. Through mechanisms such as slipped-strand mispairing/recombination between homologous repeats/polymerase slippage, genetic variation occurs in the number of bases in the homopolymeric tract or number of repeats in a tract, which can affect transcription or translation. For example, a change in the number of nucleotides located in a homopolymeric tract located in the beginning of a gene's coding region can cause a frame shift, causing premature termination of translation of the protein. Such frequent changes in the frame of translation of a gene lead to a resultant on/off switching in production of the complete protein. For more examples, see the review by Moxon et al. *(57)*.

The advent of whole genome sequencing has now permitted such phase-variable genes to be identified through computational analyses geared at identifying the homopolymeric repeats or short tandem repeats in the genome. The first report of the use of this approach was by Hood et al. *(2)*, who performed a search of the *H. influenzae* Rd genome for all possible combinations of homopolymeric repeats (two or more nucleotides), dinucleotide repeats (two or more), and tri- and tetranucleotide repeats (three or more copies of the repeat). They used the program FINDPATTERN *(58)* to identify homopolymeric and dinucleotide repeats, while BLAST was used to identify the tri- and tetranucleotide repeats and then searched sequences in the associated regions for the presence of genes. A less cumbersome approach was later developed by Saunders et al. *(59)* as an integrated system based on the ACeDB genome analysis and visualization tool that allowed the display of such motifs in the context of sequence data and other genome annotations. This tool permitted the identification of 10 novel putative phase-variable genes in the *Helicobacter pylori* 26695 genome sequence *(59)* and 52 novel candidates for phase variation in the *N. meningitidis* MC58 genome *(60)*. Note that other repeat-finding tools are now available, such as fuzznuc, of the EMBOSS (European Molecular Biology Open Software Suite) open source sequence analysis suite *(61)*.

This approach has a number of benefits, including the fact that truly novel genes involved in virulence may be identified through such analyses because the analysis is not dependent on detecting homologs of known virulence factors. Of course, genes not involved in virulence, but rather involved in conferring some other adaptive benefit, may also be identified. This type of search can also be further refined as more genomes of a particular species or genus are sequenced because then comparisons can be made of the repeat sequences to identify those repeat regions that vary their repeat number between species. Such evidence of repeat variability in a region further supports the role of that region in phase variation. Snyder et al. *(62)* capitalized on this point and performed a comprehensive bioinformatic analysis and comparison of repeat sequences in three *Neisseria* species genomes. Their analysis identified over 100 putative phase-variable genes in *Neisseria* that have strong support for their role in phase variation (i.e., there is a varying repeat number between the different genomes). This included hypothetical genes, one of which was subsequently studied further and found to play a role in competence for DNA uptake *(63)*.

Identifying Antigenic Sequences

Immunoinformatics, the application of informatics and modeling techniques to molecules of the immune system, is a rapidly expanding research area. However, because this is not the focus of this chapter, it is only touched on here. There is a growing list of resources available on the Internet (for example, see http://www.imtech.res.in/raghava/ctlpred/link.html or http://www.jenner.ac.uk/bioinfo03/), and the number of publicly accessible databases of relevance to immunologists now is in the hundreds *(2,64)*. One of the most common analyses performed involves the identification of antigenic sequences (sequences in the microbe that trigger an immune response in the host). Such sequences are of interest for their potential role in vaccines and in developing a better understanding of host–pathogen interaction.

Note that before performing any computational analysis of possible antigenic sequences, it is common first to limit the data set of peptide sequences by determining which proteins or parts of proteins are on the surface of the cell (i.e., using the protein subcellular localization tools listed in Table 1). In addition, hydrophobicity analysis may be performed to identify which regions of globular proteins are most likely exposed on the surface of the protein (i.e., exposed to solvent and so usually hydrophilic). Such analyses limit the focus to the analysis of sequences that would most likely be seen by the host immune system. Of course, if predictions of surface-exposed sequences are incorrect, all downstream analyses are affected, so care must be taken to use predictive methods that have a high accuracy for surface-exposed sequences.

Once particular protein regions are identified, the sequences can be investigated for their potential to bind major histocompatibility complex (MHC) class molecules. MHC class molecules play a critical role in the cellular immune response because they bind fragments of peptides from invading microbes. There is a limit to which specific peptide sequences the MHC will bind, hence the development of computational approaches and databases that provide information about MHC class binding to a particular peptide sequence. For such binding information, some of the most comprehensive databases include FIMM, SYFPEITHI, and JenPep, with the JenPep providing quantitative measures of binding affinity. IMGT, the ImMunoGeneTics database *(65)* is a curated information system that provides a range of resources. ANTIGENIC: EMBOSS *(61)* is a free, open source version of the program by the same name that was originally in the popular GCG package of bioinformatic analysis tools. Some programs provide simultaneous prediction of MHC binders and proteosome cleavage sites in sequence. Proteosome cleavage is an important step in the immune system's presentation of antigenic sequences from a microbe. Both proteosome cleavage and MHC binding analysis lead to the identification of potential T-cell epitopes. nHLAPred and ProPred1 are examples. BCIPep provides a curated database of B-cell epitopes reported in the literature. Currently, nHLAPred, ProPred1, BCIPep, and a range of other predictive tools and databases are accessible through http://www.imtech.res.in/raghava/ctlpred/link.html or through a mirror site at http://bioinformatics.uams.edu/mirror/mirror_imm.html.

CHARACTERIZATION OF HUMAN POLYMORPHISMS ASSOCIATED WITH INFECTIOUS DISEASE

Although most of this chapter focuses on analysis of microbial genomes for the study of microbial pathogenicity, it must be remembered that pathogenesis involves two players: the pathogen and the host.

Before Koch's landmark studies in the late 19th century elucidating the communicable nature of some diseases, such diseases as leprosy and tuberculosis were widely thought to be inherited disorders. The heritability of susceptibility to a number of infectious diseases has now been confirmed, and it is well known that infectious diseases can influence the evolution of their hosts. One of the most well-known examples is the association between sickle cell anemia and reduced susceptibility to malaria. It appears that the prevalence of the sickle hemoglobin allele in regions with endemic malaria is a result of the selective pressure that malaria exerts on the resident human populations (for a review, see ref. 66).

Several single-gene disorders have been implicated in altered susceptibility to many different infectious diseases; for example, individuals who carry mutations in the CD40 ligand are susceptible to opportunistic infections. Human leukocyte antigen (HLA) loci evolve very fast, probably as a result of selective pressure from pathogens, and polymorphisms in these HLA loci have been associated with altered susceptibility to infectious diseases, such as leprosy and tuberculosis. Several non-HLA genes have also been linked with increased susceptibility to disease: vitamin D receptor with tuberculosis, tumor necrosis factor-α with malaria, or a cytokine CD4 with human immunodeficiency virus infection *(66,67)*.

With the completion of the human genome sequence, there is now growing interest in using these genome data to perform more global analyses of genes (and their associated polymorphisms) that play a role in significantly increasing or decreasing susceptibility to an infectious disease. Such knowledge will not only lead to better understanding of the role of host genetics in microbial pathogenesis processes, but also may aid development of more appropriate therapies and vaccines.

The human genome may be directly mined for single-nucleotide polymorphisms (SNPs) that may be associated with disease, using resources such as the dbSNP database of NCBI (http://www.ncbi.nlm.nih.gov/SNP). This database is accessible through Ensembl (www.ensembl.org; *68*) and other related resources, which greatly facilitates analysis by allowing the SNPs to be viewed in the context of various genome annotations and computational analyses in a graphical display.

New SNPs not in dbSNP may be detected through the sequencing of candidate genes from both affected individuals (i.e., those who had/have the disease) and unaffected individuals (those who were likely exposed but did not succumb to disease). For the identification of new SNPs directly from such raw sequence data, the sequence chromatograms produced from automated sequencing are commonly analyzed using the Phred, Phrap, and Consed System of tools (http://www.phrap.org) coupled with PolyPhred *(69)*. Phred is a widely used base-calling program that deduces the most likely combination of basepairs reflected by a sequence chromatogram, assigning a quality score to each base call. Phrap is a sequence assembler program that generates consensus contig (contiguous) sequences from multiple overlapping chromatogram sequence "reads" and assigns a new quality score to each base in the contiguous sequences. Consed and Autofinish are Unix-based graphical editors and automated finishing programs for viewing and editing Phrap sequence assemblies. Together, these tools are widely used by many genome centers as part of their genome sequence data pipeline. However, with PolyPhred, this software package has the additional ability to find SNPs automatically in Phrap contigs, do quality calls, and add data to Phrap files to permit the visualization of the polymorphisms with Consed (Fig. 3). Perhaps one of the most valuable features of this system is the quality score for each base that is generated because this permits quantitative evaluation of the results.

SNP detection is not only relevant for the analysis of human polymorphisms, but also for analysis of the pathogen genomic sequence to detect SNPs associated with changes in virulence. As the move is increasingly toward genomic analysis of multiple strains or isolates, such methods will become increasingly used. Detection of human polymorphisms associated with susceptibility to infectious disease will also become an increasingly valuable approach for the study of microbial pathogenesis, because it will permit better understanding of the complex interplay between host and pathogen. The identification of SNPs associated with decreased susceptibility to disease is of particular interest because these SNPs may provide clues regarding disease resistance mechanisms, and the gene involved may be targets for the development of novel antiinfective therapeutics.

PATHOGEN BIOINFORMATICS:
A LOOK INTO THE FUTURE

Large-scale intelligent systems are urgently required to automate the genomic and proteomic comparisons of multiple pathogens and nonpathogens and to permit customized analyses relevant to microbial virulence and pathogenesis. For such systems, more flexible and powerful comparative genome viewers must be developed, such as that initiated by the Gbrowse project *(70)*, to permit visualization of data from the rapidly increasing number of pathogen strains sequenced. Such tools must also integrate diverse genome-based data (microarray, proteomic, gene knockout analyses) and permit the data to be viewed in a more cellular pathway–based (less genome-oriented) view. A systems biology approach is needed to permit a more holistic view of pathogens and knowledge of them. These issues are similar to those facing the analysis of genomes of multicellullar organisms, so it is hoped that the bioinformatics community will come together in tackling these issues.

However, in addition to such basic resource needs, there is also a significant need for the development of bioinformatic approaches, data vocabularies, and tools that are relatively specific to the needs of researchers studying microbial pathogens. For example, ontologies describing virulence should be developed to permit the generation of a truly useful database of virulence genes and to facilitate computational approaches for

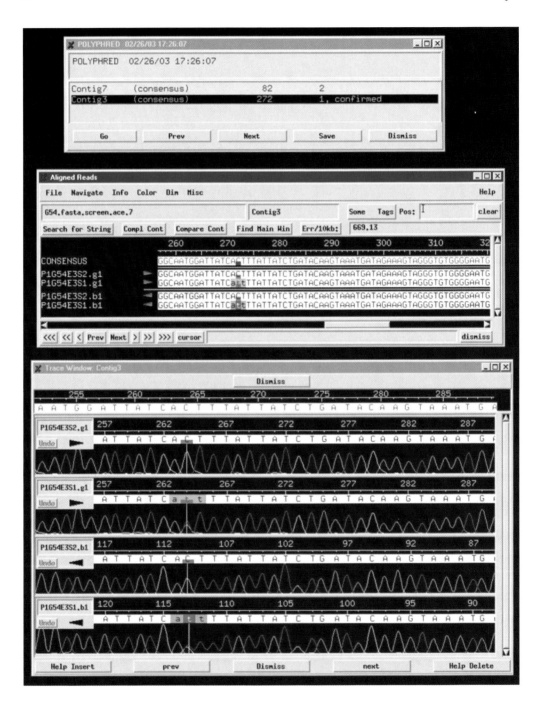

Fig. 3. Detection of human polymorphisms associated with susceptibility to a particular infectious disease can be greatly aided by the Phred, Phrap, Consed, and Polyphred collection of software. The screen shot shown illustrates the display in Consed of a polymorphism. In the Polyphred navigation window at top, contigs are listed that were created by running Phred (for base calling) and Phrap (for assembly of sequence) on sequence reads obtained from individuals affected and unaffected by disease. The alignment of sequences from multiple individuals

the identification and characterization of virulence functions and virulence genes. Such ontologies will allow better organization of thoughts surrounding "what is virulence" and empower bioinformaticists with the ability to study virulence more effectively.

Computational prediction of sequences associated with the cell surface must also be improved, particularly for Gram-negative bacteria and other related bacteria with an outer membrane. The proteins resident in such outer membrane structures have been notoriously difficult to predict and characterize computationally because of their relative lack of sequence similarity with one another (for example, see ref. *71*). Improving the prediction of such cell surface proteins and cell surface sequences will need to become increasingly more detailed as predictions move away from the whole protein level toward prediction of the subcellular localization of each individual amino acid residue in a protein. Such approaches, particularly if they calculate a probability value associated with the subcellular localization prediction, will permit more in-depth analysis of which sequences are on the cell surface. It is hoped that these will lead to a more accurate identification of protein sequences that qualify as therapeutic or vaccine candidates. Note that the development of more computational approaches for the prediction of cellular pathways that lead to the production of cell surface and secreted compounds is also required to permit the thorough evaluation of all possible cell surface and secreted targets.

Further areas to be explored in pathogen bioinformatics include the development of methods to identify proteins of similar structure and function, for instance, the development of specific computational approaches to identify host–pathogen interacting proteins and host–pathogen mimicry (for example, see ref. *72*). Bioinformatic systems that determine sequences uniquely conserved between pathogens that inhabit the same

---

permits viewing the polymorphisms in the sequences, which are detected by Polyphred and displayed in Consed. Note that Polyphred ranks the results, with 1 indicating high confidence in a polymorphism and 6 indicating low confidence. There are settings that can be manipulated to change the sensitivity and specificity of the polymorphisms detected. The user can easily browse the results using Consed by clicking on a given contig in the navigation window list to view the associated sequence alignment in the region of a detected polymorphism. In this example, a polymorphism detected in contig 3 that has a high score of 1 and is confirmed (confirmed = sequence of both strands of the DNA supports the presence of a polymorphism) is examined. The "aligned reads" window displays the sequence base calls, with uppercase letters indicating high-quality base calls and lowercase with gray background indicating lower-quality calls. Note that bases immediately surrounding a true polymorphism are usually displayed with lower-quality scores because of the way Phred produces quality scores and takes into consideration flanking base qualities. A detected polymorphism is colored according to a ranked scale, and the user can click on the region of the sequence of interest to view the "trace window," which shows the raw chromatogram trace data for the sequence regions. In this example, sequences named with .g1 and .b1 extensions represent the forward and reverse sequencing directions, respectively. The S1 sequence (P1G54E3S1) is from an individual who is heterozygous for the polymorphism (note in the traces the presence of two overlapping peaks), and the S2 sequence (P1G54E3S2) is from an individual who is homozygous. The rank score for confidence that this is a true polymorphism (vs a sequence error/artifact) is determined primarily by examining the Phred score for base call quality, as well as calculating the ratio of areas under the two peaks and comparing the peak heights with what would be expected for a hypothetical homozygous peak.

pathogenic niche are undergoing further development (J. L. Fueyo, 2003, unpublished) as are more detailed analyses of tandem repeats associated with phase-variable genes and features associated with pathogenicity islands and other horizontally transferred sequences (for example, integrons).

It should be emphasized that current pathogen bioinformatics research and development is being held up by one notable problem. Although a number of whole genome analyses have been initiated, there is still a relative lack of comprehensive studies of pathogens regarding which genes play a role in virulence. This lack of data makes it difficult to determine the accuracy of the various bioinformatic methods in predicting virulence genes because there is no clear way to determine false-positive and false-negative rates: All true positives and true negatives are not known. Complicating this fact is the notion of virulence as a "fluid" state that depends on the host state, pathogen state, infection conditions, and route of infection. However, as mentioned above, the development of ontologies for virulence may become key to dealing with this conundrum.

With more comprehensive laboratory data around the corner and many obvious analyses to be developed, there is much optimism that pathogen bioinformatics will become more quantitative and be able to play a significant role in advancing the understanding of microbial pathogens and pathogenesis. Coupled with pathogen genomics, bioinformatics should be able to perform pathogen analyses that will lead to genuinely new insights in microbial pathogenesis and the development of new therapeutic and prophylactic targets.

REFERENCES

1. World Health Organization. Report on Infectious Diseases: Removing Obstacles to Healthy Development. Atar, Switzerland: World Health Organization, 1999.
2. Hood DW, Deadman ME, Jennings MP, et al. DNA repeats identify novel virulence genes in *Haemophilus influenzae.* Proc Natl Acad Sci USA 1996; 93:11,121–11,125.
3. Nguyen L, Paulsen IT, Tchieu J, Hueck CJ, Saier MH Jr. Phylogenetic analyses of the constituents of type III protein secretion systems. J Mol Microbiol Biotechnol 2000; 2:125–144.
4. Holmes EC. Molecular epidemiology and evolution of emerging infectious diseases. Br Med Bull 1998; 54:533–543.
5. Levin BR, Lipsitch M, Bonhoeffer S. Population biology, evolution, and infectious disease: convergence and synthesis. Science 1999; 283:806–809.
6. Paine K, Flower DR. Bacterial bioinformatics: pathogenesis and the genome. J Mol Microbiol Biotechnol 2002; 4:357–365.
7. Zagursky RJ, Russell D. Bioinformatics: use in bacterial vaccine discovery. Biotechniques 2001; 31:636, 638, 640, passim.
8. Read TD, Gill SR, Tettelin H, Dougherty BA. Finding drug targets in microbial genomes. Drug Discov Today 2001; 6:887–892.
9. Altschul SF, Madden TL, Schaffer AA, et al. Gapped BLAST and PSI-BLAST: a new generation of protein database search programs. Nucleic Acids Res 1997; 25:3389–3402.
10. Altschul SF, Gish W, Miller W, Myers EW, Lipman DJ. Basic local alignment search tool. J Mol Biol 1990; 215:403–410.
11. Pearson WR, Lipman DJ. Improved tools for biological sequence comparison. Proc Natl Acad Sci USA 1988; 85:2444–2448.
12. Pearson WR. Searching protein sequence libraries: comparison of the sensitivity and selectivity of the Smith–Waterman and FASTA algorithms. Genomics 1991; 11:635–650.

13. Baxevanis AD, Ouellette BFF. Bioinformatics: A Practical Guide to the Analysis of Genes and Proteins. New York: Wiley, 2001.

14. Pertsemlidis A, Fondon JW 3rd. Having a BLAST with bioinformatics (and avoiding BLAST phemy). Genome Biol 2001; 2:REVIEWS2002.

15. Attwood TK, Bradley P, Flower DR, et al. PRINTS and its automatic supplement, prePRINTS. Nucleic Acids Res 2003; 31:400–402.

16. Apweiler R, Attwood TK, Bairoch A, et al. InterPro—an integrated documentation resource for protein families, domains and functional sites. Bioinformatics 2000; 16:1145–1150.

17. Hacker J, Bender L, Ott M, et al. Deletions of chromosomal regions coding for fimbriae and hemolysins occur in vitro and in vivo in various extraintestinal *Escherichia coli* isolates. Microb Pathog 1990; 8:213–225.

18. Censini S, Lange C, Xiang Z, et al. cag, a pathogenicity island of *Helicobacter pylori*, encodes type I–specific and disease-associated virulence factors. Proc Natl Acad Sci USA 1996; 93: 14,648–14,653.

19. Ochman H, Soncini FC, Solomon F, Groisman EA. Identification of a pathogenicity island required for *Salmonella* survival in host cells. Proc Natl Acad Sci USA 1996; 93:7800–7804.

20. Blum G, Ott M, Lischewski A. Excision of large DNA regions termed pathogenicity islands from tRNA-specific loci in the chromosome of an *Escherichia coli* wild-type pathogen. Infect Immun 1994; 62:606–614.

21. Sullivan JT, Ronson CW. Evolution of rhizobia by acquisition of a 500-kb symbiosis island that integrates into a phe-tRNA gene. Proc Natl Acad Sci USA 1998; 95:5145–5149.

22. Hentschel U, Hacker J. Pathogenicity islands: the tip of the iceberg. Microbes Infect 2001; 3: 545–548.

23. Hacker J, Blum-Oehler G, Muhldorfer I, Tschape H, et al. Pathogenicity islands of virulent bacteria: structure, function and impact on microbial evolution. Mol Microbiol 1997; 23: 1089–1097.

24. Inouye S, Sunshine MG, Six EW, Inouye M. Retronphage phi R73: an *E. coli* phage that contains a retroelement and integrates into a tRNA gene. Science 1991; 252:969–971.

25. Williams KP. Integration sites for genetic elements in prokaryotic tRNA and tmRNA genes: sublocation preference of integrase subfamilies. Nucleic Acids Res 2002; 30:866–875.

26. Lio P, Vannucci M. Finding pathogenicity islands and gene transfer events in genome data. Bioinformatics 2000; 16:932–940.

27. Karlin S. Detecting anomalous gene clusters and pathogenicity islands in diverse bacterial genomes. Trends Microbiol 2001; 9:335–343.

28. Hsiao W, Wan I, Jones SJ, Brinkman FS. IslandPath: aiding detection of genomic islands in prokaryotes. Bioinformatics 2003; 19:418–420.

29. Koski LB, Morton RA, Golding GB. Codon bias and base composition are poor indicators of horizontally transferred genes. Mol Biol Evol 2001; 18:404–412.

30. Sonnhammer ELL, Durbin R. A workbench for large scale sequence homology analysis. Comput Appl Biosci 1994; 10:301–307.

31. Florea L, Riemer C, Schwartz S, et al. Web-based visualization tools for bacterial genome alignments. Nucleic Acids Res 2000; 28:3486–3496.

32. Schwartz S, Kent WJ, Smit A, et al. Human–mouse alignments with BLASTZ. Genome Res 2003; 13:103–107.

33. Perrin A, Bonacorsi S, Carbonnelle E, et al. Comparative genomics identifies the genetic islands that distinguish *Neisseria meningitidis*, the agent of cerebrospinal meningitis, from other *Neisseria* species. Infect Immun 2002; 70:7063–7072.

34. Read TD, Salzberg SL, Pop M, et al. Comparative genome sequencing for discovery of novel polymorphisms in *Bacillus anthracis*. Science 2002; 296:2028–2033.

35. Paulsen IT, Seshadri R, Nelson KE, et al. The *Brucella suis* genome reveals fundamental similarities between animal and plant pathogens and symbionts. Proc Natl Acad Sci USA 2002; 99: 13,148–13,153.
36. Tettelin H, Masignani V, Cieslewicz MJ, et al. Complete genome sequence and comparative genomic analysis of an emerging human pathogen, serotype V *Streptococcus agalactiae*. Proc Natl Acad Sci USA 2002; 99:12391–12396.
37. Fleischmann RD, Alland D, Eisen JA, et al. Whole-genome comparison of *Mycobacterium tuberculosis* clinical and laboratory strains. J Bacteriol 2002; 184:5479–5490.
38. Perna NT, Plunkett G 3rd, Burland V, et al. Genome sequence of enterohaemorrhagic *Escherichia coli* O157:H7. Nature 2001; 409:529–533.
39. Fueyo JL. In Silico Discovery of Antimicrobial Targets. Ph.D. thesis, Philadelphia: University of Pennsylvania, 2002.
40. Bruccoleri RE, Dougherty TJ, Davison DB. Concordance analysis of microbial genomes. Nucleic Acids Res 1998; 26:4482–4486.
41. Pearson WR, Lipman DJ. Improved tools for biological sequence comparison. Proc Natl Acad Sci USA 1988; 85:2444–2448.
42. Thanassi JA, Hartman-Neumann SL, Dougherty TJ, Dougherty BA, Pucci MJ. Identification of 113 conserved essential genes using a high-throughput gene disruption system in *Streptococcus pneumoniae*. Nucleic Acids Res 2002; 30:3152–3162.
43. Dziejman M, Balon E, Boyd D, Fraser CM, Heidelberg JF, Mekalanos JJ. Comparative genomic analysis of *Vibrio cholerae*: genes that correlate with cholera endemic and pandemic disease. Proc Natl Acad Sci USA 2002; 99:1556–1561.
44. Galperin MY, Koonin EV. Searching for drug targets in microbial genomes. Curr Opin Biotechnol 1999; 10:571–578.
45. Huynen M, Dandekar T, Bork P. Differential genome analysis applied to the species-specific features of *Helicobacter pylori*. FEBS Lett 1998; 426:1–5.
46. Tatusov RL, Natale DA, Garkavtsev IV, et al. The COG database: new developments in phylogenetic classification of proteins from complete genomes. Nucleic Acids Res 2001; 29:22–28.
47. Emanuelsson O. Predicting protein subcellular localisation from amino acid sequence information. Brief Bioinform 2002; 3:361–376.
48. Gardy JL, Spencer C, Wang K, et al. PSORT-B: improving protein subcellular localization prediction for Gram-negative bacteria. Nucleic Acids Res 2003; 31:3613–3617.
49. Nair R, Rost B. Sequence conserved for subcellular localization. Protein Sci 2002; 11:2836–2847.
50. Ross BC, Czajkowski L, Hocking D, et al. Identification of vaccine candidate antigens from a genomic analysis of *Porphyromonas gingivalis*. Vaccine 2001; 19:4135–4142.
51. Pizza M, Scarlato V, Masignani V, et al. Identification of vaccine candidates against serogroup B meningococcus by whole-genome sequencing. Science 2000; 287:1816–1820.
52. Guttman DS, Vinatzer BA, Sarkar SF, Ranall MV, Kettler G, Greenberg JT. A functional screen for the type III (Hrp) secretome of the plant pathogen *Pseudomonas syringae*. Science 2002; 295:1722–1726.
53. Bannantine JP, Griffiths RS, Viratyosin W, Brown WJ, Rockey DD. A secondary structure motif predictive of protein localization to the chlamydial inclusion membrane. Cell Microbiol 2000; 2:35–47.
54. Karp PD, Paley S, Romero P. The Pathway Tools software. Bioinformatics 2002; 18(Suppl 1): S225–S232.
55. Karp PD, Riley M, Paley SM, Pellegrini-Toole A. The MetaCyc database. Nucleic Acids Res 2002; 30:59–61.
56. Kanehisa M. The KEGG database. Novartis Found Symp 2002; 247:91–101; discussion 101–103, 119–128, 244–252.

57. Moxon ER, Rainey PB, Nowak MA, Lenski RE. Adaptive evolution of highly mutable loci in pathogenic bacteria. Curr Biol 1994; 4:24–33.
58. Devereux J, Haeberli P, Smithies O. A comprehensive set of sequence analysis programs for the VAX. Nucleic Acids Res 1984; 12:387–395.
59. Saunders NJ, Peden JF, Hood DW, Moxon ER. Simple sequence repeats in the *Helicobacter pylori* genome. Mol Microbiol 1998; 27:1091–1098.
60. Saunders NJ, Jeffries AC, Peden JF, et al. Repeat-associated phase variable genes in the complete genome sequence of *Neisseria meningitidis* strain MC58. Mol Microbiol 2000; 37:207–215.
61. Rice P, Longden I, Bleasby A. EMBOSS: the European Molecular Biology Open Software Suite. Trends Genet 2000; 16:276–277.
62. Snyder LA, Butcher SA, Saunders NJ. Comparative whole-genome analyses reveal over 100 putative phase-variable genes in the pathogenic *Neisseria* spp. Microbiology 2001; 147:2321–2332.
63. Snyder LA, Saunders NJ, Shafer WM. A putatively phase variable gene (dca) required for natural competence in *Neisseria gonorrhoeae* but not *Neisseria meningitidis* is located within the division cell wall (dcw) gene cluster. J Bacteriol 2001; 183:1233–1241.
64. Brusic V, Zeleznikow J, Petrovsky N. Molecular immunology databases and data repositories. J Immunol Methods 2000; 238:17–28.
65. Lefranc MP. IMGT, the international ImMunoGeneTics database: a high-quality information system for comparative immunogenetics and immunology. Dev Comp Immunol 2002; 26:697–705.
66. Cooke GS, Hill AV. Genetics of susceptibility to human infectious disease. Nat Rev Genet 2001; 2:967–977.
67. Foster CB, Chanock SJ. Mining variations in genes of innate and phagocytic immunity: current status and future prospects. Curr Opin Hematol 2000; 7:9–15.
68. Hubbard T, Barker D, Birney E, et al. The Ensembl genome database project. Nucleic Acids Res 2002; 30:38–41.
69. Nickerson DA, Tobe VO, Taylor SL. PolyPhred: automating the detection and genotyping of single nucleotide substitutions using fluorescence-based resequencing. Nucleic Acids Res 1997; 25:2745–2751.
70. Stein LD, Mungall C, Shu S, et al. The generic genome browser: a building block for a model organism system database. Genome Res 2002; 12:1599–1610.
71. Brinkman FS, Bains M, Hancock RE. The amino terminus of *Pseudomonas aeruginosa* outer membrane protein OprF forms channels in lipid bilayer membranes: correlation with a three-dimensional model. J Bacteriol 2000; 182:5251–5255.
72. Stebbins CE, Galan JE. Structural mimicry in bacterial virulence. Nature 2001; 412:701–705.
73. Tusnady GE, Simon I. Principles governing amino acid composition of integral membrane proteins: application to topology prediction. J Mol Biol 1998; 283:489–506.
74. Krogh A, Larsson B, von Heijne G, Sonnhammer EL. Predicting transmembrane protein topology with a hidden Markov model: application to complete genomes. J Mol Biol 2001; 305:567–580.
75. Cserzo M, Wallin E, Simon I, von Heijne G, Elofsson A. Prediction of transmembrane alpha-helices in prokaryotic membrane proteins: the dense alignment surface method. Protein Eng 1997; 10:673–676.
75a. Hoffman K, Stoffel W. TMbase—a database of membrane spinning proteins segments. Biol Chem Hoppe-Seyler 1993; 374:166. http://www.ch.embnet.org/software/TMPRD_form.html.
76. Janulczyk R, Rasmussen M. Improved pattern for genome-based screening identifies novel cell wall-attached proteins in Gram-positive bacteria. Infect Immun 2001; 69:4019–4026.
77. Diederichs K, Freigang J, Umhau S, Zeth K, Breed J. Prediction by a neural network of outer membrane beta-strand protein topology. Protein Sci 1998; 7:2413–2420.

78. Jacoboni I, Martelli PL, Fariselli P, De Pinto V, Casadio R. Prediction of the transmembrane regions of beta-barrel membrane proteins with a neural network-based predictor. Protein Sci 2001; 10:779–787.

79. Martelli PL, Fariselli P, Krogh A, Casadio R. A sequence-profile-based HMM for predicting and discriminating beta barrel membrane proteins. Bioinformatics 2002; 18(Suppl 1):S46–S53.

80. Zhai Y, Saier MH Jr. The beta-barrel finder (BBF) program, allowing identification of outer membrane beta-barrel proteins encoded within prokaryotic genomes. Protein Sci 2002; 11:2196–2207.

81. Schirmer T, Cowan SW. Prediction of membrane-spanning beta-strands and its application to maltoporin. Protein Sci 1993; 2:1361–1363.

82. Wimley WC. Toward genomic identification of beta-barrel membrane proteins: composition and architecture of known structures. Protein Sci 2002; 11:301–312.

83. Horton P, Nakai K. Better prediction of protein cellular localization sites with the k nearest neighbors classifier. Proc Int Conf Intell Syst Mol Biol 1997; 5:147–152.

84. Bannai H, Tamada Y, Maruyama O, Nakai K, Miyano S. Extensive feature detection of N-terminal protein sorting signals. Bioinformatics 2002; 18:298–305.

85. Nielsen H, Engelbrecht J, Brunak S, von Heijne G. A neural network method for identification of prokaryotic and eukaryotic signal peptides and prediction of their cleavage sites. Int J Neural Syst 1997; 8:581–599.

86. Hua S, Sun Z. Support vector machine approach for protein subcellular localization prediction. Bioinformatics 2001; 17:721–728.

87. Reinhardt A, Hubbard T. Using neural networks for prediction of the subcellular location of proteins. Nucleic Acids Res 1998; 26:2230–2236.

88. Nakai K, Kanehisa M. Expert system for predicting protein localization sites in Gram-negative bacteria. Proteins 1991; 11:95–110.

89. Saleh MT, Fillon M, Brennan PJ, Belisle JT. Identification of putative exported/secreted proteins in prokaryotic proteomes. Gene 2001; 269:195–204.

90. Mott R, Schultz J, Bork P, Ponting CP. Predicting protein cellular localization using a domain projection method. Genome Res 2002; 12:1168–1174.

91. Gomez M, Johnson S, Gennaro ML. Identification of secreted proteins of *Mycobacterium tuberculosis* by a bioinformatic approach. Infect Immun 2000; 68:2323–2327.

92. Di Guilmi AM, Dessen A. New approaches towards the identification of antibiotic and vaccine targets in *Streptococcus pneumoniae*. EMBO Rep 2002; 3:728–734.

93. Wizemann TM, Heinrichs JH, Adamou JE, et al. Use of a whole genome approach to identify vaccine molecules affording protection against *Streptococcus pneumoniae* infection. Infect Immun 2001; 69:1593–1598.

94. Tettelin H, Nelson KE, Paulsen IT, et al. Complete genome sequence of a virulent isolate of *Streptococcus pneumoniae*. Science 2001; 293:498–506.

95. Adamou JE, Heinrichs JH, Erwin AL, et al. Identification and characterization of a novel family of pneumococcal proteins that are protective against sepsis. Infect Immun 2001; 69:949–958.

96. Gosink KK, Mann ER, Guglielmo C, Tuomanen EI, Masure HR. Role of novel choline binding proteins in virulence of *Streptococcus pneumoniae*. Infect Immun 2000; 68:5690–5695.

5

Bacteriophage Bioinformatics

Derrick E. Fouts

INTRODUCTION

We are becoming increasingly aware that lysogenic bacteriophages provide essential virulence and fitness factors to their hosts, affecting (among other properties), bacterial adhesion, colonization, invasion, spread, resistance to immune responses, exotoxin production, serum resistance, and resistance to antibiotics. New studies using phage-encoded enzymes that degrade the bacterial cell wall and inhibit synthesis of cell wall precursors offer renewed possibilities for phage-based pharmaceuticals. In addition, a good deal can be learned about bacterial genome plasticity and evolution by studying bacteriophage genomes.

More than 5000 bacteriophages have been classified since 1959 *(1)*; however, fewer than 3% of these have been completely sequenced and are publicly available in GenBank (Entrez Genomes: Phages; Table 1). More sequences of bacteriophage genomes are necessary for full understanding of their diversity, their full potential for genetic mobilization/exchange, and their evolution. In the meantime, bacteriophage researchers can mine an enormous, poorly explored, publicly available resource—completed bacterial genomes. This chapter is not meant to regurgitate or rehash the already well-reviewed areas of bacteriophage taxonomy *(2,3)*, evolution *(4–8)*, or comparative bacteriophage genomics *(9–18)*. Rather, the focus is on the bioinformatics of prophage identification and analysis from completed or incomplete bacterial genomes.

BACKGROUND

Bacteriophages (phages) are viruses that infect bacteria and can usually be identified by the holes or plaques that they create on a lawn of susceptible or host bacterial cells. The word *phage* is Greek for "eat"; thus, bacteriophages are "bacteria eaters." Surprisingly, bacteriophages were not the first virus to be identified; the animal and plant viruses won that race, but the bacteriophages have the largest number of descriptions recorded *(1)*. The actual discoverer of bacteriophages as viruses of bacteria is controversial, but was either the French–Canadian bacteriologist Felix d'Herelle or the English pathologist F. W. Twort in about 1915–1917 *(19,20)*. It is not known which was the first bacteriophage identified, but it either infected a dysentery-causing bacillus (*Shigella*) or an infectious enteritis-causing bacterium of locusts (a subspecies of *Enterobacter*

From: *Microbial Genomes*
Edited by: C. M. Fraser, T. D. Read, and K. E. Nelson © Humana Press Inc., Totowa, NJ

Table 1
Useful Internet Links

Description	Web address
Artemis	http://www.sanger.ac.uk/Software/Artemis/
CLUSTAL W	http://www-igbmc.u-strasbg.fr/BioInfo/ClustalW/clustalw.html
Comprehensive Microbial Resource (CMR)	http://www.tigr.org/tigr-scripts/CMR2/CMRHomePage.spl
Entrez Genomes: Phages	http://www.ncbi.nlm.nih.gov:80/genomes/static/phg.html
HMMER	http://hmmer.wustl.edu/
International Committee on Taxonomy of Viruses (ICTV)	http://www.ncbi.nlm.nih.gov/ICTVdb/
Manatee (Manual Annotation Tool, etc., etc.)	http://manatee.sourceforge.net/
Pfam	http://pfam.wustl.edu/
Phage Ecology	http://www.mansfield.ohio-state.edu/~sabedon/
Phage genomics (International Bacteriophage Genomics Group)	http://meds.queensu.ca/~ibgg/
Phage proteome	http://salmonella.utmem.edu/phage/tree/
Phage therapy	http://www.evergreen.edu/phage/
PHYLIP	http://evolution.genetics.washington.edu/phylip.html
PSI-BLAST	http://www.ncbi.nlm.nih.gov/BLAST/
REPuter	http://www.genomes.de/
TIGRFAMs	http://www.tigr.org/tigr-scripts/CMR2/find_hmm.spl?db=CMR
WU-BLAST (Washington University BLAST Archives)	http://blast.wustl.edu

aerogenes) *(19–23)*. It was not until the early 1940s that the structure of bacteriophages were observed for the first time under the electron microscope *(24–28)*.

Many important principles in bacterial genetics and molecular biology that are taken for granted today were elucidated from bacteriophage research. For example, Luria and Delbrück, using the famous fluctuation test, demonstrated that bacteria could randomly acquire mutations prior to a selective pressure with bacteriophage α (present-day T1) *(29)*. T1 is one of the seven lytic phages (T1–T7) that infect *Escherichia coli* strain B as part of the T system put together by Delbrück and the "phage group" at Cold Spring Harbor in 1944 to focus and standardize future studies on lytic bacteriophages *(28)*. One of the key experiments that showed deoxyribonucleic acid (DNA) was the genetic material, the famous Hershey–Chase experiment, used bacteriophage T2 *(30)*. The concept of *transduction*, the ability of a phage to move host genomic DNA from one host to another and a very common method of constructing bacterial strains with specific genetic backgrounds, was founded by Zinder and Lederberg using bacteriophage

PLT22 (present-day P22) *(31)*. Taking advantage of host-range mutants of T4 called rII (rapid lysis mutants type II), Benzer used deletion mapping to define the physical structure of a gene *(32)*. The genetic code was cracked and verified by studies conducted on bacteriophage T4 *(33,34)*. Many of the enzymes and techniques used routinely in molecular biology are deeply rooted in phage studies: restriction enzymes *(35)*, DNA ligase *(36,37)*, RNA ligase *(38)*, DNA polymerases *(39)*, polynucleotide kinase *(40)*, phage display *(41)*, challenge phage *(42,43)*, M13 packaging of single-stranded (ss)DNA for sequencing *(44,45)* and site-directed mutagenesis *(46,47)*, and λ cloning vectors *(48)*.

Lytic or *virulent bacteriophages* are phages that undergo only one life cycle (vegetative growth), which consists of binding to a host cell, internalization of viral nucleic acid, hijacking host metabolic machinery, making more of itself, and destroying the cell to release hundreds of newly synthesized viral particles. *Lysogenic* or *temperate phages* can undergo an additional, alternative lifestyle by integrating their DNA into the host chromosome to reside as a *prophage*, a dormant phase that is generally transcriptionally quiescent, only producing messenger ribonucleic acid (mRNA) for a repressor protein and other gene cassettes that may function in prophage fitness, referred to as *morons (7,17)*.

Bacteriophages were studied for more than 50 years before a taxonomical standard was developed *(49)*. For a review of present-day bacteriophage taxonomy, see refs. *1* and *3*. The order *Caudovirales* contains the tailed phages, which make up the bulk of known phages. A subset of this group is temperate. All of the tailed phages contain double-stranded, linear DNA, and they are divided into three families based on the morphology of their tails: *Myoviridae* phages have contractile tails like bacteriophage T2, the classic syringe model; *Siphoviridae* phages have long, noncontractile tails like phage λ; and *Podoviridae* have short tails like P22. These three families are each further divided into three morphotypes (A1 to A3 for *Myoviridae*, B1 to B3 for *Siphoviridae*, and C1 to C3 for *Podoviridae*). Each number represents the shape of the capsid: isometric, moderately elongated, and very long heads, for numbers 1 to 3, respectively. Phages are also characterized by many other properties, including type of nucleic acid genome *(3)*.

Bacteriophages are vehicles for genetic mobility of genes that can increase bacterial fitness *(7,13)*. Specifically, the role of the bacteriophage in bacterial pathogenesis of humans was appreciated and studied as early as 1927 *(50,51)*. It has been demonstrated that bacteriophage-carried genes can play a role in many aspects of bacterial virulence (adhesion, invasion, host evasion, and toxin production) *(51)*. Toxin production seems to be the most widely observed phage-derived virulence factor, with examples of phages carrying botulinum toxin *(52)*, diphtheria toxin *(53,54)*, cholera toxin *(55)*, Shiga toxins *(56,57)*, and toxic shock syndrome toxin *(58)*. PblA and PblB proteins encoded by *Streptococcus mitis* phage SM1 have been proposed to promote adhesion to human platelets, possibly leading to endocarditis *(59)*. Phages have also been implicated in enabling their bacterial hosts to possibly evade the human immune system by altering O antigen structure *(60,61)* or providing enzymes that neutralize the oxidative burst that is meant to destroy the invading bacterium *(62)*. Phages also encode enzymes that specifically play a direct role in bacterial virulence *(63–65)*.

If the well-documented roles that phages have in bacterial pathogenesis are not reason enough to expand bacteriophage research, a resurrected idea to use phages to treat human disease is gaining popularity *(66–68)*. Classic phage therapy was once an accepted means to treat certain bacterial infections prior to the invention of modern-day antibiotics. For example, phages would be isolated from the stools of an individual suffering from acute dysentery; these phages were grown to high titer and swallowed by the patient. In some cases, this was quite effective in treating the disease *(69,70)*. Bacteriophage therapy has been recently shown to cure mice bacteremic with a vancomycin-resistant *Enterococcus faecium* clinical isolate *(71)*. Cases for which phage therapy fails are likely because of improper choice of phage for the target bacterium, the ease by which bacteria can become resistant to an individual phage, and the fact that some phages actually carry virulence factors that help the bacterium to cause disease.

Medical professionals in the former Soviet Union made cocktails of different phages specific for the bacteria of interest and made preparations that did not contain bacterial contamination *(72)*. The work of Fischetti's group made the cover of *Nature*; it was a report that purified lysin protein (PlyG) from the γ-phage that infects *Bacillus anthracis* can be used for the detection and destruction of germinated *B. anthracis* cells *(73)*. This new twist on phage therapy could prove to be very useful because the frequency of resistance should be much lower than when using whole phage lysates. Lysin proteins tend to recognize components of the cell wall that are much more difficult to mutate without ill effects than typical phage receptor molecules, which can be altered by the simple addition or deletion of sugar residues or outer membrane proteins.

Another potential medical use for bacteriophage-derived products is gene therapy. Phage-derived site-specific integrases have recently been shown to function in human cells, making it theoretically possible to engineer integrases that deliver good copies of disease-causing alleles to predictable and safe regions of the human genome or other mammalian genomes *(74–78)*.

The genomics era has brought an unprecedented wealth of knowledge on the coding capacity and genome plasticity of many organisms. The contribution that bacteriophage research gave to this era has often been unappreciated and overlooked. Genomics did not begin with the sequencing of *Haemophilus influenzae* in 1995 *(79)* because the genomes of φX174 and λ had already been sequenced in 1977 and 1983, respectively *(80–83)*. The sequencing of bacteriophage genomes led to the sequencing of bacterial genomes. Ironically, bacterial genomes are being used here as a means to obtain bacteriophage genomic data in the form of integrated prophage.

FINDING PROPHAGES IN BACTERIAL GENOMES

One of the greatest challenges inhibiting accurate identification of prophage regions in bacterial genomes is the proper annotation of bacteriophage-derived genes. For instance, for a gene to be annotated with a particular functional assignment, it must meet the following criteria at The Institute for Genomic Research (TIGR): It must have a BLAST (Basic Local Alignment Search Tool) E probability value of 1×10^{-5} or less and at least 35% full-length amino acid identity to a protein whose function has been experimentally determined in the laboratory; alternatively, a protein may match above a carefully determined trusted cutoff to an equivalog hidden Markov model (HMM), which has as part of its seed a protein with a function that has been experimentally determined in the laboratory. Even if there is a very low BLAST E value and acceptable percentage identity, a gene in a putative prophage region might be annotated as "conserved hypothetical" if that protein has not been functionally characterized in the laboratory—as is the case for matches to prophages from bacterial genome projects.

There are also cases for which the gene has been very well characterized, but the reference does not appear in on-line bibliographic sources because it was published too long ago or in a journal that on-line bibliographic sources do not track. More often than not, the putative protein encoded by a gene in a prophage region does not have any significant BLAST or HMM matches and by default is labeled hypothetical. This is likely due to poor representation of characterized bacteriophage proteins in public databases and the great genetic diversity that exists in the bacteriophage gene pool. It is especially troubling when a number of open reading frames (ORFs) in a 20- to 30-kbp region are annotated as having phage or phage-like functions, but the region is not labeled as a contiguous prophage region. In other words, the local genetic structure or juxtaposition of the ORFs is overlooked, and these ORFs are not recognized as possibly within the same integrated prophage.

Before giving a detailed account of the identification of prophage regions, I define what I mean by *prophage regions*. First, it is necessary to mention that all prophage regions are "putative" prophage regions until someone is able to isolate infectious phage particles on induction. A prophage region is a cluster or stretch of genes possibly encoding proteins with bacteriophagelike functions interspersed with genes of unknown function or functions that have been known to be phage encoded, but may or may not pertain to phage replication, morphogenesis, packaging, immunity, or lysis of the host. Examination of just a few genome articles is enlightening, regarding the many different interpretations of a true prophage, a defective or cryptic prophage, and even more confusing, phagelike bacteriocins, if these are at all distinguished *(84–86)*.

A true prophage is inducible, whereas defective prophages are the uninducible remnants of formerly functional temperate bacteriophages. The designation *cryptic prophage* was historically reserved for those defective prophages that do not provide immunity against superinfecting phages *(87)*. Another category to consider is phage genes that originated from lytic/virulent phages integrated into the bacterial host genome via illegitimate recombination *(88)*. There are continuous cycles of war and peace between host and phage. The host may want to keep certain functionalities encoded on the prophage, but may not want the deadly consequences of prophage induction; therefore, there is a race against time as to which can kill the other first *(89)*.

There may be a prophage that appears functional, but has a mutation in a key gene. For instance, in the *Streptococcus pyogenes* 370.2 prophage region *(90)*, there is a mutation in the putative portal protein that presumably inactivates the prophage. Later stages of mutation via indels and recombination result in a degenerated prophage region, which is easier to label as defective. It is a little trickier to predict a *satellite prophage*, a prophage incapable of lytic development unless supplied with functional components *in trans* by a resident fully functional prophage or by an incoming phage *(9)*.

Because the size of one of the smallest functional double-stranded (ds)DNA temperate, tailed bacteriophages (*Bacillus subtilis* φ29-like phage B-103) *(91)* is about 18 kbp and some experimentally determined satellite phages from *Lactococcus lactis* are 13–15 kbp *(9)*, I have been using a potential satellite size of 10–18 kbp, which may or may not be realistic for some larger prophages. Any region greater than 18 kbp would be considered as a putative prophage until other observations discount this demarcation.

For the purpose of this study, I have not included any grouping of phage genes that is less than 10 kbp. I do not consider stand-alone "phage-like integrases" to count as a defective prophage region because it could just as well be host derived or be from some other mobile genetic element *(92)*.

Finally, other prophage regions are phage-like bacteriocins, for which the *Pseudomonas aeruginosa* R-type or F-type pyocins are an example *(93)*. These regions could easily be annotated as defective prophages because they do have many genes with similarity to known phage genes (P2 in the case of the R-type pyocin and phage λ in the case of the F-type pyocin). Phagelike bacteriocins are different in that they primarily contain genes for tail fibers and tail appendages, regulators, and lysis genes, but no capsid genes. These regions are thought to be prophages that were held hostage by the host and that lost all genes not essential for formation of tail fibers, which the host uses to kill bacteria of similar species by disrupting the membrane of the unfortunate recipient cell *(94)*. These bacteriocins are called many different names (monocin for *Listeria monocytogenes*, pestisin for *Yersinia pestis*, cerecin for *Bacillus cereus*, colicins for *E. coli*, and staphylococcin for *Staphylococcus*) *(95)*. If there is a stretch of phage-like genes that does not contain capsid genes but contains many genes with matches to tail-like genes, this region could be a putative phage-like bacteriocin of the appropriate name. Like bacteriophages, these bacteriocins have been useful for the typing/identification of pathogenic bacterial strains *(96,97)*.

Given the difficulties associated with prophage identification, it is necessary to examine each region manually. This is done by looking at the genetic organization near genes that have significant BLAST matches to known bacteriophage genes, using graphical ORF-viewing software like Artemis (Table 1) or the region display option in TIGR's Manatee annotation tool package (Table 1). The limits of the prophage region are identified by browsing 5' and 3' of the gene or genes of interest until genes with clear-cut "housekeeping" functions like glycolytic enzymes or ribosomal proteins/RNAs are found. The extended region (~1 kb) on either side of the predicted borders of the prophage region can be investigated for repeats using a tool such as REPfind, available in the publicly available REPuter package *(98)*, to search for direct repeats resulting from single crossover integration between DNA attachment sites *(99)*. One copy will be phage derived, and the other will be host derived.[1]

This method of determining the point of insertion/attachment (att) site does not work for Mu-like phages because they integrate into the genome via a transpositional mechanism, resulting in a 4- to 5-bp target site duplication *(100)*. Instead, the boundaries of Mu-like phage regions can be roughly identified by the location of functionally conserved genes. One end of the phage region will have a cI-like repressor (the Mu *c* functional equivalent), with two ORFs having similarity to transposase A and B subunits

---

[1]A good strategy to follow is to start with the Unix/Linux command line (repfind -f -l 80 -h 1 -s [filename of extended sequence]). The -f option denotes forward or direct repeats, -l is the length of the repeat in basepairs, -h is the number of allowable mismatches, and -s is to show the sequence of the repeat. To facilitate identifying a real repeat, it is a good idea to start with -l of 80, then if there are no matches, rerun the program with -l 70, then 60, and so on until you find a set of direct repeats that seems to flank the prophage region manually identified.

nearby. The other end of the region may be determined by the presence of an ORF with similarity to an adenine-specific methylase (the Mu *mom* functional equivalent) *(100)*. In the case of MuMenB from *Neisseria meningitidis* serogroup B strain MC58, which appears to be different from the four previously published Mu-like phages, the Mu-like prophage was discovered fortuitously by virtue of a disruption of NMB1077/NMB1122, an ABC transporter *(101)*. This prophage is thought to produce a 7-bp imperfect target site duplication and contain the cI-like repressor, transposase A and B subunits, but no *mom* equivalent; therefore, the boundaries of this phage could only be determined by disruption of a host gene.

If no direct repeat is found, then there are other paths to take. If there are multiple finished (closed) bacterial genome sequences available from related bacteria, disruption of identifiably conserved gene orders adjacent to the putative prophage region can be sought, as was done for putative prophage region 3 of *B. anthracis* (Fig. 1). In this example, no potential attL/R sites were identified as direct repeats; therefore, genes flanking the manually identified region were located in the published genomes of *Bacillus subtilis*, *Bacillus halodurans*, *L. monocytogenes*, and *Staphylococcus aureus*. It seemed likely that the gene order (using B. *subtilis* nomenclature) *ylbL, ylbM, ylbN*, and *rpmF* was disrupted in *B. anthracis* because of one or more prophage integration events between *ylbM* and *ylbN* (Fig. 1).

One explanation to account for the presence of *spoIIIE* (BA4067), presumably a housekeeping gene, at the 3' end of the prophage is via transduction from a related strain of *B. anthracis* or *B. cereus*, a close relative of *B. anthracis*. One piece of evidence supporting this theory is that there are two copies of *spoIIIE* in *B. anthracis* compared with only one copy in *B. subtilis* and *B. halodurans*. Since *Clostridium perfringens* bacteriophage φ3626 encodes a protein with similarity to SpoIIID *(102)*, it is tempting to speculate that these putative phage-encoded transcriptional regulators manipulate endospore development of their hosts; however, these proteins could be regulators of some unknown or uncharacterized phage functionalities. Laboratory work is needed to determine if this region is inducible and whether this putative sigma factor plays any role in fitness or virulence of *B. anthracis*.

For prophage regions that contain an identifiable integrase gene, the target site will likely be adjacent to the integrase gene *(99)*. If there happens to be a transfer RNA (tRNA) gene near the integrase gene, then this could be the target site (attB) because many integrases of the tyrosine recombinase family have tropisms for tRNA genes *(99)*. This putative target site can be used to search for another similar copy on the other side of the putative prophage region *(103)*. If the putative target site is a tRNA gene, a program such as Lowe and Eddy's tRNAscan-SE can be used to find the match *(104)*. Target sites are not always tRNA or transfer-messenger (tm)RNA. They can be conserved housekeeping genes like glucose 6-phosphate isomerase *(105)* and guanosine 5'-monophosphate (GMP) synthetase *(102)*. Perhaps it is advantageous for a bacteriophage to integrate into a conserved target, possibly increasing host range by guaranteeing a landing pad in whichever bacterial species it might encounter expressing the appropriate cell surface receptor. Because phages typically do not inactivate genes that might be essential for host viability *(103)*, it is very unlikely that one will identify an integrated prophage because it caused a frameshift in the target gene. The guanosine and cytosine

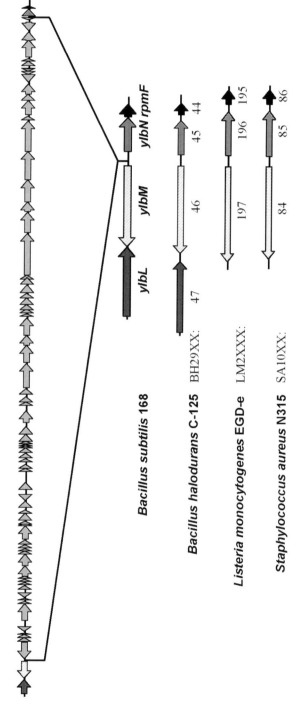

Fig. 1. Borders of prophage regions in bacterial genomes can be identified by disruption of identifiably conserved gene orders from closely related organisms. The gene order of *ylbL*, *ylbM*, *ylbN*, and *rpmF* is apparently conserved between the genomes of *Bacillus subtilis* and *Bacillus halodurans*. Broadening the scope, the gene order of *ylbM*, *ylbN*, and *rpmF* is apparently conserved among Gram-positive genomes of *B. subtilis*, *B. halodurans*, *Listeria monocytogenes*, and *Staphylococcus aureus*. The names of the corresponding ORFs are indicated under each ORF diagram.

Table 2
TIGR Role_ids That Could be Phage Associated

Rold_ids	Description
88	Cell envelope, other
89	Biosynthesis of murein sacculus and peptidoglycan
90	Biosynthesis and degradation of surface polysaccharides and lipopolysaccharides
91	Surface structures
94	Toxin production and resistance
123	2'-Deoxyribonucleotide metabolism
129	Regulatory functions, other
131	Degradation of DNA
132	DNA replication, recombination, and repair
138	Degradation of proteins, peptides, and glycopeptides
149	Adaptations to atypical conditions
152	Prophage functions
154	Transposon functions
156	Conserved hypothetical
157	Unknown function, general
175	General
183	Restriction/modification
185	Role category not yet assigned
187	Pathogenesis
261	Regulatory functions, DNA interactions
270	Disrupted reading frames
703	Enzymes of unknown specificity
704	Conserved domain

content (%G+C) or atypical nucleotide composition cannot be used reliably because the G+C content of tailed bacteriophages parallels that of their hosts *(3)*. There is one known exception; *Pseudomonas putida* putative prophages are nicely situated in atypical regions of the genome *(106)*.

Manual curation of prophage regions is laborious and very time consuming, especially when scaled up to analyze the more than 90 bacterial genomes currently maintained in TIGR's Comprehensive Microbial Resource (CMR) *(107)*. To tackle this problem, I have begun Phage_phinder, a Practical Extraction and Report Language (Perl) script developed to find prophage regions within bacterial genomes and display the output in a variety of formats. The input of this program is tab-delimited (btab) output from BLASTP 2.0MP-WashU *(108)* searches of all the protein sequences from a bacterial genome as query against a multi-FASTA protein sequence database of all the completed bacteriophage sequences from GenBank plus some manually curated prophages. Phage_phinder takes advantage of TIGR's database to determine the location of the genes that have significant ($\leq 1 \times 10^{-6}$ E value) matches in the phage BLAST-formatted database and to use TIGR gene role assignments (role_ids) *(109)* to aid in the identification of proteins likely to be host derived (Table 2).

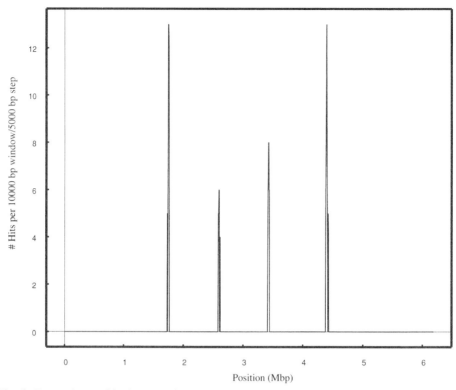

Fig. 2. Example graphical output from Phage_phinder prophage locator software. The complete genome of *Pseudomonas putida* KT2440 was analyzed with a window size of 10,000 bp and a step size of 5,000 bp *(117)*. Only windows with four or more hits are displayed to eliminate noisy background. The output is an XGRAPH (General Purpose 2-D Plotter; Table 1) input file that was saved as a PostScript file and converted to a .tif file.

Phage_phinder slides along the length of the genome and counts the number of significant hits to the phage database per window size. To eliminate noise, only those windows with at least four hits are considered worthy of further attention (Fig. 2). The point in the region that has the greatest number of hits per window is identified, and TIGR role_ids are examined for each ORF in the database 5' and 3' of the point until nonfavorable role_ids are encountered (Table 2). An example graphical output is in Fig. 2 for the genome of *P. putida* KT2440.

To get an idea just how many prophages exist in sequenced bacterial genomes, 89 of the 90 bacterial strains in the CMR were searched against the phage database, and the results were run through the Phage_phinder program. Of the 89 genomes processed, there were 141 putative prophages more than 18 kbp, totaling 5,197,329 bp or 1.85% of the bacterial genomic DNA, accounting for an average of 1.6 putative prophage per genome. This is more than the number of completed bacteriophage genomes in GenBank (72 phages > 18 kbp, totaling 3,129,029 bp) (Fig. 3). Of course, not all of these will be functional phages, but if half are functional, that is still about 70 prophages or about 80% the size of the existing phage genomic pool that can be used for comparative and phylogenetic purposes.

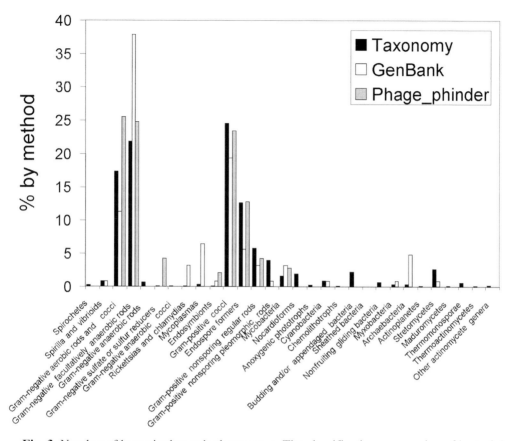

Fig. 3. Number of bacteriophages by host genus. The classification or grouping of bacterial hosts is derived from that used by Ackermann and is based on *Bergey's Manual (1,123)* The numbers of bacteriophages are represented as a percentage of the total number of phages per method (taxonomy, GenBank, or results from Phage_phinder). The data for Taxonomy were derived from Ackermann and represent the number of morphological phage descriptions in 2000 among the tailed phages only (*Myoviridae, Siphoviridae,* and *Podoviridae*) (*1*). The GenBank data were compiled from the number of completely sequenced bacteriophage genomes on Entrez Genomes: Phages as of September 16, 2002 (Table 1). The Phage_phinder data were generated from analysis of 89 completely sequence bacterial genomes and represent predicted prophages 18 kbp or larger.

Currently, a human still has to look at the data to determine whether it is likely phage related or just a cluster of genes that have a representative on at least one fully sequenced, functional phage. Problems arose when analyzing another bacterial genome in that Phage_ phinder identified an island of restriction modification genes that did not contain other "phage-keeping" genes. Other problems occur when a sequenced phage from GenBank harbors transposable elements or ABC transporter(s), resulting in many host genes erroneously marked as in a phage region when they are not.

Identification and annotation of prophage regions can be greatly facilitated by accurate identification of phage-specific proteins. The extreme sequence divergence characteristic

Table 3
HMMs for Bacteriophage Proteins

PF00959	Phage lysozyme
PF01818	Bacteriophage transcriptional regulator
PF02061	Lambda phage CIII
PF02305	Capsid protein (F protein)
PF02306	Major spike protein (G protein)
PF02316	Mu DNA-binding domain
PF02924	Bacteriophage lambda head decoration protein D
PF02925	Bacteriophage scaffolding protein D
PF03197	Bacteriophage FRD2 protein
PF03245	Bacteriophage lysis protein
PF03335	Phage tail fiber repeat
PF03354	Phage terminase, large subunit, putative
PF03374	Phage antirepressor protein
PF03406	Phage tail fiber repeat
PF03420	Prohead core protein protease, T4 family
PF03431	RNA replicase, beta-chain
PF03592	Terminase small subunit
PF03863	Phage maturation protein
PF03864	Phage major capsid protein E
PF03903	Phage T4 tail fiber
PF03906	Phage T7 tail fiber protein
PF04233	Phage Mu protein F-like protein
TIGR01446	DnaD and phage-associated domain
TIGR01537	Phage portal protein, HK97 family
TIGR01538	Phage portal protein, SPP1 family
TIGR01539	Phage portal protein, lambda family
TIGR01541	Phage tail tape measure protein, lambda family
TIGR01543	Phage prohead protease, HK97 family
TIGR01547	Phage terminase, large subunit, PBSX family
TIGR01551	Phage major capsid protein, P2 family
TIGR01554	Phage major capsid protein, HK97 family
TIGR01555	Phage-related protein, HI1409 family

of many phage protein families can make BLAST searches ineffective. Sensitivity and specificity can be greatly improved by the utilization of high-quality profile HMMs *(110,111)*.

To acquire protein sequences for the seed multiple sequence alignment used in the building of HMMs, the BLAST-formatted multiple-protein FASTA file containing sequences used as the subject for Phage_phinder BLASTP searches plus all proteins from the CMR categorized as having a putative prophage function (role_id 152) was used as a query in a BLASTP session. This "all-vs-all" search produced tab-delimited output that was fed into a Perl script that generated single-linkage clusters.

A second Perl script was created to number, annotate, and retrieve amino acid sequences for clusters that contained 10 or more members. Eighty-two clusters, containing 1793

Table 3 (Continued)

TIGR01558	Phage terminase, small subunit, putative, P27 family
TIGR01560	Uncharacterized phage protein (possible DNA packaging)
TIGR01563	Phage head–tail adaptor, putative
TIGR01592	Holin, SPP1 family
TIGR01593	Toxin secretion/phage lysis holin
TIGR01594	Phage holin, lambda family
TIGR01598	Holin, phage phi LC3 family
TIGR01600	Phage minor tail protein L
TIGR01603	Phage major tail protein, phi13 family
TIGR01606	Holin, BlyA family
TIGR01610	Phage replication protein O, N-terminal domain
TIGR01611	Phage major tail tube protein
TIGR01613	Phage/plasmid primase, P4 family, C-terminal domain
TIGR01618	Phage nucleotide-binding protein
TIGR01629	Phage/plasmid replication protein, gene II/X family
TIGR01630	Phage uncharacterized protein, C-terminal domain
TIGR01633	Phage putative tail component, N-terminal domain
TIGR01634	Phage tail protein I
TIGR01635	Phage virion morphogenesis protein
TIGR01636	Phage transcriptional regulator, RinA family
TIGR01637	Phage transcriptional regulator, ArpU family
TIGR01641	Phage putative head morphogenesis protein, SPP1 gp7 family
TIGR01644	Phage baseplate assembly protein V
TIGR01665	Phage minor structural protein, N-terminal region
TIGR01669	Phage uncharacterized protein, XkdX family
TIGR01671	Phage conserved hypothetical protein TIGR01671
TIGR01673	Phage holin, LL-H family
TIGR01674	Phage minor tail protein G
TIGR01712	Phage N-6-adenine-methyltransferase
TIGR01714	Phage replisome organizer, putative, N-terminal region
TIGR01715	Phage tail assembly protein T
TIGR01725	Phage protein, HK97 gp10 family
TIGR01760	Phage tail tape measure protein, TP901 family, core region

See Table 1 for Pfam and TIGRFAM websites.

amino acid sequences, were identified. Only 73 of these clusters were analyzed further because 9 of them already had HMMs available (Tables 1 and 3).

Initially, only those clusters that contained functionally characterized members in the conserved late gene operon responsible for packaging and head morphogenesis of tailed dsDNA bacteriophages were studied *(10,13,112)* (Fig. 4). This conserved region consists of five or more loci that encode proteins with the following functions: (a) small and large terminase subunits to recognize *pac* or *cos* sites, cleavage of concatemers of the bacteriophage genome, possibly involved in packaging of the phage genome into phage heads *(3)*; (b) portal protein to form a hole, or portal, that enables DNA passage during packaging and ejection. It also forms the junction between the phage head (capsid) and

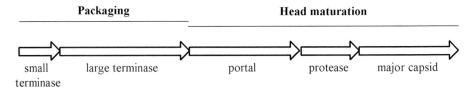

Fig. 4. Diagram of the conserved packaging and head maturation operon of tailed bacteriophages.

the tail proteins *(113)*; (c) prohead protease to cleave the major capsid protein into a mature form *(112,114)*; and (d) the major capsid or head protein that makes up the bulk of the bacteriophage capsid/head *(112)*.

For most of these loci, there are multiple clusters, corresponding to distinct protein families that carry out analogous functions, but lack obvious sequence similarity. A functionally characterized member of each of these clusters was used to initiate Position-Specific Iterated BLAST (PSI-BLAST) *(115)* (Table 1) searches to obtain a larger set of related proteins. Each set was aligned using CLUSTAL W *(116)* (Table 1). Initial alignments were trimmed, culled, realigned, or edited to improve the quality of the alignment and of sequence searches to be based on the alignment.

These manually curated alignments were fed through a pipeline for constructing and testing HMMs with the HMMER software suite *(110)* (Tables 1 and 3). Members of the better characterized families tended to be annotated correctly, while proteins that were smaller, more divergent, and less well characterized (e.g., holin, excisionase, tail accessory proteins) proved far more difficult to identify without an HMM.

How well did Phage_phinder work? Preliminary tests of Phage_phinder with *B. anthracis*, using a window size of 10,000 bp and a step size of 5,000 bp *(117)*, correctly identified the four putative prophages that were manually identified (omitting self-hits to avoid skewing the results). All four putative prophage regions found in the genome of *P. putida* were correctly identified; however, Phage_phinder called a putative conjugal transfer element in another genome a prophage region larger than 18 kbp, but correctly identified all other manually identified prophage regions.

These putative prophage regions will have to be named and classified according to the International Committee on Taxonomy of Viruses (ICTV) taxonomical system (Table 1). Currently, most putative prophage regions are named by some arbitrary methodology and classified based on BLAST matches to some previously characterized phage. Now, because of a very exciting and groundbreaking publication by Rohwer and Edwards *(118)*, the bacteriophage community has a systematic genome-based taxonomy for the classification of bacteriophages. Their method clusters bacteriophages into taxa that are consistent with existing taxonomy standards and observations from the literature.

Rohwer and Edwards *(118)* searched DNA and protein sequences for a conserved gene or motif that could be used in the construction of phylogenetic trees, but were unsuccessful. Briefly, a proteomic distance was determined through a series of steps that involved: (1) construction of sequence alignments for each BLASTP hit with an E value less than 0.1 using CLUSTAL W, (2) determination of protein distance using PROTDIST (part of PHYLIP, the Phylogeny Inference Package) with corrections for different protein lengths, (3) summation of all the length-corrected protein distances with penalties assigned for missing proteins, and (4) creation of a distance matrix in PHYLIP format. This matrix can then be used to construct a phylogenetic tree. The Perl scripts written by Edwards have been made publicly available as freeware under the GNU General Public License (Table 1).

This analysis does seem to corroborate existing taxonomy for many phage families, but for the purpose of characterizing prophages (which would fall into *Siphoviridae*, *Podoviridae*, and *Myoviridae*), a less-rigorous approach might suffice. The nomenclature they propose has three basic components:

1. Indication of the full genus and species name of the first characterized host with the Greek symbol for phi (φ) following because there are a couple of examples for which more than one phage that infects different hosts has the same name (e.g., *Mycobacterium tuberculosis* φ L5 vs *Leuconostoc oenos* φ L5).
2. Replacement of "viridae" with "phage" to indicate that the phage or prophage was classified based on computational approaches (e.g., Siphophage).
3. Naming of phages or prophages belonging to characterized subgroups with the word "-like" appended to the name of the most well studied member of the subgroup (e.g., φ29-like).

CONCLUDING REMARKS

Improved software for identification and annotation of prophage regions is needed because a wealth of untapped public information is available that can significantly contribute to phage comparative genomics, population biology, and evolutionary biology. More profile HMMs need to be constructed, particularly for underrepresented phage genes and those that are small, highly divergent, and "slip under the radar" of pairwise alignment searches. Better identification and annotation of prophage regions will greatly improve annotation of bacterial genomes because up to 7% of bacterial genomic DNA can have phage ancestry *(119)*.

Bacteriophage genomic sequences from enterics and bacteria used in the dairy industry to make yogurt and cheeses are overrepresented in the public databases. More bacteriophage genomes need to be sequenced, beginning with those bacteriophages that already have a wealth of biological information. It is disturbing that the phage that brought awareness to the nature of the transforming principle, bacteriophage T1, has not been fully sequenced and deposited in GenBank. There are many more examples of well-characterized phages that have not been fully sequenced and deposited in GenBank (e.g. many of the *S. pyogenes* phages, like A24, T12, and C1). Just as worrying, the famous bacteriophage T2 (infecting *E. coli* strain B), CTFφ (the filamentous *Vibrio cholera* phage encoding cholera toxin) *(55)*, SopEφ (the P2-like phage infecting *Salmonella typhimurium* and carrying the virulence factor SopE that is secreted by the type III secretion system) *(65)*, the γ-phage of *B. anthracis (120)*, and the *Gifsy* phages of *S. typhimurium (62,121,122)* are not in GenBank or if so are not complete or do not have separate entries. The list of phages that have importance historically or medically but have not been fully sequenced and deposited is too long to mention comprehensively in this chapter and reflects the need for bacteriophage researchers to make sequencing and depositing completed bacteriophage genomes a group effort reminiscent of that of the original phage group in the 1940s.

All of this bacteriophage and prophage genomic information needs to be stored in a publicly accessible database via the Internet. While GenBank is a great resource, not all completed bacteriophage sequences are listed on Entrez Genomes under phages (Table 1). If they are listed, the host is not clear, and the annotation is inconsistent or missing. Comparisons between phages will be greatly facilitated if prophages from bacterial genomes are included in one dedicated, curated resource.

Because most of the sequenced bacterial genomes are pathogens of either animals or plants, it is critical to have more diversity of the global estimated phage population of more than 10^{30} phages (89). Phages must be sequenced from uncultured environmental samples (ocean water, river water, polluted sites, other niches), in which much of life's diversity likely exists; this will provide understanding of the type of phages that exist in nature and the genes they carry. It is known that phages harbor and mobilize genes for bacterial virulence and fitness (7,51). Are there genes that will aid bioremediation? Bacteriophages have given us great insight into molecular mechanisms. Who knows what biological secrets are encrypted in the genomes of phages from unknown microorganisms?

ACKNOWLEDGMENTS

I would like to thank Karen E. Nelson for the invitation to author a chapter in this book. Thanks are extended to Martin Wu and Brian Haas for helping with the clustering of BLASTP btab data; Dan Haft for the building of HMMs; Scott Durkin for stimulating discussions on annotation of bacteriophages; Jonathan Eisen for making protein distance determination understandable; Robert (Bob) Koenig and Aymeric de Vazeille for help translating German into English; Bebbie Rhodes for excellent library support; Jacques Ravel, Garry Myers, and Eddy Arnold-Berkovitz for help with graphical file conversions; and Tim Read for continuous, genuine encouragement and support for my interests and efforts in pursuing bacteriophage bioinformatics at TIGR.

REFERENCES

1. Ackermann HW. Frequency of morphological phage descriptions in the year 2000. Brief review. Arch Virol 2001; 146:843–857.
2. Maniloff J, Ackermann HW. Taxonomy of bacterial viruses: establishment of tailed virus genera and the order Caudovirales. Arch Virol 1998; 143:2051–2063.
3. Ackermann HW. Tailed bacteriophages: the order *Caudovirales*. Adv Virus Res 1999; 51: 135–201.
4. Campbell A. Phage evolution and speciation. In: Calendar R (ed). The Bacteriophages. New York: Plenum, 1988, pp. 1–14.
5. Hendrix RW, Smith MC, Burns RN, Ford ME, Hatfull GF. Evolutionary relationships among diverse bacteriophages and prophages: all the world's a phage. Proc Natl Acad Sci USA 1999; 96:2192–2197.
6. Hendrix RW. Evolution: the long evolutionary reach of viruses. Curr Biol 1999; 9:R914–R917.
7. Hendrix RW, Lawrence JG, Hatfull GF, Casjens S. The origins and ongoing evolution of viruses. Trends Microbiol 2000; 8:504–508.
8. Hendrix RW. Bacteriophages: evolution of the majority. Theor Popul Biol 2002; 61:471–480.
9. Chopin A, Bolotin A, Sorokin A, Ehrlich SD, Chopin M-C. Analysis of six prophages in *Lactococcus lactis* IL1403: different genetic structure of temperate and virulent phage populations. Nucleic Acids Res 2001; 29:644–651.

10. Desiere F, Lucchini S, Brüssow H. Comparative sequence analysis of the DNA packaging, head, and tail morphogenesis modules in the temperate cos-site *Streptococcus thermophilus* bacteriophage Sfi21. Virology 1999; 260:244–253.

11. Lucchini S, Desiere F, Brüssow H. Comparative genomics of *Streptococcus thermophilus* phage species supports a modular evolution theory. J Virol 1999; 73:8647–8656.

12. Desiere F, Mahanivong C, Hillier AJ, Chandry PS, Davidson BE, Brüssow H. Comparative genomics of lactococcal phages: insight from the complete genome sequence of *Lactococcus lactis* phage BK5-T. Virology 2001; 283:240–252.

13. Desiere F, McShan WM, van Sinderen D, Ferretti JJ, Brüssow H. Comparative genomics reveals close genetic relationships between phages from dairy bacteria and pathogenic strepto-cocci: evolutionary implications for prophage–host interactions. Virology 2001; 288:325–341.

14. Brüssow H, Desiere F. Comparative phage genomics and the evolution of *Siphoviridae*: insights from dairy phages. Mol Microbiol 2001; 39:213–222.

15. Meijer WJ, Horcajadas JA, Salas M. φ29 family of phages. Microbiol Mol Biol Rev 2001; 65: 261–287.

16. Weisberg RA, Gottesmann ME, Hendrix RW, Little JW. Family values in the age of genomics: comparative analyses of temperate bacteriophage HK022. Annu Rev Genet 1999; 33:565–602.

17. Juhala RJ, Ford ME, Duda RL, Youlton A, Hatfull GF, Hendrix RW. Genomic sequences of bacteriophages HK97 and HK022: pervasive genetic mosaicism in the lambdoid bacteriophages. J Mol Biol 2000; 299:27–51.

18. Brüssow H. Phages of dairy bacteria. Annu Rev Microbiol 2001; 55:283–303.

19. Ackermann HW, Martin M, Vieu JF, Nicolle P. Felix d'Hérelle: His life and work and the foun-dation of a bacteriophage reference center. ASM News 1982; 48:346–348.

20. Duckworth DH. History and basic properties of bacterial viruses. In: Goyal SM, Gerba CP, Bitton G (eds). Phage Ecology. New York: Wiley, 1987, pp. 1–43.

21. d'Hérelle F. La coccobacille des sauterelles. Ann Inst Pasteur Paris 1914; 28:280–328.

22. d'Hérelle F. Sur un microbe invisible antagoniste des bacilles dysentériques. CR Acad Sci Paris 1917; 165:373–375.

23. Twort FW. An investigation on the nature of ultra-microscopic viruses. Lancet 1915; 2:1241–1243.

24. Ruska H. Die Sichtbarmachung der Bakteriophagen Lyse im Übermikroskop. Naturwiss 1940; 28:45–46.

25. Ruska H. Über ein neues bei der Bakteriophagen Lyse auftretendes Formelement. Naturwiss 1941; 29:367–368.

26. Pfankuch E, Kausche GA. Isolierung und übermikroskopische Abbildung eines Bakteriophages. Naturwiss 1940; 28:46.

27. Luria SE, Anderson TF. The identification and characterization of bacteriophages with the electron microscope. Proc Natl Acad Sci USA 1942; 28:127–130.

28. Anderson TF. Electron microscopy of phages. In: Cairns J, Stent GS, Watson JD (eds). Phage and the Origins of Molecular Biology. Cold Spring Harbor, NY: Cold Spring Harbor Labora-tory Press, 1992, pp. 63–78.

29. Luria SE, Delbrück M. Mutations of bacteria from virus sensitivity to virus resistance. Genetics 1943; 28:491–511.

30. Hershey AD, Chase M. Independent functions of viral protein and nucleic acids in growth of bacteriophage. J Gen Physiol 1952; 36:39–56.

31. Zinder ND, Lederberg J. Genetic exchange in *Salmonella*. J Bacteriol 1952; 64:679–699.

32. Benzer S. On the topography of genetic fine structure. Proc Natl Acad Sci USA 1961; 47: 403–415.

33. Crick FHC, Barnett L, Brenner S, Watts-Tobin RJ. General nature of the genetic code for pro-teins. Nature (London) 1961; 192:1227–1232.

34. Terzaghi E, Okada Y, Streisinger G, Emrich J, Inouye M, Tsugita A. Change of a sequence of amino acids in phage T4 lysozyme by acridine-induced mutations. Proc Natl Acad Sci USA 1966; 56:500–507.

35. Linn S, Arber W. Host specificity of DNA produced by *Escherichia coli*, X. In vitro restriction of phage fd replicative form. Proc Natl Acad Sci USA 1968; 59:1300–1306.

36. Weiss B, Richardson CC. Enzymatic breakage and joining of deoxyribonucleic acid, I. Repair of single-strand breaks in DNA by an enzyme system from *Escherichia coli* infected with T4 bacteriophage. Proc Natl Acad Sci USA 1967; 57:1021–1028.

37. Weiss B, Jacquemin-Sablon A, Live TR, Fareed GC, Richardson CC. Enzymatic breakage and joining of deoxyribonucleic acid. VI. Further purification and properties of polynucleotide ligase from *Escherichia coli* infected with bacteriophage T4. J Biol Chem 1968; 243:4543–4555.

38. Silber R, Malathi VG, Hurwitz J. Purification and properties of bacteriophage T4-induced RNA ligase. Proc Natl Acad Sci USA 1972; 69:3009–3013.

39. De Waard A, Paul AV, Lehman IR. The structural gene for deoxyribonucleic acid polymerase in bacteriophages T4 and T5. Proc Natl Acad Sci USA 1965; 54:1241–1248.

40. Panet A, van de Sande JH, Loewen PC, et al. Physical characterization and simultaneous purification of bacteriophage T4 induced polynucleotide kinase, polynucleotide ligase, and deoxyribonucleic acid polymerase. Biochemistry 1973; 12:5045–5050.

41. Smith GP. Filamentous fusion phage: novel expression vectors that display cloned antigens on the virion surface. Science 1985; 228:1315–1317.

42. Benson N, Sugiono P, Bass S, Mendelman LV, Youderian P. General selection for specific DNA-binding activities. Genetics 1986;114:1–14.

43. MacWilliams MP, Celander DW, Gardner JF. Direct genetic selection for a specific RNA-protein interaction. Nucleic Acids Res 1993; 21:5754–5760.

44. Salser W, Fry K, Brunk C, Poon R. Nucleotide sequencing of DNA: preliminary characterization of the products of specific cleavages at guanine, cytosine, or adenine residues (bacteriophage M13–ribosubstitution–DNA polymerase I–electrophoresis–two-dimensional fingerprinting). Proc Natl Acad Sci USA 1972; 69:238–242.

45. Schreier PH, Cortese R. A fast and simple method for sequencing DNA cloned in the single-stranded bacteriophage M13. J Mol Biol 1979; 129:169–172.

46. Kunkel TA. Rapid and efficient site-specific mutagenesis without phenotypic selection. Proc Natl Acad Sci USA 1985; 82:488–492.

47. Kunkel TA, Roberts JD, Zakour RA. Rapid and efficient site-specific mutagenesis without phenotypic selection. Methods Enzymol 1987; 154:367–382.

48. Chauthaiwale VM, Therwath A, Deshpande VV. Bacteriophage lambda as a cloning vector. Microbiol Rev 1992; 56:577–591.

49. Lwoff A, Horne R, Tournier P. A system of viruses. Cold Spring Harbor Symp Quant Biol 1962; 27:51–55.

50. Frobisher M, Brown J. Transmissible toxicogenicity of streptococci. Bull. Johns Hopkins Hosp 1927; 41:167–173.

51. Wagner PL, Waldor MK. Bacteriophage control of bacterial virulence. Infect Immun 2002; 70:3985–3993.

52. Fujii N, Oguma K, Yokosawa N, Kimura K, Tsuzuki K. Characterization of bacteriophage nucleic acids obtained from *Clostridium botulinum* types C and D. Appl Environ Microbiol 1988; 54:69–73.

53. Holmes RK, Barksdale L. Genetic analysis of tox+ and tox− bacteriophages of *Corynebacterium diphtheriae*. J Virol 1969; 3:586–598.

54. Uchida T, Gill DM, Pappenheimer AM. Mutation in the structural gene for diphtheria toxin carried by temperate phage β. Nat New Biol 1971; 233:8–11.

55. Waldor MK, Mekalanos JJ. Lysogenic conversion by a filamentous phage encoding cholera toxin. Science 1996; 272:1910–1914.

56. Huang A, de Grandis S, Friesen J, et al. Cloning and expression of the genes specifying Shiga-like toxin production in *Escherichia coli* H19. J Bacteriol 1986; 166:375–379.

57. Recktenwald J, Schmidt H. The nucleotide sequence of Shiga toxin (Stx) 2e-encoding phage φP27 is not related to other Stx phage genomes, but the modular genetic structure is conserved. Infect Immun 2002; 70:1896–1908.

58. Betley MJ, Mekalanos JJ. Staphylococcal enterotoxin A is encoded by phage. Science 1985; 229:185–187.

59. Bensing BA, Rubens CE, Sullam PM. Genetic loci of *Streptococcus mitis* that mediate binding to human platelets. Infect Immun 2001; 69:1373–1380.

60. Wright A. Mechanism of conversion of the *Salmonella* O antigen by bacteriophage ε^{34}. J Bacteriol 1971; 105:927–936.

61. Guan S, Bastin DA, Verma NK. Functional analysis of the O antigen glucosylation gene cluster of *Shigella flexneri* bacteriophage SfX. Microbiology 1999; 145(Pt 5):1263–1273.

62. Figueroa-Bossi N, Bossi L. Inducible prophages contribute to *Salmonella* virulence in mice. Mol Microbiol 1999; 33:167–176.

63. Hynes WL, Ferretti JJ. Sequence analysis and expression in *Escherichia coli* of the hyaluronidase gene of *Streptococcus pyogenes* bacteriophage H4489A. Infect Immun 1989; 57:533–539.

64. Sako T, Sawaki S, Sakurai T, Ito S, Yoshizawa Y, Kondo I. Cloning and expression of the staphylokinase gene of *Staphylococcus aureus* in *Escherichia coli*. Mol Gen Genet 1983; 190: 271–277.

65. Mirold S, Rabsch W, Rohde M, et al. Isolation of a temperate bacteriophage encoding the type III effector protein SopE from an epidemic *Salmonella typhimurium* strain. Proc Natl Acad Sci USA 1999; 96:9845–9850.

66. Fischetti VA. Phage antibacterials make a comeback. Nat Biotechnol 2001; 19:734–5.

67. Nelson D, Loomis L, Fischetti VA. Prevention and elimination of upper respiratory colonization of mice by group A streptococci by using a bacteriophage lytic enzyme. Proc Natl Acad Sci USA 2001; 98:4107–4112.

68. Loeffler JM, Nelson D, Fischetti VA. Rapid killing of *Streptococcus pneumoniae* with a bacteriophage cell wall hydrolase. Science 2001; 294:2170–2172.

69. Stone R. Bacteriophage therapy. Stalin's forgotten cure. Science 2002; 298:728–731.

70. Summers WC. Bacteriophage therapy. Annu Rev Microbiol 2001; 55:437–451.

71. Biswas B, Adhya S, Washart P, et al. Bacteriophage therapy rescues mice bacteremic from a clinical isolate of vancomycin-resistant *Enterococcus faecium*. Infect Immun 2002; 70:204–210.

72. Chernomordik AB. Bacteriophages and their therapeutic-prophylactic use. Med Sestra 1989; 48:44–47.

73. Schuch R, Nelson D, Fischetti VA. A bacteriolytic agent that detects and kills *Bacillus anthracis*. Nature 2002; 418:884–889.

74. Le Y, Gagneten S, Tombaccini D, Bethke B, Sauer B. Nuclear targeting determinants of the phage P1 cre DNA recombinase. Nucleic Acids Res 1999; 27:4703–4709.

75. Groth AC, Olivares EC, Thyagarajan B, Calos MP. A phage integrase directs efficient site-specific integration in human cells. Proc Natl Acad Sci USA 2000; 97:5995–6000.

76. Schagen FH, Rademaker HJ, Cramer SJ, et al. Towards integrating vectors for gene therapy: expression of functional bacteriophage MuA and MuB proteins in mammalian cells. Nucleic Acids Res 2000; 28:E104.

77. Olivares EC, Hollis RP, Calos MP. Phage R4 integrase mediates site-specific integration in human cells. Gene 2001; 278:167–176.

78. Stoll SM, Ginsburg DS, Calos MP. Phage TP901-1 site-specific integrase functions in human cells. J Bacteriol 2002; 184:3657–3663.

79. Fleischmann RD, Adams MD, White O, et al. Whole-genome random sequencing and assembly of *Haemophilus influenzae* Rd. Science 1995; 269:496–512.

80. Sanger F, Air GM, Barrell BG, et al. Nucleotide sequence of bacteriophage phi X174 DNA. Nature 1977; 265:687–695.

81. Sanger F, Coulson AR, Friedmann T, et al. The nucleotide sequence of bacteriophage phiX174. J Mol Biol 1978; 125:225–246.

82. Daniels DL, Sanger F, Coulson AR. Features of bacteriophage lambda: analysis of the complete nucleotide sequence. Cold Spring Harb Symp Quant Biol 1983; 47(Pt 2):1009–1024.

83. Sanger F, Coulson AR, Hong GF, Hill DF, Petersen GB. Nucleotide sequence of bacteriophage lambda DNA. J Mol Biol 1982; 162:729–773.

84. Kunst F, Ogasawara N, Moszer I, et al. The complete genome sequence of the Gram-positive bacterium *Bacillus subtilis*. Nature 1997; 390:249–256.

85. Glaser P, Frangeul L, Buchrieser C, et al. Comparative genomics of *Listeria* species. Science 2001; 294:849–852.

86. Perna NT, Plunkett G 3rd, Burland V, et al. Genome sequence of enterohaemorrhagic *Escherichia coli* O157:H7. Nature 2001; 409:529–533.

87. Campbell AM. Cryptic prophages. In: Neidhardt FC, Curtiss R III, Ingraham JL, et al. (eds). *Escherichia coli* and *Salmonella* Cellular and Molecular Biology. Washington, DC: American Society for Microbiology Press, 1996, pp. 2041–2046.

88. Read TD, Brunham RC, Shen C, et al. Genome sequences of *Chlamydia trachomatis* MoPn and *Chlamydia pneumoniae* AR39. Nucleic Acids Res 2000; 28:1397–1406.

89. Brüssow H, Hendrix RW. Phage genomics: small is beautiful. Cell 2002; 108:13–16.

90. Ferretti JJ, McShan WM, Ajdic D, et al. Complete genome sequence of an M1 strain of *Streptococcus pyogenes*. Proc Natl Acad Sci USA 2001; 98:4658–4663.

91. Pečenková T, Benes V, Pačes J, Vlček C, Pačes V. Bacteriophage B103: complete DNA sequence of its genome and relationship to other *Bacillus* phages. Gene 1997; 199:157–163.

92. Osborn MA, Böltner D. When phage, plasmids, and transposons collide: genomic islands, and conjugative- and mobilizable-transposons as a mosaic continuum. Plasmid 2002; 48:202–212.

93. Nakayama K, Takashima K, Ishihara H, et al. The R-type pyocin of *Pseudomonas aeruginosa* is related to P2 phage, and the F-type is related to lambda phage. Mol Microbiol 2000; 38: 213–231.

94. Uratani Y, Hoshino T. Pyocin R1 inhibits active transport in *Pseudomonas aeruginosa* and depolarizes membrane potential. J Bacteriol 1984; 157:632–636.

95. Daw MA, Falkiner FR. Bacteriocins: nature, function and structure. Micron 1996; 27:467–479.

96. Rampling A, Whitby JL. Preparation of phage-free pyocin extracts for use in the typing of *Pseudomonas aeruginosa*. J Med Microbiol 1972; 5:305–312.

97. Jones LF, Zakanycz JP, Thomas ET, Farmer JJ 3rd. Pyocin typing of *Pseudomonas aeruginosa*: a simplified method. Appl Microbiol 1974; 27:400–406.

98. Kurtz S, Schleiermacher C. REPuter: fast computation of maximal repeats in complete genomes. Bioinformatics 1999; 15:426–427.

99. Zhao S, Williams KP. Integrative genetic element that reverses the usual target gene orientation. J Bacteriol 2002; 184:859–860.

100. Morgan GJ, Hatfull GF, Casjens S, Hendrix RW. Bacteriophage Mu genome sequence: analysis and comparison with Mu-like prophages in *Haemophilus*, *Neisseria* and *Deinococcus*. J Mol Biol 2002; 317:337–359.

101. Masignani V, Giuliani MM, Tettelin H, Comanducci M, Rappuoli R, Scarlato V. Mu-like prophage in serogroup B *Neisseria meningitidis* coding for surface-exposed antigens. Infect Immun 2001; 69:2580–2588.

102. Zimmer M, Scherer S, Loessner MJ. Genomic analysis of *Clostridium perfringens* bacterio-phage φ3626, which integrates into *guaA* and possibly affects sporulation. J Bacteriol 2002; 184:4359–4368.

103. Williams KP. Integration sites for genetic elements in prokaryotic tRNA and tmRNA genes: sublocation preference of integrase subfamilies. Nucleic Acids Res 2002; 30:866–875.

104. Lowe TM, Eddy SR. tRNAscan-SE: a program for improved detection of transfer RNA genes in genomic sequence. Nucleic Acids Res 1997; 25:955–964.

105. Shimizu-Kadota M, Kiwaki M, Sawaki S, Shirasawa Y, Shibahara-Sone H, Sako T. Insertion of bacteriophage phiFSW into the chromosome of *Lactobacillus casei* strain Shirota (S-1): characterization of the attachment sites and the integrase gene. Gene 2000; 249:127–134.

106. Nelson KE, Weinel C, Paulsen IT, et al. Complete genome sequence and comparative analysis of the metabolically versatile *Pseudomonas putida* KT2440. Environ Microbiol 2002; 4:799–808.

107. Peterson JD, Umayam LA, Dickinson T, Hickey EK, White O. The Comprehensive Microbial Resource. Nucleic Acids Res 2001; 29:123–125.

108. Altschul SF, Gish W, Miller W, Myers EW, Lipman DJ. Basic Local Alignment Search Tool. J Mol Biol 1990; 215:403–410.

109. Riley M. Functions of the gene products of *Escherichia coli*. Microbiol Rev 1993; 57:862–952.

110. Eddy SR. Profile hidden Markov models. Bioinformatics 1998; 14:755–63.

111. Wang IN, Smith DL, Young R. Holins: the protein clocks of bacteriophage infections. Annu Rev Microbiol 2000; 54:799–825.

112. Duda RL, Martincic K, Hendrix RW. Genetic basis of bacteriophage HK97 prohead assembly. J Mol Biol 1995; 247:636–647.

113. Casjens S, Hendrix R. Control mechanisms in dsDNA bacteriophage assembly. In: Calendar R (ed). The Bacteriophages. New York: Plenum Press, 1988, pp. 15–91.

114. Black LW, Showe MK, Steven AC. Morphogenesis of the T4 head. In: Karam J (ed). Molecular Biology of Bacteriophage T4. Washington, DC: American Society for Microbiology Press, 1994, pp. 518–558.

115. Altschul SF, Madden TL, Schaffer AA, et al. Gapped BLAST and PSI-BLAST: a new generation of protein database search programs. Nucleic Acids Res 1997; 25:3389–3402.

116. Thompson JD, Higgins DG, Gibson TJ. CLUSTAL W: improving the sensitivity of progressive multiple sequence alignment through sequence weighting, position-specific gap penalties and weight matrix choice. Nucleic Acids Res 1994; 22:4673–4680.

117. Takami H, Nakasone K, Takaki Y, et al. Complete genome sequence of the alkaliphilic bacterium *Bacillus halodurans* and genomic sequence comparison with *Bacillus subtilis*. Nucleic Acids Res 2000; 28:4317–4331.

118. Rohwer F, Edwards R. The Phage Proteomic Tree: a genome-based taxonomy for phage. J Bacteriol 2002; 184:4529–4535.

119. Simpson AJ, Reinach FC, Arruda P, et al. The genome sequence of the plant pathogen *Xylella fastidiosa*. The *Xylella fastidiosa* Consortium of the Organization for Nucleotide Sequencing and Analysis. Nature 2000; 406:151–157.

120. Brown ER, Cherry W. Specific identification of *Bacillus anthracis* by means of a variant bacteriophage. J Infect Dis 1955; 96:34–39.

121. Figueroa-Bossi N, Coissac E, Netter P, Bossi L. Unsuspected prophage-like elements in *Salmonella typhimurium*. Mol Microbiol 1997; 25:161–173.

122. Figueroa-Bossi N, Uzzau S, Maloriol D, Bossi L. Variable assortment of prophages provides a transferable repertoire of pathogenic determinants in *Salmonella*. Mol Microbiol 2001; 39: 260–271.

123. Holt JG (ed). Bergey's Manual of Systematic Bacteriology. Baltimore, MD: Williams & Wilkins, 1984.

III CORE FUNCTIONS

6

Comparative Microbial Metabolism

Karen E. Nelson

INTRODUCTION

The completion and subsequent availability of the genome sequence of the free living bacterium *Haemophilus influenzae (1)* in 1995 opened the field of microbial genomics. Since that year, more than 100 microbial genomes have been completely sequenced and published, and at least another 300 are estimated to be in progress worldwide (www. tigr.org). In the early stages of the development of the field of microbial genomics, choices for whole genome sequencing were clearly geared toward organisms of medical importance such as *H. influenzae (1)* and *Mycoplasma genitalium (2)*. This was undoubtedly because the characterization of the major pathogens was anticipated to increase understanding of their basic biology and increase opportunities for the identification of antimicrobial targets *(3)*.

The initial focus on pathogenic species later changed to include the sequencing of microbes of agricultural, environmental, evolutionary, and biotechnological importance (for review, see ref. *4*). In addition to revealing how bacterial species may initiate and cause disease *(5–7)*, the sequencing of microbial genomes has given insight into lateral gene transfer in microbial species *(8)*, genome rearrangements between closely related species *(7,9–11)*, and potential environmental applications of some of these microbial species *(12)*. The ability to sequence microbial genomes completely has also allowed entrance to avenues that give basic biochemical information about the organism in question and about entire microbial populations and communities.

This chapter on microbial metabolism is devoted to a description of how genomic sequencing has enabled increased understanding of the basic biology, biochemistry, and physiology of microbial species and how comparisons within and across microbial genomes are giving insight into previously uncharacterized biological information for every new species sequenced. Several examples are drawn from published and ongoing genome projects.

DATA MINING OF MICROBIAL GENOME SEQUENCES

Other chapters in this volume address methodologies currently used for constructing genome libraries and for generating a complete microbial genome sequence. As such,

From: *Microbial Genomes*
Edited by: C. M. Fraser, T. D. Read, and K. E. Nelson © Humana Press Inc., Totowa, NJ

techniques for sequencing genomes are not revisited here. Once the genome is assembled and all physical and sequencing gaps closed, bioinformatic analyses become essential for interpretation of the species biology. Bioinformatic analyses allow identification of all open reading frames (ORFs), as well as such other features including transfer RNA (tRNA), ribosomal RNA (rRNA), repeat sequences, and the like in the genome. These analyses often extend to the identification of intergenic regions, nucleotide biases, origins of replication, putative regions of horizontal gene transfer, insertion sequence elements, and plasmids.

Effective automated gene identification can be accomplished with programs that employ hidden Markov models (HMMs) *(13–15)* or interpolated Markov models such as the Gene Locator and Interpolated Markov Modeler (Glimmer; *14*), with biological name assignments and functions made by a combination of computer programs and human curation. Basic Local Alignment Search Tool (BLAST) *(16,17)* or FASTA searches against sequence databases and comparisons with homologous families of proteins, including HMMs, Pfams, and Clusters of Orthologous Groups (COGs), aid in the process of making functional predictions. For more detail on genome annotation, please see Chapter 3.

A closer evaluation of the genome sequence accompanied by detailed analyses and an examination of existing literature can allow reconstruction of physiological profiles of the organism and prediction of biochemical pathways, which may not have been previously identified. Often, there may be steps in pathways that cannot be identified in the initial round of annotation and curation. By making HMMs or alignments of sequences for steps that have been identified in other organisms and searching these alignments against the genome sequence of interest, a putative assignment can often be made to fill gaps in a pathway. Again, it should be reiterated that, at the end of the annotation phase of a genome project, a significant proportion of ORFs remain as either conserved hypothetical or hypothetical proteins, which are either shared with another species or unique to the organism in question, respectively. These ORFs for which functional assignments cannot be made, undoubtedly contribute to the biology of the organism, but based on current limitations in analysis and understanding of microbial diversity, the putative roles that these elements play in the cell are yet to be determined.

Although it is possible to obtain metabolic maps from various sources for many biochemical pathways, few software tools are currently available to assist in the reconstruction of physiological profiles of microbial species directly from genomic data. For viewing metabolic maps, publicly available tools include the Kyoto Encyclopedia of Genes and Genomes (also known as KEGG; http://www.genome.ad.jp/kegg/kegg2.html), the WIT (What Is There?) project (http://wit.mcs.anl.gov/WIT2/), and EcoCyc (http://www.ecocyc.org/).

KEGG acts primarily as a bioinformatics resource for interpreting the utilities of organisms based on genome information. As of 2003, the KEGG database contained information on 132 organisms, representing 481,325 genes and 5,415 chemical reactions.

The WIT project tries to produce metabolic reconstructions based on sequence, biochemical, and phenotypic data for various genomes. WIT contains 25 reconstructions in different stages of completion. For the organisms that have biochemical reconstructions, a list of the pathways that have been identified in the genome and a table that relates the ORFs to the proposed functional roles are presented.

EcoCyc is devoted to the biochemical machinery of *Escherichia coli* and is based primarily on available *E. coli* genome sequences. This database has become an electronic reference source for *E. coli*, as well as for other microbial species that have pathways present in *E. coli*. The Pathway/Genome Navigator user interface in EcoCyc enables the visualization of genes in *E. coli*, as well as pathways for a range of biochemical reactions. EcoCyc also has descriptions of all known signal-transduction pathways of *E. coli*, as well as operons, promoters, transcription factors, and transcription-factor binding sites. This database appears to contain the most complete description of the genetic network of any single organism in electronic format.

The Enzymes and Metabolic Pathways (EMP) database (http://www.empproject.com/) is a comprehensive resource of biochemical data that incorporates many aspects of enzymology and metabolism. This database includes references to about 15,000 experimental journal publications to support any descriptions. Information in this database includes cell cultivation conditions, enzyme and reaction, enzyme kinetics, and thermodynamics, as well as physical chemistry. The EMP database contains more than 3,000 metabolic diagrams useful for mathematical simulation of metabolic networks and metabolic design.

Finally, the University of Minnesota Biocatalysis/Biodegradation Database (UM-BBD; http://umbbd.ahc.umn.edu/) is an electronic resource of microbial biocatalytic reactions and biodegradation pathways primarily for xenobiotic and chemical compounds; therefore, it provides information primarily on reactions of importance to biotechnology. Reactions and pathways include information on starting and intermediate chemical compounds, the organisms that transform the compounds, the enzymes, and the genes.

As yet, no single tool is capable of automatically or reliably reconstructing the physiological profile of a cell based on genome data. The main problem with all current tools for analyzing metabolic data is that they depend on name assignments rather than on the actual sequence; as a result, mistakes such as transitive error (see Chapter 3) are incorporated into these predictions. To alleviate this problem and to generate predictions with a high degree of accuracy at The Institute for Genomic Research (TIGR), metabolic and transport profiles are manually reconstructed (see Chapter 7; Figs. 1 and 2). Each pathway and gene assignment is manually checked to ensure that the gene and name calls are accurate. When a step in a pathway cannot be reliably identified, the genome sequence is searched with available sequences for that particular missing enzyme from other species to designate a putative ORF that may be responsible for that particular reaction.

As mentioned, a significant proportion (sometimes as much as 40 to 50%) of the ORFs in each microbial genome is considered to be hypothetical or conserved hypothetical proteins. As such, it is obvious that a significant amount of microbial species biology remains to be elucidated. The magnitude of the possible diversity that exists is evident when it is considered that more than 99% of microbial species remain unidentified. Regardless, with the available sequence data, biochemical profiles for the majority of species sequenced at TIGR have been successfully reconstructed. These represent both prokaryotic and eukaryotic species isolated from a range of environments and include extremophiles *(8,18)*, soil-dwelling aromatic degraders *(12)*, photosynthetic species *(19)*, and many pathogenic species (Gill et al., unpublished manuscript; *5,7,20*).

Analyses of some species, such as the soil-dwelling remediator *Pseudomonas putida* *(12)*, have revealed a far greater number of metabolic pathways for the conversion of

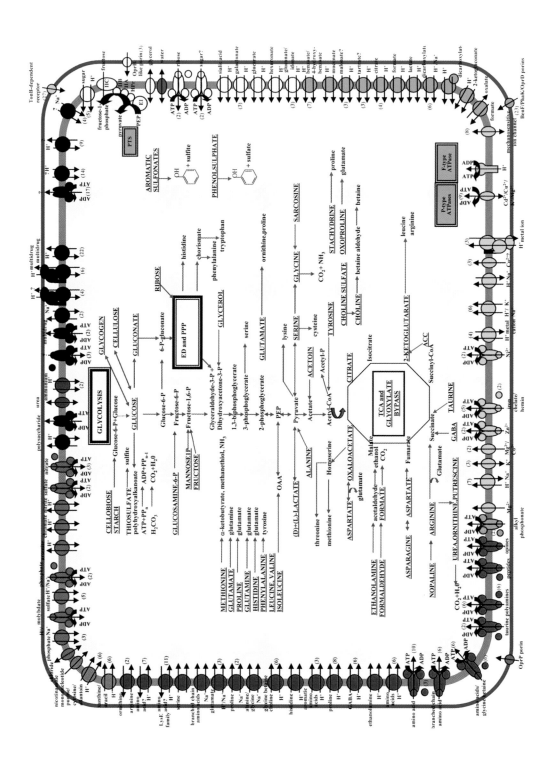

atypical compounds than previously identified (Fig. 1). *Caulobacter crescentus* was sequenced primarily to gain insights into cell cycle control and regulation *(21,22)*, but genome analysis revealed the presence of unsuspected pathways, such as the β-keto-adipate pathway for the metabolism of aromatic compounds. Similar analyses of the *Streptococcus pneumoniae* genome *(7)* highlighted the ability of this organism to import and metabolize a range of sugars, many of which had not been previously identified as potential substrates. *S. pneumoniae* type 4 is the causative agent of life-threatening diseases, including pneumonia, bacteremia, and meningitis, with high morbidity and mortality rates among young children and elderly people. It is also the leading cause of otitis media and sinusitis *(23)*. Capsular types are considered major virulent determinants across pneumococcal strains and are also known to differ significantly. Detailed analysis of the genome sequence allowed the reconstruction of a pathway for biosynthesis of the type 4 capsule and for the degradation and metabolism of host polysaccharides (*see* Figure in Paulsen Chapter 7).

Analysis of the *Thermotoga maritima* genome *(8)* revealed numerous pathways for the metabolism of plant compounds, including cellulose and xylan, as well for the metabolism of a range of sugars. The bacterium also has a significant number of transporters devoted to the import of polysaccharides and oligopeptides. Based on the predictions made from the *T. maritima* genome sequence, this bacterium (and close relatives such as *Thermotoga neopolitana*; Fig. 2) is being investigated for its potential to produce hydrogen gas as a renewable energy source (S. Van Otengham, personal communication, 2003). Recent studies described a method of producing 20–30% hydrogen gas by *T. neapolitana* in batch experiments of relatively short duration (50–250 h).

Porphyromonas gingivalis is one of the many bacterial species that reside in the oral cavity, where microorganisms in the supragingival plaque are exposed to the host's dietary intake. Anaerobic species such as *P. gingivalis* that occupy the subgingival region are not exposed to the dietary fraction, but are exposed to the host tissue proteins and metabolic end products from other microbial species *(24)*. In an attempt to characterize further the biology of *P. gingivalis*, a whole genome sequencing project was initiated at TIGR and completed in 2002 *(25)*. Although glucose utilization by *P. gingivalis* is known to be poor *(26)* the sequenced strain does contain putative ORFs for all enzymes of the glycolytic pathway. In addition, there are ORFs for a putative glucose/galactose transporter and glucose kinase. Four putative ORFs for the pentose phosphate pathway were identified, and it is possible that this pathway is used to generate precursor metabolites during anaerobic growth. Aspartate, asparagine, and glutamine are readily utilized by this bacterium *(24,26)*, and pathways have been identified from the genome sequence that suggest many additional amino acids can be utilized. In total, 44 peptidases could be identified, as well as enzymes for the degradation of complex amino sugars.

The results from genome analysis also suggest that the major fermentation end products of *P. gingivalis* include propionate, butyrate, isobutyrate, isovalerate, actetate, ethanol,

Fig. 1. (*Opposite page*) Metabolic reconstruction of pathways present in the *Pseudomonas putida* genome. Reproduced with permission from Environmental Microbiology. Nelson KE, Weinel C, Paulsen IT, et al. Complete genome sequence and comparative analysis of the metabolically versatile *Pseudomonas putida* KT2440. Environmental Microbiology 2003; 5:

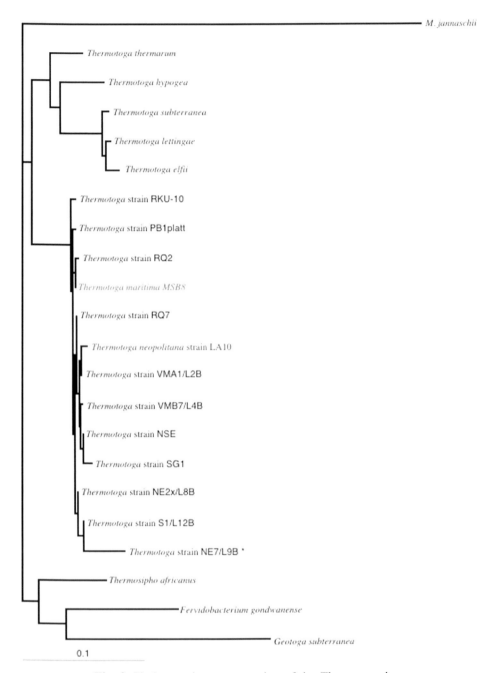

Fig. 2. Phylogenetic representation of the *Thermotogales*.

...l. Some of these fermentation end-products have been shown to be toxic to
...d may result in cell death, disruption of immunocyte activity, and cytokine
These end products may therefore act as virulence agents against the

EXAMPLES FROM EUKARYOTIC SPECIES:
PLASMODIUM FALCIPARUM

As another example of recent analysis of completed genome sequences, reconstruction of biochemical pathways of *Plasmodium falciparum (29)* (Fig. 3) has given tremendous insight into the biology of this organism and has allowed the tentative identification of novel drug targets. From the genome sequence, all of the enzymes necessary for a functional glycolytic pathway, as well as candidate genes for all but one of the pentose phosphate pathways, could be identified. The presence of a phosphoenolpyruvate carboxylase suggests that *P. falciparum* may cope with a drain of intermediates from the tricarboxylic acid (TCA) cycle by using this enzyme to replenish oxaloacetate from cytosolic phosphoenolpyruvate and bicarbonate.

Biochemical, genetic, and chemotherapeutic data suggest that malaria and other apicomplexan parasites possess the ability to synthesize chorismate from erythrose 4-phosphate and phosphoenolpyruvate via the shikimate pathway *(30,31)*. Apart from chorismate synthase, the genes for the enzymes in the pathway could not be identified with any certainty from the genome sequence.

Finally, the malaria parasite utilizes hemoglobin from the host cytoplasm as a food source, hydrolyzing globin and releasing heme, which is detoxified in the form of hemazoin. It was unclear whether *de novo* synthesis occurs using imported host enzymes or using the parasite's own enzymes. From analysis of the genome sequence, orthologs could be identified for every enzyme in the pathway except for uroporphyrinogen-III synthase.

GENOMIC APPROACHES TO CULTIVATION STUDIES

The significant value of reconstructions of microbial physiological profiles is even more evident when the true substrate utilization patterns are realized from the genome predictions. The complete genome sequence of *Deinococcus radiodurans (32)*, for example, was used to identify the key nutritional constituents that could restore growth of the bacterium in nutritionally limiting radioactive environments *(33)*. To develop a synthetic minimal medium, the authors tested combinations of different amounts of carbohydrates, amino acids, salts, and vitamins in both liquid media and solid media. This allowed the identification of the minimal nutrient constituents and concentrations of these nutrients necessary for substantial growth. In addition to a metabolizable carbon source, growth of *D. radiodurans* was dependent on exogenous amino acids and a vitamin; sulfur-rich amino acids together with nicotinic acid were particularly effective at supporting growth. It was also realized that there was no specific amino acid combination necessary because many different combinations of amino acids supported growth. One factor that strongly influenced the extent of growth was the total amino acid concentration in the growth medium, not the composition of the amino acid pool. It is anticipated that similar techniques will be applied to figuring out growth requirements for a number of microbial species currently being sequenced but for which cultivation still is difficulty or there is an inability to cultivate, such as *Dehalococcoides ethenogenes*, which currently requires microbial extracts to support growth, and *Epulopiscium*, which is still uncultured.

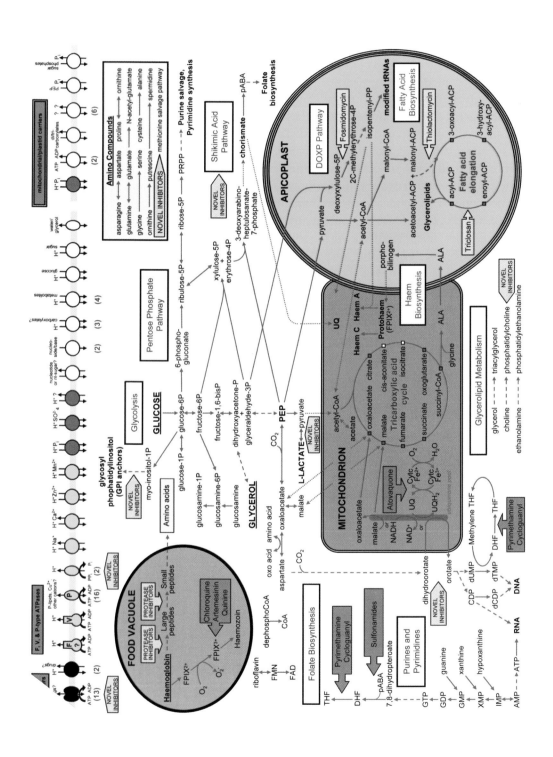

Epulopiscium spp are intestinal symbionts of certain species of surgeonfish. These heterotrophs are among the largest known bacteria and are capable of reaching up to 600 µm in length *(34)*. Despite increased understanding of this species, it still cannot be cultured in the laboratory. Phylogenetic analyses have revealed that *Epulopiscium* spp belong to the low guanine plus cytosine (G+C) Gram-positive bacteria *(34)*.

A distinguishing feature of this group of bacteria is their unusual ability to produce multiple offspring internally. Daughter cell production in *Epulopiscium* may represent the next step in the evolution of a novel form of cellular reproduction *(35)*, with this bacterium capable of producing multiple viviparous offspring intracellularly. Even though this bacterium still cannot be cultured in the laboratory, the genome sequence stands to provide important information on the unusual ability of *Epulopiscium* to produce live multiple offspring internally, as well as the changes that occurred early in the transition from the prokaryotic cell to eukaryotic cell. In addition, it is anticipated that metabolic reconstruction and successful growth of this organism based on the functionally annotated genome will be possible.

Because of significant interest in this organism, TIGR recently initiated a whole genome sequencing project for *Epulopiscium*, with an overarching goal to use the genome sequence to decipher major metabolic pathways that are present in this organism and as such identify a list of putative substrates that can be used for successful cultivation of this unusual species in the laboratory. Generation of sufficient deoxyribonucleic acid (DNA) for sequencing of *Epulopiscium* was made possible, in part, by the development of bacteriophage φ29 DNA polymerase as a tool for multiply primed rolling circle amplification of DNA directly from cells or plaques *(36,37)*; (Repli-G™, Molecular Staging, New Haven, CT; TempliPhi™, Amersham Biosciences, Piscataway, NJ). Substrate uptake by *Epulopiscium* will be investigated by incubation of samples with ^{14}C-labeled substrates. Substrate disappearance from the medium and end product formation from the bacterial species under testing can also be determined with gas chromatographic–mass spectrometric analysis. It is anticipated that the ability to reconstruct the growth medium of a species based on its sequenced genome will have applications to successful growth of the numerous uncultured species based on information contained in their genomes.

In addition to substrate utilization analysis on *Epulopiscium*, comparative genome analyses will be conducted with the genomes of endospore-forming bacteria to identify conserved genes that may be involved in offspring production across all these species. Among the endospore formers for which genome sequences are currently available are *Bacillus subtilis*, *Bacillus anthracis*, *Bacillus halodurans*, *Clostridium acetobutylicum*, and *Clostridium perfringens*. Genome projects for *Clostridium difficile*, *Clostridium botulinum*, *Clostridium tetani*, *Clostridium thermocellum*, *Clostridium* sp BC1, *Bacillus cereus*, and *Bacillus stearothermophilus*) are nearing completion.

Fig. 3. (*Opposite page*) Metabolic reconstruction of the pathways present in the *Plasmodium falciparum* genome. Reproduced with permission from *Nature*. Gardner MJ, Hall N, F··· E, et al. Genome sequence of the human malaria parasite *Plasmodium falciparum*. Nᵒᵗ 419(6906):498–511.

COMPARATIVE METABOLIC PREDICTIONS
AS A TOOL FOR UNDERSTANDING ECOSYSTEMS

The genomes that have been sequenced to date represent a few hundred microbial species that in almost all cases have been grown in pure culture; they represent a limited picture of true species and physiological diversity. The reality, however, is that natural ecosystems in general are driven by communities of microorganisms with significant interspecies associations. These organisms are responsible for biogeochemical cycles, element recycling, the degradation of pollutants, the conversion of biomass, and the maintenance of health in the human body and in animal systems.

Deciphering the genetic information of species in diverse ecosystems can currently be achieved by sequencing genomic libraries that are created directly from environmental DNA (metagenomics). The ability to analyze large sections of genomic DNA that have been created from bacterial artificial chromosome (BAC) and fosmid libraries created directly from environmental samples has been one of the major reasons for the success of this methodology. DeLong and coworkers successfully used this technique to construct libraries and sequence DNA generated from planktonic marine samples. This group pioneered the sequencing of large fragments of DNA to characterize genes from uncultivated species. In the first of a series of reports, they were able to analyze a 40-kbp genome fragment from a planktonic marine archaeon *(38)* that had been cloned into a fosmid library. By using large fragments that have a ribosomal sequence present as a marker, a fragment associated with a species can be characterized rather than just analyzing a piece of DNA derived from an unknown source.

DeLong and coworkers subsequently constructed BAC libraries that had an average insert size of 80 kbp, with a maximum insert size of 155 kbp *(39)*. This group demonstrated the value and utility of BAC-based libraries for providing genetic information on uncultivated species, with the analysis of these large fragments providing information on gene composition, gene organization, and the functional roles of these genes. At the same time these fragments have a phylogenetic link to define the closest relatives of the uncultivated species. In one of the studies published by this group, a novel bacteriorhodopsin was identified through the sequencing and characterization of a BAC derived from a previously uncultured organism *(40)*.

This technology has been extended to the analysis of other environments with a vast number of uncultivable inhabitants, such as soils, the human oral cavity, the human gastrointestinal tract, and the Sargasso Sea; the goal is generate more information on these uncultivable species and the genomic potential of these natural populations. In general, the basic technique for analyzing the genetic potential of these species is based on filtering and retrieving all the prokaryotic cells in the samples; constructing small, medium, and large insert libraries; and generating enough sequence for a predetermined of coverage from these environments. The major challenge for some of these environments appears to be the successful assembly of all the sequence data so that large pieces of DNA that contain operons or links to phylogenetic markers can be ideal would be to regenerate complete microbial genomes of previously from these samples, but it remains to be seen if current assembly his magnitude of a challenge.

sed as an example, humans are dependent on the activities of that inhabit multiple environmental niches in and on the human body for survival. Although many of these organisms comprise the endogenous microbial flora that exist in a dynamic and mutually beneficial relationship with the human host, others are opportunistic pathogens that can cause both chronic infections and life-threatening diseases. The current view of microbial host interactions is extremely limited by the fact that the overwhelming majority of microbes resist cultivation in the laboratory and are dependent on molecular approaches for identification and characterization.

The impact of this vast number of uncharacterized microbes on human health and disease is obviously significant and remains to be elucidated. Many chronic diseases, such as Crohn's disease and Kawasaki's disease, possess features of infectious disease despite no demonstrated presence of a microbial pathogen. The identification and characterization of these microbial populations will undoubtably establish links between these microorganisms and infectious and chronic diseases, their roles in development of the immune system, and their overall impact on human evolution *(41)*. A comprehensive survey of microorganisms and their genomic content in the human body would go a long way to help elucidate the role that these species play in this environment.

Members of the oral microbial community play key roles in maintaining the oral health of individuals and as causative agents responsible for the onset of periodontal diseases. A thorough survey of cultivable and uncultivable microorganisms and their genomic diversity in both healthy and diseased individuals will (1) enable identification of all the bacterial and archaeal species within the oral cavity and (2) identify genes within this microbial population that may encode virulence factors or proteins involved in biofilm formation, colonization, and antimicrobial resistance. Such a broad-range gene discovery approach will ultimately facilitate the development of therapeutic agents for control of oral and systemic diseases caused by this diverse microbial population. Overall, the structured community of heterogeneous bacterial species within the oral biofilm plays a role in preventing colonization by exogenous bacterial species and controlling the state of oral health and disease in the host.

While the approach based on 16S rDNA has revealed a great deal of bacterial and archaeal diversity in samples from different environments, this sampling technique only allows an estimate of the microbial species diversity and the populations that are present, but does not give any insight into the genetic diversity of the population. What are the genes that are present in these environments that have not been characterized from any of the completely sequenced genomes or other sequences currently available in public databases? As such, for completely new environments, the physiological role that is being played by individual species identified by 16S rDNA sequencing cannot be defined. Some level of analogy may be drawn by relative closeness to other species based on 16S rDNA sequences, but it is becoming increasingly apparent from whole genome sequencing that species that appear closely related from 16S rDNA sequences may have tremendous differences in genome composition *(42)*. Without being able to cultivate these organisms, the option of sequencing and analyzing large DNA fragments retrieved directly from the environment of choice becomes more attractive. A deeper understanding of the genetic information present in these ecosystems and their interactions will increase understanding of how communities control the overall processes. By having the entire genetic complements of these environments, expression analysis can be used to monitor how communities change in response to different environmental pressures.

It can be envisioned that, with detailed bioinformatic analysis, these entire communities can be treated in a manner similar to the annotation of the sequence data generated from a single microbial genome. As such, a single database can be constructed that contains all the predicted ORFs, their putative annotation, and biological role categories. By examining the role assignments, it is possible to identify all the biochemical pathways that may be present in a given environment. Just as a photorhodopsin was identified by Beja and coworkers in the study based on the analysis of a single fosmid clone *(40)*, an unprecedented amount of biochemical and physiochemical data will be obtained from the sequencing of complex environments. It can be imagined, for example, that analysis of the soil metagenome will yield many new pathways for the catabolism of ring-based/aromatic compounds and novel antibiotics, as well as pathways for the synthesis of secondary metabolites. The human gastrointestinal tract metagenome should yield significant information on the metabolic potential of the species that inhabit the gastrointestinal tract, their fermentation potential, and the end products that they are capable of producing. This in turn will have implications for human gastrointestinal tract function, human health, and the potential for these species to cause diseases. The metagenome of a healthy individual, can be compared with the metagenome of a diseased individual (that is, for diseases related to gastrointestinal tract function), allowing the identification of possible microbial species and factors responsible for the onset of disease.

THE FUTURE IN TERMS OF MICROBIAL METABOLISM BASED ON GENOME SEQUENCES

Beyond employing bioinformatic tools to interpret the physiological profile of an organism based on a genome sequence, the physiology of a microbial species can be addressed in detail through the use of functional techniques such as microarrays and proteomic technologies. DNA microarrays allow the identification of genes that are turned on or off under different growth conditions, and comparative genome hybridization (CGH) studies can be used to identify the extent of genomic diversity across related species.

Data from studies that have investigated gene regulation in microbial species under various growth conditions are currently available. For example, *Pasteurella multocida* was examined for the effect of nutrient limitation on gene expression *(43)*. Microarrays were used to compare gene expression during growth in rich and minimal media. Expression patterns of 669 genes were detectably altered in this study. The majority of the genes altered were expressed at higher levels in rich medium. Genes with increased expression in minimal medium included those encoding amino acid biosynthesis and transport systems, outer membrane proteins, and heat shock proteins. *Neisseria meningitidis* serogroup B gene regulation during interaction with human epithelial cells has been followed using microarrays *(44)*. In this study, it was found that contact between host and cell induced expression of 347 genes. Genes that were upregulated included transporters of iron, chloride, amino acids, and sulfate, many virulence factors, and the entire pathway of sulfur-containing amino acids.

Microarrays have also been used successfully to monitor gene expression in environments with mixed microbial populations. Dennis and workers constructed a microarray to which they applied total RNA from pure and mixed cultures containing the 2,4-dichlorophenoxyacetic acid (2,4-D)–degrading bacterium *Ralstonia eutropa*, and the inducing agent 2,4-D *(45)*. Gene induction of two of five 2,4-D catabolic genes from populations as low as 10^5 cells/mL was detectable against a background of 10^8 cells/mL, demonstrating the successful utility of microarrays for gene expression studies in complex ecosystems.

Taroncher-Oldenburg *(46)* and coworkers developed a microarray method for the detection and quantification of genes involved in the nitrogen cycle. Hybridization patterns differed between sediment samples from two locations in the Choptank River/Chesapeake Bay area. This finding implied that there were important differences in the makeup of the denitirifer communities in this environmental gradient. Again, this study demonstrated the successful utility of functional gene microarrays for environmental applications.

Similarly, CGH have been used successfully to identify differences in the metabolic potential of closely related species. The analysis of the complete *T. maritima* genome *(8)* suggested that this bacterium was capable of extensive gene transfer with other microbial species in its environment. Subsequent biochemical studies that employed techniques including gene amplification with degenerate polymerase chain reaction primers, and subtractive hybridization have also suggested extensive genomic diversity, as well as gene exchange in this genus *(47,48)*.

Nesbo and coworkers *(47)* conducted a study with 16 strains of *Thermotoga* and other related members of the *Thermotogales*; they investigated the distribution of two of the many predicted "archaeal-like" genes based on the analysis of the complete genome sequence. The genes investigated were those that encoded the large subunit of glutamate synthetase and myo-inositol 1P synthase. The distribution patterns of these two genes showed that they had been acquired from multiple archaeal lineages during the divergence of the *Thermotogales*, to the exclusion of other bacterial species.

In a subsequent study, Nesbo and colleagues *(48)* used suppressive subtractive hybridization techniques to identify genes present in different *Thermotoga* strains that do not have homologs in the sequenced *Thermotoga* genome. Their studies focused on *Thermotoga* sp strain RQ2, which differs from the sequenced *T. maritima* MSB8 by only 0.3% in the 16S rRNA gene, and numerous differences between these two strains could be identified.

At TIGR, we have used CGH on a range of *Thermotoga* strains isolated from numerous locations throughout the world and donated to us by Drs. Karl Stetter and Robert Huber at the University of Regensburg, Germany, to obtain further details on the extent of gene transfer in this genus. Preliminary results support that *T. maritima* MSB8 shares a high level of genome conservation with *Thermotoga* sp RQ2, with which it shares 99.7% identity in the small-subunit rRNA sequence. At least 7% (129 ORFs) in the MSB8 genome do not have homologous sequences in the RQ2 genome. These include 45 hypothetical proteins, 13 conserved hypothetical proteins, and 23 (18% of total) devoted to transport. Of these 129, only 18 occur as single ORFs, and the remaining correspond to islands that range in size from 2 kb to 38 kb. Many of these islands invariably correspond to complete pathways for the metabolism of sugars that are absent from the RQ2 genome. Among pathways absent from RQ2 are those for the metabolism of tagatose, pectin, ribose, and glycerol. Using CGH, major operonic differences (absence or presence in different strains) have been identified that correlate with different geographic locations from which these strains were obtained and likely reflect the availability of different substrates in these different locations.

It can be envisioned that suppressive subtractive techniques coupled with microarray technology can be used to identify major differences in the microbial populations of different ecosystems, such as comparing the gastrointestinal tract populations between healthy and diseased individuals. White and coworkers have successfully used suppressive subtractive hybridization to identify major differences in the populations of microbial species derived from the rumen of cows fed different diets (B. White, personal communication, August 2003). These differences likely reflect the varying populations that have adapted in this environment to different incoming substrates.

METABOLIC MODELING OF SINGLE GENOMES AND COMPLETE ENVIRONMENTS

Schilling and coworkers *(49)* have gone to extreme lengths to use available genome and biochemical data to predict the metabolic networks of a number of microbial species. They described the limitations in ability to predict cellular behavior in the absence of kinetic constants. They proposed that, in the absence of kinetic information, it is still possible to assess the theoretical capabilities of integrated cellular processes such as metabolism and to examine the feasible metabolic flux distributions under a steady-state assumption. The steady-state analysis is based on the constraints imposed on the metabolic network, and the steady-state analysis of metabolic networks based on the mass balance constraints is known as flux balance analysis (FBA). In one of a series of articles, they used available biochemical literature, the annotated genome sequence data, and strain-specific information to formulate an organism scale *in silico* representation of the metabolic capabilities of *E. coli* MG1655. FBA then was used to assess metabolic capabilities subject to these constraints, leading to qualitative predictions of growth performance. Eventually, such predictions will be extremely useful for the analysis of microbial biochemistry in mixed ecosystems.

CONCLUDING PERSPECTIVES

At the individual microbe level, close to 40% of each genome remains as hypothetical or conserved hypothetical proteins. The amount of data generated to date demands that high-throughput methodologies will have to be developed for the efficient analysis of these data sets. This includes high-throughput proteomics, gene expression, and protein–protein interaction studies. It is anticipated that these types of methodologies will further extend into an analysis of microbial communities, for which new techniques for studying the complex communities need to be developed so the associations between the previously cultivated and the estimated greater than 99% of uncultivated species can be evaluated. As mentioned, microarrays are rapidly becoming standard laboratory tools for investigating gene expression under different conditions and for looking at the presence and absence of genes in different stains or species related to a reference genome. Their applicability extends to many areas of microbial research, including microbial physiology, pathogenesis, epidemiology, ecology, phylogeny, and pathway engineering.

Recent work has shown that, by addressing the level of nutrient requirements of individual species and by trying to mimic the concentrations of these nutrients in their natural environments, some previously uncultivated species can be successfully cultivated. Zengler and coworkers described a high-throughput cultivation method by which they used the encapsulation of cells in gel microdroplets for microbial cultivation under low nutrient flux conditions, followed by flow cytometry to detect microdroplets that contained microcolonies *(50)*. They showed that this technique can successfully be applied to multiple environments. They pointed out that, although progress in obtaining genetic material from uncultivated species is being made, cultivation will ultimately be necessary if a comprehensive understanding of these species is desired. Advances in culturing previously uncultivated species implies the continued generation of data that will have to be deciphered in terms of microbial biochemical and physiological potential. It is obvious that these predictions need to be proven on a large scale, and available technologies such as Biolog Plates™ (Haywood, CA) will need to be enhanced to incorporate new substrates identified from complete genome analysis.

Despite the massive advances of genomics in the field of microbiology, it is grounding to realize that no single prokaryote has been completely deciphered such that all gene functions are known. This fact and the other challenges outlined in this chapter will continue to motivate the application of genomic-scale methods to the elucidation of microbial metabolism.

REFERENCES

1. Fleischmann RD, Adams MD, White O, et al. Whole-genome random sequencing and assembly of *Haemophilus influenzae* Rd. Science 1995; 269:496–512.
2. Fraser CM, et al. The minimal gene complement of *Mycoplasma genitalium*. Science 1995; 270: 397–403.
3. Hoffman SL, Rogers WO, Carucci DJ, Venter JC. From genomics to vaccines: malaria as a model system. Nat Med 1998; 4:1351–1353.
4. Nelson KE, Paulsen IT, Heidelberg JF, Fraser CM. Status of genome projects for nonpathogenic bacteria and archaea. Nat Biotechnol 2000; 18:1049–1054.
5. Tettelin H, Saunders NJ, Heidelberg S, et al. Complete genome sequence of *Neisseria meningitidis* serogroup B strain MC58. Science 2000; 287:1809–1815.
6. Tettelin H, Masignani V, Cieslewicz MJ, et al. Complete genome sequence and comparative genomic analysis of an emerging human pathogen, serotype V *Streptococcus agalactiae*. Proc Natl Acad Sci USA 2002; 99:12,391–12,396.
7. Tettelin H, Nelson, KE, Pausen IT, et al. Complete genome sequence of a virulent isolate of *Streptococcus pneumoniae*. Science 2001; 293:498–506.
8. Nelson KE, Clayton RA, Gill SR, et al. Evidence for lateral gene transfer between Archaea and bacteria from genome sequence of *Thermotoga maritima*. Nature 1999; 399:323–329.
9. Read TD, et al. Genome sequences of *Chlamydia trachomatis* MoPn and *Chlamydia pneumoniae* AR39. Nucleic Acids Res 2000; 28:1397–1406.
10. Read TD, Myers GS, Brunham RC, et al. Genome sequence of *Chlamydophila caviae* (*Chlamydia psittaci* GPIC): examining the role of niche-specific genes in the evolution of the Chlamydiaceae. Nucleic Acids Res 2003; 31:2134–2147.
11. Paulsen IT, Seshadri R, Nelson KE, et al. The *Brucella suis* genome reveals fundamental similarities between animal and plant pathogens and symbionts. Proc Natl Acad Sci USA 2002; 99: 13,148–13,153.

12. Nelson KE, Weinel C, Paulsen T, et al. Complete genome sequence and comparative analysis of the metabolically versatile *Pseudomonas putida* KT2440. Environ Microbiol 2002; 4:799–808; erratum in Environ Microbiol 2003; 5:630.

13. Delcher AL, Harmon D, Kasif S, White O, Salzberg SL. Improved microbial gene identification with Glimmer. Nucleic Acids Res 1999; 27:4636–4641.

14. Eddy SR. Noncoding RNA genes. Curr Opin Genet Dev 1999; 9:695–699.

15. Henderson J, Salzberg S, Fasman KH. Finding genes in DNA with a hidden Markov model. J Comput Biol 1997; 4:127–141.

16. Altschul SF, Gish W, Miller W, Myers EW, Lipman DJ. Basic local alignment search tool. J Mol Biol 1990; 215:403–410.

17. Altschul SF, Madden TL, Schaffer AA, et al. Gapped BLAST and PSI-BLAST: a new generation of protein database search programs. Nucleic Acids Res 1997; 25:3389–3402.

18. Klenk HP, Clayton RA, Tomb JF, et al. The complete genome sequence of the hyperthermophilic, sulphate-reducing archaeon *Archaeoglobus fulgidus*. Nature 1997; 390:364–370.

19. Eisen JA, Nelson KE, Paulsen IT, et al. The complete genome sequence of *Chlorobium tepidum* TLS, a photosynthetic, anaerobic, green-sulfur bacterium. Proc Natl Acad Sci USA 2002; 99: 9509–9514.

20. Tomb JF, White O, Kerlavage AR, et al. The complete genome sequence of the gastric pathogen *Helicobacter pylori*. Nature 1997; 388:539–547.

21. Nierman WC, Feldblyum N, Laub MT, et al. Complete genome sequence of *Caulobacter crescentus*. Proc Natl Acad Sci USA 2001; 98:4136–4141.

22. Laub MT, McAdams HH, Feldblyum T, Fraser CM, Shapiro L. Global analysis of the genetic network controlling a bacterial cell cycle. Science 2000; 290:2144–2148.

23. Paton JC, Andrew PW, Boulnois GJ, Mitchell TJ. Molecular analysis of the pathogenicity of *Streptococcus pneumoniae*: the role of pneumococcal proteins. Annu Rev Microbiol 1993; 47: 89–115.

24. Shah HN, Williams RAD. Utilization of glucose and amino acids by *Bacteroides intermedius* and *Bacteroides gingivalis*. Curr Microbiol 1987; 15:241–246.

25. Nelson KE, Fleischmann RD, DeBoy RT, et al. The complete genome sequence of the oral pathogenic bacterium *Porphyromonas gingivalis* strain W83. J. Bacteriol 2003; 185:5591–5601.

26. Takahashi N, Sato T, Yamada T. Metabolic pathways for cytotoxic end product formation from glutamate- and aspartate-containing peptides by *Porphyromonas gingivalis*. J Bacteriol 2000; 182:4704–4710.

27. Niederman R, Brunkhorst B, Smith S, Weinreb RN, Ryder MI. Ammonia as a potential mediator of adult human periodontal infection: inhibition of neutrophil function. Arch Oral Biol 1990; 35(Suppl):205S–209S.

28. Niederman R, Zhang J, Kashket S. Short-chain carboxylic-acid-stimulated, PMN-mediated gingival inflammation. Crit Rev Oral Biol Med 1997; 8:269–290.

29. Gardner MJ, Hall N, Fung E, et al. Genome sequence of the human malaria parasite *Plasmodium falciparum*. Nature 2002; 419:498–511.

30. Roberts CW, Roberts I, Lyons RE, et al. The shikimate pathway and its branches in apicomplexan parasites. J Infect Dis 2002; 185(Suppl 1):S25–S36.

31. Roberts F, Roberts CW, Johnson JJ, et al. Evidence for the shikimate pathway in apicomplexan parasites. Nature 1998; 393:801–805.

32. White O, Eisen JA, Heidelberg JF, et al. Genome sequence of the radioresistant bacterium *Deinococcus radiodurans* R1. Science 1999; 286:1571–1577.

33. Venkateswaran A, McFarlan SC, Ghosal D, et al. Physiologic determinants of radiation resistance in *Deinococcus radiodurans*. Appl Environ Microbiol 2000; 66:2620–2626.

34. Angert ER, Clements KD, Pace NR. The largest bacterium. Nature 1993; 362:239–241.

35. Angert ER, Losick RM. Propagation by sporulation in the guinea pig symbiont *Metabacterium polyspora*. Proc Natl Acad Sci USA 1998; 95:10,218–10,223.
36. Dean FB, Nelson JR, Giesler TL, Lasken RS. Rapid amplification of plasmid and phage DNA using phi 29 DNA polymerase and multiply-primed rolling circle amplification. Genome Res 2001; 11:1095–1099.
37. Dean FB, Hosono S, Fang L, et al. Comprehensive human genome amplification using multiple displacement amplification. Proc Natl Acad Sci USA 2002; 99:5261–5266.
38. Stein JL, Marsh TL, Wu KY, Shizuya H, DeLong EF. Characterization of uncultivated prokaryotes: isolation and analysis of a 40-kilobase-pair genome fragment from a planktonic marine archaeon. J Bacteriol 1996; 178:591–599.
39. Beja O, Suzuki MT, Koonin EV, et al. Construction and analysis of bacterial artificial chromosome libraries from a marine microbial assemblage. Environ Microbiol 2000; 2:516–529.
40. Beja O, Aranind L, Koenin EV, et al. Bacterial rhodopsin: evidence for a new type of phototrophy in the sea. Science 2000; 289:1902–1906.
41. Relman DA, Falkow S. The meaning and impact of the human genome sequence for microbiology. Trends Microbiol 2001; 9:206–208.
42. Perna NT, Plunkett G 3rd, Burland V, et al. Genome sequence of enterohaemorrhagic *Escherichia coli* O157:H7. Nature 2001; 409:529–533.
43. Paustian ML, May BJ, Kapur V. Transcriptional response of *Pasteurella multocida* to nutrient limitation. J Bacteriol 2002; 184:3734–3739.
44. Grifantini R, Bartolini E, Muzzi A, et al. Previously unrecognized vaccine candidates against group B meningococcus identified by DNA microarrays. Nat Biotechnol 2002; 20:914–921.
45. Dennis P, Edwards EA, Liss SN, Fulthorpe R. Monitoring gene expression in mixed microbial communities by using DNA microarrays. Appl Environ Microbiol 2003; 69:769–778.
46. Taroncher-Oldenburg G, Griner EM, Francis CA, Ward BB. Oligonucleotide microarray for the study of functional gene diversity in the nitrogen cycle in the environment. Appl Environ Microbiol 2003; 69:1159–1171.
47. Nesbo CL, L'Haridon S, Stetter KO, Doolittle WF. Phylogenetic analyses of two "archaeal" genes in *Thermotoga maritima* reveal multiple transfers between archaea and bacteria. Mol Biol Evol 2001; 18:362–375.
48. Nesbo CL, Nelson KE, Doolittle WF. Suppressive subtractive hybridization detects extensive genomic diversity in *Thermotoga maritima*. J Bacteriol 2002; 184:4475–4488.
49. Schilling CH, Covert MW, Famili I, Church GM, Edwards JS, Palson BO, et al. Genome-scale metabolic model of *Helicobacter pylori* 26695. J Bacteriol 2002; 184:4582–4593.
50. Zengler K, Toledo G, Rappe M, et al. Cultivating the uncultured. Proc Natl Acad Sci USA 2002; 99:15,681–15,686.

7

Genomic Analysis of Membrane Transport

Ian T. Paulsen, Katherine H. Kang,
Mark E. Hance, and Qinghu Ren

INTRODUCTION

Membrane transporters are essential for the movement of substrates across the cell membrane, which acts as a permeability barrier. Transporters enable the acquisition of organic nutrients, efflux of toxic compounds, ion homeostasis, environmental sensing, energy production, and other important cellular functions. Emphasizing the importance of transporters in the bacterial lifestyles, between 3 and 15% of open reading frames (ORFs) in bacterial genomes are predicted to encode membrane transport proteins (1).

Membrane transporters mediate solute transport via several different energy coupling mechanisms (Fig. 1; for review, see ref. 2). Primary active transporters use chemical or solar energy to drive transport, with adenosine triphosphate (ATP) the most commonly used energy source. Secondary active transporters use chemiosmotic energy in the form of a proton, sodium or other ion, or solute gradient to drive transport. Channel proteins allow the energy-independent passage of small molecules or ions. Phosphotransferase systems (PTS) transport and concomitantly phosphorylate their sugar substrates using phosphoenolpyruvate as both energy source and phosphate donor.

Cytoplasmic membrane transporters typically consist of at least one membrane-localized protein with multiple transmembrane-spanning α-helical segments. This has led to membrane transport systems being difficult to study experimentally. Their hydrophobic nature and solubility only in the presence of detergents typically make them recalcitrant to purification and crystallization. Thus, there is a paucity of high-resolution, three-dimensional structural data for membrane transporters, although significant progress has been made using high-throughput approaches to obtain structures of transporters such as the *Mycobacterium tuberculosis* mechanosensitive MscL channel (3), *Streptomyces lividans* KcsA potassium channel (4), calcium P-type ATPase from skeletal muscle sarcoplasmic reticulum (5), *Escherichia coli* ATP-driven lipid flippase MsbA (6), and proton-driven multidrug efflux transporter AcrB (7).

The difficulties in applying traditional experimental approaches for studying membrane transport make genomic/bioinformatic analyses an attractive option. More than 140 microbial organisms have had their genomes sequenced approximately 300 publically funded microbial genome sequencing projects are currently under way around the

From: *Microbial Genomes*
Edited by: C. M. Fraser, T. D. Read, and K. E. Nelson © Humana Press Inc., Totowa, NJ

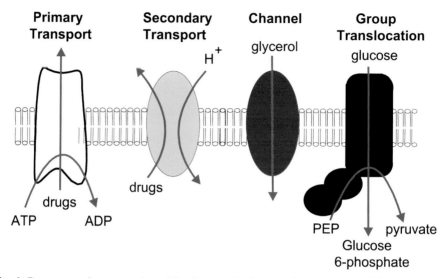

Fig. 1. Representative examples of the four main classes of transporters. Primary transporter; *Lactococcus lactis* LmrP multidrug efflux pump; secondary transporter; *Staphylococcus aureus* QacA multidrug efflux transporter; channel; *Escherichia coli* GlpF glycerol channel; and group translocation; *E. coli* PtsG/Crr glucose PTS transporter.

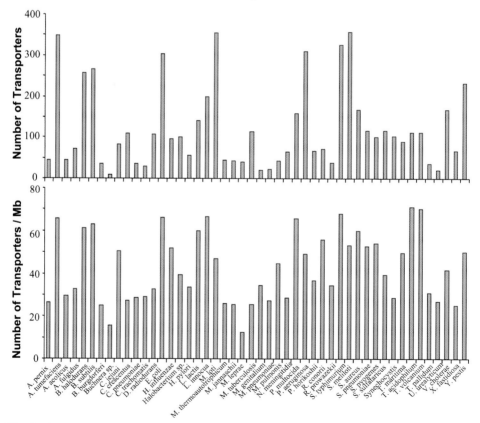

Fig. 2. (A) Number of predicted cytoplasmic membrane transporters across selected sequenced microorganisms. **(B)** Number of predicted cytoplasmic membrane transporters per megabase of genome sequence for selected sequenced microorganisms.

world *(8)* (http://www.tigr.org/tdb/mdb/mdbcomplete.html). Moreover, these genomic efforts cover a broad range of microbial organisms from different phylogenetic groupings, allowing comparative genomic analyses across a diverse range of organisms and lifestyles.

GENOMIC ANALYSIS OF MEMBRANE TRANSPORTERS

We have previously undertaken systematic genome-wide comparisons of the transporter gene content across a range of sequenced organisms *(1,9)*, and similar studies have been undertaken on specific genomes *(10)*. A complete catalogue of predicted membrane transporters and their likely substrate specificities has been compiled for each completely sequenced organism (http://www.membranetransport.org). This classification and compilation effort is continuing, and we are developing a relational database to enable more sophisticated querying of these data.

All of the identified transporters were classified according to the Transporter Classification (TC) system for membrane transport proteins and families *(9,11*; http://www. biology.ucsd.edu/~msaier/transport/). The TC system uses information on transporter function and protein phylogeny to classify all known types of transporters. Each transporter classified by the TC system has a five-digit/letter designation as follows: V.W.X. Y.Z. V (a number) corresponds to the transporter class (i.e., channel, secondary transporter, primary transporter, or group translocator); W (a letter) corresponds to the transporter subclass, which in the case of primary active transporters refers to the energy source used to drive transport; X (a number) corresponds to the transporter family; Y (a number) corresponds to the subfamily in which a transporter is found; and Z corresponds to the substrate or range of substrates transported. Thus, the well-characterized *E. coli* LacY lactose permease *(12)* is represented in the TC system as 2.A.1.5.1, where 2 indicates it is a secondary transporter, A indicates it is a uniporter/symporter/antiporter, 1 indicates it belongs to the major facilitator superfamily (MFS), 5 indicates it belongs to an oligosaccharide symporter subfamily within the MFS, and 1 indicates it is a lactose/proton symporter. More than 150 families of transporters are represented in the TC system.

The number of predicted cytoplasmic membrane transporters varies over an approximately 40-fold range across sequenced microorganisms (Fig. 2A). When the number of transporters is compared relative to the microorganism's genome size (Fig. 2B), the ratio of transporters/megabase varies over only a 2- to 3-fold range, indicating that the number of transporters present in microorganisms is partially dependent on its genome size.

Several trends can be discerned now that it is possible to conduct such comparative analyses across a broad range of organisms. Soil/plant-associated organisms such as *Streptomyces coelicolor, Pseudomonas aeruginosa, Mesorhizobium loti, Agrobacterium tumefaciens*, and *Bacillus subtilis* appear to be overrepresented in transporters relative to their genome sizes, as do some organisms related to the gastrointestinal tract, such as *E. coli* and *Campylobacter jejuni*. Many of these organisms also have large numbers of transporters in absolute terms. This probably reflects the versatility of these organisms and their exposure to a wide range of different substrates in their natural environments; in the case of organisms such as *P. aeruginosa*, it explains their ability to flourish in a broad diversity of different environments.

In contrast, most intracellular pathogens or symbionts, such as *Mycobacterium leprae*, *Mycoplasma* spp, and *Buchnera* sp possess relatively few transporters/megabase. This probably reflects their lifestyle limitations and the more static nature of their external environments. Most of the sequenced archaea also are underrepresented in transporters relative to their genome size, in some cases because of their largely autotrophic rather than heterotrophic metabolism. The exceptions to this generalization about archaea are *Thermoplasma acidophilum* and *Thermoplasma volcanium*, which have a high number of transporters relative to their genome size, probably reflecting their sugar-scavenging lifestyles.

INDIVIDUAL ORGANISM TRANSPORTER PROFILES

The complete genome sequence of an organism allows reconstruction of detailed models of transport and metabolism based on bioinformatics predictions. Figure 3 shows an example of such an analysis of metabolism and transport for the respiratory tract pathogen *Streptococcus pneumoniae (13)*. Other examples of detailed models of transport and metabolism of the eukaryotic parasite *Plasmodium falciparum* and the bacterium *Pseudomonas putida* can be seen in the chapter on metabolism.

The precise substrate specificity for a significant percentage of the transporters could not be predicted with confidence (these are seen in Fig. 3 either with question marks for predicted specificity or with a "generic" prediction such as "amino acid" rather than a specific prediction). Nevertheless, a good correlation can be observed between the predicted transporters and the metabolic pathways, suggesting that this type of analysis provides a good overview of the capabilities of the organism.

In the case of *S. pneumoniae*, over 30% of its transporters were predicted to be for carbohydrates, one of the highest percentages observed in any sequenced organism *(13)*. In particular, it utilizes 21 distinct PTS-type sugar transporters, as well as multiple ATP binding cassette (ABC) superfamily sugar transporters. This was consistent with the presence of multiple sugar catabolic pathways that fed directly into the glycolytic pathway, as well as pathways for pentitols via the pentose phosphate pathway.

Fig. 3. (*Opposite page*) Model of transport and metabolism of *Streptococcus pneumoniae*. Pathways for energy production, metabolism of organic compounds, and capsular biosynthesis are shown. Transporters are grouped by substrate specificity as follows: inorganic cations (green), inorganic anions (pink), carbohydrates/carboxylates (yellow), amino acids/peptides/amines/ purines and pyrimidines (red), and drug efflux and other (black). Question marks indicate uncertainty about the substrate transported. Export or import of solutes is designated by the direction of the arrow through the transporter. The energy-coupling mechanisms of the transporters are also shown: Solutes transported by channel proteins are shown with a double-headed arrow; secondary transporters are shown with two arrowed lines, indicating both the solute and coupling ion; ATP-driven transporters are indicated by the ATP hydrolysis reaction; and transporters with an unknown energy coupling mechanism are shown with only a single arrow. Components of transporter systems that function as multisubunit complexes that were not identified are outlined with dotted lines. When multiple homologous transporters with similar substrate predictions exist, the number of that type of transporter is indicated in parentheses. Systematic gene numbers (SPxxxx) are indicated next to each pathway or transporter; those separated by a dash represent a range of consecutive genes. (Reprinted with permission from ref. *13*. Copyright 2000 American Association for the Advancement of Science.)

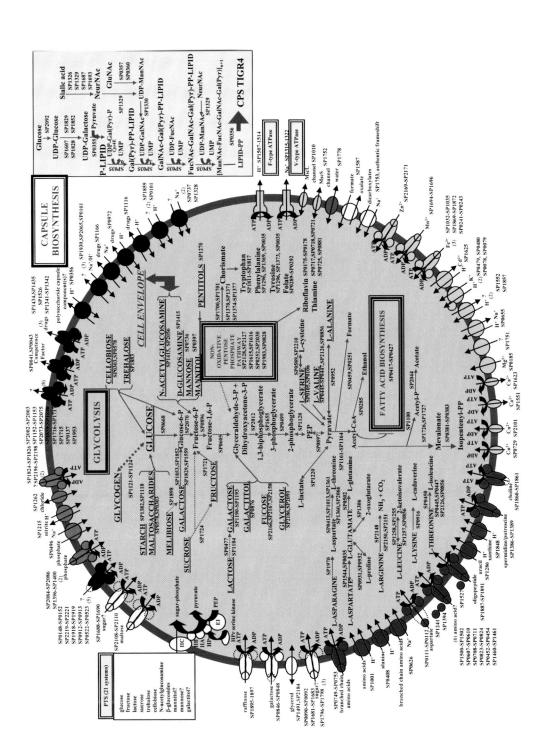

Furthermore, *S. pneumoniae* encodes a range of extracellular enzymes, such as *N*-acetylglucosaminidases, α- and β-galactosidases, endoglycosidases, hydrolases, hyaluronidases, and neuraminidases, probably enabling the breakdown of carbohydrate polymers such as host murein, glycolipids, and hyaluronic acid to their constituent sugars. Therefore, the wealth of sugar and polyol transporters in *S. pneumoniae* probably reflects an ability to catabolize host sugar polymers, providing sustenance and at the same time damaging host tissues and facilitating colonization. *Streptococcus pneumoniae* may also be able to use its sugar transporters and catabolic pathways to recycle consituents of its sugar polysaccharide capsule.

Other sequenced respiratory tract pathogens, such as *Neisseria meningitidis* and *Haemophilus influenzae*, have a relative paucity of sugar transporters and instead appear to prefer carboxylates and other carbon compounds *(1)*. This suggests that *S. pneumoniae* may inhabit a distinct microenvironment in the respiratory tract. Other streptococci also appear to be highly reliant on sugar metabolism and transport.

Streptococcus pneumoniae also possesses transporters for a range of amino acids, polyamines, purines, and pyrimidines. Only two transporters for carboxylates were identified, one of which has an authentic frameshift mutation in the TIGR4 strain of *S. pneumoniae*. The lack of carboxylate transporters in contrast to the abundance of sugar transporters correlates with the lack of a tricarboxylic acid (TCA) cycle and the fermentative lifestyle of this facultative anaerobe. *Streptococcus pneumoniae* has a relatively limited repertoire of inorganic ion transporters, including manganese and zinc transporters, which have been reported to play a role in virulence *(14,15)*. Three iron chelate ABC transporters and three phosphate ABC transporters are probably involved in iron and phosphate scavenging in the host and may be associated with virulence.

The overview of transport and metabolism of *S. pneumoniae* provides an example of the types of insight and overall picture it is possible to obtain from such bioinformatics analyses. It should be emphasized that the cross-correlation of predictions of transport and metabolic capabilities together with previous physiological information from the literature greatly enhance this type of analysis (see Chapter 6).

Qualitative and quantitative *in silico* modeling of the transport and metabolic capabilities of an organism are now becoming feasible. One example is the incorporation of membrane transporters into the EcoCyc database of *E. coli* metabolism *(16,17)* (Paulsen, IT and Karp, PD, 2003, unpublished data). The known and predicted transporters of *E. coli* have been represented in EcoCyc, and a detailed description of each transporter is provided based on the primary literature. The structured representation of transporters and metabolic pathways in the EcoCyc database enables complex queries of the overlap between transport and metabolism and qualitative modeling of these processes.

Another example is the reconstruction of the metabolic and transport networks of *Helicobacter pylori* 26695 *(18)* and *H. influenzae* *(19,20)*. In this case, constraint-based modeling, extreme pathway analysis, and flux balance analysis were used to analyze the properties of the *in silico* model and to provide predictions of the minimum substrate requirements for growth and the potential substrates capable of providing the bulk carbon requirements. *In silico* gene deletion and simulation of the growth of these "mutants" in different media allow for the identification of putatively essential genes. In some instances, the essentiality of these genes has been verified by in vitro gene deletion studies *(18)*.

COMPARISON OF SUBSTRATE SPECIFICITIES/BIOENERGETICS

The overall substrate specificities and bioenergetics of the complete set of predicted transporters can be compared across different genomes. Previous analyses have shown the usage of transporter energy-coupling mechanisms appears to reflect the overall metabolism and bioenergetics of an organism *(1,9)*. For instance, organisms that lack a TCA cycle and an electron transfer chain, such as *Mycoplasma* spp, spirochetes, *Thermotoga maritima*, and *S. pneumoniae*, are largely reliant on ATP-driven transporters, presumably because they generate their proton motive force via indirect methods. Photosynthetic organisms such as *Synechocystis* PCC6803, *Synechococcus* WH8102, and *Chlorobium tepidum* also have a preponderance of ATP-driven transporters, probably because of their ability to synthesize an ATP pool via photosynthesis.

In some cases, the rationale for preferring a particular class of transporters is not yet clear; for instance, a group of α-proteobacteria, including *A. tumefaciens*, *M. loti*, *Sinorhizobium meliloti*, and *Brucella suis*, is highly reliant on ATP-driven transporters, yet there is no clear bioenergetic reason why this would be the case. One possible explanation is that primary transporters frequently have a higher affinity for their substrates than secondary transporters, so these organisms have a particular need for high-affinity transport *(21)*.

Analyses of the overall substrate specificities of membrane transporters across sequenced organisms have shown highly idiosyncratic differences, which probably largely reflect the availability and abundance of various substrates in their natural habitats *(1)*. For instance, obligate intracellular parasites such as sequenced *Rickettsia* and *Chlamydia* species have fairly limited transport capabilities, with very little capacity for transport of free sugars, but a variety of transporters for amino acids and nucleotides, which they are probably able to scavenge from their hosts. In contrast, as discussed here, *S. pneumoniae* and other streptococci have a relatively large percentage of transporters for carbohydrates, as do some other organisms such as *T. maritima*, which can break down and utilize a wide variety of complex polysaccharides of plant origin, and *Enterococcus faecalis*, for which the sugar transporters facilitate the fermentation of nonabsorbed sugars by *E. faecalis* in the gastrointestinal tract.

Rhizobial organisms, such as *A. tumefaciens*, *M. loti*, and *S. meliloti*, all share a very large contingent of transporters for amino acids, peptides, and sugars, probably reflecting the nutrient-rich environment of the rhizosphere. Other soil/plant-associated organisms, such as *B. subtilis*, *P. aeruginosa*, and *P. putida*, also have very broad transport capabilities, although the last two largely lack transporters for carbohydrates and have broader capacities for uptake of other carbon compounds, including aromatic substrates, particularly in the case of *P. putida*. Multidrug transporter genes are also prevalent in soil/plant-associated bacteria and overrepresented in the genomes of some intracellular pathogens, such as *Coxiella burnetii* and *Rickettsia prowazekii*, possibly facilitating efflux of host-produced antimicrobial peptides *(22)*.

TRANSPORTER COMPARISONS
BETWEEN CLOSELY RELATED SPECIES/STRAINS

As the genome sequencing field matures, it is becoming possible to undertake comparative genomic analyses at a different level: between closely related bacterial species or strains. For instance, the complete genome sequences of *B. suis (23)* and *Brucella meli-*

tensis (23a) have now been published. These pathogenic bacteria both are causative agents of the human/animal disease brucellosis *(24)*, but differ with respect to their virulence (*B. melitensis* is more virulent than *B. suis* in humans) and host preference (swine are the preferred host for *B. suis* and goats or sheep are the preferred host for *B. melitensis*).

Despite these phenotypic differences, these two genomes share a very high level of conserved gene order or synteny. They share an absolutely conserved backbone of more than 3.2 Mb over their 3.31-Mb genomes *(23)*. A small number of unique regions or "islands" were identified in each of the two genomes; many of these islands appeared to be generated by phage integration events. A total of 42 genes specific for *B. suis* and 32 specific for *B. melitensis* were identified; the majority were of unknown function *(23)*. Two of the *B. suis*–unique islands include predicted ABC amino acid transporter genes, and one of these islands appears to be caused by a deletion in *B. melitensis* as there are fragments of the ABC amino acid transporter gene cluster remaining. The presence of these two ABC amino acid transporters in *B. suis* probably explains the capacity of this organism, but not *B. melitensis*, to utilize ornithine, citrulline, arginine, and lysine. Whether these differences play any role in the host preferences or virulence remains to be explored.

The exploration of the complete genome sequences of *Brucella abortus* (Halling et al., unpublished) amd *Brucella ovis* (Paulsen et al., unpublished) are now under way, so it should soon be possible to expand these observations to a broader range of species in the *Brucella* genera. Similar analyses are now possible for specific species or genera for which multiple genome sequences are available, such as *E. coli* and related enterobacteria, various streptococcal and chlamydial species, *M. tuberculosis* strains, and so on. We envisage that such comparisons between closely related strains/species should become more and more informative as the volume of sequence data continues to grow.

PHYLOGENETIC/PHYLOGENOMIC ANALYSES OF MEMBRANE TRANSPORTERS

Another type of comparative genomic analysis is to examine in detail specific types or families of transporters across all sequenced genomes using a phylogenomic approach *(25)* to reconstruct the evolutionary history of a family and to investigate instances of gene loss, lateral transfer, and gene duplications or expansions.

Previous phylogenetic studies of transporter families have revealed that substrate specificity is a highly conserved evolutionary trait; that is, transporters with similar substrate specificities tend to cluster together *(1,26–29)*. An example of this phenomenon is demonstrated in the phylogenetic tree of representative members of the MFS displayed in Fig. 4. Phylogenetic analyses of the MFS have identified 34 distinct subfamilies in this superfamily *(28,29)*, of which 15 are represented in the tree displayed in Fig. 4. The MFS currently contains literally thousands of identified members, and together with the ABC superfamily, is one of the two largest families of transporters in nature.

The MFS is very functionally diverse and includes transporters for sugars, carboxylates, amino acids, aromatic compounds, drugs, inorganic anions and cations, and other diverse compounds, although they all share the common feature of being secondary transporters *(28)*. Most of the clusters or subfamilies in the tree in Fig. 4 are specific for a particular class of substrate (e.g., drugs, monosaccharides, nucleosides, etc.). One of the families

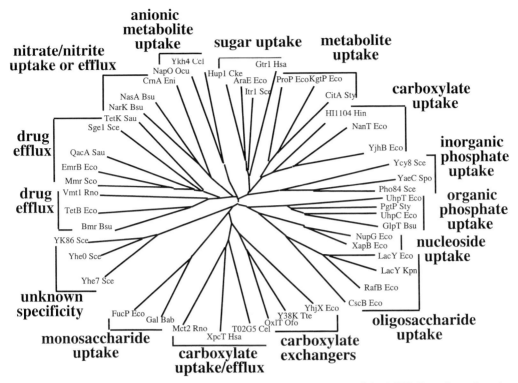

Fig. 4. Unrooted phylogenetic tree of representative members of the MFS. Protein and strain abbreviations are as per Pao et al. *(28)* and Saier et al. *(29).* Of the known 34 subfamilies of the MFS, 15 are represented in this tree and are labeled with their known substrate specificities.

shown does not include any characterized members, and the one cluster labeled metabolite uptake shows a somewhat broader diversity of substrate specificity. Thus, phylogenetic analysis provides an aid for functional predictions, at least of the class of substrate, for a novel transporter.

However, making precise predictions for transporters by bioinformatic analysis may be more problematic. For instance, single basepair substitutions can modify the substrate specificity of the *E. coli* LacY lactose transporter to maltose, arabinose, or other sugars *(30–33)* or change the drug specificity of the *Staphylococcus aureus* QacA *(34)* and *B. subtilis* Bmr *(35)* multidrug efflux transporters. This suggests that it may be relatively easy for transporters to modulate their substrate specificity in a class of compounds. However, there is no small number of changes that can be introduced into LacY to turn it into a multidrug transporter. The ability to transport an entirely different class of compounds appears to have evolved infrequently over an evolutionary timescale. The phylogenetic analysis of the separate subfamilies of the MFS suggests that the ability to transport sugars arose and was maintained on at least three distinct occasions, while the ability to transport other classes of substrates, such as nucleosides or inorganic phosphate, arose and was maintained on only a single occasion during the evolution of the MFS.

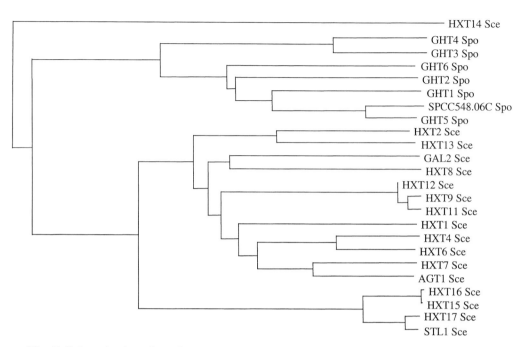

Fig. 5. Selected subsection of a larger phylogenetic tree of fungal sugar transporters (Greenberg and Paulsen, unpublished data). Sugar transporters from *Saccharomyces cerevisiae* (Sce) and *Schizosaccharomyces pombe* (Spo) are shown.

Detailed phylogenomic analysis of particular families of transporters in sequenced genomes is also a valuable aid for functional predictions. Figure 5 shows part of a phylogenetic tree of known and predicted sugar transporters from the yeasts *Saccharomyces cerevisiae* and *Schizosaccharomyces pombe*. It is clear from this phylogenetic tree that the majority of the *S. cerevisiae* HXT transporters and the *S. pombe* GHT transporters were caused by independent expansions of this family after the evolutionary divergence of these two yeasts. Thus, probably none of these yeast transporters shown in Fig. 5 are orthologs of each other; in this case, there should be hesitancy in extrapolating precise functional information between the sugar transporters of these two organisms.

FUNCTIONAL GENOMIC ANALYSES OF MEMBRANE TRANSPORT

The wealth of genome sequence data is making functional genomic analyses, such as microarray expression analyses and large-scale gene knockout or expression studies, an increasingly attractive option for investigating biological questions. To date, mostly small-scale systematic gene knockout or overexpression studies have used particular families or types of transporters in different microorganisms.

One example of such an approach was the systematic insertional disruption of 30 putative multidrug transporter genes in *E. faecalis (36)*. These 30 transporter mutants were screened for susceptibility to 28 different antimicrobial drugs or compounds, and four ABC transporter genes were identified with disruption that led to significantly increased sensitivity to at least one antimicrobial agent; hence, these probably encode drug efflux transporters. An analogous study in *S. cerevisiae* undertook single and multiple deletions of yeast ABC drug transporters and regulators and screened the mutant collection against a bank of antifungal compounds *(37)*. This analysis revealed that many of these transporters had overlapping specificities and generated a drug-hypersensitive strain with multiple gene deletions.

Homologous and heterologous expression studies have also been used to characterize novel transport. For instance, Nishino and Yamaguchi *(38)* cloned 37 putative drug efflux transporters from *E. coli* and screened their drug-resistance phenotypes in an *E. coli* drug-sensitive mutant. Of the 37 putative drug efflux transporters, 20 conferred increased resistance to one or more antimicrobial agents. These 20 drug transporters included 7 novel drug transporters and 6 drug transporters with specificities that were broader than previously recognized. Similarly, heterologous expression in *E. coli* has been utilized to characterize multiple SMR family multidrug efflux genes from a range of pathogenic bacteria *(39)*.

Small-scale functional genomic analyses of transporters have focused not only on drug efflux transporters. For instance, systematic insertional disruption was used to characterize the predicted amino acid transporter genes in *Synechocystis* PCC6803 *(40)*. We are currently employing a combination of heterologous expression, gene knockout, and microarray analysis to characterize the transporter complement of the marine cyanobacterium *Synechococcus* WH8102 (Palenik, Brahamsha, and Paulsen, unpublished data).

Microarray expression analyses have started to produce a wealth of data that should shed light on the function and regulation of transporter genes. Several examples are described here from *E. coli* microarray expression studies. Microarray expression analysis of *E. coli* constitutively expressing the MarA global regulator identified two efflux systems, YadGH and YdeA, with gene expression that was affected, in addition to the multidrug efflux genes *acrAB* and *tolC (41)*. Microarray expression analysis of mutants of the 36 *E. coli* two-component signal transduction genes identified a variety of *E. coli* transporters that are subject to regulation via two-component systems *(42)*. Expression analysis of wild-type *E. coli* and a mutant with a knockout in the leucine-responsive regulatory protein, which plays an important role in regulating expression of stationary phase genes, identified a number of transporters upregulated in stationary phase, under leucine-responsive regulatory protein control, including transporters for osmoprotectants *(43)*. Array studies of *E. coli* strains with either a knockout mutation or overexpression of the *evgA* response regulator gene revealed they regulate a number of acid-resistance genes and the drug efflux pump genes *emrK* and *yhiUV (44)*. Microarray expression analyses have also provided insight into the regulation of transporters in other bacteria, such as *B. subtilis (45)*, *M. tuberculosis (46)*, *Shewanella oneidensis (47)*, and *Erwinia chrysanthemi (48)*.

Large-scale knockout studies have also started to shed a little light on the function and essentiality of membrane transporter genes. Transposon mutagenesis of *Mycoplasma genitalium* and *Mycoplasma pneumoniae*, which have very small genomes, has been used to identify a minimum required gene set, suggesting only 265–330 genes were essential in *M. genitalium (49)*. Knockouts were obtained in several transporters, including both predicted ABC drug efflux pumps and the PTS fructose uptake system.

A very large collection of *S. cerevisiae* mutants has been constructed with knockouts in 96% of the predicted yeast ORFs and with unique deoxyribonucleic acid sequences acting as "molecular bar codes" *(50)*. Examination of the 15% of homozygous mutants that displayed a slow-growth phenotype in rich medium revealed most of the subunits of the F- and V-type ATPases, as well as selected other transporters. A variety of other transporters that were required for optimal growth under different environmental conditions was also identified.

CONCLUSIONS

One of the challenges of the "genomic era" is to make full use of the vast volumes of data that are being generated from genome sequencing, microarray experiments, and high-throughput functional genomics. In terms of membrane transport, the twin goals of understanding the complete complement of transporters in an organism and of being able to model *in silico* the transport and metabolism of an organism are closer. The recent advances in structure determination for membrane transport proteins are also yielding insights at the molecular level into substrate recognition and transport of membrane transport systems. However, much work still remains for comprehensive understanding of membrane transport; even in *E. coli*, almost half of the predicted membrane transport systems have yet to be characterized experimentally.

REFERENCES

1. Paulsen IT, Nguyen L, Sliwinski MK, Rabus R, Saier MH Jr. Microbial genome analyses: comparative transport capabilities in 18 prokaryotes. J Mol Biol 2000; 301:75–100.
2. Saier MH Jr. Vectorial metabolism and the evolution of transport systems. J Bacteriol 2000; 182:5029–5035.
3. Chang G, Spencer RH, Lee AT, Barclay MT, Rees DC. Structure of the MscL homolog from *Mycobacterium tuberculosis*: a gated mechanosensitive ion channel. Science 1998; 282: 2220–2226.
4. Doyle DA, Morais Cabral J, Pfuetzner RA, et al. The structure of the potassium channel: molecular basis of K+ conduction and selectivity. Science 1998; 280:69–77.
5. Toyoshima C, Nakasako M, Nomura H, Ogawa H, et al. Crystal structure of the calcium pump of sarcoplasmic reticulum at 2.6 A resolution. Nature 2000; 405:647–655.
6. Chang G, Roth CB. Structure of MsbA from *E. coli*: a homolog of the multidrug resistance ATP binding cassette (ABC) transporters. Science 2001; 293:1793–1800.
7. Murakami S, Nakashima R, Yamashita E, Yamaguchi A. Crystal structure of bacterial multidrug efflux transporter AcrB. Nature 2002; 419:587–593.
8. Nelson KE, Paulsen IT, Heidelberg JF, Fraser CM. Status of genome projects for nonpathogenic bacteria and archaea. Nat Biotechnol 2000; 18:1049–1054.
9. Paulsen IT, Sliwinski MK, Saier MH Jr. Microbial genome analyses: global comparisons of transport capabilities based on phylogenies, bioenergetics and substrate specificities. J Mol Biol 1998; 277:573–592.
10. Meidanis J, Braga MD, Verjovski-Almeida S. Whole-genome analysis of transporters in the plant pathogen *Xylella fastidiosa*. Microbiol Mol Biol Rev 2002; 66:272–299.
11. Saier MH Jr. A functional-phylogenetic classification system for transmembrane solute transporters. Microbiol Mol Biol Rev 2000; 64:354–411.
12. Kaback HR, Sahin-Toth M, Weinglass AB. The kamikaze approach to membrane transport. Nat Rev Mol Cell Biol 2001; 2:610–620.
13. Tettelin H, Nelson KE, Paulsen IT. Complete genome sequence of a virulent isolate of *Streptococcus pneumoniae*. Science 2001; 293:498–506.

14. Jakubovics NS, Smith AW, Jenkinson HF. Expression of the virulence-related Sca (Mn2+) permease in *Streptococcus gordonii* is regulated by a diphtheria toxin metallorepressor-like protein ScaR. Mol Microbiol 2000; 38:140–153.

15. Dintilhac A, Alloing G, Granadel C, Claverys JP. Competence and virulence of *Streptococcus pneumoniae*: Adc and PsaA mutants exhibit a requirement for Zn and Mn resulting from inactivation of putative ABC metal permeases. Mol Microbiol 1997; 25:727–739.

16. Karp PD, Riley M, Saier M, Paulsen IT, Paley SM, Pellegrini-Toole A. The EcoCyc and Meta Cyc databases. Nucleic Acids Res 2000; 28:56–59.

17. Karp PD, Riley M, Saier M. The EcoCyc Database. Nucleic Acids Res 2002; 30:56–58.

18. Schilling CH, Covert MW, Famili I, Church GM, Edwards JS, Palsson BO. Genome-scale metabolic model of *Helicobacter pylori* 26695. J Bacteriol 2002; 184:4582–4593.

19. Edwards JS, Palsson BO. Systems properties of the *Haemophilus influenzae* Rd metabolic genotype. J Biol Chem 1999; 274:17,410–17,416.

20. Schilling CH, Palsson BO. Assessment of the metabolic capabilities of *Haemophilus influenzae* Rd through a genome-scale pathway analysis. J Theor Biol 2000; 203:249–283.

21. Wood DW, Setubal JC, Kaul R, et al. The genome of the natural genetic engineer *Agrobacterium tumefaciens* C58. Science 2001; 294:2317–2323.

22. Paulsen IT, Lewis K. Microbial multidrug efflux: introduction. J Mol Microbiol Biotechnol 2001; 3:143–144.

23. Paulsen IT, Seshadri R, Nelson KE, et al. The *Brucella suis* genome reveals fundamental similarities between animal and plant pathogens and symbionts. Proc Natl Acad Sci USA 2002; 99: 13,148–13,153.

23a. Del Vecchio VG, Kapatral V, Redkar RJ, et al. The genome sequence of the facultative intracellular pathogen *Brucella melitensis*. Proc Natl Acad Sci USA 2002; 99:443–448

24. Smith LD, Ficht TA. Pathogenesis of *Brucella*. Crit Rev Microbiol 1990; 17:209–230.

25. Eisen JA, Hanawalt PC. A phylogenomic study of DNA repair genes, proteins, and processes. Mutat Res 1999; 435:171–213.

26. Jack DL, Paulsen IT, Saier MH. The amino acid/polyamine/organocation (APC) superfamily of transporters specific for amino acids, polyamines and organocations. Microbiology 2000; 146 (Pt 8):1797–1814.

27. Paulsen IT, Brown MH, Skurray RA. Proton-dependent multidrug efflux systems. Microbiol Rev 1996; 60:575–608.

28. Pao SS, Paulsen IT, Saier MH Jr. Major facilitator superfamily. Microbiol Mol Biol Rev 1998; 62:1–34.

29. Saier MH Jr, Eng BH, Fard S, et al. Phylogenetic characterization of novel transport protein families revealed by genome analyses. Biochim Biophys Acta 1999; 1422:1–56.

30. King SC, Wilson TH. Characterization of *Escherichia coli* lactose carrier mutants that transport protons without a cosubstrate. Probes for the energy barrier to uncoupled transport. J Biol Chem 1990; 265:9645–9651.

31. Goswitz VC, Brooker RJ. Isolation of lactose permease mutants which recognize arabinose. Membr Biochem 1993; 10:61–70.

32. Varela MF, Brooker RJ, Wilson TH. Lactose carrier mutants of *Escherichia coli* with changes in sugar recognition (lactose vs melibiose). J Bacteriol 1997; 179:5570–5573.

33. King SC, Wilson TH. Identification of valine 177 as a mutation altering specificity for transport of sugars by the *Escherichia coli* lactose carrier. Enhanced specificity for sucrose and maltose. J Biol Chem 1990; 265:9638–9644.

34. Paulsen IT, Brown MH, Littlejohn TG, Mitchell BA, Skurray KA. Multidrug resistance proteins QacA and QacB from *Staphylococcus aureus*: membrane topology and identification of residues involved in substrate specificity. Proc Natl Acad Sci USA 1996; 93:3630–3635.

35. Klyachko KA, Schuldiner S, Neyfakh AA. Mutations affecting substrate specificity of the *Bacillus subtilis* multidrug transporter Bmr. J Bacteriol 1997; 179:2189–2193.

36. Davis DR, McAlpine JB, Pazoles CJ, et al. *Enterococcus faecalis* multi-drug resistance transporters: application for antibiotic discovery. J Mol Microbiol Biotechnol 2001; 3:179–184.

37. Rogers B, Decottignies A, Koloczkowski M, Carvajal E, Balzi E, Goffeau A. The pleitropic drug ABC transporters from *Saccharomyces cerevisiae*. J Mol Microbiol Biotechnol 2001; 3: 207–214.

38. Nishino K, Yamaguchi A. Analysis of a complete library of putative drug transporter genes in *Escherichia coli*. J Bacteriol 2001; 183:5803–5812.

39. Ninio S, Rotem D, Schuldiner S. Functional analysis of novel multidrug transporters from human pathogens. J Biol Chem 2001; 276:48,250–48,256.

40. Quintero MJ, Montesinos ML, Herrero A, Flores E. Identification of genes encoding amino acid permeases by inactivation of selected ORFs from the *Synechocystis* genomic sequence. Genome Res 2001; 11:2034–2040.

41. Barbosa TM, Levy SB. Differential expression of over 60 chromosomal genes in *Escherichia coli* by constitutive expression of MarA. J Bacteriol 2000; 182:3467–3474.

42. Oshima T, Arba H, Masuda Y. Transcriptome analysis of all two-component regulatory system mutants of *Escherichia coli* K-12. Mol Microbiol 2002; 46:281–291.

43. Tani TH, Khodursky A, Blumenthal RM, Brown PO, Mathews RG. Adaptation to famine: a family of stationary-phase genes revealed by microarray analysis. Proc Natl Acad Sci USA 2002; 99: 13,471–13,476.

44. Masuda N, Church GM. *Escherichia coli* gene expression responsive to levels of the response regulator EvgA. J Bacteriol 2002; 184:6225–6234.

45. Britton RA, Eichenberger P, Gonzalez-Pastor JE, et al. Genome-wide analysis of the stationary-phase sigma factor (sigma-H) regulon of *Bacillus subtilis*. J Bacteriol 2002; 184:4881–4890.

46. Wilson M, DeRisi J, Kristensen HH, et al. Exploring drug-induced alterations in gene expression in *Mycobacterium tuberculosis* by microarray hybridization. Proc Natl Acad Sci USA 1999; 96:12,833–12,838.

47. Beliaev AS, Thompson DK, Fields MW, et al. Microarray transcription profiling of a *Shewanella oneidensis* etrA mutant. J Bacteriol 2002; 184:4612–4616.

48. Okinaka Y, Yans CH, Perra NJ, Keen NT. Microarray profiling of *Erwinia chrysanthemi* 3937 genes that are regulated during plant infection. Mol Plant Microbe Interact 2002; 15:619–629.

49. Hutchison CA, Peterson SN, Gill SR, et al. Global transposon mutagenesis and a minimal *Mycoplasma* genome. Science 1999; 286:2165–2169.

50. Giaever G, Chu AM, Ni L, et al. Functional profiling of the *Saccharomyces cerevisiae* genome. Nature 2002; 418:387–391.

8

Genomics-Based Analysis of the Bacterial Cell Cycle

Michael T. Laub, Harley H. McAdams, and Lucy Shapiro

INTRODUCTION

The advent of whole genome sequencing and the technologies that enable global analysis of the genome are revolutionizing microbiology. These global analyses are leading to the identification of patterns, phenomena, and mechanisms that would be more difficult, if not impossible, to identify by studying genes or proteins one at a time. The mechanics of bacterial cell cycle control, the subject of this chapter, is one area for which a combination of genomics with molecular genetics, biochemistry, and cell biology is leading to rapid progress. The molecular mechanisms in the temporal regulation of cell cycle progression have received extensive attention in a number of eukaryotic systems, but are less well understood in prokaryotes, owing partly to the lack of suitable model systems.

However, the Gram-negative bacterium *Caulobacter crescentus* is proving to be a tractable system for addressing these issues in bacteria. This experimental system has well-established genetics and cell biology; perhaps most important, it is relatively easy to obtain populations of synchronized *Caulobacter* cells. Experiments with synchronized cells facilitate precise analysis of temporal events during cell cycle progression. The complete genome sequence *(1)* for *Caulobacter* has been determined, opening the door to genomic analysis of the bacterial cell cycle. This chapter discusses recent studies using DNA microarrays that have dramatically increased the understanding of the genetic network and molecular mechanisms that drive bacterial cell cycle progression.

CAULOBACTER CRESCENTUS: A MODEL SYSTEM FOR STUDYING THE BACTERIAL CELL CYCLE

Progression through the life cycle of *C. crescentus* (Fig. 1) requires precisely ordered execution of a series of morphological, metabolic, and regulatory events. *Caulobacter* divides asymmetrically into two distinct progeny cells, a motile "swarmer" cell and a sessile "stalked" cell. Cells in the G1 phase have a single polar flagellum and several polar pili. These motile swarmer cells do not initiate replication of their single circular chromosome. In response to signals that are not yet understood, swarmer cells differentiate into stalked cells by shedding their polar flagellum, retracting their polar pili, and constructing a stalk at that same pole. The stalk is a tubular extension of the cell

From: *Microbial Genomes*
Edited by: C. M. Fraser, T. D. Read, and K. E. Nelson © Humana Press Inc., Totowa, NJ

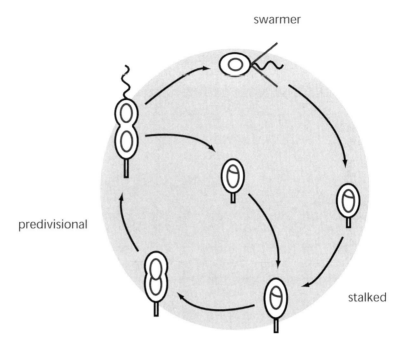

Fig. 1. Cell cycle progression and the formation of two distinct progeny cells in *Caulobacter*. Cell types (swarmer, stalked, predivisional) are listed around the outside next to the corresponding stages of the cell cycle. The circle or theta structure inside each cell represents the replication state of a given cell type.

envelope with a holdfast substance at the tip that allows the cell to adhere to various surfaces. The initiation of deoxyribonucleic acid (DNA) replication occurs coincident with the swarmer-to-stalked cell differentiation and can be viewed as the G1–S transition. At this point, the swarmer progeny is at the same point in the cell cycle as the stalk progeny immediately after the parent cell division. As the stalked cell proceeds through S phase, an unidentified mechanism establishes an asymmetrically placed cell division site approximately 0.6 of the cell length from the stalked pole by starting construction of a FtsZ ring *(2)* and initiation of cytokinesis. Next, the pinched predivisional cells begin construction of a new flagellum at the pole opposite the stalk. After the completion of chromosome replication and segregation, cell division is completed to produce the daughter cells with their differing morphologies and distinct cell fates.

   This asymmetric division is exploited for synchronizing *Caulobacter* cells; density centrifugation is used to separate the smaller, motile swarmer cells away from the more buoyant stalked and predivisional cells. The resulting highly purified subpopulation of swarmer cells then proceeds synchronously through the cell cycle. Even at its fastest generation time (approximately 90 min), the *Caulobacter* chromosome is replicated once and only once per cell cycle, with no overlapping replication phases as in *Escherichia coli* and many other bacterial species. In sum, *Caulobacter* cells are synchronizable and show distinct G1 and S phases; together, these properties make *Caulobacter* a powerful model system for studying the bacterial cell cycle.

SEQUENCING OF THE *CAULOBACTER* GENOME

The complete 4.0-Mb sequence of the *C. crescentus* chromosome has been determined *(1)*. While the annotation and organization of this genome sequence provide insights into the entire α-class of proteobacteria, the challenge is to determine what it reveals about the regulation of bacterial cell cycle progression and the establishment of cellular asymmetry.

Homology searches can identify some genes predicted to be involved in specific cell cycle events. This sort of analysis is most successful for genes involved in well-characterized processes such as DNA replication or chromosome segregation. However, it is often equally informative, if not more so, to identify which genes are not present in the genome rather than which ones are present. For example, the genes *minCD* and their homologs have been shown to be central players in cell division site selection in *E. coli* and *Bacillus subtilis (3–6)*. However, no *minCD* homologs are found in the *C. crescentus* sequence, suggesting that an alternative mechanism involving a distinct set of genes, is used to determine the asymmetric placement of the cell division site in this organism.

Although the genome sequence is informative for such comparisons and analyses, sequence analysis alone cannot identify the molecular mechanisms that regulate the cell cycle. However, the availability of the complete sequence enables the use of a wide range of new genomic tools to elucidate the regulatory network driving cell cycle progression. Foremost among these tools are DNA microarrays. Whole genome DNA microarrays for *Caulobacter* have allowed the first global analysis of the transcriptional network controlling a bacterial cell cycle.

MAPPING THE CELL CYCLE
GENETIC NETWORK WITH DNA MICROARRAYS

Early work on cell cycle progression and cell differentiation in *Caulobacter* identified the presence of temporally regulated transcription *(7)*. This was followed by extensive characterization of the transcriptional control of flagellum biogenesis that occurs in the *Caulobacter* predivisional cell (reviewed in ref. *8*). The polar flagellum is built from the interior of the cell toward the exterior. The membrane-embedded motor and basal body are built first, followed by export and assembly of the hook subunits, and finally export and polymerization of the flagellin subunits of the flagellar filament. The more than 40 genes required for flagellar biogenesis are organized in a four-tier transcriptional hierarchy with the order of expression of genes in this cascade reflecting their order of assembly at the nascent swarmer pole. The basal body genes (class II) are transcribed and expressed first, while genes with products that are required later in assembly, such as genes for the hook (class III) and filament (class IV) proteins, are expressed last. In addition, each class of flagellar genes includes genes that encode trans-acting factors that activate the next class of genes, thereby coupling order of assembly to order of expression.

These previous studies demonstrated that *Caulobacter* uses regulated transcription to control cell cycle events. However, these studies utilized time-consuming and labor-intensive gene-by-gene experiments. A complete genome sequence and the availability of *Caulobacter* DNA microarrays afforded the first opportunity to conduct a comprehensive assessment of genes with messenger ribonucleic acids (mRNAs) that are expressed in a cell cycle-dependent fashion *(9)*. In this first global study, a large population of G1-phased swarmer cells were isolated and then allowed to proceed synchronously through their 150-min cell cycle, with RNA samples collected every 15 min. Each of these RNA samples was then compared on a microarray to a common reference, producing cell cycle expression profiles for 2966 genes. Using a modified Fourier technique, computational analysis of the resulting expression profile data set identified about 550 mRNAs that changed in a cyclical, or cell cycle-dependent, manner. The temporal expression profiles of these genes were then clustered in two ways: First, they were clustered based on their temporal patterns of expression (Fig. 2), revealing that there is a continuum of differential gene expression during the cell cycle, rather than discrete phases. Second, the expression profiles were clustered based on the corresponding gene's functional classification and predicted cellular role.

Figure 3A shows the expression profiles for those genes known or predicted to play a role in flagellum biogenesis. On a coarse level, these genes are all expressed late in the cell cycle, during the predivisional stage, which is precisely when the organism is building the new polar flagellum. However, these profiles also have enough resolution to discern their previously known order of transcription, confirming the colinearity of flagellar gene expression with order of assembly (Fig. 3A). These profiles fall into four distinct temporal clusters, recapitulating the previously established class I–IV designations (Fig. 3A) *(9)*. Similar timing of the *E. coli* flagellar transcriptional hierarchy was confirmed by an entirely different approach *(10)*. In that study, the promoters of 14 individual flagellar genes were each fused to the reporter gene *gfp* (encoding green fluorescent protein [GFP]). Induction of these flagellar genes, as measured by GFP fluorescence, was also colinear with the order of assembly and revealed precise ordering of gene expression, parallel to that seen in *Caulobacter*.

In *Caulobacter*, the expression profiles for genes associated with pili biogenesis and cell division also exhibit a hierarchical pattern of expression, suggesting that the proper assembly of these multiprotein structures also involves ordered gene expression. The construction of polar pili in *Caulobacter* is known to require expression of at least three operons that are adjacent on the chromosome *(11)*. The temporally ordered expression of these genes (Fig. 3B) suggests that, like the flagellar genes, they are expressed in an order that reflects their order of assembly and usage during the pili construction process in late predivisional cells. The predicted membrane-embedded portions of these type IV-related pili are induced first, peaking at approximately 100 min into the cell cycle. This is followed, after a 15- to 30-min lag, by maximal expression of the prepilin peptidase and then finally by expression of the pilin subunit itself. These expression profiles suggest the following testable model for temporal control of pili biogenesis in *Caulobacter*: The pilin-anchoring subunits are expressed first to provide a site at the cell pole for pilus assembly in the cell envelope. This is followed by expression of the prepilin peptidase *cpaA*, which must be present to process the prepilin subunit. The prepilin subunit gene *pilA* is expressed last, and then PilA is immediately cleaved by the CpaA peptidase to generate pilin monomers that are polymerized into the membrane-anchored pilus.

Similarly, the cell division genes (five were found to be cell cycle regulated: *ftsZ, ftsI, ftsW, ftsQ,* and *ftsA*) are expressed in an ordered fashion, again arguing that temporal ordering of transcription is important for proper assembly into a subcellular structure. Supporting this hypothesis, it is known that assembly and localization of FtsZ to the nascent cell division site is required for subsequent assembly and activity of FtsQ and FtsA at the cytokinetic ring *(12,13)*; consistently, *ftsZ* expression peaks in stalked cells, well before induction of *ftsQ* and *ftsA* in the late predivisional cell.

In addition to structural and morphological genes, the grouping of cell cycle-regulated gene expression profiles based on functional classification reveals a wide range of other processes that seem to be temporally controlled, at least in part, by staged gene expression. Notably, nearly all aspects of DNA replication require genes that were identified as cell cycle regulated. Genes involved in DNA replication initiation are maximally expressed in swarmer cells, presumably to ensure the availability of their gene products at the G1–S transition. This is followed by expression at the G1–S transition of genes required for replication elongation, nucleotide synthesis, and DNA repair. These in turn are followed by peak expression of genes known or predicted to play roles in chromosome decatenation and segregation. In sum, these patterns of expression suggest that the *Caulobacter* cell controls the timing of various aspects of DNA replication at least partly by transcriptional regulation.

The expression profiles of cell cycle-regulated genes showed that *Caulobacter* cells generally express genes immediately before or coincident with the time in the cell cycle when they are needed. Microarray-based analysis of gene expression during cell cycle progression in yeast and mammalian cells has demonstrated that eukaryotes also employ a similar approach to control of the cell cycle *(14–16)*. For example, in *Saccharomyces cerevisiae* the DNA replication initiation genes are also maximally expressed in the G1 phase, immediately before the initiation of the S phase. While these genes share no strong homology with any genes in *Caulobacter*, the induction of replication initiation genes in the G1 phase in both organisms has apparently been selected as advantageous to the operation of the cell cycle.

Given this paradigm of expressing genes immediately before or during the time of use, examination of cell cycle-regulated transcripts by functional category suggests additional processes that may be executed in a cell cycle-dependent fashion. Among these are three sets of genes encoding subunits of the ribosome, RNA polymerase, and the nicotinamide adenine dinucleotide (NADH) dehydrogenase complex of oxidative respiration. The induction of these three sets of genes at approximately the same time (immediately after the G1–S transition) suggests that stalked and predivisional cells may have a higher metabolic requirement than swarmer cells, and that the transition to a replication-competent state is accompanied by a more general transition to a metabolically active state.

The wild-type gene expression profiles may also help assign function to genes with a function that has not been predicted by sequence analysis. Nearly 260 genes with cell cycle-dependent expression profiles have no obvious predicted function. Both the timing of expression of these genes and the functional identity of genes with which they are coregulated may help pinpoint their function or at least generate testable hypotheses. Expression profiles may also help assign new or more precise function to some genes that already are annotated with a predicted function. For instance, there are 10 genes expressed in swarmer and early stalked cells that are predicted, based on sequence analysis, to encode cell wall synthesis enzymes. The biogenesis of the polar stalk in

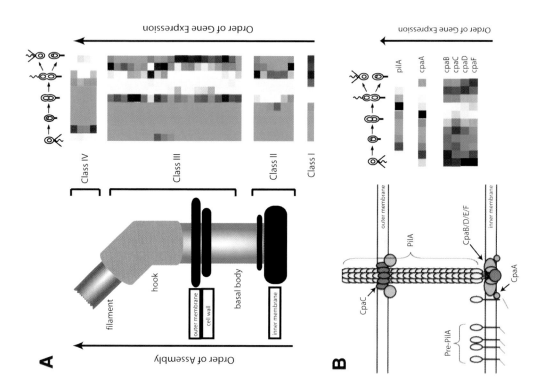

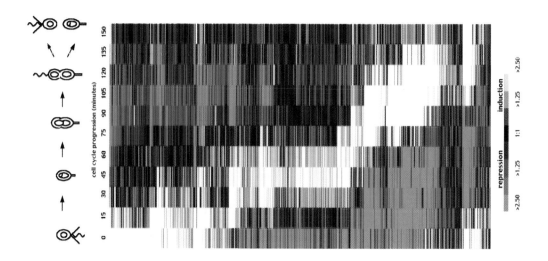

Caulobacter is known to require localized cell envelope synthesis at the transition of swarmer to stalked cell. The subset of cell wall metabolism genes that are induced during this G1 and early S period are thus logical candidates for participation in construction of the stalk.

WHY DIFFERENTIAL GENE EXPRESSION?

The discovery and cataloging of more than 550 cell cycle-regulated transcripts raises a major and largely unanswered question: Why are these genes differentially expressed during the cell cycle? As discussed for the flagellar and pili biogenesis genes and the cell division machinery, temporally regulated transcription may relate to coordinated assembly of large subcellular structures. In addition, there are genes with products that are harmful to the cell when present at the wrong time. For example, misexpression or constitutive expression of the *Caulobacter* gene *ccrM*, which encodes an essential DNA methyltransferase, leads to extremely sick cells, suggesting that regulated transcription is important to confine its activity to a precise stage of the cell cycle *(17)*. Another factor driving temporally controlled transcription is that the cell may be saving energy and resources by only transcribing, and presumably translating, genes immediately before or during the time when they are needed. This "just-in-time" design principle in biological systems may allow cells to operate with increased efficiency, greater robustness, and hence better long-term fitness.

Another possible rationale for *Caulobacter*'s use of cell cycle-regulated transcription is to offset the asymmetric distribution of certain gene products at the time of cell division, a well-documented phenomenon in the asymmetrically dividing *Caulobacter* *(18–20)*. For example, consider the histidine kinase CckA, which has been shown, on the protein level, to be present at approximately constant levels during the cell cycle *(20)*. Fluorescent microscopy of GFP-tagged CckA showed that CckA is predominantly in the swarmer cell after cell division. Interestingly, the *cckA* mRNA is cell cycle regulated, peaking in stalked cells *(9)*. The increased transcription of *cckA* in stalked cells may help offset the asymmetric distribution of CckA to swarmer cells, thereby maintaining the constant levels seen by Western blot analysis. The histidine kinase PleC is also present in all cell types, but is localized predominantly to the swarmer pole at cell division; like *cckA*, the *pleC* gene shows dramatically increased mRNA levels in stalked cells. Finally, the mechanism of localization may be coupled to cell stage-specific transcription. In budding yeast, in some instances the localization of a bud-neck protein can only occur in a specific cell type because the mechanism of localization, involving directed secretion to the bud-neck, requires *de novo* transcription and protein produc-

Fig. 2. Cell cycle-regulated gene expression. Expression profiles for the 553 *Caulobacter* genes identified as cell cycle regulated are shown using the color coding scheme at the bottom to indicate relative RNA levels. Each gene's expression profile runs from left to right, with the timing of cell cycle progression indicated in minutes and with a diagram at the top. (Reprinted with permission from ref. 9. Copyright 2000 American Association for the Advancement of Science.)

Fig. 3. Control of multiprotein structure assembly by ordered transcription. (**A**) Expression profiles for genes required for flagellum assembly are shown next to a schematic of the polar *Caulobacter* flagellum. The arrows on the left and right indicate the timing of flagellum assembly and flagellar gene expression, which are colinear (see text). (**B**) Expression profiles for pili biogenesis genes are shown next to a schematic of a pilus. The order of expression of pilus genes suggests that the timing of expression of these genes, like the flagellar genes, may be critical to proper assembly of the organelle.

tion *(21)*. Thus, in yeast cell type-specific or cell cycle-regulated transcription can be required for proper localization; a similar situation may occur in *Caulobacter*, causing the expression of certain genes to be cell cycle regulated.

DISSECTING THE REGULATORY NETWORK

The identified cell cycle-regulated genes in *Caulobacter* include 34 two-component signal transduction genes and 5 RNA polymerase-associated sigma factors. Presumably, the cell cycle-dependent expression of these regulatory genes produces a cell cycle-dependent expression pattern in the genes they regulate. The two-component genes, some of which are already known to play critical roles in governing cell cycle progression, are of particular interest. Two-component signal transduction genes, encompassing response regulators and histidine kinases, are a major form of internal cell signaling in the bacterial kingdom and are found in plants and some eukaryotes *(22)*. The two-component proteins are best known for their function as signaling factors involved in adaptive responses to environmental changes. However, a *Caulobacter* response regulator, *ctrA*, was identified and shown to play an essential role in controlling cell cycle progression *(23)*.

The *ctrA* gene was originally identified as a temperature-sensitive, loss-of-function allele, *ctrA401^{ts} (23)*. The *ctrA401^{ts}* mutant grows normally and appears morphologically intact at the permissive temperature of 30°C, but becomes nonmotile, stalkless, and filamentous after a shift to the restrictive temperature of 37°C. In addition, *ctrA401^{ts}* loses viability at the restrictive temperature, indicating that the *ctrA* gene is essential. CtrA was shown to control the expression of genes involved in a wide range of cell cycle-regulated events, including DNA methylation, cell division, and flagellum synthesis *(23)*. In addition, CtrA directly represses DNA replication initiation by binding to at least five sites in the origin of replication *(24)*.

Given the crucial role that CtrA plays in governing many cell cycle events, it is no surprise that CtrA is itself controlled at multiple levels (Fig. 4). During the transition from swarmer to stalked cell, CtrA protein is rapidly degraded, in a process dependent on the ClpXP protease *(25)*, thereby allowing replication to begin. Subsequently, the stalked cell downregulates *ctrA* degradation, allowing newly transcribed CtrA to accumulate. A burst of transcription from promoter P1 drives initial production of CtrA. As levels increase, CtrA blocks additional expression from P1, but dramatically stimulates production from promoter P2 *(26)*. This positive feedback leads to a rapid increase in CtrA levels in predivisional cells. CtrA, which must be phosphorylated to function as a transcription factor, exhibits a cell cycle-dependent phosphorylation pattern matching the CtrA protein profile *(18,27)*. Even when CtrA is rendered resistant to its usual degradation at the transition from swarmer to stalked cell, its cell cycle regulated phosphorylation maintains cell cycle-regulated activity *(18)*.

Global analyses have recently been used to identify the genes and cell cycle events controlled by the CtrA master regulator. First, DNA microarrays were used to identify all genes dependent on CtrA for normal levels of expression *(9,28)*. RNA was isolated from a *ctrA^{ts}* strain at the permissive temperature of 30°C and also 4 h after shifting to the restrictive temperature, immediately before the strain began losing viability. Comparison of these RNA samples on microarrays, after elimination of genes responding merely to the temperature change, led to the identification of approximately 200 genes with expression that is dependent on *ctrA* and that are cell cycle regulated in wild-type cells.

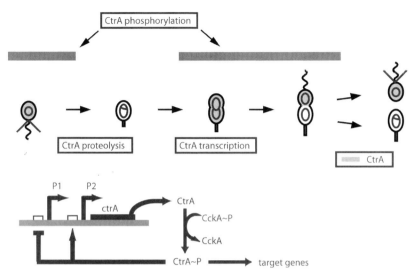

Fig. 4. CtrA activity during the cell cycle is controlled on multiple levels. The level of active CtrA is controlled during cell cycle progression by at least three mechanisms, described in detail in the text. Transcription of the *ctrA* gene is driven by two promoters, P1 and P2, as schematically shown at the bottom. The activation of P1 in the stalked cell drives the initial production of CtrA. As CtrA accumulates to higher levels, it feeds back negatively on P1, but positively on P2, leading to rapid production of CtrA. Proteolysis of CtrA (cell types with CtrA protein are shaded gray) in stalked cells and in the stalked half of predivisional cells prevents CtrA binding of the origin of replication and hence allows the initiation of DNA replication initiation in stalked cells. Even in cells that constitutively express an allele of CtrA that is resistant to degradation, *ctrAΔ3Ω*, cell cycle-regulated phosphorylation (shown as dark gray bars at the top) leads to cell cycle-regulated fluctuations in CtrA activity.

However, expression profiling experiments alone cannot distinguish between those genes directly regulated and those indirectly regulated. One way to discriminate between direct and indirect targets is to search the predicted, upstream regulatory region of each of these CtrA-dependent genes *(9)* for the previously identified consensus CtrA binding site. This approach is unsatisfactory regardless of the algorithm used to identify sites; even sites that match perfectly to a consensus may not bind CtrA in vivo; conversely, CtrA might bind in vivo to sites that do not match the consensus binding site pattern. An alternative approach, called *location analysis (29,30)*, offers a way to map experimentally, on a genome-wide level, the in vivo binding sites of transcription factors such as CtrA *(28)*.

In the CtrA location analysis experiment, formaldehyde was added to mid-log phase *Caulobacter* cells to crosslink CtrA (and all DNA-binding proteins) to the DNA to which it is bound in vivo. Shearing of this crosslinked chromosomal DNA was followed by immunoprecipitation with an anti-CtrA antibody enriched for CtrA-bound pieces of DNA. After reversing the crosslinking, this enriched DNA sample was compared to unenriched chromosomal DNA by competitive hybridization on DNA microarrays containing spots specific for each *Caulobacter* intergenic region. Of 116 enriched, CtrA-bound intergenic regions, 55 flanked genes or operons that were identified as cell cycle regulated and CtrA dependent for normal expression; these genes comprise the

CtrA cell cycle regulon. Taking into account multigene operons, this regulon of direct CtrA targets contains 95 genes. The list includes all but one of the genes identified previously by other experimental methods as direct CtrA targets.

Genes in this CtrA cell cycle regulon fall into two broad functional categories: polar morphogenesis and essential cell cycle processes. Many of the genes involved in flagellum and pili biogenesis and chemotaxis are directly regulated by CtrA, with maximal expression of these genes occurring in predivisional cells after CtrA has accumulated to high levels. In addition to activating expression of these polar morphogenesis genes, CtrA regulates transcription of a number of essential genes, including the DNA methyltransferase *ccrM* and five genes required for proper cell wall assembly and cell division, *ftsZ*, *ftsQ*, *ftsA*, *ftsW*, and *murG*.

In addition, CtrA activates expression of the *clpP* gene that encodes a component of the essential ClpXP protease required for cell stage-specific degradation of CtrA itself *(25)*. CtrA is also required for expression of the response regulator DivK, which is involved in triggering turnover of CtrA at the transition of swarmer to stalked cell *(25a)*. Together, these data suggest that, as CtrA builds up to high levels in late predivisional cells, it may help activate its own destruction by turning on the genes *clpP* and *divK*. This model, however, does not explain why CtrA is only degraded in the stalked half of the late predivisional cell. Other mechanisms, likely not transcriptionally based, must control this spatial aspect of CtrA proteolysis. Regardless of the spatial component, it appears that this bacterial cell cycle regulator is responsible for triggering its own destruction, as is the case for key eukaryotic cell cycle regulators such as the cyclin Clb2 in *S. cerevisiae (16,31,32)*.

The CtrA regulon contains at least 14 regulatory genes. These genes are now candidates for control of additional cell cycle events and may ultimately help link other pathways to the CtrA-controlled subnetwork (Fig. 5) of the cell cycle regulatory network. Expression profiling and global location analysis of these other regulatory factors is needed to complete the identification of the transcriptional network that drives *Caulobacter* cell cycle progression. Such a systematic approach has already been taken for 12 cell cycle regulatory factors in the yeast *S. cerevisiae (33)*. Location analysis of these 12 factors showed that they form a serially connected loop of transcriptional activators that helps drive the regular oscillations of the cell cycle in that organism. It is expected that bacteria such as *Caulobacter* have similar transcriptionally based loops in their cell cycle regulatory circuitry.

In addition to the 55 cell cycle-regulated operons directly controlled by CtrA, there are approximately as many that are dependent on CtrA for normal expression, but that were not identified as CtrA bound in the global location analysis. Many of these genes are likely indirect targets of CtrA (Fig. 5). For instance, the expression of genes encoding ribosomal proteins and the NADH dehydrogenase complex is strongly affected by loss of CtrA function, but the intergenic regions upstream of these genes were not enriched in the CtrA immunoprecipitation experiments.

In sum, CtrA is a major hub of the genetic network driving cell cycle progression in *Caulobacter* (Fig. 5). This single master regulator controls a surprisingly large number of genes that coordinate a wide range of cell cycle and morphological processes. However, although the understanding of CtrA's regulatory role has increased substantially, many questions remain: How does CtrA activate the expression of genes at widely dis-

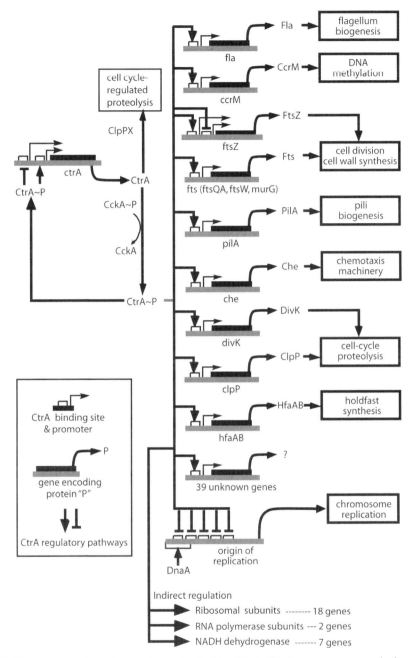

Fig. 5. Genes and cell cycle events controlled by CtrA. CtrA acts as a transcription factor to control directly a wide range of cell cycle events (listed in boxes on the right). DNA microarray analyses, as described in the text, have led to the rapid identification of the CtrA regulon.

parate times, a range spanning more than half of the cell cycle? While CtrA appears to control, directly or indirectly, nearly a quarter of the cell cycle–regulated genes in *Caulobacter*, what controls the remaining three-quarters (approximately 350 genes) that are apparently CtrA independent?

The answers to these questions likely involve the activity of additional, as yet uncharacterized, regulatory molecules. These additional regulators may be cell cycle-regulated themselves. As described here, included among the more than 550 cell cycle-regulated genes are 34 two-component signal transduction genes, 5 sigma factors, and 10 other transcription factors. All of these genes are now candidates for controlling sets of cell cycle-regulated genes, and they are all being characterized.

BEYOND TRANSCRIPTIONAL PROFILING

While DNA microarrays are driving the rapid generation of new data, and hence new insights, into cell cycle progression in bacteria, there are obvious limitations to this technology. First, there are a number of other levels of regulation that help control the cell cycle and that cannot be studied by DNA microarrays. This includes posttranslational modifications such as phosphorylation, controlled proteolysis, regulated protein–protein interaction, and subcellular localization, all of which are known in specific instances to function in regulation of the *Caulobacter* cell cycle.

One recent study has begun documenting, on a global level, changes in protein synthesis and stability during the cell cycle *(34)*. The complete cataloging of changes in protein levels, phosphorylation state, and protein localization as a function of cell cycle progression will be critical to the development of an integrated model of the *Caulobacter* cell.

However, even the collection of all such data will not illuminate all aspects of cell cycle regulation in *Caulobacter*. A critical missing component will be the assessment of function by means of genetic perturbation and subsequent phenotypic characterization. Global survey methods, such as DNA microarrays or most proteomic methods, must ultimately be coupled with methods for genome-wide analysis of function.

Several techniques have been developed and applied for analyzing function on a global level in yeast: massive, near-saturating insertional mutagenesis *(35)* and comprehensive in-frame deletions *(35–37)*. Pools of in-frame deletions have been used to identify all genes necessary for normal progression through the yeast cell cycle in minimal media *(37)*. Surprisingly, this list of genes showed little, if any, correlation with the list of genes differentially expressed in minimal media relative to rich media or differentially expressed during the cell cycle, highlighting the need to combine a global analysis of function with global expression analyses. The systematic mapping of the genetic network driving cell cycle progression in *Caulobacter* will ultimately require such a combination of techniques.

With genomics leading the way to a global, integrated view of the bacterial cell cycle, new questions and challenges will arise. Of particular interest will be a dissection of the similarities and differences in regulatory strategies used by eukaryotes and prokaryotes to control cell cycle progression. For example, do bacteria use checkpoints and surveillance mechanisms to control cell cycle transitions in the same way that eukaryotes do *(38)*? How are multiple signals, both intracellular and extracellular, integrated into the functioning of the cell cycle machinery? How do cells couple morphological changes with the mechanics of the cell cycle? While conservation at the sequence level is not expected for regulatory genes in the two kingdoms, there is likely to be conservation of design principles and regulatory mechanisms. In fact, identifying these conserved design principles and the genetic architectures that have been repeatedly selected by evolution will be easier by parallel study, with the aid of genomics, of both bacterial and eukaryotic cell cycles.

REFERENCES

1. Nierman WC, Feldblyum TV, Laub MT, et al. Complete genome sequence of *Caulobacter crescentus*. Proc Natl Acad Sci USA 2001; 98:4136–4141.
2. Fukuda A, Iba H, Okada Y. Stalkless mutants of *Caulobacter crescentus*. J Bacteriol 1977; 131: 280–287.
3. Levin PA, Shim JJ, Grossman AD. Effect of *minCD* on FtsZ ring position and polar septation in *Bacillus subtilis*. J Bacteriol 1998; 180:6048–6051.
4. Marston AL, Thomaides HB, Edwards DH, Sharpe ME, Errington J. Polar localization of the MinD protein of *Bacillus subtilis* and its role in selection of the mid-cell division site. Genes Dev 1998; 12:3419–3430.
5. Marston AL, Errington J. Selection of the midcell division site in *Bacillus subtilis* through MinD-dependent polar localization and activation of MinC. Mol Microbiol 1999; 33:84–96.
6. de Boer PA, Crossley RE, Rothfield LI. A division inhibitor and a topological specificity factor coded for by the minicell locus determine proper placement of the division septum in *E. coli*. Cell 1989; 56:641–649.
7. Newton A. Role of transcription in the temporal control of development in *Caulobacter crescentus*. Proc Natl Acad Sci USA 1972; 69:447–451.
8. Gober JW, England JC. Regulation of flagellum biosynthesis and motility in Caulobacter. In: Brun YV, Shimkets LJ (eds). Prokaryotic Development. Washington, DC: ASM Press, 2000, pp. 319–39.
9. Laub MT, McAdams HH, Feldblyum T, Fraser CM, Shapiro L. Global analysis of the genetic network controlling a bacterial cell cycle. Science 2000; 290:2144–2148.
10. Kalir S, McClure J, Pabbaraju K, et al. Ordering genes in a flagella pathway by analysis of expression kinetics from living bacteria. Science 2001; 292:2080–2083.
11. Skerker JM, Shapiro L. Identification and cell cycle control of a novel pilus system in *Caulobacter crescentus*. Embo J 2000; 19:3223–3234.
12. Sackett MJ, Kelly AJ, Brun YV. Ordered expression of *ftsQA* and *ftsZ* during the *Caulobacter crescentus* cell cycle. Mol Microbiol 1998; 28:421–434.
13. Wortinger M, Sackett MJ, Brun YV. CtrA mediates a DNA replication checkpoint that prevents cell division in *Caulobacter crescentus*. Embo J 2000; 19:4503–4512.
14. Cho RJ, Campbell MJ, Winzeler EA, et al. A genome-wide transcriptional analysis of the mitotic cell cycle. Mol Cell 1998; 2:65–73.
15. Cho RJ, Huang M, Campbell MJ, et al. Transcriptional regulation and function during the human cell cycle. Nat Genet 2001; 27:48–54.
16. Spellman PT, Sherlock G, Zhang MQ, et al. Comprehensive identification of cell cycle-regulated genes of the yeast *Saccharomyces cerevisiae* by microarray hybridization. Mol Biol Cell 1998; 9:3273–3297.
17. Zweiger G, Marczynski G, Shapiro L. A *Caulobacter* DNA methyltransferase that functions only in the predivisional cell. J Mol Biol 1994; 235:472–485.
18. Domian IJ, Quon KC, Shapiro L. Cell type-specific phosphorylation and proteolysis of a transcriptional regulator controls the G1-to-S transition in a bacterial cell cycle. Cell 1997; 90: 415–424.
19. Wheeler RT, Shapiro L. Differential localization of two histidine kinases controlling bacterial cell differentiation. Mol Cell 1999; 4:683–694.

20. Jacobs C, Domian IJ, Maddock JR, Shapiro L. Cell cycle-dependent polar localization of an essential bacterial histidine kinase that controls DNA replication and cell division. Cell 1999; 97:111–120.

21. Lord M, Yang M C, Mischke M, Chant J. Cell cycle programs of gene expression control morphogenetic protein localization. J Cell Biol 2000; 151:1501–1512.

22. Loomis WF, Kuspa A, Shaulsky G. Two-component signal transduction systems in eukaryotic microorganisms. Curr Opin Microbiol 1998; 1:643–648.

23. Quon KC, Marczynski GT, Shapiro L. Cell cycle control by an essential bacterial two-component signal transduction protein. Cell 1996; 84:83–93.

24. Quon KC, Yang B, Domian IJ, Shapiro L, Marczynski GT. Negative control of bacterial DNA replication by a cell cycle regulatory protein that binds at the chromosome origin. Proc Natl Acad Sci USA 1998; 95:120–125.

25. Jenal U, Fuchs T. An essential protease involved in bacterial cell-cycle control. Embo J 1998; 17:5658–5669.

25a. Hung DY, Shapiro L. A signal transduction protein cues proteolytic events critical to *Caulobacter* cell cycle progression. Proc Natl Acad Sci USA 2002; 99:13,160–13,165.

26. Domian IJ, Reisenauer A, Shapiro L. Feedback control of a master bacterial cell-cycle regulator. Proc Natl Acad Sci USA 1999; 96:6648–6653.

27. Quon KC. Thesis: Temporal control during the *Caulobacter crescentus* cell cycle [doctoral thesis]. Stanford, CA: Stanford University, 1996.

28. Laub MT, Chen SL, Shapiro L, McAdams HH. Genes directly controlled by CtrA, a master regulator of the *Caulobacter* cell cycle. Proc Natl Acad Sci USA 2002; 99:4632–4637.

29. Iyer VR, Horak CE, Scafe CS, Botstein D, Snyder M, Brown PO. Genomic binding sites of the yeast cell-cycle transcription factors SBF and MBF. Nature 2001; 409:533–538.

30. Ren B, Robert F, Wyrick JJ, et al. Genome-wide location and function of DNA binding proteins. Science 2000; 290:2306–2309.

31. Chen KC, Csikasz-Nagy A, Gyorffy B, Val J, Novak B, Tyson JJ. Kinetic analysis of a molecular model of the budding yeast cell cycle. Mol Biol Cell 2000; 11:369–391.

32. Yeong FM, Lim HH, Padmashree CG, Surana U. Exit from mitosis in budding yeast: biphasic inactivation of the Cdc28-Clb2 mitotic kinase and the role of Cdc20. Mol Cell 2000; 5:501–511.

33. Simon I, Barnett J, Hannett N, et al. Serial regulation of transcriptional regulators in the yeast cell cycle. Cell 2001; 106:697–708.

34. Grunenfelder B, Rummel G, Vohradsky J, Roder D, Langen H, Jenal U. Proteomic analysis of the bacterial cell cycle. Proc Natl Acad Sci USA 2001; 98:4681–4686.

35. Ross-Macdonald P, Sheehan A, Roeder GS, Snyder M. A multipurpose transposon system for analyzing protein production, localization, and function in *Saccharomyces cerevisiae*. Proc Natl Acad Sci USA 1997; 94:190–195.

36. Shoemaker DD, Lashkari DA, Morris D, Mittmann M, Davis RW. Quantitative phenotypic analysis of yeast deletion mutants using a highly parallel molecular bar-coding strategy. Nat Genet 1996; 14:450–456.

37. Winzeler EA, Shoemaker DD, Astromoff A, et al. Functional characterization of the *S. cerevisiae* genome by gene deletion and parallel analysis. Science 1999; 285:901–906.

38. Hartwell LH, Weinert TA. Checkpoints: controls that ensure the order of cell cycle events. Science 1980; 246:629–634.

IV THE EVOLUTION OF MICROBIAL GENOMES

9

A Brief History of Views
of Prokaryotic Evolution and Taxonomy

Lorraine Olendzenski, Olga Zhaxybayeva, and J. Peter Gogarten

We are living in an era of prokaryotic systematics, one in which we believe we have the tools and knowledge to decipher the natural phylogenetic relationships of all organisms, particularly the prokaryotic Bacteria and Archaea. This was not always true, and 30 years hence, we may have a different understanding of phylogenetic relationships of these organisms. Certainly, the enormous increase in available genomic data will contribute to the view of natural relationships among the prokaryotes and is affecting the current understanding of prokaryotic evolution.

TOWARD A NATURAL SYSTEM OF PROKARYOTES:
EARLY DELINEATION OF MEMBERS AND RELATIONSHIPS

If the goal of phylogeny is to present a system that mirrors the presumed natural relationships of organisms and reconstruct the evolution of lineages, in the case of prokaryotic microorganisms this goal has been historically problematic. Prior to the advent of genetic approaches to prokaryotic systematics, the earliest schemes were based on the limited data workers were able to collect from light microscopy. Following the discovery of the microbial world by Leeuwenhoek, early workers tended to name and describe observed forms without regard to existing names and previous descriptions. Few tried to represent natural relationships (or phylogeny) among organisms, although they worked under the assumption that a natural system of bacteria was possible because they adopted the same structure of nomenclature used for animals and plants (orders, families, genera, etc.) (*see* e.g., *1*).

Because microscopy was the major tool for observation, most descriptions were based on shape, behavior, and habitat and were made by people looking at primarily aquatic natural habitats *(2,3)*. Haeckel *(4)* established a group he called the Moneres as a distinct subgroup of the Protista. The work of Cohn *(5–7)* was notable in that he recognized that the different shapes and sizes of organisms he observed probably represented different species rather than different life stages of the same organism. Cohn recognized the earlier work of Ehrenberg and Dujardin as a valid foundation for a new taxonomic classification of bacteria *(2,8,9)*. He classified bacteria into four groups based on cell shape and felt that the bacteria (as then known) were related to the plants through the blue-green algae and that the two should be classified together. In the first half of the 20th

From: *Microbial Genomes*
Edited by: C. M. Fraser, T. D. Read, and K. E. Nelson © Humana Press Inc., Totowa, NJ

century, Orla-Jensen proposed using physiological characters such as metabolic by-products, sugar fermentation, and temperature ranges for growth to create a comprehensive taxonomic system based on evolutionary relationships *(3,10)*.

In 1941, Stanier and van Neil *(11)* presented a system of classification for the Kingdom Monera composed of Myxophyta (cyanobacteria) and Schizomycetae (all other bacteria) with a proposed phylogeny. The subdivisions of the Schizomycetae contained three main groups (Eubacteriae, which included photosynthetic bacteria, actinomycetes, and a wide variety of unicellular bacteria; Myxobacteriae; and Spirocheatae). They also suggested a number of additional appendices (i.e., groups containing *Leptothrix, Crenothrix, Achromatium*, and *Pasteuria* and *Hyphomicrobium*). Important taxonomic characteristics were cell shape and insertion of flagella. Following earlier suggestions, organisms such as the sulfur-oxidizing *Beggiatoa* were classified with the Myxophyta (cyanobacteria) as colorless forms because of their distinctively similar morphology *(11)*.

Stanier and van Neil *(11)* argued against using physiology as a phylogenetic character because it led to the splitting of natural groups based on morphological characters, and it forced organisms into assemblages in which they did not belong. Deciding which physiologies are ancestral or derived is also problematic. Despite their long work studying bacterial evolutionary relationships, by 1946, van Neil seemed to have changed his mind and abandoned all attempts toward phylogeny *(12)*. It became widely accepted that the lack of characteristics, particularly the lack of distinguishing morphological characteristics visible by the light microscope of the time, made phylogenies of prokaryotes impossible.

PROKARYOTIC TAXONOMIES: EMPIRICAL APPROACHES

Prior to the advent of modern molecular phylogenetic methods, the majority of schemes of prokaryotic taxonomy were determinative; that is, they were designed for determining the identity of the organism only, without attempting to address presumed evolutionary relationships. The early work of Chester, who noted that the lack of organized taxon descriptions and classification schemes made identification and recognition of new bacterial species virtually impossible *(13)*, greatly influenced the development of the first *Bergey's Manual of Determinative Biology*. This work, the "bible" of bacterial taxonomy used by most workers, went through nine successful editions and has been substituted with *Bergey's Manual of Systematic Bacteriology*, which is now in its second edition.

The *Bergey's Manual of Determinative Bacteriology* was initiated in 1923 by the Society of American Biologists and subsequently directed by the independent Bergey's Manual Trust, founded in 1936. Regulation of bacterial taxonomy occurred with the creation of the International Committee on Systematic Bacteriology formed under the auspices of the International Association of Microbiological Sciences *(3)*. This committee enacted major reforms that shape the current system of taxonomy and names in place today for the prokaryotes. They published an Approved List of Bacterial Names *(14)*; set a new starting date for bacterial names as of January 1, 1980, to replace those of May 1, 1753 (the date of Linneaus's *Species Plantarum*); provided definition in the Bacterial Code of valid and invalid publication of names *(15)*; and freed up names not on the Approved List for future use *(3)*.

The long history of *Bergey's Manual* attests to its popularity and usefulness as a guide for workers seeking to identify bacteria. The advent of using computers to analyze large amounts of phenotypic data (numerical taxonomy or Adansonian method; *16*) improved the validity of phenotypic identification and reinforced the determinative approach. All available data (or a statistically significant proportion of it) can be used to calculate coefficients of similarity between strains, and organisms are grouped on the basis of phenotypic similarities into taxospecies *(17,18)*. These empirical methods do not take into account the modes of origin of the observed resemblances *(19)* and are in no way phylogenetic. Instances of convergence or acquisition of traits through horizontal transfer would tend to group organisms together, but obscure their evolutionary histories.

THE CONCEPT OF A BACTERIUM

When *Bergey's* was founded, the nature of bacterial cells was not well understood. Discussions of bacterial (*Schizomycetes*) classification took place at botanical congresses. The seminal work of Stanier and van Neil *(20)* defined the uniqueness of bacteria as a group. With advances in electron microscopy, the fundamental differences between prokaryotic and eukaryotic cells became clear. Bacteria (as they were then circumscribed) lacked a nuclear membrane and internal organelles. They also had distinctive cell walls. As would become significant many years later with the discovery of the Archaea, the bacteria were initially defined by exclusion, that is, by negative characteristics rather than by a defining set of unifying characteristics *(21)*.

GENETIC AND EPIGENETIC APPROACHES: NEW TOOLS FOR TAXONOMY AND PHYLOGENY

With the understanding that DNA was the genetic material of the cell and thus controlled phenotype, molecular biological and genetic methods became available that could be applied to phylogenetic relationships of organisms. This led to renewed interest on the part of microbiologists in an evolutionary taxonomy of bacteria based on the use of genetic techniques *(22)*. Mean DNA base composition, reported in mole percent guanine and cytosine (G+C) residues provided a rough measure for grouping organisms and helped delineate closely related organisms. Widely differing G+C percentages could be used to separate groups definitively. Although not used as a primary diagnostic character today, base composition is still included in the demarcation of prokaryotic groups *(23)*.

Techniques of nucleic acid hybridization, measured as the thermal denaturation temperature, also yield a rough quantitative expression of the degree of overall relatedness between organisms under comparison *(22,24)*. Single-strand DNA from two different bacteria can be incubated together and heteroduplexes allowed to form; the percentage of unpaired bases in the heteroduplex DNA is an indication of degree of divergence. The number of unpaired bases can be approximated by comparing thermal stability of the heteroduplex to the homoduplex DNA (ΔT_m). Duplex formation does not occur if mismatch is greater than 20% *(18)*. Immunological comparison of proteins using quantitative determination of cross-reactivity also helped to define closely related organisms *(19,22)*. Genetic relatedness was supported by the work of Arnheim and colleagues

(25), who demonstrated correlation between immunological resemblance and sequence homology.

All of these techniques solved many problems at lower phylogenetic levels (e.g., species, family, or phylum), but none was useful for determining relationships at the highest taxonomic levels or for inferring relationships among phyla. These advances encouraged Stanier to envision the future achievement of a "fragmented hierarchy," a genetically based classification in which recognized homology groups would remain unlinked at the higher levels, yet grouped together as prokaryotes *(22)*.

PROKARYOTIC PHYLOGENY: THE MOLECULAR FOUNDATION

The suggestions of Zuckerkandl and Pauling *(26)* that the history of life could be recorded in the sequences of nucleic acids and proteins provided a foundation for a revolution in microbial taxonomy. The first molecules used for the construction of phylogenetic trees were amino acid sequences of cytochrome c and ferredoxin *(27,28)*. Schwartz and Dayhoff *(29)* used them to construct phylogenies of prokaryotes and eukaryotes and used these data to strengthen the arguments that the eukaryotic organelles, mitochondria, and plastids were in fact derived from free-living bacteria *(30)*.

The use of oligonucleotide cataloging of RNA molecules by Fox, Woese and colleagues *(31–33)* launched a new era in approaches to phylogeny and provided a measure by which all organisms could be compared. Small-subunit RNA molecules were ubiquitous in distribution, performed the same function in all organisms, interacted with many other components, and so were thought not to be easily horizontally transferred; they contained both highly conserved and highly variable regions that could be compared among organisms *(21)*.

At the time this work began, the sequencing of whole RNA molecules was unfeasible. Rather, many small fragments of each 16S RNA were obtained by digesting with ribonuclease T1 and sequencing, cataloging, and comparing the sequences among groups *(34)*. Comparisons of RNA catalogs could be used for constructing phylogenies; early results emerging from the changing bacterial phylogeny necessitated a change in bacterial taxonomy at the highest level.

The methanogenic bacteria were found to form a coherent group that was distinctly different from and distantly related to all other bacteria previously examined *(35)*. Concurrently, the observation that methanogens and halophilic bacteria did not contain peptidoglycan in their cell walls supported the idea that they were very different from all other known bacteria *(36)*. This led to the realization that there were two fundamentally different types of cells with prokaryotic organization: the Eubacteria and Archaebacteria, now know as Bacteria and Archaea *(33,37,38)*.

Subsequently, sequencing of whole rDNA molecules (i.e., the DNA encoding ribosomal RNA [rRNA]) became routine, allowing larger amounts of sequence information to be compared among organisms, using a variety of techniques for phylogenetic inference *(39)*. The ability to amplify rRNA coding genes, either through direct cloning of environmental DNA or using the polymerase chain reaction, allows analyses of rRNAs not only from cultured organisms, but also from environmental samples *(40–42)*. With the isolation of environmental rRNA sequences that did not match any cultured organisms came the realization that the diversity of the prokaryotes had been vastly under-

estimated *(43)*. Fewer than 5000 species of prokaryotes have been cultured and offici-
ally named; however, it is generally believed that only between 1 and 10% of existing
prokaryotic diversity has been cultured *(44)*.

Routine sequencing of rRNA from cultures and environmental samples has led to the
explosion of the rRNA database *(45)* that is still ongoing. A universal phylogeny of three
domains is recognized: Bacteria, Archaea, and Eucarya *(38)*. Prompted by this progress,
Bergey's Manual of Systematic Bacteriology adopted a taxonomic approach based on
16S ribosomal (rRNA) sequences for determination of phyla. The second edition lists 2
phyla of Archaea and 23 phyla of Bacteria *(46)* (Table 1). The overlap between the phyla
recognized by 16S rRNA methods and organisms previously classified in a strictly
determinative approach is illustrated in Fig. 2 of ref. *46*.

Current practice considers 16S rRNAs with more than 3% divergence as separate
species corresponding to an overall genomic sequence similarity of less than 70% and
a melting point difference (ΔT_m) of 5°C or more (see ref. *44* for a recent review). However, in some instances, this approach is problematic. Because of the conserved nature
of 16S rRNA, some distinct species have identical or nearly identical rRNA *(47)*, whereas
in other instances a single species can have divergent rRNA genes *(48,49)*. Furthermore, single organisms contain multiple rRNA operons; in a few cases, these were shown
to be more than 5% divergent *(50–53)*.

ROOTING THE TREE OF LIFE

The first universal trees based on 16S rRNA were unrooted; when examining all of
life, there is no outgroup that can be used to root the phylogeny. Schwarz and Dayhoff
(29) suggested that sequences of paralogous or duplicated protein-coding genes could
be used to solve this problem. Phylogenies of one paralog could be used as an outgroup
to the other paralog. Applying this method to the proton-pumping adenosine triphos-
phatases *(54)*, elongation factors EF-Tu and EF-G *(55–57)*, signal recognition particles
(58), and amino acyl transfer RNA (tRNA) synthetases *(59)* gives a tree in which the root
occurs between the Bacteria on one side and the Archaea and Eukaryotes on the other.
Archaea share a number of traits, primarily related to genome structure, translation, and
transcription, with Eukaryotes *(60–62)*. However, they also have genes best described as
prokaryotic in character, leading to the suggestion that this lineage is chimeric *(63,64)*.

SHAKING THE MOLECULAR CORE/TREE

As more molecular data became available, different gene trees were reconstructed and
compared to rRNA trees. Some were in agreement with rRNA trees; others were not.
One explanation for the incongruence of phylogenetic trees constructed using different
markers is horizontal gene transfer (HGT) *(65)*. The exchange of genetic information
is now recognized as a major factor in microbial evolution *(66,67)*. The formation of
operons has been attributed to frequent gene transfer *(68)*, and gene acquisition, often
in the form of genomic islands, is recognized as an important factor that allows organisms to occupy new ecological niches *(69,70)*. The high rate of horizontal exchange
within a species has been suggested as a criterion for defining microbial species analogous to the biological species concept *(68,71)*. Population studies support this reasoning
for at least some species—those for which within-species gene phylogenies differ in

Table 1
Commonly Used Classification Schemes of Bacterial Phyla

Bergey's Manual (23)	Representatives of different phyla *(43,99)*	National Center for Biotechnology Information Taxonomy Database, June 2002 *(100)*
Bacteria		
BI. Aquificae	Aquificales	Aquificae
BII. Thermotogae	Thermotogales	Thermotogae
BIII. Thermodesulfobacteria	Thermodesulfobacterium	Thermodesulfobacteria
BIV. "Deinococcus-Thermus"	Thermus/Deinococcus	Thermus/Deinococcus group
BV. Chrysiogenetes	—	Chrysiogenetes
BVI. Chloroflexi	Green nonsulfur bacteria	Chloroflexi
BVII. Thermomicrobia	—	Thermomicrobia
BVIII. Nitrospirae	Nitrospira	Nitrospirae
BIX. Deferribacteres	Synergistes Flexistipes	Deferribacteres
BX. Cyanobacteria	Cyanobacteria	Cyanobacteria
BXI. Chlorobi	Chlorobiaceae	Bacteroidetes/Chlorobi group
BXX. Bacteroidetes	Bacteroidetes/Cytophaga	
BXII. Proteobacteria	Proteobacteria	Proteobacteria (purple bacteria and relatives)
BXIII. Firmicutes	Low C+G Gram-positive	Firmicutes
BXIV. Actinobacteria	Actinomycetales	
BXV. Planctomycetes	Planctomycetales	Planctomycetales
BXVI. Chlamydiae	Chlamydia	Chlamydiales/Verrucomicrobia group
BXXII. Verrucomicrobia	Verrucomicrobium	
BXVII. Spirochaetes	Spirochaetes Leptospira	Spirochaetales
BXVIII. Fibrobacteres	Fibrobacter	Fibrobacter/Acidobacteria group
BXIX. Acidobacteria	Acidobacterium	
BXXI. Fusobacteria	Fusobacteria	Fusobacteria
BXXIII. Dictyoglomus	Dictyoglomus	Dictyoglomus group
—	—	Dehalococcoides group
	Environmental phylotypes from hot springs: OP1, OP3-OP10, EM19[a]	
Archaea		
AI. Crenarchaeota	Crenarchaeota	Crenarchaeota
AII. Euryarchaeota	Euryarchaeota	Euryarchaeota
—	Korarchaeota	Korarchaeota
—	—	Nanoarchaeota

[a]Each phylotype represents a separate phylum.

topology, whereas interspecies gene phylogenies appear more congruent to each other *(71,72)*.

However, gene transfer is not limited to within-species transfer and can even occur across domains. In addition to genes under sporadic selection *(68)*, HGT affects housekeeping genes *(73)* and functions involved in information processing *(67,74–76)*. The widespread occurrence of HGT has been amply documented through the analyses of completely sequenced microbial genomes *(63,77–79)*. The complexity hypothesis states that informational genes, because of their highly cooperative gene products, are less frequently transferred than operational genes *(80)*; however, there does not appear to be a core of genes that is never subject to horizontal transfer *(81,82)*.

MOSAIC NATURE OF rRNA AND GENOME CONTENT

Because of their central role in translation and the complexity of the ribosome, it was traditionally thought that the likelihood of an organism being able to use an rRNA from another organism was very low. However, functioning ribosomes can be reconstituted from components from different organisms *(83–86)*, ribosomal operons of an organism can be replaced under laboratory conditions with those from another species *(87)*, divergent rRNA operons can coexist in the same genome *(50,51)*, and mosaic rRNA operons showing extensive recombination have been observed and have been demonstrated to function *(51,52,88)*. Could it be that the congruence between phylogenies based on gene content *(89–93)* and rRNA phylogenies is because of the mosaic nature of both the genome and the rRNA *(94)*?

Over long periods of time, even low rates of HGT lead to a mosaic genome in which different parts of the genome reflect different histories. The potential difference between a molecular or gene tree and the organismal evolution has been widely recognized; the latter is often assumed to be netlike or reticulate *(65,73)*.

TREES, WEBS, AND LINES OF DESCENT

Even in the presence of rampant HGT, an organismal line of descent can be defined over very short time intervals as the majority consensus of genes passed on by vertical inheritance (i.e., from mother to daughter cells during cell division). It appears that the transfer between divergent species usually only involves the transfer of a few genes *(64,80,95)*. The above definition of an organismal lineage thus is applicable in most instances, provided that the time interval under consideration is sufficiently short. However, reliable genome content information is only available for present-day organisms and not for infinitely many points along past genealogies.

Congruence between phylogenies for different molecular markers can be a reflection of the organismal lineage through which these genes were transmitted mainly via vertical inheritance. However, HGT itself has the potential to create congruent molecular phylogenies (see Fig. 1); furthermore, these congruent molecular phylogenies do not necessarily reflect organismal phylogeny. The mosaic nature of genomes *(81,82)* makes it difficult, and in many instances probably impossible, to reconstruct the organismal lineages.

SUMMARY

Darwin's work on the origin of species *(96)* and the neo-Darwinian synthesis (see e.g., ref. *97*) suggested that, over long periods of time, organismal evolution can be described as a strictly bifurcating tree. A taxonomy reflecting this tree of life is the goal of modern systematics *(98)*. Initially, microbial taxonomy had to be based on so few characters that the goal of a natural systematic system was unattainable. The introduction of small-

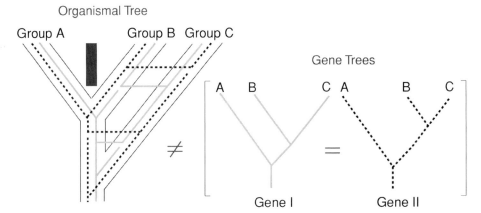

Fig. 1. How horizontal gene transfer (HGT) influences the ability to reconstruct organismal evolution. The gray and the dotted lines represent two gene trees. The solid black line encompassing them represents organismal evolution (see text for discussion). Group A is a more recently branching taxonomic unit; however, after its separation from group B, it occupies a niche or develops a physiology that drastically reduces the frequency with which it shares genes with groups B and C (vertical black bar). In contrast, groups B and C continue to exchange genes. As a result, the molecular phylogenies of the separate gene trees (I and II) differ from the organismal phylogeny. According to this scenario, a taxonomy that is based on molecular data reflects HGT frequency rather than the vertical transmission of genes, accumulation of mutations, and subsequent divergence of lineages. In particular, the deep-branching lineage in the gene trees (A) is one that participates in HGT less frequently rather than one that evolved earlier.

subunit rRNA as a tool in microbial taxonomy *(33)* by Woese and Fox led many microbiologists to assume that the concepts of animal and plant taxonomy could be extended to the realm of prokaryotes. "Until the advent of molecular sequencing, bacterial evolution was not a subject that could be approached experimentally" *(21)*. The hope was that more sequence data would allow an even more precise placement of species onto the universal tree of life. Molecular data, particularly 16S rRNA and whole genome data, are valuable tools for classifying microorganisms. Groups defined on the basis of molecular sequence data usually are also united by shared biochemical, physiological, and structural characteristics. However, the recognition of HGT as an important force in microbial evolution and the discovery of the mosaic nature of microbial genomes have initiated a reassessment of microbial evolution *(94)*. The extent to which taxonomic groupings reflect shared ancestry or preferred HGT remains under debate.

ACKNOWLEDGMENTS

The work on this chapter was supported through the NASA Exobiology program and NASA Astrobiology Institute at Arizona State University, Tempe. We thank Ed Leadbetter, W. Ford Doolittle, and Otto Kandler for providing useful references.

REFERENCES

1. Müller OF. Animacula Infusoria Fluviatilia et Marina, quae Detexit, Systematice Descripsit et ad Vivum Delineari Curavit, Hauniae: Typis N Mölleri, 1786.

2. Ehrenberg CG. Die Infusionsthierchen als vollkommene Organismen. Leipzig: Leopold Voss, 1838.
3. Murray RGE, Holt JG. The history of Bergey's Manual. In: Garrity GM (ed). Bergey's Manual of Systematic Bacteriology. Volume One. The Archaea and Deeply Branching and Phototrophic Bacteria. New York: Springer, 2001, pp. 1–13.
4. Haeckel E. Generelle Morphologie der Organismen. Berlin: Georg Reimer, 1866.
5. Cohn F. Untersuchungen über Bakterien. Beitr Biol Pflanz 1872; 1:127–224.
6. Cohn F. Untersuchungen über Bakterien II. Beitr Biol Pflanz 1875; 1:141–207.
7. Brock TD. Milestones in Microbiology, 1556 to 1940. Washington, DC: ASM Press, 1998.
8. Dujardin F. Histoire Naturelle des Zoophytes. Paris: Roret, 1841.
9. Drews G. The roots of microbiology and the influence of Ferdinand Cohn on microbiology of the 19th century. FEMS Microbiol Rev 2000; 24:225–249.
10. Orla-Jensen S. Die Hauptlinien des natürlichen Bakterien-systems. Zentbl Bakteriol Parasitenkd Infektkrankh Hyg Abt II 1909; 22:97–98 and 305–346.
11. Stanier RY, van Neil CB. The main outlines of bacterial classification. J Bacteriol 1941; 42: 437–466.
12. van Neil CB. The classification and natural relationships of bacteria. Cold Spring Harb Symp Quant Biol 1946; 11:285–301.
13. Chester FD. A Manual of Determinative Bacteriology. New York: Macmillan, 1901.
14. Skerman VBD, McGowan V, Sneath PHA. Approved list of bacterial names. Int J Syst Bacteriol 1980; 30:225–420.
15. Lapage SP, Sneath PHA, Lessel J, Skerman VBD, Seeliger HPR, Clark WA. International Code for Nomenclature of Bacteria, 1976 revision. Washington, DC: American Society for Microbiology, 1975.
16. Sneath PHA, Sokal RR. Numerical taxonomy. Nature 1962; 193:855–860.
17. Sokal RR. Typology and empiricism in taxonomy. J Theor Biol 1962; 3:230–267.
18. Brenner DJ, Staley JT, Krieg NR. Classification of procaryotic organisms and the concept of bacterial speciation. In: Garrity GM (ed). Bergey's Manual of Systematic Bacteriology. Volume One. The Archaea and Deeply Branching and Phototrophic Bacteria. New York: Springer, 2001, pp. 27–31.
19. Marmur J, Falcow S, Mandel M. New approaches to bacterial taxonomy. Ann Rev Microbiol 1963; 17:329–372.
20. Stanier RY, van Neil CB. The concept of a bacterium. Arch Mikrobiol 1962; 42:17–35.
21. Woese CR. Bacterial evolution. Microbiol Rev 1987; 51:221–271.
22. Stanier RY. Toward an evolutionary taxonomy of the bacteria. In: Pérez-Miravete A, Peláez D (eds). Recent Advances in Microbiology, 10th International Congress for Microbiology, Mexico: Associacion Mexicana De Microbiologia; 1971, pp. 595–604.
23. Garrity G, (ed). Bergey's Manual of Systematic Bacteriology. New York: Springer, 2001.
24. Marmur J, Doty P. Thermal renaturation of DNA. J Mol Biol 1961; 3:584–594.
25. Arnheim N, Prager EM, Wilson AC. Immunological prediction of sequence differences among proteins. Chemical comparison of chicken, quail, and phesant lysozymes. J Biol Chem 1969; 244:2085–2094.
26. Zuckerkandl E, Pauling L. Evolutionary divergence and convergence in proteins. In: Bryson V, Vogel HJ (eds). Evolving Genes and Proteins. New York: Academic Press, 1965, pp. 97–166.
27. Margoliash E. Primary structure and evolution of cytochrome c. Proc Natl Acad Sci USA 1963; 50:672–679.
28. Fitch WM, Margoliash E. Construction of phylogenetic trees. Science 1967; 155:279–284.
29. Schwartz RM, Dayhoff MO. Origins of prokaryotes, eukaryotes, mitochondria, and chloroplasts. Science 1978; 199:395–403.

30. Sagan L. On the origin of mitosing cells. J Theor Biol 1967; 14:255–274.
31. Zablen L, Woese CR. Procaryote phylogeny IV: concerning the phylogenetic status of a photo-synthetic bacterium. J Mol Evol 1975; 5:25–34.
32. Woese CR, Fox GE, Zablen L, et al. Conservation of primary structure in 16S ribosomal RNA. Nature 1975; 254:83–86.
33. Woese CR, Fox GE. Phylogenetic structure of the prokaryotic domain: the primary kingdoms. Proc Natl Acad Sci USA 1977; 74:5088–5090.
34. Fox GE, Peckman KJ, Woese CR. Comparative cataloging of 16S ribosomal ribonucleic acid: molecular approach to prokaryotic systematics. Int J Syst Bacteriol 1977; 27:44–57.
35. Fox GE, Magrum LJ, Balch WE, Wolfe RS, Woese CR. Classification of methanogenic bacteria by 16S ribosomal RNA characterization. Proc Natl Acad Sci USA 1977; 74:4537–4541.
36. Kandler O, Hippe H. Lack of peptidoglycan in the cell walls of *Methanosarcina barkeri*. Arch Microbiol 1977; 113:57–60.
37. Woese CR, Fox GE. The concept of cellular evolution. J Mol Evol 1977; 10:1–6.
38. Woese CR, Kandler O, Wheelis ML. Towards a natural system of organisms: proposal for the domains Archaea, Bacteria, and Eucarya. Proc Natl Acad Sci USA 1990; 87:4576–4579.
39. Swofford DL, Olsen GJ, Waddell PJ, Hillis DM. Phylogenetic inference. In: Hillis DM, Moritz C, Mable BK (ed). Molecular Systematics. Sunderland, MA: Sinauer, 1996, pp. 407–514.
40. Giovannoni SJ, Britschgi TB, Moyer CL, Field KG. Genetic diversity in Sargasso Sea bacterioplankton. Nature 1990; 345:60–63.
41. Schmidt TM, DeLong EF, Pace NR. Analysis of a marine picoplankton community by 16S rRNA gene cloning and sequencing. J Bacteriol 1991; 173:4371–4378.
42. Hugenholtz P, Pace NR. Identifying microbial diversity in the natural environment: a molecular phylogenetic approach. Trends Biotechnol 1996; 14:190–197.
43. Pace NR. A molecular view of microbial diversity and the biosphere. Science 1997; 276:734–740.
44. Rossello-Mora R, Amann R. The species concept for prokaryotes. FEMS Microbiol Rev 2001; 25:39–67.
45. Maidak BL, Cole JR, Lilburn TG, et al. The RDP-II (Ribosomal Database Project). Nucleic Acids Res 2001; 29:173–174.
46. Garrity GM, Holt JG. The road map to the manual. In: Garrity GM (ed). Bergey's Manual of Systematic Bacteriology 2nd ed. New York: Springer, 2001, pp. 119–116.
47. Probst A, Hertel C, Richter L, Wassill L, Ludwig W, Hammes WP. *Staphylococcus condimenti* sp nov, from soy sauce mash, and *Staphylococcus carnosis* (Schleifer and Fisher 1982) subsp utilis subsp nov. Int J Syst Bacteriol 1998; 48:651–658.
48. Fox GE, Wisotzkey JD, Jurtshuk P Jr. How close is close: 16S rRNA sequence identity may not be sufficient to guarantee species identity. Int J Syst Bacteriol 1992; 42:166–170.
49. Martinez-Murcia AJ, Benlloch S, Collins MD. Phylogenetic interrelationships of members of the genera *Aeromonas* and *Plesiomonas* as determined by 16S ribosomal DNA sequencing: lack of congruence with results of DNA–DNA hybridizations. Int J Syst Bacteriol 1992; 42: 412–421.
50. Yap WH, Zhang Z, Wang Y. Distinct types of rRNA operons exist in the genome of the actinomycete *Thermomonospora chromogena* and evidence for horizontal transfer of an entire rRNA operon. J Bacteriol 1999; 181:5201–5209.
51. Mylvaganam S, Dennis PP. Sequence heterogeneity between the two genes encoding 16S rRNA from the halophilic archaebacterium *Haloarcula marismortui*. Genetics 1992; 130:399–410.
52. Dennis PP, Ziesche S, Mylvaganam S. Transcription analysis of two disparate rRNA operons in the halophilic archaeon *Haloarcula marismortui*. J Bacteriol 1998; 180:4804–4813.
53. Amann G, Stetter KO, Llobet-Brossa E, Amann R, Anton J. Direct proof for the presence and expression of two 5% different 16S rRNA genes in individual cells of *Haloarcula marismortui*. Extremophiles 2000; 4:373–376.

54. Gogarten JP, Kibak H, Dittrich P, et al. Evolution of the vacuolar H+-ATPase: implications for the origin of eukaryotes. Proc Natl Acad Sci USA 1989; 86:6661–6665.
55. Iwabe N, Kuma K, Hasegawa M, Osawa S, Miyata T. Evolutionary relationship of Archae-bacteria, Eubacteria, and Eukaryotes inferred from phylogenetic trees of duplicated genes. Proc Natl Acad Sci USA 1989; 86:9355–9359.
56. Cammarano P, Palm P, Creti R, Ceccarelli E, Sanangelantoni AM, Tiboni O. Early evolutionary relationships among known life forms inferred from elongation factor EF-2/EF-G sequences: phylogenetic coherence and structure of the archaeal domain. J Mol Evol 1992; 34:396–405.
57. Baldauf SL, Palmer JD, Doolittle WF. The root of the universal tree and the origin of eukaryotes based on elongation factor phylogeny. Proc Natl Acad Sci USA 1996; 93:7749–7754.
58. Gribaldo S, Cammarano P. The root of the universal tree of life inferred from anciently duplicated genes encoding components of the protein-targeting machinery. J Mol Evol 1998; 47:508–516.
59. Brown JR, Doolittle WF. Root of the universal tree of life based on ancient aminoacyl–tRNA synthetase gene duplications. Proc Natl Acad Sci USA 1995; 92:2441–2445.
60. Marsh TL, Reich CI, Whitelock RB, Olsen GJ. Transcription factor IID in the Archaea: sequences in the *Thermococcus celer* genome would encode a product closely related to the TATA-binding protein of eukaryotes. Proc Natl Acad Sci USA 1994; 91:4180–4184.
61. Langer D, Hain J, Thuriaux P, Zillig W. Transcription in Archaea: similarity to that in Eucarya. Proc Natl Acad Sci USA 1995; 92:5768–5772.
62. Keeling PJ, Doolittle WF. Archaea: narrowing the gap between prokaryotes and eukaryotes. Proc Natl Acad Sci USA 1995; 92:5761–5764.
63. Koonin EV, Mushegian AR, Galperin MY, Walker DR. Comparison of archaeal and bacterial genomes: computer analysis of protein sequences predicts novel functions and suggests a chimeric origin for the Archaea. Mol Microbiol 1997; 25:619–637.
64. Olendzenski L, Hilario E, Gogarten JP. Horizontal gene transfer and fusing lines of descent: the Archaebacteria—a Chimera? In: Syvanen M, Kado C (eds). Horizontal Gene Transfer. London: Chapman and Hall, 1998, pp. 349–362.
65. Gogarten JP. The early evolution of cellular life. Trends Ecol Evol 1995; 10:147–151.
66. Doolittle WF. Phylogenetic classification and the universal tree. Science 1999; 284:2124–2129.
67. Woese CR, Olsen GJ, Ibba M, Soll D. Aminoacyl-tRNA synthetases, the genetic code, and the evolutionary process. Microbiol Mol Biol Rev 2000; 64:202–236.
68. Lawrence JG, Roth JR. Selfish operons: horizontal transfer may drive the evolution of gene clusters. Genetics 1996; 143:1843–1860.
69. Hacker J, Kaper JB. Pathogenicity islands and the evolution of microbes. Ann Rev Microbiol 2000; 54:641–679.
70. Perna NT, Plunkett G 3rd, Burland V, et al. Genome sequence of enterohaemorrhagic *Escherichia coli* O157: H7. Nature 2001; 409: 529–533.
71. Dykhuizen DE, Green L. Recombination in *Escherichia coli* and the definition of biological species. J Bacteriol 1991; 173:7257–7268.
72. Feil EJ, Holmes EC, Bessen DE, et al. Recombination within natural populations of pathogenic bacteria: short-term empirical estimates and long-term phylogenetic consequences. Proc Natl Acad Sci USA 2001; 98:182–187.
73. Hilario E, Gogarten JP. Horizontal transfer of ATPase genes—the tree of life becomes a net of life. Biosystems 1993; 31:111–119.
74. Ibba M, Bono JL, Rosa PA, Soll D. Archaeal-type lysyl–tRNA synthetase in the Lyme disease spirochete *Borrelia burgdorferi*. Proc Natl Acad Sci USA 1997; 94:14383–14388.
75. Gogarten JP, Olendzenski L. Orthologs, paralogs and genome comparisons. Curr Opin Genet Dev 1999; 9:630–636.
76. Olendzenski L, Liu L, Zhaxybayeva O, Murphey R, Shin DG, Gogarten JP. Horizontal transfer of archaeal genes into the Deinococcaceae: detection by molecular and computer-based approaches. J Mol Evol 2000; 51:587–599.

77. Nelson KE, Clayton RA, Gill SR, et al. Evidence for lateral gene transfer between Archaea and Bacteria from genome sequence of *Thermotoga maritima*. Nature 1999; 399:323–329.

78. Bult CJ, White O, Olsen GJ, et al. Complete genome sequence of the methanogenic archaeon, *Methanococcus jannaschii*. Science 1996;273:1058–1073.

79. Deckert G, Warren PV, Gaasterland T, et al. The complete genome of the hyperthermophilic bacterium *Aquifex aeolicus*. Nature 1998; 392:353–358.

80. Jain R, Rivera MC, Lake JA. Horizontal gene transfer among genomes: the complexity hypothesis. Proc Natl Acad Sci USA 1999; 96:3801–3806.

81. Zhaxybayeva O, Gogarten J. Bootstrap, Bayesian probability and maximum likelihood mapping: exploring new tools for comparative genome analyses. BMC Genomics 2002; 3:4.

82. Nesbø CL, Boucher Y, Doolittle WF. Defining the core of nontransferable prokaryotic genes: the euryarchaeal core. J Mol Evol 2001; 53:340–350.

83. Bellemare G, Vigne R, Jordan BR. Interaction between *Escherichia coli* ribosomal proteins and 5S RNA molecules: recognition of prokaryotic 5S RNAs and rejection of eukaryotic 5S RNAs. Biochimie 1973; 55:29–35.

84. Nomura M, Traub P, Bechmann H. Hybrid 30S ribosomal particles reconstituted from components of different bacterial origins. Nature 1968; 219:793–799.

85. Wrede P, Erdmann VA. Activities of *B. stearothermophilus* 50S ribosomes reconstituted with prokaryotic and eukaryotic 5S RNA. FEBS Lett 1973; 33:315–319.

86. Daya-Grosjean L, Geisser M, Stoffler G, Garret RA. Heterologous protein-RNA interactions in bacterial ribosomes. FEBS Lett 1973; 37:17–20.

87. Asai T, Zaporojets D, Squires C, Squires CL. An *Escherichia coli* strain with all chromosomal rRNA operons inactivated: complete exchange of rRNA genes between bacteria [see comments]. Proc Natl Acad Sci USA 1999; 96:1971–1976.

88. Wang Y, Zhang Z, Ramanan N. The actinomycete *Thermobispora bispora* contains two distinct types of transcriptionally active 16S rRNA genes. J Bacteriol 1997; 179:3270–3276.

89. Snel B, Bork P, Huynen MA. Genome phylogeny based on gene content [see comments]. Nat Genet 1999; 21:108–110.

90. Fitz-Gibbon ST, House CH. Whole genome-based phylogenetic analysis of free-living microorganisms. Nucleic Acids Res 1999; 27:4218–4222.

91. Lin J, Gerstein M. Whole-genome trees based on the occurrence of folds and orthologs: implications for comparing genomes on different levels. Genome Res 2000; 10:808–818.

92. House CH, Fitz-Gibbon ST. Using homolog groups to create a whole-genomic tree of free-living organisms: an update. J Mol Evol 2002; 54:539–547.

93. Tekaia F, Lazcano A, Dujon B. The genomic tree as revealed from whole proteome comparisons. Genome Res 1999; 9:550–557.

94. Olendzenski L, Zhaxybayeva O, Gogarten JP. What's in a tree? Does horizontal gene transfer determine microbial taxonomy? In: Seckbach J (ed). Symbiosis. Dordrecht, The Netherlands: Kluwer, 2002, pp. 63–78.

95. Nesbø CL, L'Haridon S, Stetter KO, Doolittle WF. Phylogenetic analyses of two "archaeal" genes in *Thermotoga maritima* reveal multiple transfers between Archaea and Bacteria. Mol Biol Evol 2001; 18:362–375.

96. Darwin C. On the Origin of Species by Means of Natural Selection, or the Preservation of Favoured Races in the Struggle for Life. London: John Murray, 1859.

97. Futuyma DJ. Evolutionary Biology. Sunderland, MA: Sinauer Associates, 1986.

98. Hennig W. Phylogenetic Systematics. Urbana: University of Illinois Press, 1966.

99. Barns SM, Delwiche CF, Palmer JD, Pace NR. Perspectives on archaeal diversity, thermophily and monophyly from environmental rRNA sequences. Proc Natl Acad Sci USA 1996;93:9188–9193.

100. Wheeler DL, Chappey C, Lash AE, et al. Database resources of the National Center for Biotechnology Information. Nucleic Acids Res 2000; 28:10–14.

10

How Bacterial Genomes Change

Timothy D. Read and Garry S. A. Myers

INTRODUCTION

Microbial genomes are inherently dynamic. They acquire mutations and undergo gene deletions, acquisitions, and rearrangements over time-scales that encompass few to countless many generations. Some of these genomic alterations survive and become fixed in progeny, potentially setting a new evolutionary course. The development of microbial molecular genetics in the last quarter of the 20th century and the more recent advent of comparative genomics have provided most of the current knowledge of genome-scale variation. Despite the cost and effort in obtaining high-quality complete sequences, the assemblage of A's, T's, G's, and C's that make up a genome sequence is only a snapshot, a frozen moment in time. This does not diminish the value of genome sequences for their insight into microbiology. However, there is much extra value to be extracted from sequence analysis if the mechanisms of genome dynamics are understood.

This chapter discusses the various factors influencing bacterial genomes as an aid to understanding the nature of the evolutionary process. While studies of pathogenic eubacteria are the source of many of the examples given (reflecting the current bias in whole genome sequence funding), they do illustrate general principles of bacterial genome dynamics. However, it is both apparent and interesting that the same processes for change do not operate across all genomes.

SELECTION, DRIFT, AND GENOME CHANGE

In discussing microbial genome change, there must be awareness of the effect of selection on any observations. Some mutations may truly be "neutral," having no effect on the overall fitness of the organism. On the other hand, nonsynonymous mutations in genes present as a single copy, and encoding proteins essential for cellular function may result in lethal phenotypes and thus will not survive to be detected. Likewise, mutations that increase or decrease fitness may be detected at greater or lesser frequency than changes that actually occur. Random genetic drift (increase or decrease of the frequency of a given mutation in a population due to chance alone) is another factor to be considered when examining genome changes. Drift often plays a major role in small populations, for example, causing the fixing of deleterious mutations following a population bottleneck. Examination of bacterial genomes from sequestered populations with little

From: *Microbial Genomes*
Edited by: C. M. Fraser, T. D. Read, and K. E. Nelson © Humana Press Inc., Totowa, NJ

or no opportunity or facility for lateral gene transfer (LGT), such as the insect endosymbionts *Buchnera (1)* and *Wigglesworthia glossinidia (2)*, shows that these bacteria are directly affected by random genetic drift, which contributes to the observed extreme genome reduction in these organisms *(3)*.

Mutations may include larger-scale changes than point mutations, but these must still be considered in terms of selection or drift. Capsule genes of many human pathogens, which have been shown to be among the most variable regions (following studies on closely related genomes; see refs. *4* and *5*), are an example of the need for careful consideration of the role of selection. It is possible that these loci are a focus for high-frequency deletion and reintroduction of capsule genes owing to multiple flanking insertion sequences (ISs). Alternatively, selection for novel capsule structure to evade host immune responses may increase the frequency at which capsule loci changes are fixed.

Loss of larger regions modulated by the pressure of selection has also been demonstrated in the evolution of pathogenic *Shigella* from *Escherichia coli* through "pathoadaptive mutations" *(6)*. Cadaverine, the product of lysine decarboxylase (encoded by *cadA*), blocks the action of *Shigella* plasmid-encoded enterotoxins (acquired by horizontal transfer) and is considered an antivirulence factor. A variety of *cadA*-inactivating alterations, including insertion sequences, phage insertion, and other gene rearrangements, have been identified in the various *Shigella* species *(6)*. The convergent evolution of these pathoadaptive mutations suggests the environmental niche (in this case, host tissues) allowed the selection of clones with increased fitness and virulence through loss of *cadA* functionality and hence enterotoxin activity.

These examples show that genomic change can be effected through a variety of often-nonexclusive pathways. To explore these pathways in this chapter, we consider some of the better-known processes of genome change.

POINT MUTATIONS

Point mutations, usually caused by inaccurate repair of damaged nucleotides, are a driving force of genome change. Typically, using exponentially growing *E. coli* cells, the rate of spontaneous point mutation has been estimated as 5×10^{-10} per base per generation *(7)*. Although early models assumed an even distribution of point mutations, there are certain biases across bacterial genomes that may be the result of differential mutation or selection.

One such bias is the increased frequency of transition mutations (A<>G or C<>T) over transversions, with the increase caused by the likelihood that repair processes will preserve the ring structure of the damaged nucleotide. Comparison of very similar genomes (Table 1) shows an average of about a two- to fourfold greater occurrence of transitions. Another bias consistently reported is guanine and cytosine (G+C) skew, the preponderance of purine nucleotides on the leading replication strand *(8)*. G+C skew (as well as operon skew, the finding that certain nucleotide "words" are unevenly distributed; 9) is a powerful predictive analysis for the origins of replication and the replication termination regions of genomes. Other biases include increased frequency of point mutation with distance from the replication origin *(10)*; increased frequency at redundant third positions in codons within open reading frames *(11)*, and decreased likelihood of mutations in highly expressed genes *(12)*.

Table 1
Comparison of Transition and Transversion
Occurrence From Selected Closely Related Genome Pairs[a]

	Transitions			Transversions				
	T<>C	A<>G	Total	C<>A	C<>G	T<>A	T<>G	Total
Chlamydophila pneumoniae[b]	137	138	275	16	9	9	18	52
Brucella spp.[c]	128	152	280	52	64	29	52	197
Buchnera aphidicola[d]	283	230	513	118	35	420	136	709
Streptococcus pneumoniae[e]	7,068	7,025	14,093	1,603	725	1,572	1,652	5,552
Helicobacter pylori[f]	25,721	25,566	51,287	4,295	2,895	2,729	4,298	14,217

[a]Calculated using MUMmer (MUMsize = 20; http://www.tigr.org/software/mummer/); forward strand only.
[b]*C. pneumoniae* AR39 vs *C. pneumoniae* CWL029.
[c]*B. suis* 1330 vs *B. melitensis* 16M.
[d]*B. aphidicola* Ap vs *B. aphidicola* Sg.
[e]*S. pneumoniae* R6 vs *S. pneumoniae* TIGR4.
[f]*H. pylori* J99 vs *H. pylori* 26695.

It has been proposed that the genomic mutation rate has evolved to be as low as possible *(13)*, which is corroborated by the presence of numerous bacterial systems that prevent the appearance of mutations (i.e., deoxyribonucleic acid [DNA] repair) *(14)*. Bacteria differ in the proficiency and specificity of their DNA repair systems, which may result in different patterns of change on the genomic scale. In this regard, it is interesting to note how mutations in bacterial genes can themselves affect mutation frequency.

One example is the finding that certain pathogenic bacteria acquire defects in their DNA mismatch repair systems in the process of adapting to new environments *(15)*. Natural isolates with subpopulations possessing increased mutation rates are frequently observed (for example, *Pseudomonas aeruginosa* in cystic fibrosis lung infections; *16*). These mutants have a markedly higher frequency of mutation because of error-prone repair of DNA mismatches. For pathogenic *E. coli*, this may also provide a selective advantage for high-frequency mutational change when the organism needs to adapt rapidly to a new ecological niche in a host *(17)*. Reversions to mismatch repair proficiency will occur after the organism has undergone a period of mutagenic adaptation. Recombination with an intact gene is one possible mechanism for reversion, explaining the observed degree of mosaicism and also why phylogenetic reconstructions of repair genes from Gram-negative bacteria do not corroborate with trees built with typical housekeeping genes *(18)*.

The external environment is an important factor influencing the nature and rate of changes in the bacterial genome sequence. However, in most cases, these "extrinsic" effects are far less studied than the intrinsic mechanisms detailed here. When considering environmental effects, it is worth noting that some of the earliest studies in microbial genetics demonstrated an enhanced rate of *E. coli* DNA damage and mutation in the presence of mutagens such as ultraviolet light and nitrosoguanadine. It is also well known that bacteria inhabiting environments that cause higher rates of DNA damage, for instance, high temperatures (*Pyrococcus*), and ultraviolet light and desiccation (*Deinococcus*), have adaptations that allow them to repair DNA lesions and breakages with higher efficiency compared to *E. coli*. However, even for well-studied bacteria, the effects of environmental damage in the complex niches typically occupied (for instance, soil or the human intestine) and the temperatures and pH typically faced are not clear-cut. Alongside the mutagenic potential of the environment, the effect on gene expression of DNA repair systems must also be considered. For instance, are mismatch repair systems expressed at the same rate in aerobic and anaerobic environments?

INSERTIONS, DELETIONS, AND CONVERSIONS CAUSED BY HOMOLOGOUS RECOMBINATION, SLIPPAGE, AND ILLEGITIMATE RECOMBINATION

Insertions and deletions (termed *indels* when the history of the event is not assumed) are an important source of genome sequence change. Repeat sequences in the genome are important foci for indel events. Bacterial repeats are generally divided into two types: low-complexity repeats (tandem repeats) and longer repeats. Low-complexity repeats consist of small oligonucleotides, ranging from mono- to pentanucleotide in size, head-to-tail repeated many times. Longer repeats include transposable elements, large tandem repeats, and spaced repeats. A number of mechanisms have been proposed for the production of tandem repeats, including slipped stand mispairing, unequal cross-over via homologous recombination, rolling circle, and circle excision with reinsertion *(19)*. Nucleotide composition bias has been shown to have a strong influence on the rate of tandem repeat creation *(20)*.

Once thought to be lacking in prokaryotes because of their more compact genomes, large tandem repeats are hypothesized to arise through duplication mechanisms similar to those acting on eukaryotic long repeats by using low-complexity repeats as primers *(20,21)*. Moreover, large repeats can play a role in antigenic variation strategies and can represent a major fraction of some genomes, for example, the reduced genomes of the *Mycoplasma*. In *Mycoplasma genitalium*, repeats of a three-gene operon encoding the adhesin MgPa makes up more than 4% of the genome *(22)*. Sequence similarity of these repeats ranges from 78 to 90%, sufficiently close for homologous recombination, yet sufficiently distant to represent a complete gene conversion once recombined. Another example of the biological importance of large tandem repeats is the *tyrP* gene of *Chlamydophila pneumoniae*, which encodes an important aromatic amino acid transport protein. Strains isolated from around the world have one to three tandem repeats of the gene, and there appears to be an association of two or three repeats in respiratory isolates and one repeat unit in cardiovascular disease isolates (Fig. 1; R. Belland, personal communication, 2003).

Of the low-complexity repeats, the class termed variable numbers of tandem repeats (VNTR) has been developed into useful epidemiological and forensic typing schemes in bacteria such as *Bacillus anthracis*, for which nucleotide variation between strains is very limited *(23,24)*. Indeed, polynucleotide runs (repeats with a unit number of 1) are the most common VNTR and typically are among the most variable features of bacterial genomes. A VNTR of 35 (nt) in the *B. anthracis* Ames genome has been shown to increase or decrease in size by 1 nucleotide unit at a frequency of about 10^{-4} per gen-

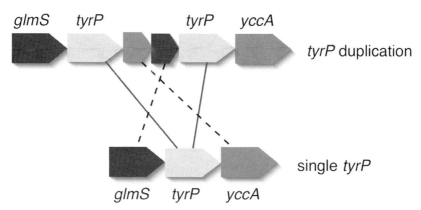

Fig. 1. Duplication of the *C. pneumoniae tyrP* locus correlates to disease type. Multiple *tyrP* genotypes are associated with respiratory infections; the single-copy genotype is associated with cardiovascular conditions. Remnants of *glmS* and *yccA* and distribution of single-nucleotide polymorphisms in *tyrP* indicate the single-copy genotype is ancestral. *glmS*, glucosamine-fructose-6-phosphate aminotransferase; *tyrP*, tyrosine/tryptophan-specific transport protein; *yccA*, integral membrane protein (possible permease). (Data kindly provided by R. Belland.)

eration (P. Keim, personal communication, 2003). The larger the repeat unit, the less frequently changes in repeat number seem to occur.

In some cases, VNTRs may influence gene functions. For instance, in the *Haemophilus influenzae* genome, there are numerous genes containing short VNTRs with a nucleotide repeat unit length that is not a multiple of 3 at their 5′ end. Similar strategies for gene regulation by hypervariable sequence have been reported for the pathogens *Campylobacter jejuni (25)*, *Neisseria meningzae (26)*, and *Streptococcus pneumoniae (5)*. VNTR unit changes can result in a frameshift mutation that negates gene function or alternatively could allow a frameshifted gene to be translated correctly. In another example of phase variation in *H. influenzae*, expression of the pilus structure is controlled by bidirectional promoters separated by a poly-AT VNTR *(27)*. When the VNTR has 9 units, the spacing between the promoters allows efficient expression of pilus genes. When there are any other numbers of units, the pilus is not produced by the cell.

Comparative genome studies indicate that shorter direct repeats (sometimes as few as 7 bases) are commonly sites for indels involving generally fewer than 1000 nt of intervening sequence *(28,29)*. However, the size of the repeat sequences is generally accepted as too short to allow RecA-mediated homologous recombination through crossover formation and Holliday junction resolution. The frequency of these events decreases with distance between the direct repeats and also with smaller repeats. Two basic mechanisms for this type of illegitimate recombination have been proposed. The first involves incorrect annealing of the lagging strand in replication (*slip-strand mispairing* or *slippage*). An alternative model involves annealing of short homologous single strands produced by the action of exonucleases *(30)*.

Gene conversion, the unbalanced exchange of sequence homologs via recombination. This has been best characterized as a mechanism for genome change in the mating-type switching system of yeast *(31)*. Gene conversion allows concerted evolution (maintenance

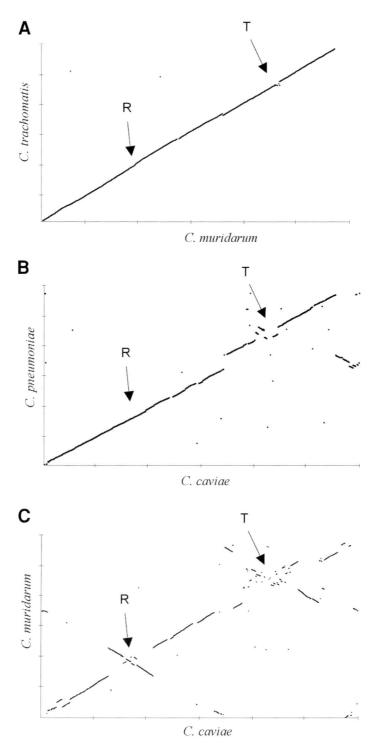

Fig. 2. (I) Comparison of the Chlamydiaceae showing breaking of chromosomal synteny around the replication origin and terminus regions. Each dot represents the most significant protein hit. R and T indicate the origins of replication and termination, respectively. (**A**) *Chlamydia trachomatis* vs *Chlamydia muridarum*. (**B**) *Chlamydia pneumoniae* vs *Chlamydia caviae*. (**C**) *Chlamydia muridarum* vs *Chlamydia caviae*. (Reproduced with permission from ref. *38*.)

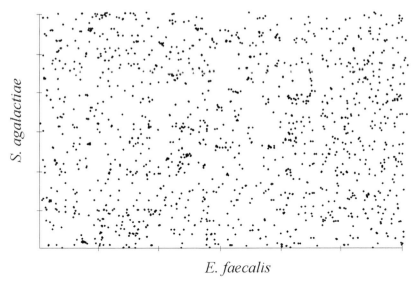

E. faecalis

Fig. 2. (II) Comparison of *Enterococcus faecalis* vs *Streptococcus agalactiae* demonstrating no large-scale chromosomal synteny.

of high-level similarity between genes over extended time periods) of multigene families in bacterial genomes *(32,33)*. In *Neisseria gonorrhoeae*, recombination between expressed pilus subunit genes and up to 19 silent copies of the genes in different loci allow antigenic variation of this surface-exposed structure *(34)*. Detailed analysis of the intermediate pilus gene sequences suggests in this case that gene conversion involves both homologous recombination and RecA-independent exchanges between shorter homologous sequences *(34)*.

INVERSIONS AROUND CHROMOSOME REPLICATION AND TERMINATION ORIGINS

Symmetrical recombinations around the replication origin and replication terminus regions represent intrinsic changes occurring on a whole genome scale (Fig. 2). This phenomenon was originally noted in the Enterobacteriaceae and, with insights from comparative genomic sequencing, is widely observed in bacteria *(35,36)*. A plausible explanation is that double-strand breaks that occur through errors in the processing of the bidirectional replication fork promote recombination between DNA strands at sites approximately equidistant from the origin *(37)*. The frequency of these symmetrical inversions appears to be greater at the terminus than at the origin *(36,38)*, perhaps because of a higher likelihood of DNA break formation at the end of replication.

Another feature of many genomes discovered from comparative genomics studies is the partitioning of genes that have globally conserved functions near the origin, with a higher concentration of hypotheticals and putatively laterally acquired genes near the terminus *(38–41)*. There may be several overlapping reasons for this: higher mutation rate at the terminus, lower expression level, or higher frequency of insertion of exogenous DNA. Housekeeping genes may be selected to move away from the replication terminus if there is a higher rate of mutation. An imbalance of housekeeping genes at

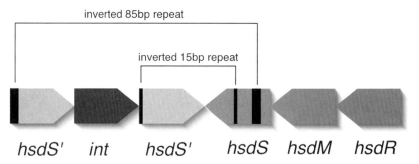

Fig. 3. Organization of the polymorphic type I restriction enzyme (*hsdS*) operon of *Streptococcus pneumoniae* TIGR4. Genes marked *hsdS'* are partial *hsdS* genes (pseudogenes). *int* is an integrase gene, *hsdSMR* are the specificity, modification and restriction subunits respectively. The inverted 85 bp repeat and inverted 15 bp repeats are marked. Clones were found that were fusions of the *hsdS* and the *hsdS'* pseudogenes with the boundary being either the 85 or 15 bp repeat.

the origin and nonessential genes at the terminus may cause further exacerbation of bias in the rate of fixation of mutations across the genome. The surfeit of conserved genes at the origin may also lead to selection against replication-directed symmetrical inversion, which may account for the lower number of these events observed at the origin compared to the terminus. This issue illustrates the potential complexities in consideration of the multiple factors that influence bacterial genome dynamics.

The distinctive X pattern when the positions of putative orthologous genes from two closely related genomes are plotted does not occur for all bacteria, for instance, *Enterococcus faecalis* and *Streptococcus agalactiae* (Fig. 2). In this case, there is very little conserved gene order, probably owing to the high number of recombination events that have occurred in *E. faecalis (42)*, which has many IS elements, prophages, and integrated plasmids.

PROGRAMMED GENOMIC CHANGE

The example of antigenic variation in *N. gonorrhoeae* pilus by gene conversion *(34)* represents a type of high-frequency switching event that may also be characterized as a "programmed" change. Programmed changes are not unique events initiated by errors in DNA repair and replication or the action of "selfish" DNA elements; rather, the ability to produce these events generally imparts a selective advantage for the bacterium by enabling response to frequently occurring environmental variation.

Many of the recognized programmed changes involve site-specific recombination systems. Site-specific recombinases catalyze DNA exchanges at specific target sites. In many of the well-studied systems, recombination causes an inversion between two or more target sites found close together in the genome.

One example is the modulation in the specificity subunit of a type I restriction system in *S. pneumoniae (5)*. A site-specific recombinase, situated next to the *hsdS*, *hsdM*, and *hsdR* genes, which respectively encode the specificity, methylation, and restriction subunits of the tripartite enzyme can cause inversion at two sites in the *hsdS* subunit and nearby *hsdS* pseudogenes (Fig. 3). This system, discovered by the variant sequence clones found in random shotgun sequencing, gives four possible nucleotide sequences for the specificity sequences; presumably, these variant genes allow different target specificity for the restriction enzyme, which may be an adaptation to protect against virus infection. Similar systems are found in *Mycoplasma pulmonis (43,44)*. Other well known examples include the *Salmonella hin* flagellar phase variation system, the *Moraxella piv* pilus variation system *(45)*, and the multiple outer surface protein variations in *Bacteroides (46)*.

With the increased understanding of genomes gained with sequencing studies, preprogrammed changes may be shown to be ubiquitous. Other examples include the types of gene and protein expression events caused by slip-strand mispairing *(47)*. It is also possible that selfish elements may be coopted to control certain features of the biology of the bacterium. For example, in *Bacillus subtilis*, the excision of *skin* prophage reconstitutes the *sigK* gene and is a critical event in sporulation *(48)*.

MOBILE GENETIC ELEMENTS AND GENOMIC REARRANGEMENT

Mobile genetic elements in a genome may be an important source of intrinsic genome change. ISs are very commonly encountered mobile elements in bacterial genomes. A basic IS element consists of a site-specific recombinase (transposase) gene and flanking DNA sequence, usually bound by inverted repeats that mark the sites for recombination *(49)*. Often, there are short direct repeats flanking the insertion site; these are generated through repeats of short, single-strand DNA ends created during transposition.

A key feature of IS elements is their ability to transpose ("jump") between different sites within genomes. IS elements may be mobile as individual units or as complex transposons containing multiple units. In some cases, mobile insertion cassettes consisting of DNA sequence flanked by inverted repeats may be transposed *in trans (50)*.

Several bacterial genomes contain multiple identical IS sequences suggestive of recent expansion through nonconservative transposition *(42,51)*. In such cases, the multiple IS elements become targets for homologous recombination, which can lead to rearrangements such as deletions, inversions, and insertions. Transposition mechanisms and specificity of target DNA sites vary enormously across different families of elements. IS7 and IS30, for instance, have very distinct specificity for insertion at a particular sequence *(52,53)*. At the other end of the scale, IS1 seems to have a preference for A and T nucleotides at the insertion site *(54)*. Target specificity may be an adaptation to long-term survival within specific hosts by avoiding insertion in sequences critical to the survival of the host. For example, of the 84 IS elements in the *S. pneumoniae* genome, only 2 are inserted in genes of known function *(5)*.

Other, less-studied mobile elements include introns, retrons, and inteins. Bacteria contain both type I and type II introns that can be spliced from transcribed ribonucleic acid (RNA) molecules *(55)*. On the other hand, intein elements are spliced from translated proteins. Retrons are small genetic elements that can replicate through an encoded reverse transcriptase *(56)*. Many bacterial genomes also contain a complex structure of repeats (examples are BOX elements in *S. pneumoniae [5]*, Correia elements in *Neisseria [57,58]*, and REP elements in *E. coli [59]*). These repeats are often dispersed in several genomic locations on different strains, suggesting some ability to move around the genome, although the mechanisms for maintenance and expansion of these elements is generally much less well understood than for ISs and introns.

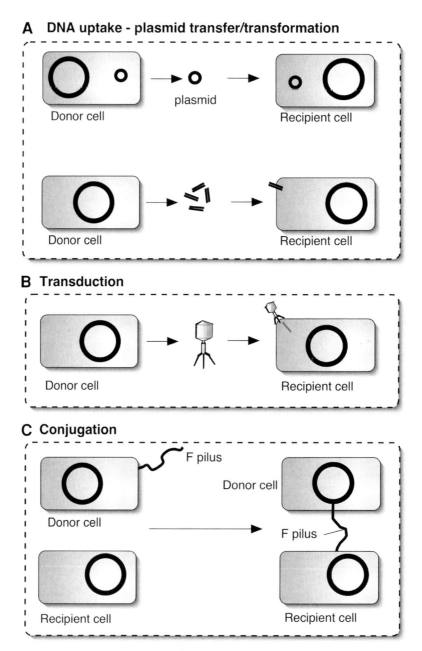

Fig. 4. Mechanisms for the acquisition of exogenous DNA by lateral gene transfer (LGT).

ACQUISITION OF EXOGENOUS DNA BY LATERAL TRANSFER

A major source of extrinsic genome change is the ability of certain bacteria to acquire DNA from other organisms, so-called LGT *(60)*. Three basic mechanisms for exogenous DNA acquisition by LGT are transformation bacteriophage-mediated transfer (transduction), and conjugation (Fig. 4). *Transformation* is the process by which bacteria acquire naked DNA from the environment. This phenomenon was discovered by Avery and coworkers in 1944 *(61)* in *S. pneumoniae* and led to the realization that DNA was the hereditary material in cells. Certain bacterial species have "competence" genes that facilitate binding and uptake, generally of DNA with specific sequence signatures. The best-described competence systems are found in *Bacillus, Streptococcus, Neisseria,* and *Haemophilus (62,63)*. Following uptake, exogenous DNA can be rescued by recombination-mediated integration into the genome. It may be no coincidence that some of these species have a high rate of recombination *(64–67)*. Even without specific DNA uptake genes, some bacteria may still be competent for acquisition of environmental DNA. The discovery by Mandel and Higa that *E. coli* was transformable in the presence of calcium chloride *(68)*, for example, was important for the early development of molecular cloning techniques.

The process by which self-transmissible and mobilizable plasmids move between sometimes quite unrelated bacteria is termed *conjugation*. As is often the case, the Enterobacteriaceae provide the best-understood models, exemplified by the well-studied F plasmid *(69)*. Conjugative transfer of the F plasmid is initiated by nicking of the double-strand DNA molecule, followed by rolling circle replication to generate a leading strand. The single-strand DNA is transferred from the donor cell through a plasmid-encoded type IV secretion apparatus, which forms a bridge into the plasmid-free recipient. PilE, also encoded by the plasmid, serves to stabilize interactions between bacterial cells to increase DNA transfer efficiency. Once in the recipient, the linear single-strand DNA circularizes and can undergo replication to reconstitute the original double-strand molecule. There is a class of smaller, "mobilizable" plasmids that lack their own conjugation genes, but contain the key DNA targets that allow them to be transferred by the machinery encoded by larger plasmids in the same cell *(69)*.

Numerous laboratory studies have shown that conjugative DNA transfer can take place between phylogenetically very separated organisms, perhaps most startlingly, *E. coli* and yeast *(70)*. Most conjugative plasmids are unable to be maintained stably in a broad range of hosts. This feature can lead to a "suicide" DNA transfer, by which the plasmid replicon is lost from the recipient shortly after conjugation, but short segments may be integrated into the new genome by the process of transposition or recombination. A further plot development is *retrotransfer*, by which the newly arrived plasmid undergoes conjugative transfer with the donor cell *(71)*. In this manner, DNA sequences transferred from the recipient chromosome to the immigrant plasmid can end up in the original donor cell; effectively, plasmid-containing bacteria use retrotransfer as a "pickpocket" of genes from their neighbors.

Another well-studied conjugation system is the Ti plasmids of *Agrobacterium tumefaciens*, which can transfer a self-mobilizable DNA segment to plant genomes using an apparatus similar to that of the F plasmids *(72)*. In many cases, plasmids found outside the Enterobacteriaceae may be self-mobilizable, but contain few, if any, F-like conjugation determinants. For example, the enterococcal pheromone-induced plasmids encode a set of gene products that enable the host cell to sense the presence of exogenous chromosomally encoded pheromone secreted by plasmid-less cells *(73)*. In response to pheromone detection, conjugation functions are activated that transfer the plasmid to the plasmid-less strain. Other functions on the plasmid also prevent self-induction in response to the cells' own pheromone. Thus, this form of mobilization is a complex bacterial behavior controlled by sophisticated cell–cell signaling.

There are numerous families of these plasmids, each responding to different phero-mones, enabling rapid dissemination of bacterial traits to other enterococci. As recent genome studies have shown, pheromone-induced plasmids and conjugative transposons were key elements in the acquisition of vancomycin resistance *(42)*.

With characteristics reminiscent of phages, transposons, and conjugative plasmids, conjugative transposons have been generally found to date only in Gram-positive ge-nomes *(74,75)*. These elements excise from host genomes to form circular double-strand intermediates, which then transfer conjugatively via a rolling circle mechanism and inte-grate at random locations in the recipient genome.

Bacteriophages offer another route for incorporation of exogenous DNA into bacterial genomes *(60,75)*. Many phages have the ability to integrate into genomes (lysogeny) as an alternative to killing their hosts after infection. On integration, the prophage represses most transcription of its genes to enter a dormant state. Phages can subsequently revive following derepression (often caused by perturbation of the host cell), when they excise from the genome and enter the lytic cycle, usually killing the host, and produce numer-ous viral progeny. Phages generally exhibit much narrower host specificity than conjuga-tive plasmids. This is probably because of the requirements for attachment to a specific receptor on the outside of the bacterium, target sites in the genome for recombinase-mediated integration, and the necessity for coevolution of the phage and bacterial gene regulation systems.

The excision and integration of phages (and the gain or loss of plasmids) can repre-sent the most significant difference in gross gene content between closely related bacte-ria *(76)*. Often, lysogenic phages contain important genes that confer advantages to their host, such as toxins [in *Cholera (77)* and *Clostridium (78)*, for example] and restriction enzymes. Phages also serve to mobilize genomic DNA. The best-known mechanism is transduction, by which prophage excision results in incorporation of neighboring host genomic DNA through illegitimate recombination and packaging into the viral capsule. Subsequent infection of the mutant phage results in homologous recombination-medi-ated integration of the nonphage DNA into the new host. In the absence of transduc-tion, infecting phages recombine with genomic DNA if they contain DNA sequences of sufficient identity. It also must be remembered that prophages that are bracketed by *att* sites and lie dormant in a bacterial genome can acquire genes through transposition or insertion in the same manner as any other piece of DNA. This newly acquired DNA sequence will be packaged and mobilized on the reactivation of the phage.

Comparative sequencing has led to the discovery of islands of inserted genes that appear to insert and move as a unit. Originally called pathogenicity islands *(79)* because the first examples contained genes essential for virulence in pathogenic enterobacterial strains (see Chapter 4), they are now more correctly termed *genomic islands* because the genes contained in these regions are responsible for a much wider range of pheno-types. Features associated with the first islands discovered included insertion near trans-fer RNA (tRNA) genes, integrase genes and flanking inverted repeats. However, recent genome sequencing has revealed a much more complex and interesting picture, with many islands having features of phages, transposons, conjugative plasmids, and conju-gative transposons *(75,80)*.

In considering how exogenous DNA may change microbial genomes, it is important to take into account the barriers that can impede the efficiency of acquisition. These may be primarily ecological: Many organisms, such as intracellular bacteria and extremophiles, may only have a very limited complement of donors or recipient organisms in their environment. Poor or nonexistent expression of key genes will also limit elements such as phages and IS elements. As discussed above, the nonexpression of replication genes in other hosts limits the host range of many bacterial plasmids. "Promiscuous" plasmids such as the IncP incompatibility group, with an extraordinarily large host range, have overcome this problem through specific adaptations *(81)*. A significant barrier to homologous recombination between distantly related bacteria is the specificity of the host mismatch repair systems. The more mismatched bases in the heteroduplex formed by the recombining strands, the less efficiently recombination occurs, as was shown in *E. coli/ Salmonella* exchanges *(82)*. Thus, many recombination-mediated mechanisms will be confined within species or highly conserved DNA regions. Another barrier to uptake of parasitic DNA suggested by Lawrence et al. may be some bias toward deletion in recombination between repeat sequences *(83)*. This may explain the relatively rapid removal of pseudogenes in many organisms and why bacterial genomes do not continue to expand in size.

Restriction/modification systems are often cited as a bacterial defense strategy against bacteriophage and plasmid infection, although other roles are also plausible *(84)*. A potential problem exists in this defense strategy in that if the bacteria possess only a single restriction/modification system, escape of methylated bacteriophage could lead to elimination of the entire bacterial population. Many plasmids and bacteriophages possess systems to promote their evasion of restriction *(85)*. The genome sequence of *N. meningitidis (86)* reveals two strategies to avoid this scenario. First, over 20 restriction/modification systems are reported in the genome. In addition, 6 of these systems have associated repeat sequences subject to phase variation by slipped-strand mispairing *(84)*. These two strategies ensure that the *N. meningitidis* population is always heterogeneous for restriction/modification phenotype.

Repeats are also associated with control of restriction/modification genes in other bacteria, including phase variation control of a methyltransferase (*mod*) in *H. influenzae* Rd *(87)* and invertible element control of the restriction/modification *hsd* genes (Fig. 3). Such repeats could also possibly be involved in long-distance homologous recombination. Moreover, restriction/modification systems are also likely to play a role not only in inactivating bacteriophage DNA, but also in processing incoming DNA into fragments large enough for recombination, enabling potentially useful segments to be recombined independent of possible deleterious flanking regions *(84)*.

The integration of extraneous DNA sequences may itself be another selection for genome change. If a foreign gene provides a useful phenotype to the cell (i.e., antibiotic resistance), there may be a process of amelioration *(88)* by which there is selection for mutations that give the gene a similar codon usage bias as other efficiently expressed native open reading frames. Other pressures for amelioration of newly acquired DNA may be in avoiding target sites for restriction enzymes or adjustment in the purine content of the gene owing to its position of the sequence relative to the replication fork. In the majority of cases, it is likely that foreign DNA will not encode a function that will maintain its selection and will therefore begin to deteriorate, accumulating point mutations and mutations. IS elements may be able to prolong their survival in their new host genome even though they may not encode a selectable function or by replication through intragenomic transposition. Thus, the presence of an IS element may be the trace of an ancient LGT event, long after the plasmid, phage, or transforming DNA that originally bore the element has been lost from the genome.

INTEGRONS: PROGRAMMED EXOGENOUS GENE TRANSFER?

Could bacteria have evolved mechanisms that allow them to take specific sequences from invading DNA or allow plasmids to acquire genes from chromosomes they contact? This seems to be the raison dêtre of integron gene capture systems *(89)*. Integrons are recombination cassettes usually found on plasmids and transposons. The main portion of the integron consists of an integrase gene, the target for site-specific recombination (classically a 59-bp sequence) and promoter for expression of genes in the cassette. Genes with an adjacent 59-bp sequence can be "captured" by integrase-mediated circularization and integration events. The integrase can also work in reverse, excising gene cassettes from the integron. In this manner, naturally occurring integrons consist of a highly variable complex of genes; one of the best examples is found in the megaplasmid of the *Vibrio cholerae* genome *(90,91)*. Recent evidence also indicates that integron/gene cassettes are both widespread and diverse in the broader environment and may represent a vast metagenomic template from which LGT and thus bacterial genome evolution are facilitated *(92)*.

GENOMICS IN THE FOURTH DIMENSION

The new era of comparative sequencing will provide the static single-genome sequence with a time dimension, yielding insights into the dynamics of bacterial evolution. Key will be finding not only the mechanisms, but also the relative rates of changes that affect the whole genome. The nature of these genome dynamics will likely reflect the lifestyle and selection pressures faced by the organism. A deeper knowledge of the different rates of genome changes might also lead eventually to more sophisticated models for phylogenetic reconstruction.

Two of the different patterns that have emerged from early comparative genomics are found in obligate intracellular parasites such as *Chlamydia* and *Rickettsia* and commensal organisms such as *Neisseria* and *Enterococcus*. The obligate intracellular pathogens appear to have evolved through reductive evolution from free-living ancestors. The process of gene loss through deletions and the relative lack of acquisition of new sequences has had a tendency of removing repetitive sequences from the genomes, including IS elements *(93)*. In the case of the *Chlamydia*, this certainly may indicate near-minimal stable gene content in which insertions, inversion, and deletions almost inevitably have deleterious outcomes.

Suyama and Bork *(36)* noted that the rate of nucleotide change in intracellular pathogens appears to be higher relative to the rate of replication origin-dependent inversions in other bacteria. One possible explanation is that the DNA repair system of the bacteria is more efficient at repairing double-strand breaks at the replication fork. There appears to be little evidence for recent uptake of foreign DNA, presumably owing to the privileged intracellular niche occupied by the organisms as well as their paucity of mechanisms for LGT. However, there are "plasticity zones," located at the presumed recombination hot spot of the replication termination region, where there may be gene exchanges between chlamydiae.

In contrast, the rules for survival of the commensal bacteria produce quite different genome dynamics. These organisms exist and compete in a characteristically diverse microbial environment. In addition, they must evade the ever-changing response of the host immune system. Typically, these bacteria have more repeats, IS elements, and site-specific recombination systems. The frequency of inversions and deletions compared to point mutations is much higher than for intracellular bacteria, as is the appearance of newly acquired genes in the form of phages, plasmids, and pathogenicity islands.

Another research area for which knowledge of the mechanisms, rates, and factors affecting genome change will be extremely significant is the nascent science of genomics-based microbial forensics. It is currently possible to obtain a "draft" sequence of a microbial genome within days. The future holds the likelihood of even faster, cheaper Sanger sequencing reactions and possibly alternative technologies (such as microarray-based resequencing) playing a role in easy generation of complete genome sequence data from multiple bacterial strains from a criminal or epidemiological investigation. A forerunner of this type of project was the draft sequencing of a *B. anthracis* strain recovered from the 2001 bioterror attack in the United States *(29)*. Comparison of the bioterror Ames strain sequence to the chromosome of a laboratory Ames strain revealed 11 authentic single-nucleotide polymorphisms. It turned out that these polymorphisms were unique to the laboratory strain, which had been subjected to chemical mutagenesis and elevated temperatures to cure resident pXO1 and pXO2 virulence plasmids.

Studies involving the sequencing of multiple very closely related genomes would allow identification of lineage-specific markers that will enable strain "pedigrees" to be determined. Information about mutation rates is needed to understand the significance of the findings. For instance, what does it mean if two genomes are sequenced and it is found that they are identical at every nucleotide? What is the maximum number of generations that likely separate them? Are changes found within regions of the genome that accumulate mutations at a relatively higher rate? The changes discovered may also provide clues about the recent history of the strains. A strain grown for multiple generations in a laboratory may accumulate frameshift or nonsynonymous mutations in genes that may be essential in a natural isolate (for example, the laboratory Ames strain of *B. anthracis* has nonsynonymous mutations in two genes encoding global regulators, *relA* and *spo0A; 29*). The relative rates of different types of changes (single-nucleotide substitution, tandem duplication, transposition, etc.) may also be indicative of the environments in which the strains were grown.

With the avalanche of genomes appearing in public databases, genome dynamism will be more readily detectable. Use of these data to build accurate models of how genomes change is a challenge that will have a great impact on the biology of the future.

REFERENCES

1. Tamas I, Klasson L, Canback B, et al. Fifty million years of genomic stasis in endosymbiotic bacteria. Science 2002; 296:2376–2379.
2. Akman L, Yamashita A, Watanabe H, et al. Genome sequence of the endocellular obligate symbiont of tsetse flies, *Wigglesworthia glossinidia*. Nat Genet 2002; 32:402–407.

3. Wernegreen JJ. Genome evolution in bacterial endosymbionts of insects. Nat Rev Genet 2002; 3:850–861.

4. Tettelin H, Masignani V, Cieslewicz MJ, et al. Complete genome sequence and comparative genomic analysis of an emerging human pathogen, serotype V *Streptococcus agalactiae*. Proc Natl Acad Sci USA 2002; 99:12391–12396.

5. Tettelin H, Nelson KE, Paulsen IT, et al. Complete genome sequence of a virulent isolate of *Streptococcus pneumoniae*. Science 2001; 293:498–506.

6. Day WA Jr, Fernandez RE, Maurelli AT. Pathoadaptive mutations that enhance virulence: genetic organization of the *cadA* regions of *Shigella* spp. Infect Immun 2001; 69:7471–7480.

7. Drake JW, Charlesworth B, Charlesworth D, Crow JF. Rates of spontaneous mutation. Genetics 1998; 148:1667–1686.

8. Lobry JR. Asymmetric substitution patterns in the two DNA strands of bacteria. Mol Biol Evol 1996; 13:660–665.

9. Salzberg SL, Salzberg AJ, Kerlavage AR, Tomb JF. Skewed oligomers and origins of replication. Gene 1998; 217:57–67.

10. Mira A, Ochman H. Gene location and bacterial sequence divergence. Mol Biol Evol 2002; 19: 1350–1358.

11. Wright F, Bibb MJ. Codon usage in the G+C-rich *Streptomyces* genome. Gene 1992; 113:55–65.

12. Sharp PM, Li WH. Codon usage in regulatory genes in *Escherichia coli* does not reflect selection for "rare" codons. Nucleic Acids Res 1986; 14:7737–7749.

13. Kimura M. On the evolutionary adjustment of spontaneous mutation rates. Genet Res, 1967; 23–34.

14. Giraud A, Radman M, Matic I, Taddei F. The rise and fall of mutator bacteria. Curr Opin Microbiol 2001; 4:582–585.

15. Taddei F, Matic I, Godelle B, Radman M. To be a mutator, or how pathogenic and commensal bacteria can evolve rapidly. Trends Microbiol 1997; 5:427–428; discussion 428–429.

16. Oliver A, Canton R, Campo P, Baquero F, Blazquez J. High frequency of hypermutable *Pseudomonas aeruginosa* in cystic fibrosis lung infection. Science 2000; 288:1251–1254.

17. Taddei F, Radman M, Maynard-Smith J, Toupance B, Gouyon PH, Godelle B. Role of mutator alleles in adaptive evolution. Nature 1997; 387:700–702.

18. Denamur E, Lecointre G, Darlu P, et al. Evolutionary implications of the frequent horizontal transfer of mismatch repair genes. Cell 2000; 103:711–721.

19. Romero D, Palacios, R. Gene amplification and genomic plasticity in prokaryotes. Annu Rev Genet 1997; 31:91–111.

20. Achaz G, Rocha EP, Netter P, Coissac E. Origin and fate of repeats in bacteria. Nucleic Acids Res 2002; 30:2987–2994.

21. Levinson G, Gutman GA. Slipped-strand mispairing: a major mechanism for DNA sequence evolution. Mol Biol Evol 1987; 4:203–221.

22. Fraser CM, Gocayne JD, White O, et al. The minimal gene complement of *Mycoplasma genitalium*. Science 1995; 270:397–403.

23. Schupp JM, Klevytska AM, Zinser G, Price LB, Keim P. *vrrB*, a hypervariable open reading frame in *Bacillus anthracis*. J Bacteriol 2000; 182:3989–3997.

24. Keim P, Price LB, Klevitska AM, et al. Multiple-locus variable-number tandem repeat analysis reveals genetic relationships within *Bacillus anthracis*. J Bacteriol 2000; 182:2928–2936.

25. Parkhill J, Wren BW, Mungall K, et al. The genome sequence of the food-borne pathogen *Campylobacter jejuni* reveals hypervariable sequences. Nature 2000; 403:665–668.

26. Saunders NJ, Jeffries AC, Peden JF, et al. Repeat-associated phase variable genes in the complete genome sequence of *Neisseria meningitidis* strain MC58. Mol Microbiol 2000; 37:207–215.

27. van Ham SM, van Alphen L, Mooi FR, van Putten JP. The fimbrial gene cluster of *Haemophilus influenzae* type b. Mol Microbiol 1994; 13:673–684.

28. Michel B. Illegitimate recombination in bacteria. In: Charlebois R (ed). Organization of the Prokaryotic Genome. Washington, DC: ASM, 1999, pp. 129–150.

29. Read TD, Salzberg SL, Pop, M, et al. Comparative genome sequencing for discovery of novel polymorphisms in *Bacillus anthracis*. Science 2002; 296:2028–2033.

30. Conley EC, Saunders VA, Saunders JR. Deletion and rearrangement of plasmid DNA during transformation of *Escherichia coli* with linear plasmid molecules. Nucleic Acids Res 1986; 14: 8905–8917.

31. Haber JE. Mating-type gene switching in *Saccharomyces cerevisiae*. Annu Rev Genet 1998; 32: 561–599.

32. Pride DT, Blaser MJ. Concerted evolution between duplicated genetic elements in *Helicobacter pylori*. J Mol Biol 2002; 316:629–642.

33. Hashimoto JG, Stevenson BS, Schmidt TM. Rates and consequences of recombination between rRNA operons. J Bacteriol 2003; 185:966–972.

34. Howell-Adams B, Seifert HS. Molecular models accounting for the gene conversion reactions mediating gonococcal pilin antigenic variation. Mol Microbiol 2000; 37:1146–1158.

35. Eisen JA, Heidelberg JF, White O, Salzberg SL. Evidence for symmetric chromosomal inversions around the replication origin in bacteria. Genome Biol 2000; 1:1–9.

36. Suyama M, Bork P. Evolution of prokaryotic gene order: genome rearrangements in closely related species. Trends Genet 2001; 17:10–13.

37. Tillier ER, Collins RA. Genome rearrangement by replication-directed translocation. Nat Genet 2000; 26:195–197.

38. Read TD, Myers GS, Brunham RC, et al. Genome sequence of *Chlamydophila caviae* (*Chlamydia psittaci* GPIC): examining the role of niche-specific genes in the evolution of the Chlamydiaceae. Nucleic Acids Res 2003; 31:2134–2147.

39. Read TD, Peterson SN, Tourasse N, et al. The genome sequence of *Bacillus anthracis* Ames and comparison to closely related bacteria. Nature 2003; 432:81–86.

40. Salama N, Guillemin K, McDaniel TK, Sherlock G, Tompkins L, Falkow S. A whole-genome microarray reveals genetic diversity among *Helicobacter pylori* strains. Proc Natl Acad Sci USA 2000; 97:14668–14673.

41. Lecompte O, Ripp R, Puzos-Barbe V, et al. Genome evolution at the genus level: comparison of three complete genomes of hyperthermophilic archaea. Genome Res 2001; 11:981–993.

42. Paulsen IT, Banerjei L, Myers GS, et al. Role of mobile DNA in the evolution of vancomycin-resistant *Enterococcus faecalis*. Science 2003; 299:2071–2074.

43. Gumulak-Smith J, Teachman A, Tu AH, Simecka JW, Lindsey JR, Dybvig K. Variations in the surface proteins and restriction enzyme systems of *Mycoplasma pulmonis* in the respiratory tract of infected rats. Mol Microbiol 2001; 40:1037–1044.

44. Sitaraman R, Denison AM, Dybvig K. A unique, bifunctional site-specific DNA recombinase from *Mycoplasma pulmonis*. Mol Microbiol 2002; 46:1033–1040.

45. Tobiason DM, Lenich AG, Glasgow AC. Multiple DNA binding activities of the novel site-specific recombinase, Piv, from *Moraxella lacunata*. J Biol Chem 1999; 274:9698–9706.

46. Krinos CM, Coyne MJ, Weinacht KG, Tzianalbos AO, Kasper DL, Conestack LE. Extensive surface diversity of a commensal microorganism by multiple DNA inversions. Nature 2001; 414:555–558.

47. Hood DW, Deadman ME, Jennings MP, et al. DNA repeats identify novel virulence genes in *Haemophilus influenzae*. Proc Natl Acad Sci USA 1996; 93:11,121–11,125.

48. Sato T, Harada K, Kobayashi Y. Analysis of suppressor mutations of *spoIVCA* mutations: occurrence of DNA rearrangement in the absence of site-specific DNA recombinase SpoIVCA in *Bacillus subtilis*. J Bacteriol 1996; 178:3380–3383.

49. Chandler M, Mahillon J. Insertion sequence revisited. In: Lambowitz AM (ed). Mobile DNA II. Washington, DC: ASM, 2002, pp. 305–366.

50. Chen Y, Braathen P, Leonard C, Mahillon J. MIC231, a naturally occurring mobile insertion cassette from *Bacillus cereus*. Mol Microbiol 1999; 32:657–668.

51. Lawrence JG, Ochman H, Hartl DL. The evolution of insertion sequences within enteric bacteria. Genetics 1992; 131:9–20.

52. Peters JE, Craig NL. Tn7: smarter than we thought. Nat Rev Mol Cell Biol 2001; 2:806–814.

53. Olasz F, Kiss J, Konig P, Buzas Z, Stalder R, Arber W. Target specificity of insertion element IS30. Mol Microbiol 1998; 28:691–704.

54. Zerbib D, Gamas P, Chandler M, Prentki B, Bass S, Galas D. Specificity of insertion of IS1. J Mol Biol 1985; 185:517–524.

55. Belfort M, Reaban ME, Coetzee T, Dalgaard JZ. Prokaryotic introns and inteins: a panoply of form and function. J Bacteriol 1995; 177:3897–3903.

56. Lampson B, Inouye M, Inouye S. The msDNAs of bacteria. Prog Nucleic Acid Res Mol Biol 2001; 67:65–91.

57. Liu SV, Saunders NJ, Jeffries A, Rest RF. Genome analysis and strain comparison of correia repeats and correia repeat-enclosed elements in pathogenic *Neisseria*. J Bacteriol 2002; 184: 6163–6173.

58. Correia FF, Inouye S, Inouye M. A 26-base-pair repetitive sequence specific for *Neisseria gonorrhoeae* and *Neisseria meningitidis* genomic DNA. J Bacteriol 1986; 167:1009–1015.

59. Meyer BJ, Schottel JL. Characterization of *cat* messenger RNA decay suggests that turnover occurs by endonucleolytic cleavage in a 3' to 5' direction. Mol Microbiol 6, 1095–1104 (1992).

60. Ochman H, Lawrence JG, Groisman EA. Lateral gene transfer and the nature of bacterial innovation. Nature 2000; 405:299–304.

61. Avery OT, MacLeod CM, McCarty M. Studies on the chemical nature of the substance inducing transformation of pneumococcal types—induction of transformation by a deoxyribonucleic-acid fraction isolated from pneumococcus type-III. J Exp Med 1944; 79:137–158.

62. Lorenz MG, Wackernagel W. Bacterial gene transfer by natural genetic transformation in the environment. Microbiol Rev 1994; 58:563–602.

63. Dubnau D. DNA uptake in bacteria. Annu Rev Microbiol 1999; 53:217–244.

64. Vazquez JA, Berron S, O'Rourke M, et al. Interspecies recombination in nature: a meningococcus that has acquired a gonococcal PIB porin. Mol Microbiol 1995; 15:1001–1007.

65. Suerbaum S, Achtman M. Evolution of *Helicobacter pylori*: the role of recombination. Trends Microbiol 1999; 7:182–184.

66. Feil EJ, Maiden MC, Achtman M, Spratt BG. The relative contributions of recombination and mutation to the divergence of clones of *Neisseria meningitidis*. Mol Biol Evol 1999; 16: 1496–1502.

67. Meats E, Feil EJ, Stringer S, et al. Characterization of encapsulated and noncapsulated *Haemophilus influenzae* and determination of phylogenetic relationships by multilocus sequence typing. J Clin Microbiol 2003; 41:1623–1636.

68. Mandel M, Higa A. Calcium-dependent bacteriophage DNA infection. J Mol Biol 1970; 53: 159–162.

69. Zechner EL, de la Cruz F, Eisenbrandt R, et al. Conjugative-DNA transfer processes. In: Thomas CM (ed). The Horizonal Gene Pool. Amsterdam: Harwood, 2000, pp. 87–174.

70. Bates S, Cashmore AM, Wilkins BM. IncP plasmids are unusually effective in mediating conjugation of *Escherichia coli* and *Saccharomyces cerevisiae*: involvement of the *tra2* mating system. J Bacteriol 1998; 180:6538–6543.

71. Szpirer C, Top E, Couturier M, Mergeay M. Retrotransfer or gene capture: a feature of conjugative plasmids, with ecological and evolutionary significance. Microbiology 1999; 145:3321–3329.

72. Christie PJ, Vogel JP. Bacterial type IV secretion: conjugation systems adapted to deliver effector molecules to host cells. Trends Microbiol 2000; 8:354–360.

73. Clewell DB. Bacterial sex pheromone-induced plasmid transfer. Cell 1993; 73:9–12.
74. Salyers AA, Shoemaker NB, Stevens AM, Li LY. Conjugative transposons: an unusual and diverse set of integrated gene transfer elements. Microbial Rev 1995; 59:579–590.
75. Osborn AM, Boltner D. When phage, plasmids, and transposons collide: genomic islands, and conjugative- and mobilizable-transposons as a mosaic continuum. Plasmid 2002; 48:202–212.
76. Banks DJ, Beres SB, Musser JM. The fundamental contribution of phages to GAS evolution, genome diversification and strain emergence. Trends Microbiol 2002; 10:515–521.
77. Waldor MK, Mekalanos JJ. Lysogenic conversion by a filamentous phage encoding cholera toxin. Science 1996; 272:1910–1914.
78. Fujii N, Oguma K, Yokosawa N, Kimura K, Tsuzuki K. Characterization of bacteriophage nucleic acids obtained from *Clostridium botulinum* types C and D. Appl Environ Microbiol 1988; 54:69–73.
79. Hacker J, Blum-Oehler G, Muhldorfer I, Tschape H. Pathogenicity islands of virulent bacteria: structure, function and impact on microbial evolution. Mol Microbiol 1997; 23:1089–1097.
80. Osborn AM, da Silva Tatley FM, Steyn LM, Pickup RW, Saunders JR. Mosaic plasmids and mosaic replicons: evolutionary lessons from the analysis of genetic diversity in IncFII-related replicons. Microbiology 2000; 146:2267–2275.
81. Perlin MH. The subcellular entities a.k.a. plasmids. In: Yasbin RE (ed). Modern Microbial Genetics. New York: Wiley-Liss, 2002, pp. 507–560.
82. Matic I, Rayssiguier C, Radman M. Interspecies gene exchange in bacteria: the role of SOS and mismatch repair systems in evolution of species. Cell 1995; 80:507–515.
83. Lawrence JG, Hendrix RW, Casjens S. Where are the pseudogenes in bacterial genomes? Trends Microbiol 2001; 9:535–540.
84. Blumenthal RM, Cheng X. Restriction-modification systems. In: Yasbin RE (ed). Modern Microbial Genetics. New York: Wiley-Liss, 2002, pp. 177–225.
85. Wilkins BM. Plasmid promiscuity: meeting the challenge of DNA immigration control. Environ Microbiol 2002; 4:495–500.
86. Tettelin H, Saunders NJ, Heidelberg J, et al. Complete genome sequence of *Neisseria meningitidis* serogroup B strain MC58. Science 2000; 287:1809–1815.
87. De Bolle X, Bayliss CD, Field D, et al. The length of a tetranucleotide repeat tract in *Haemophilus influenzae* determines the phase variation rate of a gene with homology to type III DNA methyltransferases. Mol Microbiol 2000; 35:211–222.
88. Lawrence JG, Ochman H. Amelioration of bacterial genomes: rates of change and exchange. J Mol Evol 1997; 44:383–397.
89. Hall RM, Collis CM. Mobile gene cassettes and integrons: capture and spread of genes by site-specific recombination. Mol Microbiol 1995; 15:593–600.
90. Clark CA, Purins L, Kaewrakon P, Manning PA. VCR repetitive sequence elements in the *Vibrio cholerae* chromosome constitute a mega-integron. Mol Microbiol 1997; 26:1137–1138.
91. Heidelberg JF, Eisen JA, Nelson WC, et al. DNA sequence of both chromosomes of the cholera pathogen *Vibrio cholerae*. Nature 2000; 406:477–483.
92. Holmes AJ, Gillings MR, Nield BS, Mabbutt BC, Nevalainen KM, Stokes HW. The gene cassette metagenome is a basic resource for bacterial genome evolution. Environ Microbiol 2003; 5:383–394.
93. Mira A, Ochman H, Moran NA. Deletional bias and the evolution of bacterial genomes. Trends Genet 2001; 17:589–596.

11

Concepts of Bacterial
Biodiversity for the Age of Genomics

Frederick M. Cohan

INTRODUCTION

A full accounting of ecological diversity in the bacterial world will no doubt require genome-based and sequence-based approaches. Because only a small fraction of bacterial species is currently cultivable, the best hope of identifying the full complement of bacterial biodiversity is from sequence surveys of genes that can be amplified from natural bacterial habitats (1,2). In addition, genome- and sequence-based approaches are uncovering ecological diversity even within the most familiar (and cultivated) species: Ecologically distinct populations within named species are being discovered as sequence clusters even when there is ignorance of their ecology (3–5); populations are also being discovered as clusters based on the content of their genomes (6,7). Beyond discovery of ecologically distinct populations, genomic approaches promise to elucidate the ecological functioning of each community member (8,9) and how the various populations partition the environment and manage to coexist. The role that horizontal transfer has played in allowing bacteria to occupy new ecological niches (10) can be discovered. Further, the donors of horizontal transfer events (11) can be identified, and the genetic and ecological barriers to gene transfer can be discovered (12).

For each of these goals now made accessible by the genomic revolution, it is critical to identify ecologically distinct populations of strains and to recognize which strains are ecologically interchangeable and thus members of the same population. For example, consider future investigations into the role of horizontal transfer in fostering invasion of new ecological niches. A difference in the genes present in two ecologically distinct populations could represent a horizontal transfer event responsible for the populations' ecological divergence. On the other hand, horizontal transfer events that distinguish two ecologically interchangeable strains of the same population would be ecologically meaningless craters on the chromosome brought about by the meteors of horizontal transfer (13). Here, I describe recent approaches for discovering and classifying the ecological diversity of bacteria, with the aim of providing a sound ecological basis for studies of genome content and expression.

We may first turn to bacterial systematics for guidance in classifying strains into ecologically distinct groups. In the systematics of bacteria, organisms fall into clusters of

From: *Microbial Genomes*
Edited by: C. M. Fraser, T. D. Read, and K. E. Nelson © Humana Press Inc., Totowa, NJ

phenotypically, genetically, and ecologically similar organisms, with large gaps between clusters *(14–16)*. In bacteria, as in other organisms, these clusters are the recognized, named species. Bacterial species were originally discerned as phenotypic clusters, usually based on presence vs absence of metabolic capabilities (see Chapter 9) *(4,16)*. Later, as molecular techniques became available, they have been incorporated into species demarcation by calibrating each to provide the familiar clusters yielded by phenotypic analyses *(4)*. For example, whole genome deoxyribonucleic acid (DNA)–DNA hybridization has become a principal criterion for distinguishing species; this is based on the observation that groups whose genomes anneal for 70% or more of the chromosome correspond to the phenotypic clusters of yore *(17,18)*. More recently, 16S ribosomal RNA (rRNA) sequence similarity has demarcated species; this is based on the observation that 16S rRNA sequence divergence greater than 2.5% usually corresponds to different species, although some distinct species show less than this divergence *(19)*.

Do these named bacterial species, discernible by phenotypic and whole genome and sequence similarity, correspond to ecologically distinct groups, and do they correspond to anything resembling species for other kinds of organisms (e.g., animals and plants)? These are the critical questions as we attempt to use sequence and genomic data to glean information about biodiversity and the ecological functioning of a microbial community.

In the world of frequently sexual higher eukaryotes, species are understood to have special dynamic properties that guarantee a high degree of homogeneity within a species, as well as heterogeneity between species *(20)*. Most fundamentally, species have the property that genetic diversity within a species is constrained by a force of cohesion *(4,21,22)*. In the case of the highly sexual eukaryotes, the ability to exchange genes is understood as the primary force impeding divergence *(23,24)*. A second universal property is that different species are irreversibly separate; once populations reach a critical threshold of divergence, they become free to diverge without bound *(25,26)*. In the case of the highly sexual eukaryotes, this threshold is most likely the point of reproductive isolation *(23)*. Finally, while members of a single species are ecologically interchangeable, different species can coexist by partitioning resources as well as the conditions at which they thrive *(27)*. As discussed here, these dynamic properties of species apply beyond the plants and animals to groups with peculiar sexual characteristics, including the bacteria.

There is a growing consensus among microbial evolutionary geneticists that the named species of bacterial systematics do not exhibit the special dynamic properties of species. First, bacterial systematics has demonstrated a great deal of metabolic and presumably ecological diversity within a typical species *(13,28,29)*. Second, for decades DNA–DNA hybridization studies in systematics have revealed an enormous level of genomic diversity within named species *(30)*, which has recently been corroborated by genomic sequencing of two or more strains from each of several species *(31–36)*. Third, multilocus sequence typing (MLST) has revealed the existence of multiple sequence clusters within named species *(3,37,38)*, and these are likely to correspond to ecologically distinct populations *(5,37,39)*. Finally, natural history studies of ecological and sequence variation in the environment have revealed the existence of multiple ecologically distinct populations that are similar enough in sequence to be subsumed within a single named species *(40–43)*.

It is thus clear that an inventory of ecological diversity in the bacterial world must go beyond counting named species. There is less consensus, however, as to whether each ecologically distinct group within a named species has the dynamic properties ascribed to species. Here, I present several contrasting, contemporary views on the nature of biological diversity: the biological species concept as applied to bacteria *(39,44,45)*, the ecotype concept *(4,46)*, and the species-less concept of bacterial diversity *(12)*. Because these concepts begin with somewhat different assumptions about the nature of genetic exchange in bacteria, I review the properties of bacterial genetic exchange and the consequences of different rates of genetic exchange on bacterial population dynamics.

THE CHARACTER OF GENETIC EXCHANGE IN BACTERIA

The Rarity of Genetic Exchange

Bacteria can reproduce clonally for an indefinite number of generations, with pure clonality punctuated occasionally by recombination events, when a short segment from a donor replaces the homologous segment in a recipient. The rates of bacterial recombination in nature have been estimated by "retrospective" approaches utilizing surveys of sequence or allozyme variation in natural populations. The rationale is that low recombination rates can be inferred when alleles at different loci show high degrees of association among individuals (i.e., linkage disequilibrium) or when different DNA segments yield congruent phylogenies *(47)*.

These retrospective approaches have shown that recombination in most bacterial species occurs in a gene segment at about the rate of mutation or somewhat higher *(48–50)*. For example, *Staphylococcus aureus* is among the most clonal of bacterial taxa, and a gene segment undergoes recombination at a rate three times lower than mutation *(51, 52)*; *Neisseria meningitidis* is one of the most frequently recombining bacterial taxa, and here a gene segment undergoes recombination about 3.6 times more frequently than by mutation *(37)*.

Because a given recombination event effects many more nucleotides than does a point mutation, the rate at which recombination affects a given nucleotide can be up to 80 times greater than the rate of mutation *(37)*. This result has fueled the notion that recombination is not really rare in bacteria, and that models of bacterial evolution that depend on rare recombination are not valid *(12,45)*. However, recent sequence-based estimates of recombination yield essentially the same recombination rates per gene segment obtained by earlier allozyme-based approaches (i.e., with recombination occurring at a rate less than 10 times that of mutation). As discussed here, this rarity of recombination allows natural selection to have a profound effect on genetic diversity within a bacterial population.

Promiscuity of Genetic Exchange

While bacteria recombine only rarely, they are not fussy about their choice of partners in genetic exchange. Bacteria can undergo homologous recombination with organisms that differ by as much as 25% in their DNA sequence *(53–56)*.

There are, nevertheless, some important constraints on genetic exchange between divergent bacteria, including the requirement that recipient and donor share vectors of recombination (for phage- and plasmid-mediated recombination) and microhabitats *(57,58)*. Also, homologous recombination is limited by molecular constraints on integration of divergent donor sequences. Recombination requires the ends of the donor segment to match the recipient's homolog nearly perfectly, and this is unlikely when the donor and recipient are highly divergent *(54,59,60)*. In addition, mismatch repair systems tend to reverse integration when they detect nucleotide mismatches between recipient and donor *(55,61)*; mismatch repair has been shown to play a major part in sexual isolation in the Enterobacteriaceae *(55,61)*, although not in Gram-positive bacteria *(53,59)*.

Recombination in bacteria is not limited to the transfer of homologous segments. In heterologous recombination, bacteria can "capture" new gene loci and gene operons from other organisms, sometimes from organisms that are extremely distantly related *(62)*. Genomic analyses have recently shown that a sizable fraction (frequently 5–10%) of genomes of bacterial species has typically been acquired from other species *(63)*.

While horizontal transfer has clearly made a substantial mark on bacterial genomes, it should not necessarily be concluded that horizontal transfer occurs at a high rate, especially on a per capita (i.e., per individual cell) basis. Lawrence *(45)* estimated that successful horizontal transfer events have occurred in *Escherichia coli* at a rate of 6 to 7 per million years. Assuming even a modest population size of 10^{12}, this leads to a per capita, per generation rate of successful horizontal transfer events of 7×10^{-22} (assuming 1 division per hour) *(64)*. Even if horizontal transfers succeed at a rate as low as 1 in 1 million, the per capita rate of all horizontal transfer events (whether successful or unsuccessful) would be 7×10^{-16}.

EVOLUTIONARY CONSEQUENCES OF RARE, BUT PROMISCUOUS, GENETIC EXCHANGE

Introduction of Adaptive DNA from Other Taxa

Genetic exchange in bacteria is clearly frequent enough to effect transfer of adaptive alleles from one species to another. For example, antibiotic resistance alleles have spread across species of *Neisseria* and *Streptococcus*, replacing antibiotic-sensitive alleles through homologous recombination *(65)*. Also, heterologous recombination has been frequent enough to have introduced hundreds of gene loci into a typical bacterial genome *(63)*. Interspecific transfer of adaptive DNA, whether an allele or a novel operon, is the least challenging feat for recombination in that extremely low rates of recombination can accomplish it. As calculated in the preceding section, the history of horizontal transfer in *E. coli* could have been accomplished by a per capita recombination rate of 7×10^{-16}, even if only 1 in 1 million horizontal transfer events were adaptive. Introduction of adaptive DNA requires so little recombination simply because only a single recombination event is required to introduce an adaptive gene into a lineage; once the gene is present, natural selection can increase the abundance of the newly adaptive genotype. Moreover, if an acquired gene allows invasion of a new niche, the recombinant genotype is especially likely to be successful *(12,66)*.

Effects of Recombination on Neutral Sequence Diversity

Maiden and coworkers developed MLST, a sequence-based system for classifying strains into clusters called *clonal complexes (3)*. MLST is based on sequencing genes for seven housekeeping proteins whose diversity is assumed to be neutral in fitness; also, the proteins are assumed not to be involved in niche-specific adaptations and should be interchangeable between ecologically distinct populations. Therefore, MLST data can be used to view empirically the effect of recombination on neutral sequence diversity.

MLST has shown recombination to impact the sequence-based phylogeny of strains. For example, the sequence-based phylogeny for strains of *N. meningitidis* varies depending on the gene *(67)*: There is no one organismal phylogeny, but each strain is truly a composite of genes from throughout the named species and from outside as well *(12)*.

As I discussed, MLST also shows recombination has greater impact on neutral sequence diversity than mutation, owing to recombination occurring at about the rate of mutation per gene, with each recombination event involving hundreds of nucleotides or more.

There is, however, one realm in which recombination does not impact neutral sequence diversity. This is the ability to use MLST to classify strains into clonal complexes *(3)*. Here, the evolutionary distance between strains is quantified as the number of loci that are different, with two strains scored as different for a locus whether they differ by one nucleotide substitution or by scores of nucleotides (possibly because of a recombination event). Strains are then classified into clonal complexes: All strains that are identical to a particular central strain at five or more (in some cases, six or more) of the seven loci are deemed members of a clonal complex.

In development of the MLST approach, Maiden and coworkers found that the clonal complexes obtained with 7 loci were the same as obtained with up to 11 loci, and the subset of 7 loci chosen did not affect the classification of strains into complexes *(3)*. While a minority of loci may be recombinant in any given strain, the complexes appear to be a robust signal classifying the diversity of strains. Even *N. meningitidis*, the most frequently recombining of the species studied, yields robust clonal complexes. As will be discussed the MLST clonal complexes empirically correspond to ecologically distinct groups, and there is a theoretically compelling reason for this correspondence.

Effect of Recombination on Adaptive
Divergence Between Ecologically Distinct Populations

In the highly sexual world of animals and plants, divergence between two closely related populations requires sexual isolation between the populations *(23)*. If recombination between animal or plant populations were to proceed at the same rate as within populations, recombination would rapidly eliminate any adaptive divergence between populations.

In contrast, because recombination in bacteria is so rare, recurrent recombination between bacterial species cannot hinder their divergence *(46)*. Suppose, for example, that two populations are ecologically specialized on different substrates, and that alleles at several loci are responsible for the adaptive divergence. This model implies a cost of recombination: If the multilocus genotype *ABCDE* confers one population's adaptations, and genotype *abcde* confers the other population's adaptation, then the fitness of a recombinant genotype at these critical niche-determining loci (e.g., *Abcde*) would be reduced to $1 - s$ (where s is the intensity of selection against recombinants).

I have previously shown that the equilibrium frequency of a maladaptive foreign allele is c_b/s, where c_b is the rate of recombination between populations *(46)*. Given that the rate of recombination within bacterial populations is already quite rare ($c_w \approx 10^{-6}$, per gene segment), the frequency of maladaptive foreign alleles is expected to be negligible, even if the rate of recombination between populations were as high as recombination within populations. (A similar argument applies if the basis of ecological divergence is the presence vs absence of horizontally transferred gene loci.) Thus, while the evolution of sexual isolation is an important milestone in the origin of animal and plant species, it is irrelevant to the evolution of adaptive divergence in the bacterial world (4,46).

Effect of Recombination on Diversity Within a Population

It has long been understood that natural selection will purge all genetic diversity from an entirely asexual population in a process known as *periodic selection (68)*. In the absence of recombination, any adaptive mutant and its clonal descendants eventually replace the other cells of the population, thus extinguishing genetic diversity at all loci.

Frequent recombination can clearly quash the diversity-purging effect of periodic selection. If the adaptive mutation recombines into another genetic background, then the entire genome of the recipient is saved from extinction; alternatively, if a segment from a strain lacking the adaptive mutation should recombine into a strain with the adaptive mutation, then that segment (only) will be saved from extinction. This quashing of periodic selection is the most difficult challenge for recombination: Extremely frequent recombination, with recombination nearly obligately tied to sex, is required to prevent the purging of diversity (69,70) (Fig. 1). When the intensity of periodic selection is strong (i.e., fitness advantage for the adaptive mutation is 10%), each bout of periodic selection purges nearly all diversity within a population. Over recombination rates typically observed in nature (from 0.3 to 3.6 times the mutation rate), periodic selection purges all but 0.001–0.2% of the sequence diversity (Fig. 1). Over more modest selection intensity (i.e., fitness advantage of 1%), periodic selection purges all but 0.02–2% of sequence diversity over naturally occurring recombination rates. Thus, even if recombination rates in bacteria were orders of magnitude greater than estimated, recombination would be ineffective in quashing the diversity-purging effects of periodic selection.

Lawrence (45) argued that periodic selection does not have a significant role in reducing genetic diversity in bacterial populations. His argument is based in part on a sequence survey by Guttman and Dykhuizen that demonstrated a periodic selection event in *E. coli (71)*. Guttman and Dykhuizen demonstrated that a segment of at least 30 kb near *gapA* was anomalously homogeneous throughout *E. coli*, whereas the rest of the chromosome showed much greater diversity. Lawrence claimed that this observation proves periodic selection has an insignificant effect on bacterial diversity genomewide, much like a selective sweep within an animal population, owing to rampant recombination.

Majewski and I previously pointed out that Guttman and Dykhuizen's (71) original interpretation of the periodic selection is flawed because it assumes that one adaptive mutant (or recombinant) would out-compete all other variants in *E. coli (72)*. This is clearly impossible owing to what both bacterial systematists (16) and evolutionary geneticists (12,73,74) understand about the tremendous ecological diversity in *E. coli* or in most other named species: Because a named species contains strains adapted to distinct niches, one adaptive mutant could not out-compete the rest of the species' diversity. Also, as noted, theory shows that periodic selection would purge nearly all of a bacterial

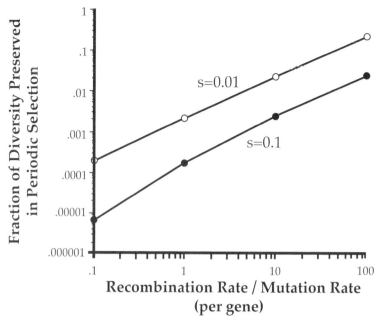

Fig. 1. The relationship between recombination rate and the diversity-purging effect of periodic selection over different intensities of selection *s* favoring the adaptive mutation. The ratios of recombination rate to mutation rate seen in the figure reflect the range of ratios typically observed in nature. The ordinate is based on the mean fraction of a cell's genome at the end of periodic selection that is not descended from the genome of the original mutant. These results are based on a Monte Carlo simulation. (Used with permission from ref. *70*, Landes Bioscience.)

population's diversity, even if recombination was substantially more frequent than typically observed *(69,70)* (Fig. 1). The correct interpretation of Guttman and Dykhuizen's *(71)* data requires a more nuanced view of ecological diversity in bacteria, as I describe next.

MODELS OF ECOLOGICAL DIVERSITY IN BACTERIA

Bacterial systematics has largely been based on demarcation of species as phenotypic and genetic clusters *(16)*, but the last decade has seen the introduction of several concepts of bacterial diversity based on the dynamics of bacterial evolution.

Species Concepts Based on Genetic Exchange and Sexual Isolation

The biological species concept of Mayr *(23,24)* may be credited for infusing evolutionary theory into systematics. The biological species concept changed zoologists' and botanists' views of what a species should represent: A species is not merely a cluster of similar organisms, but rather a fundamental unit of ecology and evolution, with certain dynamic properties. In the case of Mayr's biological species concept, a species is viewed as a group of organisms whose divergence is opposed by recombination between them *(21–23)*.

Dykhuizen and Green *(39)* proposed classifying bacteria into species according to the biological species concept, demarcating bacterial species as groups of strains that recombine within the group, but not with strains from other such groups. They suggested a phylogenetic approach to demarcation; the phylogeny of members of a single species would be expected to vary from gene to gene, while the phylogeny of members of different species would be congruent across genes.

This approach has been criticized for not taking into consideration the intrinsic promiscuity of genetic exchange in bacteria *(4,12,46)*: Bacteria do exchange genes both within and between the clusters recognized as named species *(12,65,75–77)*. Generally, phylogenetic evidence for recombination is more common within named bacterial species than between them, but only if the donor species are not included in the phylogeny *(12)*. When donor species are included, horizontally transferred alleles become evident in a phylogeny of multiple species.

Two recent concepts of bacterial diversity explain the evolutionary origin of sexual isolation in bacteria. In Lawrence's *(45)* "speciation without species" model begins with two populations that are ecologically distinct owing to modest differences in their sets of acquired genes. Any interpopulation genetic exchange that includes a gene involved in the populations' adaptive divergence is strongly selected against. Thus, regions of the chromosome near these genes are protected from the homogenizing effects of genetic exchange, and neutral sequence divergence will be allowed to accumulate in these regions. This in turn reduces the efficiency of successful recombination in these regions *(44,54)*. As each population evolves further into its respective niche, more genetic changes accumulate throughout the chromosome, and each such change protects its neighboring region of the chromosome. Eventually, the entire chromosome is protected from recombination, and the consequent sequence divergence impedes genetic exchange anywhere on the chromosome.

I believe that Lawrence's model is correct in that any gene responsible for adaptive divergence will lead to lower successful genetic exchange in the flanking region and will thereby accelerate neutral sequence divergence. However, it is not clear that this local prevention of genetic exchange is necessary for eventual sequence divergence between ecologically distinct populations. This is because neutral sequence divergence between populations is intrinsically a self-accelerating process in which any random neutral sequence divergence (in any part of the genome, whether "protected" from recombination or not) tends to increase sexual isolation in that region, and this increased sexual isolation results in lower recombination, which further increases sequence divergence, and so on. I have previously shown that the positive feedback between sequence divergence and sexual isolation in *Bacillus* results eventually in unbounded neutral sequence divergence between ecologically distinct populations *(78)*. Nevertheless, I agree with Lawrence *(45)* that the accumulation of niche-specific genes throughout the chromosome will speed up the process.

The "molecular keys to speciation" concept notes that the adaptive changes allowing a niche invasion can be promoted by modulation of the mismatch repair system *(44,55)*. Because mismatch repair is the primary agent of sexual isolation in some bacteria (e.g., the Enterobacteriaceae) *(55)*, a debilitated mismatch repair system more readily allows uptake of potentially adaptive genes from other species. However, once a population becomes adapted to a new environmental challenge, mismatch repair deficiencies can be disadvantageous because they yield high mutation rates *(79)*. The population may then regain a functional mismatch repair system, often by horizontal transfer *(80)*. In the time that an incipient species is lacking a fully functional mismatch repair system, it rapidly acquires sequence divergence from its parental population, and this acquired sequence divergence contributes to sexual isolation when mismatch repair is reinstated *(44)*.

The biological species concept motivates these dynamic models for the evolution of sexual isolation, as well as Dykhuizen and Green's *(39)* plan to classify strains by their recombination history: If groups of bacteria that recombine at a high rate can be identified, then species can be identified; if the accumulation of sexual isolation between populations can be understood, then the origins of species can be understood. However, as I have discussed here, recurrent recombination between bacterial species cannot hinder their divergence *(12,46)*. At the moment that two populations acquire ecologically distinct traits, there is already too little recombination between them to threaten the integrity of their separate adaptations. Moreover, recombination cannot prevent the accumulation of further niche-specific adaptations in each. The evolution of sexual isolation is irrelevant to the evolution of permanent divergence in the bacterial world, so the biological species concept is a red herring for bacterial systematics *(4)*.

While accurately predicting the course of increase in sexual isolation, the models of Lawrence *(45)* and Vulic et al. *(44)* are not necessary for understanding the origin of bacterial species. As I discuss next, the quintessential step in the origin of bacterial species is the ecological divergence of populations.

Species Concept Based on Periodic Selection

I have previously defined a bacterial "ecotype" with respect to the fate of an adaptive mutant (or recombinant): An *ecotype* is a set of strains using the same or similar ecological niche, such that an adaptive mutant from within the ecotype out-competes to extinction all other strains from the ecotype; an adaptive mutant cannot, however, drive to extinction strains from other ecotypes *(4,46,81)* (Fig. 2). Thus, an ecotype is the set of strains whose diversity is purged through periodic selection favoring each adaptive mutant. Periodic selection is expected to be a powerful force of cohesion within a bacterial ecotype in that it recurrently resets the genetic diversity to near zero.

When two populations become ecologically distinct, they may each undergo their own private periodic selection events. At this point, natural selection favoring an adaptive mutant purges the diversity only within the mutant's own population. This is a milestone toward forming new species; such populations are now irreversibly separate because periodic selection cannot prevent further divergence and, as I have explained, neither can recombination *(46)*.

Bacterial ecotypes, as defined by the domains of periodic selection, are expected to share the fundamental properties of species: Ecotypes are each subject to an intense force of cohesion, periodic selection, which recurrently purges diversity within an ecotype; divergence between ecotypes is irreversible; ecotypes are expected to form distinct phenotype- and sequence-based clusters (as discussed in the next section on predictions); and bacterial ecotypes are ecologically distinct *(4,81)*. According to the ecotype concept, a true species in the bacterial world may be understood as an evolutionary lineage bound together by ecotype-specific periodic selection *(4)*.

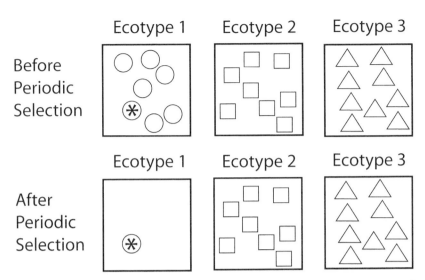

Fig. 2. Periodic selection purges the diversity within, but not between, ecotypes. Each individual symbol represents an individual organism, and the distance between symbols represents the sequence divergence between organisms. The asterisk represents an adaptive mutation in one individual of ecotype 1. Once populations are divergent enough to escape one another's periodic selection events (as separate ecotypes), those populations are free to diverge permanently. (Used with permission from ref. *81a.*)

Bacterial Diversity Without Species

Finally, in one view, bacterial taxa with the dynamic attributes of species may not exist *(12)*. This model assumes that recombination is too frequent to allow purging of diversity by periodic selection, such that there is no force of cohesion within a bacterial population. Without cohesion, there are no species *(4,12,22)*. Gogarten et al. *(12)* suggest that, owing to frequent recombination, sequence data will not be a reliable marker for ecologically distinct populations, and sequence data will not provide a true indication of the phylogenetic relationships among such populations.

PREDICTIONS OF THE ECOTYPE AND SPECIES-LESS CONCEPTS OF DIVERSITY

Predictions on Correspondence Between Ecotypes and Sequence Clusters

The species-less concept makes no specific predictions about the correspondence between sequence diversity and ecologically distinct populations, only that recombination will make sequence diversity an unreliable indicator of an organism's history and ecological characteristics *(12)*. In contrast, the ecotype concept provides a rationale for sequence-based discovery of bacterial diversity *(4,5)*. Because periodic selection purges diversity within, but not between, ecotypes, each bacterial ecotype is expected eventually to be identifiable as a sequence cluster, distinct from all closely related ecotypes. Palys et al. *(5)* showed that, under the recombination rates typical of bacteria, the average sequence divergence between ecotypes should greatly exceed that within ecotypes. Nevertheless, recombination may result in misdiagnosis of an occasional strain to a donor's ecotype, especially if strain diagnosis is based on the sequence of a single gene.

The MLST approach of Maiden and coworkers *(3)* appears to yield a more robust approach to ecotype discovery and classification than the method of Palys et al. *(5)*, and as will be discussed here, it also is based on the diversity-purging effects of periodic selection. These clonal complexes generated by MLST are robust with respect to recombination in that they are only rarely affected by the choice of loci *(3)*.

I previously hypothesized that the clonal complexes of MLST are ecotypes, and that the diversity-purging effect of periodic selection causes a correspondence between clonal complexes and ecotypes *(4,49,50)*. Because periodic selection is recurrently purging the diversity within an ecotype, ecotypes are expected to accumulate only a limited level of sequence diversity between periodic selection events. It may be speculated (and later tested) that ecotypes typically have only enough time between periodic selection events to accumulate divergence at 1 or 2 loci, at most, of 7, whether by mutation or recombination. This would justify MLST's 5/7 and 6/7 criteria for including strains within a clonal complex. In general, it would be expected that frequently recombining bacteria would diverge at more loci between periodic selection events, compared to rarely recombining bacteria, for which nearly all divergence accumulates simply through mutation. In any case, the hypothesis that MLST clusters correspond to ecotypes appears reasonable and should be rigorously tested.

In contrast, the species-less concept of diversity does not acknowledge the diversity-purging effect of periodic selection and so denies any mechanism for ecologically distinct populations to be visualized as sequence clusters. No mechanism is provided even for the existence of multiple discrete sequence clusters within named species.

Consider next how one can rigorously test whether MLST clonal complexes correspond to ecotypes, each dominated by periodic selection, or whether these groups are ecologically meaningless, as expected under the species-less concept.

Predictions Regarding Ecological Distinctness

The ecotype concept predicts that sequence clusters should be ecologically distinct (with the caveat of genotypes, discussed in a later section). One test of this prediction is that each sequence cluster might be associated with a different microhabitat. Indeed, several studies have shown that very closely related sequence clusters form a series of discrete populations occupying different ecological niches *(40,41,43)*. For example, Ramsing et al. *(41)* found that very closely related sequence clusters of *Synechococcus* in Yellowstone's hot springs are distributed at different depths, with different light conditions.

Another direct test of ecological distinctness, at least for pathogens, is that the various putative ecotypes should have different disease-causing properties. MLST was originally designed for the purpose of diagnosing unknown strains of pathogenic species into ecologically distinct populations; indeed, several of the MLST clusters have been shown to correspond to populations with different virulence and transmission characteristics *(3)*.

Also, genomic approaches can test whether putative ecotypes are ecologically distinct. Using either subtractive hybridization or microarray technology, strains can be

Small Population Size   Large Population Size
       and   and
Frequent Recombination   Rare Recombination

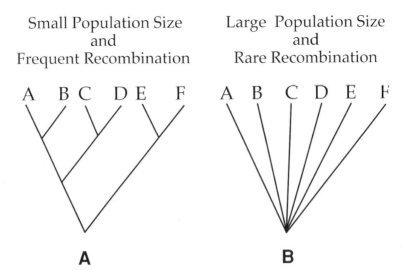

Fig. 3. The phylogenetic signatures of populations with diversity that is controlled by periodic selection vs genetic drift. (Used with permission from ref. *4*.)

assayed for the sets of genes they contain. The ecological distinctness of putative ecotypes will be supported when members of the same putative ecotype tend to share nearly all their genes, and members of different putative ecotypes share many fewer genes *(6)*. For example, Salama et al. *(7)* found that strains of *Helicobacter pylori* resolved clearly into clusters on the basis of the genes they contain. Unfortunately, these genome-content clusters cannot be compared to putative ecotypes demarcated by MLST clonal complexes because *H. pylori* is the one taxon known to recombine at too high a rate to yield robust clonal complexes *(82)*. In addition, messenger ribonucleic acid (mRNA) microarray approaches can address whether different putative ecotypes have different patterns of gene expression for the genes they share *(13,83)*. These genomic approaches are particularly promising in that they should yield information about the nature of the ecological differences between ecotypes *(6)*.

Phylogenetic Predictions

Sequence diversity within an ecotype is assumed to be limited largely by periodic selection and not by genetic drift, so nearly all strains randomly sampled from an ecotype should trace their ancestries directly back to the adaptive mutant that caused and survived the last selective sweep *(4,70)*. Thus, the phylogeny of an ecotype should be consistent with a star clade with only one ancestral node, such that all members of an ecotype are equally closely related to one another (Fig. 3). In contrast, a population with sequence diversity that is limited by genetic drift would have a phylogeny with many nodes.

In a strictly asexual ecotype, a sequence-based phylogeny would yield a perfect star clade, with only minor exceptions due to homoplasy (convergent nucleotide substitutions in different lineages) *(4)*. However, in an ecotype subject to modest rates of recombination, particularly with other ecotypes, the sequence-based phylogeny can deviate significantly from a perfect star. Libsch and I have developed a computer simulation (Star) to determine how closely a phylogeny based on multilocus sequencing should resemble a star clade *(4,84)*. Taking into account a taxon's mutation and recombination parameters, the Star simulation determines the likelihood that the phylogeny of strains from a single ecotype would have only one significant node (i.e., a perfect star) vs two, three, four, or more significant nodes.

It was found that, for *S. aureus*, which is among the least frequently recombining bacteria *(51,52)*, an ecotype should almost never have more than one node *(4,84)*. In *N. meningitidis*, which is among the most frequently recombining bacteria *(37)*, the phylogeny of an ecotype is expected to have either one or two significant nodes, but almost never three or more.

Let us consider how well the clonal complexes of MLST fit Star's phylogenetic predictions. I have tested whether the phylogenies of each of the 10 clonal complexes found within *N. meningitidis* were consistent with the Star simulation's expectations for a single ecotype *(4)*. The phylogenies of all but 1 clonal complex were found to contain 1 or 2 nodes, as expected given this taxon's recombination parameters. Similarly, all but 1 of the 26 clonal complexes within *S. aureus* contained only 1 significant node, consistent with the expectation for a single ecotype in this taxon. The pooled strains of most pairs of clonal complexes contained 2 significant nodes, indicating that pairs of complexes represented 2 ecotypes. However, 3 exceptional clonal complexes, when pooled together, contained only 1 significant node among them, suggesting that they are members of the same ecotype. I previously argued that, in *S. aureus*, a more stringent criterion for inclusion within a clonal complex (e.g., 6/7 identical loci) might yield a more perfect match between clonal complexes and the phylogenetic expectation for an ecotype *(4)*.

In summary, the MLST clonal complexes appear to pass two tests of their correspondence with ecotypes: At least some are known to be ecologically distinct, and the phylogenies of the clonal complexes are generally consistent with the expectations for a single ecotype.

Prediction of Private Periodic Selection Within Each Ecotype

The ecotype concept predicts that each putative ecotype should show evidence of having undergone its own private periodic selection events. To explore tests of this prediction, let us begin with Guttman and Dykhuizen's *(71)* sequence-based evidence for periodic selection within *E. coli*: Most genes fell into four major sequence clusters, but in the chromosomal region near *gapA*, all strains were anomalously homogeneous in sequence. As I have discussed, Guttman and Dykhuizen's *(71)* interpretation that an adaptive mutant (or recombinant) out-competed all other variants within *E. coli* is unlikely. The strains of *E. coli* are ecologically diverse, so a single adaptive mutant within *E. coli* could not out-compete the entire species; moreover, even if all of *E. coli* strains were a single ecotype, periodic selection would be expected to purge nearly all diversity over the entire chromosome (Fig. 1).

Majewski and I *(72)* proposed the adapt globally, act locally model to explain anomalous homogeneity around a small chromosomal region, as seen in *gapA* of *E. coli*. We proposed that there may be multiple ecotypes within *E. coli* (perhaps corresponding to the four major sequence clusters, or smaller subclusters), and that the adaptive mutation around *gapA* was generally useful for all of the ecotypes of *E. coli*. We proposed that the adaptive mutation first caused a purging of diversity within its original ecotype and was

even when the phylogenetic relationships among ecotypes are not. This is why sequence diversity can help us identify ecologically distinct populations.

RECOMMENDATIONS FOR THE SCIENCE OF GENOMICS

With thanks to genomics, bacteriologists are now poised to make enormous progress in discovering and characterizing bacterial diversity: Ecological diversity can be discovered in terms of the sets of genes that are contained within the genome and in terms of the expression levels of genes that are shared; the genes responsible for invasions of new niches can even be identified, as can the donor sources of each gene *(6)*. However, before embarking on this great adventure, we should make a point of focusing on those genomic differences that determine differences in ecological niche. Efforts should be made, therefore, to try to distinguish strains that are ecologically distinct from those that are not.

To this end, genomic approaches should be used to test for the existence of ecotypes, as defined here, and to hone sequence-based approaches to discovering ecotypes. I suggest that we first assign strains to putative ecotypes based on multilocus sequence clustering (using MLST) and then test whether these putative ecotypes fit the ecological, phylogenetic, and genomic expectations of a single ecotype. Tests up to this point have shown good correspondence between MLST clonal complexes and ecotypes, and further tests will be important. There are several important contributions that genomics can make toward validating the ecotype concept: use of microarray approaches to show that genomic patterns of gene expression differ between putative ecotypes, use of subtractive hybridization and microarrays to identify gene acquisitions that might suggest the nature of ecological differences between putative ecotypes, and genomic comparisons to show that putative ecotypes have undergone their own periodic selection events.

How should genomics proceed to characterize diversity once we are confident that sequence-based approaches can successfully classify organisms into ecotypes, and that there are many ecotypes contained within a typical named species? The ecotype concept should allow organization of the apparent chaos of genomic diversity within a named species by shifting the focus from the diversity of genes among strains of a named species *(73,74)* to the genomic differences among ecotypes. It should be recognized that variation in gene content and gene expression within a putative ecotype is likely to represent merely the random changes occurring within a population between periodic selection events. However, any genomic differences that correspond to different ecotypes could represent the changes responsible for invasion of new ecological niches and sustained coexistence among populations. These are the genomic differences that matter, and these are the differences that should demand attention.

ACKNOWLEDGMENTS

I am grateful to Michael Dehn for many suggestions that improved the clarity of the manuscript. This research was supported by National Science Foundation grant DEB-9815576 and by research funds from Wesleyan University.

REFERENCES

1. DeLong, E, Pace N. Environmental diversity of bacteria and archaea. Syst Biol 2001; 50:470–478.
2. Ward DM, Weller R, Bateson MM. 16S rRNA sequences reveal numerous uncultured microorganisms in a natural community. Nature 1990; 345:63–65.

3. Maiden MC, Bygraves JA, Feil E, et al. Multilocus sequence typing: a portable approach to the identification of clones within populations of pathogenic microorganisms. Proc Natl Acad Sci USA 1998; 953140–3145.

4. Cohan FM. What are bacterial species? Annu Rev Microbiol 2002; 56:457 487.

5. Palys T, Nakamura LK, Cohan FM. Discovery and classification of ecological diversity in the bacterial world: the role of DNA sequence data. Int J Syst Bacteriol 1997; 47:1145–1156.

6. Joyce EA, Chan K, Salama NR, Falkow S. Redefining bacterial populations: a post-genomic reformation. Nat Rev Genet 2002; 3:462–473.

7. Salama N, Guillemin K, McDaniel TK, Sherlock G, Tompkins L, Falkow S. A whole-genome microarray reveals genetic diversity among *Helicobacter pylori* strains. Proc Natl Acad Sci USA 2000; 97:14668–14673.

8. Hihara Y, Kamei A, Kanehisa M, Kaplan A, Ikeuchi M. DNA microarray analysis of cyanobacterial gene expression during acclimation to high light. Plant Cell 2001; 13:793–806.

9. Harrington CA, Rosenow C, Retief J. Monitoring gene expression using DNA microarrays. Curr Opin Microbiol 2000; 3:285–291.

10. Nesbo CL, Nelson KE, Doolittle WF. Suppressive subtractive hybridization detects extensive genomic diversity in *Thermotoga maritima*. J Bacteriol 2002; 184:4475–4488.

11. Sandberg R, Winberg G, Branden CI, Kaske A, Ernberg I, Coster J. Capturing whole-genome characteristics in short sequences using a naive Bayesian classifier. Genome Res 2001; 11: 1404–1409.

12. Gogarten JP, Doolittle WF, Lawrence JG. Prokaryotic evolution in light of gene transfer. Mol Biol Evol 2002; 19:2226–2238.

13. Feldgarden M, Byrd N, Cohan FM. Gradual evolution in bacteria: evidence from *Bacillus* systematics. Int J Syst Bacteriol 2003; 149:3565–3573.

14. Sneath PH. Future of numerical taxonomy. In: Goodfellow M, Jones D, Priest F (eds). Computer-Assisted Bacterial Systematics. Orlando, FL: Academic, 1985, pp. 415–431.

15. Rossello-Mora R, Amann R. The species concept for prokaryotes. FEMS Microbiol Rev 2001; 25:39–67.

16. Goodfellow M, Manfio GP, Chun J. Towards a practical species concept for cultivable bacteria. In: Claridge MF, Dawah HA, Wilson MR (eds). Species: The Units of Biodiversity. London: Chapman and Hall, 1997.

17. Johnson J. Use of nucleic-acid homologies in the taxonomy of anaerobic bacteria. Int J Syst Bacteriol 1973; 23:308–315.

18. Wayne LG, Brenner DJ, Colwell RR, et al. Report of the Ad Hoc Committee on reconciliation of Approaches to Bacterial Systematics. Int J Syst Bacteriol 1987; 37:463–464.

19. Stackebrandt E, Goebel BM. Taxonomic note: a place for DNA:DNA reassociation and 16S rRNA sequence analysis in the present species definition in bacteriology. Int J Syst Bacteriol 1994; 44:846–849.

20. de Queiroz K. The general lineage concept of species, species criteria, and the process of speciation. In: Howard DJ, Berlocher SH (eds). Endless Forms: Species and Speciation. Oxford, UK: Oxford University Press, 1998, pp. 57–75.

21. Meglitsch P. On the nature of species. Syst Zool 1954; 3:491–503.

22. Templeton A. The meaning of species and speciation: a genetic perspective. In: Otte D, Endler J (eds). Speciation and Its Consequences. Sunderland, MA: Sinauer, 1989, pp. 3–27.

23. Mayr E. Animal Species and Evolution. Cambridge, MA: Belknap Press of Harvard University Press, 1963.

24. Mayr E. Systematics and the Origin of Species from the Viewpoint of a Zoologist. New York: Columbia University Press, 1944.

25. Simpson G. Principles of Animal Taxonomy. New York: Columbia University Press, 1961.

26. Wiley E. The evolutionary species concept reconsidered. Syst Zool 1978; 27:17–26.

27. Eldredge N. Unfinished Synthesis: Biological Hierarchies and Modern Evolutionary Thought. New York: Oxford University Press, 1985.

28. Yohalem DS, Lorbeer JW. Intraspecific metabolic diversity among strains of *Burkholderia cepacia* isolated from decayed onions, soils, and the clinical environment. Antonic Van Leeuwenhoek 1994; 65:111–131.

29. Logan NA, Berkeley RC. Identification of *Bacillus* strains using the API system. J Gen Microbiol 1984; 130(Pt 7):1871–1882.

30. Lan R, Reeves PR. Gene transfer is a major factor in bacterial evolution. Mol Biol Evol 1996; 13:47–55.

31. Edwards RA, Olsen GJ, Maloy SR. Comparative genomics of closely related Salmonellae. Trends Microbiol 2002; 10:94–99.

32. Parkhill J, Achtman M, James KD, et al. Complete DNA sequence of a serogroup A strain of *Neisseria meningitidis* Z2491. Nature 2000; 404:502–506.

33. Perna NT, Plunkett G 3rd, Burland V, et al. Genome sequence of enterohaemorrhagic *Escherichia coli* O157:H7. Nature 2001; 409:529–533.

34. Read TD, Brunham RC, Shen C, et al. Genome sequences of *Chlamydia trachomatis* MoPn and *Chlamydia pneumoniae* AR39. Nucleic Acids Res 2000; 28:1397–1406.

35. Tettelin H, Saunders NJ, Heidelberg J, et al. Complete genome sequence of *Neisseria meningitidis* serogroup B strain MC58. Science 2000; 287:1809–1815.

36. Alm RA, Ling LS, Moir DT, et al. Genomic-sequence comparison of two unrelated isolates of the human gastric pathogen *Helicobacter pylori*. Nature 1999; 397:176–180.

37. Feil EJ, Maiden MC, Achtman M, Spratt BG. The relative contributions of recombination and mutation to the divergence of clones of *Neisseria meningitidis*. Mol Biol Evol 1999; 16: 1496–1502.

38. Feil EJ, Smith JM, Enright MC, Spratt BG. Estimating recombinational parameters in *Streptococcus pneumoniae* from multilocus sequence typing data. Genetics 2000; 154:1439–1450.

39. Dykhuizen DE, Green L. Recombination in *Escherichia coli* and the definition of biological species. J Bacteriol 1991; 173:7257–7268.

40. Rocap G, Distel DL, Waterbury JB, Chisholm SW. Resolution of *Prochlorococcus* and *Synechococcus* ecotypes by using 16S-23S ribosomal DNA internal transcribed spacer sequences. Appl Environ Microbiol 2002; 68:1180–1191.

41. Ramsing NB, Ferris MJ, Ward DM. Highly ordered vertical structure of *Synechococcus* populations within the one-millimeter-thick photic zone of a hot spring cyanobacterial mat. Appl Environ Microbiol 2000; 66:1038–1049.

42. Ward DM. A natural species concept for prokaryotes. Curr Opin Microbiol 1998; 1:271–277.

43. Beja O, Koonin E, Aravind L, et al. Comparative genomic analysis of archaeal genotypic variants in a single population and in two different oceanic provinces. Appl Environ Microbiol 2002; 68:335–345.

44. Vulic M, Lenski RE, Radman M. Mutation, recombination, and incipient speciation of bacteria in the laboratory. Proc Natl Acad Sci USA 1999; 96:7348–7351.

45. Lawrence JG. Gene transfer in bacteria: speciation without species? Theor Popul Biol 2002; 61: 449–460.

46. Cohan FM. The effects of rare but promiscuous genetic exchange on evolutionary divergence in prokaryotes. Am Naturalist 1994; 143:965–986.

47. Posada D. Evaluation of methods for detecting recombination from DNA sequences: empirical data. Mol Biol Evol 2002; 19:708–717.

48. Smith JM, Smith N, O'Rourke M, Spratt BG. How clonal are bacteria? Proc Natl Acad Sci USA 1993; 15:4384–4388.

49. Cohan FM. Clonal structure: an overview. In: Pagel M (ed). Encyclopedia of Evolution. Vol. 1. New York: Oxford University Press, 2002, pp. 159–161.

50. Cohan FM. Population structure and clonality of bacteria. In: Pagel M (ed). Encyclopedia of Evolution. Vol. 1. New York: Oxford University Press, 2002, pp. 161–163.

51. Feil EJ, Cooper JE, Grundman H, et al. How clonal is *Staphylococcus aureus?* J Bacteriol 2003; 185:3307–3316.

52. Enright MC, Robinson DA, Randle G, Feil EJ, Grundmann H, Spratt BG. The evolutionary history of methicillin-resistant *Staphylococcus aureus* (MRSA). Proc Natl Acad Sci USA 2002; 99:7687–7692.

53. Majewski J, Zawadzki P, Pickerill P, Cohan FM, Dowson CG. Barriers to genetic exchange between bacterial species: *Streptococcus pneumoniae* transformation. J Bacteriol 2000; 182: 1016–1023.

54. Majewski J, Cohan F. M. DNA sequence similarity requirements for interspecific recombination in Bacillus. Genetics 1999; 153:1525–1533.

55. Vulic M, Dionisio F, Taddei F, Radman M. Molecular keys to speciation: DNA polymorphism and the control of genetic exchange in enterobacteria. Proc Natl Acad Sci USA 1997; 94:9763–9767.

56. Duncan KE, Ferguson N, Kimura K, Zhou X, Istock CA. Fine-scale genetic and phenotypic structures in natural populations of *Bacillus subtilis* and *Bacillus licheniformis*: important implications for bacterial evolution and speciation. Evolution 1994; 48:2002–2025.

57. Majewski J. Sexual isolation in bacteria. FEMS Microbiol Lett 2001; 199:161–169.

58. Cohan FM. Sexual isolation and speciation in bacteria. Genetica 2002; 116:359–370.

59. Majewski J, Cohan FM. The effect of mismatch repair and heteroduplex formation on sexual isolation in Bacillus. Genetics 1998; 148:13–18.

60. Rao BJ, Chiu SK, Bazemore LR, Reddy G, Radding CM. How specific is the first recognition step of homologous recombination? Trends Biochem Sci 1995; 20:109–113.

61. Rayssiguier C, Thaler DS, Radman M. The barrier to recombination between *Escherichia coli* and *Salmonella typhimurium* is disrupted in mismatch-repair mutants. Nature 1989; 342:396–401.

62. Doolittle WF. Lateral genomics. Trends Cell Biol 1999; 9:M5–M8.

63. Ochman H, Lawrence JG, Groisman EA. Lateral gene transfer and the nature of bacterial innovation. Nature 2000; 405:299–304.

64. Licht TR, Christensen BB, Krogfelt KA, Molin S. Plasmid transfer in the animal intestine and other dynamic bacterial populations: the role of community structure and environment. Microbiology 1999; 145(Pt 9):2615–2622.

65. Maynard Smith JM, Dowson CG, Spratt BG. Localized sex in bacteria. Nature 1991; 349: 29–31.

66. Arthur W. Mechanisms of Morphological Evolution: A Combined Genetic, Developmental and Ecological Approach. New York: Wiley, 1984, pp. 182–186.

67. Feil EJ, Holmes EC, Bessen DE, et al. Recombination within natural populations of pathogenic bacteria: short-term empirical estimates and long-term phylogenetic consequences. Proc Natl Acad Sci USA 2001; 98:182–187.

68. Atwood KC, Schneider LK, Ryan FJ. Periodic selection in *Escherichia coli*. Proc Natl Acad Sci USA 1951; 37:146–155.

69. Cohan FM. Genetic exchange and evolutionary divergence in prokaryotes. Trends Ecol Evol 1994; 9:175–180.

70. Cohan FM. Periodic selection and ecological diversity in bacteria. In: Nurminsky D (ed). Selective Sweep. Georgetown, TX: Landes Bioscience, in press.

71. Guttman DS, Dykhuizen DE. Detecting selective sweeps in naturally occurring *Escherichia coli*. Genetics 1994; 138:993–1003.

72. Majewski J, Cohan FM. Adapt globally, act locally: the effect of selective sweeps on bacterial sequence diversity. Genetics 1999; 152:1459–1474.

73. Lan R, Reeves PR. Intraspecies variation in bacterial genomes: the need for a species genome concept. Trends Microbiol 2000; 8:396–401.

74. Boucher Y, Nesbo CL, Doolittle WF. Microbial genomes: dealing with diversity. Curr Opin Microbiol 2001; 4:285–289.

75. Ravin A. The origin of bacterial species: genetic recombination and factors limiting it between bacterial populations. Bacteriol. Rev 1960; 24:201–220.

76. Ravin A. Experimental approaches to the study of bacterial phylogeny. Am Nat 1963; 97:307–318.

77. Linz B, Schenker M, Zhu P, Achtman M. Frequent interspecific genetic exchange between commensal Neisseriae and *Neisseria meningitidis*. Mol Microbiol 2000; 36:1049–1058.

78. Cohan FM. Does recombination constrain neutral divergence among bacterial taxa? Evolution 1995; 49:164–175.

79. Giraud A, Matic I, Tenaillon O, et al. Costs and benefits of high mutation rates: adaptive evolution of bacteria in the mouse gut. Science 2001; 291:2606–2608.

80. Denamur E, Lecointre G, Darlu P, et al. Evolutionary implications of the frequent horizontal transfer of mismatch repair genes. Cell 2000; 103:711–721.

81. Cohan FM. Bacterial species and speciation. Syst Biol 2001; 50:513–524.

81a. Cohan FM. The role of genetic exchange in bacterial evolution. ASM News 1996; 62:631–636.

82. Suerbaum S, Smith JM, Bapumia K, et al. Free recombination within *Helicobacter pylori*. Proc Natl Acad Sci USA 1998; 95:12619–12624.

83. Gibson G. Microarrays in ecology and evolution: a preview. Mol Ecol 2002; 11:17–24.

84. Cohan FM, Libsch J. Sequence-based evidence for numerous ecologically distinct and irreversibly separate taxa within bacterial species.

85. Papke T, Ramsing NB, Bateson MM, Ward DM. Geographical isolation in hot spring cyanobacteria. Environ Microbiol 2003; 5:650–659.

12

Coevolution of Symbionts and Pathogens With Their Hosts

Robert A. Feldman

INTRODUCTION

Many microbes live in close association with metazoans. These associations can be obligate, as is the case for subcellular organelles and most human intracellular pathogens, or they can be commensal, almost peripheral, neither positively nor negatively affecting microbe or host. Typically, microbial–host associations are derived from a long-term coevolutionary history, resulting in intimate physiological and regulatory dependencies between microbe and host. Genomics is playing a central role in deciphering the genetic basis of microbial–host associations by identifying microbial pathogenicity genes and genetic regulatory systems modulated by host interactions. In the near future, genomic techniques will lead to understanding of the patterns of gene expression and physiological coupling of microbes and hosts. These studies in turn will lead to better understanding of the coevolutionary histories of microbial–host associations.

In this chapter, I describe some of the mechanisms underlying microbial–host coevolution, including the origin, persistence, and maintenance of these relationships. Then, I describe various types of microbial–host associations, starting with subcellular organelles, then progressing to microbes associated with plants, insect symbionts, marine invertebrates, and humans. Finally, I discuss the uncultivated human microflora and the promise genomics has for its characterization.

MECHANISMS OF COEVOLUTION: PERSISTENCE, MAINTENANCE, AND SELECTION

Symbiont Acquisition and Transmission Strategies

Symbionts are established in and transmitted by their hosts through three mechanisms. These are horizontal (from established adults to new recruits), vertical (parent to offspring), and reinfection (environmental). The consequence of each mode of symbiont transmission can be reflected in phylogenetic congruence or noncongruence of gene phylogenies between symbionts and hosts (Fig. 1).

Under horizontal transmission, the symbiont's phylogeny should be interpretable through a phylogeographic signal that may be independent of host phylogeny. Horizontal transmission of endosymbionts occurs in the deep-sea Vestimentiferan tubeworms

From: *Microbial Genomes*
Edited by: C. M. Fraser, T. D. Read, and K. E. Nelson © Humana Press Inc., Totowa, NJ

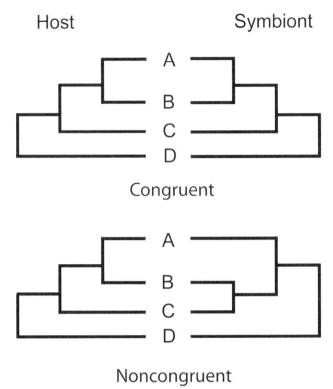

Fig. 1. Congruent and noncongruent phylogenies. Gene trees calculated from genes from a hypothetical symbiont and host. In the top pair of trees, the symbiont and host phylogenies are congruent by virtue of a shared coevolutionary history. In the bottom pair, symbiont and host phylogenies are different, suggesting they have not undergone the same evolutionary history.

and is evident by noncongruent phylogenetic trees drawn from the host worm and symbiont *(1)*.

Under the vertical mode, transmission is from parents to offspring through gametes. Vertical transmission should be reflected in congruent host–symbiont phylogenies by virtue of a shared coevolutionary history. Vertical transmission gives rise to the cospeciation patterns (congruent phylogenies) between deep sea clams and their chemoautotrophic bacteria *(2)*.

The mechanisms of the reinfection mode of symbiont transmission are more difficult to decipher, but phylogenetic trees of symbionts should show a strong phylogeographic component.

Cellular Communication

A wide variety of plant and animal pathogens use the type III secretion systems for delivering effector proteins into hosts to assist in host cell entry. The type III secretion system is made up of more than 20 proteins phylogenetically related to flagellar apparatus machinery *(3)*. The mechanism of type III secretion is not well understood, but it is known to involve formation of a needlelike complex serving as a hollow conduit through which type III secreted proteins are injected *(3)*. Once inside the host cytoplasm, the proteins have the capacity to modulate host cellular functions. Type III secretion systems occur in the vertically transmitted mutualistic endosymbiont of the grain weevil (*Soldalis glossinidius*), presumably originating from a single acquisition event over 50 million years ago *(4)*.

Quorum sensing in bacteria is the regulation of gene expression in response to population density *(5)*. Quorum sensing works through autoinduction, by which the bacteria environmentally sense and monitor their population density by producing a diffusible compound (an autoinducer) that accumulates in the environment as the population grows *(5)*. The gene expression system controlling quorum sensing is well known from the free-living marine bacterium *Vibrio harveyi* as a two-quorum sensing system, each comprised of a specific sensor and autoinducer that function in parallel to control density-dependent expression of bioluminescence *(6)*. The *V. harveyi*–type system 1 autoinducer (AI-1, a homoserine lactone) and genes involved in its production (*luxL* and *luxM*) are well known *(5–7)*. In *V. harveyi*, *Escherichia coli*, and *Salmonella typhimurium*, autoinduction for system 2 is controlled by a recently discovered highly homologous gene called *luxS (6)*. A more complete understanding of the similarities and differences between these quorum sensing systems promises to provide new methods of antibiotic control of pathogens.

Genome Reduction and Mutation Rate

Microbial symbionts and pathogens live in a metabolically static, yet metabolite-rich, environment. For a pathogen living in a constant environment fueled by nutrients from its host, many of its genes become useless and are lost. In general, a symbiotic or pathogenic lifestyle usually leads to loss of gene pathways and overall genome reduction compared to free-living counterparts. Many of the gene functions that are lost are compensated for by the host. Further genome reduction leads to further loss of metabolic flexibility as microbes become irreversibly restricted to an obligate pathogen lifestyle. In the extreme, obligate intracellular parasites have the smallest known genomes *(8)*. Loss of deoxyribonucleic acid (DNA) repair genes and a strong A + T bias in base composition can also accelerate mutation rates.

Small population sizes of symbionts and pathogens can lead to elevated rates of evolution. Due to restricted habitats, repeated bottlenecks, and extremely small effective population sizes, more random mutations will drift to fixation in symbionts as compared to free-living relatives *(9,10)*. These rates can approach amino acid substitution rates twice as high as for free-living counterparts *(11)*. The elevated mutation rate of pathogens creates higher levels of sequence-based genomic diversity that can be used to establish recent genetic origins of pathogens and potentially to identify epidemiology of pathogen transmission *(12)*. This is particularly pertinent in light of the recent biowarfare threats, but also has relevance in characterizing infectious disease outbreaks.

TYPES OF MICROBIAL–HOST ASSOCIATIONS

Subcellular Organelles: Mitochondria and Chloroplasts

It is generally accepted that modern-day plastid and mitochondrial symbioses with eukaryotic cells arose through host acquisition of bacteria that were once free living. Plastids (photosynthesizing subcellular organelles) and mitochondria (the energy-burning cell respiration organelles) represent perhaps the most ancient and highly developed of all known types of symbioses *(13)*. The highly reduced genomes of these organelles and transfer of essential genes and gene control to the nucleus are perhaps the strongest proof of a mutually beneficial arrangement *(14)*. Indeed, it is difficult to imagine how life on earth would have evolved without the ancient marriage of these bacteria with their hosts. The key evolutionary questions regarding the origin of plastid and mitochondrial acquisition are no longer whether they have occurred, but rather how many times did it happen, and who precisely was the host organism?

For primary plastids, molecular phylogenetic evidence is mounting in support of a monophyletic origin event between cyanobacteria and eukaryotic host cells that led to modern-day green plants, red algae, and glaucophytes *(14,15)*. Moreira et al. *(15)* performed phylogenetic analysis of nuclear amino acid sequence from the elongation factor 2 and the RNA polymerase II genes and showed that the red algae and green plants form a monophyletic clade within the eukaryotes. This conclusion is also supported by other nuclear molecular phylogenetic studies using small- and large-subunit ribosomal ribonucleic acid (rRNA) sequences *(16)* and a study of four nuclear genes (α-tubulin, β-tubulin, actin, and elongation factor 1-α) in a combined amino acid phylogenetic study *(17)*. The number of secondary plastid symbiotic acquisition events (ingestion by a host cell of a cell that already contained plastids) is uncertain *(14)*.

The rRNA phylogenies suggest that eukaryotic mitochondria are derived from an α-proteobacteria-like ancestor *(18)*. The modern-day descendants evolutionarily closest to the original mitochondrial donor are a group of intracellular parasites from the rickettsial subdivision, including the genera *Rickettsia*, *Anaplasma*, and *Ehrlichia* *(19)*. There is relatively strong evidence for a monophyletic origin of the mitochondria, including the findings that phylogenetic trees calculated from rRNA genes and protein sequences are congruent, and that the gene content of all mitochondrial DNAs (mtDNAs) is a subset of the largest known mitochondrial genome *(19)*.

Mitochondrial genome sizes and gene content are highly variable. The smallest mitochondrial genome (less than 6 kb) is found in the malaria parasite *Plasmodium* and the largest (366,924 bp) in the flowering plant *Arabidopsis* *(19)*. The mitochondrion of the flagellated protozoan *Reclinomonas americana* has the largest gene content (97 genes) of any mitochondrial genome sequenced to date *(20)*. The *R. americana* mitochondrial genome contains all of the protein-coding genes found in all the other mitochondrial genomes in addition to 18 novel ones, a 5S rRNA gene, an RNAse P gene, and a bacterial-type RNA polymerase gene *(20)*. Thus, the *R. americana* mitochondrial genome represents the most bacteria-like mitochondrial genome *(19,20)*.

As a result of endosymbiotic lifestyle, subcellular organelles have undergone high levels of genome loss and transfer of gene functions to the host nucleus. These genes and gene functions tend to be the "housekeeping" genes, those involved in activities such as amino acid biosynthesis, nucleoside biosynthesis, and anaerobic glycolysis and regulation *(19)*. Modern-day transfer of genes from the animal mitochondria to the nucleus appears to have stopped, presumably because of the different genetic codes used by the nuclear and mitochondrial genomes. In flowering plants, for which the mitochondrial and nuclear genome use the same genetic code, gene transfer from mitochondrion to nucleus appears to be an ongoing and dynamic process *(21,22)*. Transfer of chloroplast

DNA to the nucleus appears to be a common occurrence and can be visualized directly in the laboratory *(23)*.

Intracellular compartmentalization, translocation, and gene product sharing appear to have reached an evolutionary pinnacle in the chromophyte algae. Chromophytes arose by secondary symbiogenesis of a red or green algae (already containing chloroplasts and a nucleus), with a nonphotosynthetic flagellated host contributing a nucleus, endomembranes, and mitochondria *(24,25)*. In the evolutionary line leading to the cryptomonads, the red algal endosymbiont nucleus has undergone miniaturization, become vestigial, and is called a *nucleomorph (25)*. Modern-day cryptomonads rely on four genomes (chloroplast, nucleus, nucleomorph, and mitochondrion) cooperatively transferring gene products through a complex intertwined intracellular membrane network *(25)*.

Plant Microbial Pathogens and Symbionts

The α-subdivision of the proteobacteria contains bacterial species that tend to form close physiological associations with eukaryotic cells. Forming a tight cluster within this group are the rhizobia (legume symbionts causing nitrogen fixation), agrobacteria (plant pathogens), and the rickettsias (intracellular animal pathogens) *(26)*.

Agrobacterium tumefaciens is a plant pathogen and the causative agent of crown gall disease; it can transfer and integrate a defined segment of DNA from its genome into that of a eukaryotic host cell. In nature, this occurs when *Agrobacterium* detects small molecules released by plant cells growing in a plant wound *(27)*. The molecules induce *A. tumefaciens* Ti-plasmid encoded *vir* genes to express a set of genes leading to export of transferred DNA (T-DNA) to the plant cell. The T-DNA integrates into the plant cell host genome randomly. Expression of T-DNA genes alters plant growth hormone levels, leading to formation of plant gall tumors. Expression of the T-DNA genes also releases bacterial growth nutrients called opines. The genome of *A. tumefaciens* was sequenced by two groups independently and reported simultaneously *(27,28)*. The *A. tumefaciens* genome is 5.67 Mb and consists of four elements: a 2.841-Mb circular chromosome, a 2.075-Mb linear chromosome, and the two plasmids pAtC58 and pTiC58, which are 542.8 kb and 214.2 kb, respectively *(27,28)*. The telomeres of the linear chromosome are covalently closed, possibly by hairpin loops *(28)*. The genes involved in plant cell transformation and tumorigenesis are dispersed on all four genomic components *(27,28)*. *Agrobacterium tumefaciens* and all of the other rhizobial genomes sequenced to date show high numbers of ABC transporter genes, perhaps reflecting the high-affinity uptake systems necessary for acquisition of nutrients in highly competitive soil environments *(28)*.

Sinorhizobium meliloti is an α-proteobacterial nitrogen-fixing symbiont of alfalfa. A member of the rhizobia, it infects roots and produces nodules where bacterial endosymbionts fix (i.e., make biologically available) nitrogen within the plant cells *(29)*. The bacteria and plant communicate during nodule development and establish metabolic sharing networks for the bacteria to receive plant carbon compounds in exchange for bacterial fixed nitrogen *(30)*. The *S. meliloti* genome is comprised of a 3.65-Mb chromosome and 1.35-Mb pSymA and 1.68-Mb pSymB megaplasmids *(30)*. The combined 6.7 Mb genome encodes 6204 protein-coding open reading frames (ORFs) *(30)*. Nitrogen metabolism genes are clustered on pSymA, while pSymB contains genes involved in transport of small molecules. The genome of *S. meliloti* was compared to the rhizobial genome of *Mesorhizobium loti (31)*. Only 35% of the *M. loti* genes have orthologs in *S. meliloti*, and the genetic information carried by the tripartite genome of *S. meliloti* is dispersed in *M. loti (30)*, strongly suggesting that the rhizobia differ significantly in gene content and genome organization. However, the circular chromosomes of *A. tumefaciens* and *S. meliloti* show high levels of synteny, supporting the notion that the circular chromosomes are the descendants of the original α-Proteobacteria-like ancestors *(27,30)*.

Insect Microbial Symbionts

Insects contain "bacterial menageries," many of which are well studied *(34)*. Most of these symbionts are phylogenetically classified in the γ-proteobacteria, within the Enterobacteriaceae, and are closely related to *E. coli*. The symbionts occur in the insect suborder Sternorrhyncha (Hemiptera), which includes psyllids, aphids, mealybugs, and whiteflies *(33)*. The best studied of these systems is the highly mutualistic intracelluar aphid symbionts of the genus *Buchnera*. The closest ancestor of *Buchnera* is *E. coli*, and it seems likely that the *Buchnera* genome evolved from an *E. coli* like ancestor largely through genome reduction *(34,35)*.

Key features of the *Buchnera aphidicola* genome are its small size (640,681 bp) and presence of genes for biosynthesis of amino acids essential for the host, but not those for nonessential amino acids *(35)*. *Buchnera* lacks genes for cell surface receptors, regulator genes, and genes involved in defense, suggesting that it is a completely symbiotic and "resident" genome *(35,36)*. A comparison of two *B. aphidicola* genomes from different hosts (*Aphidicola pisum* and *Schizaphis graminum*) revealed extreme genome stability; that is, no rearrangements (including inversions, translocations, duplications, and gene acquisitions) were found in either lineage, spanning a divergence time of 50–70 million years *(37)*. The genomes are similar in size (641.5 Kb, 640.7 Kb), protein-coding gene number (545, 564), and guanine and cytosine content (G+C) (26.2, 26.3) *(37)*. The remarkable similarity of these genomes suggests that genome reduction in *Buchnera* had already occurred in the common ancestor of these two symbionts before they diverged *(34)*. However, a study using pulse field gel electrophoresis (PFGE) genome sizing of nine *Buchnera* genomes collected from five subfamilies of aphid hosts showed *Buchnera* genome sizes varied from 670 to 450 Kb *(38)*. Clearly, more work is needed in comparative *Buchnera* genomics.

MARINE INVERTEBRATE MICROBIAL SYMBIOSES

Archaeal Symbionts of Sponges

Crenarchaeal symbionts were first discovered living in the common marine sponge *Axinella mexicana*, found off the coast of California *(39)*. The symbionts are phylogenetically related to hyperthermophilic Crenarchaea, but occur at mesophilic oceanic temperatures and can be grown in sponges kept in aquaria at temperatures of 10°C. This temperature is more than 60°C lower than for any of the other crenarchaeotes in laboratory culture collections *(39)*. The discovery of this symbiont helped establish the ubiquity of cold-to-moderate temperature crenarchaeotes in the environment. The symbionts were named *Cenarchaeum symbiosum (39)* and remain uncultivated outside of sponge host tissues. Since the *C. symbiosum* discovery, more Archaea have been discovered living as symbionts of sponges *(40)* and other invertebrates *(41)*.

Genomic diversity of *Cenarchaeum* communities was assessed using the phylogenetic anchoring technique developed by Stein et al. *(42)*. This technique involves constructing environmental DNA libraries of large (initially about 40 Kb, but now over 150 Kb) *(43)* insert clones, then probing the libraries for genes of interest, typically rRNA genes. Clones containing interesting rRNA genes are then fully sequenced, and the genes flanking the rRNA gene anchor are found using ORF-finding programs. The ORFs are compared against databases (BLAST [Basic Local Alignment Search Tool] is the most popular method) *(44)*, and then inferences are made about phylogenetic relationships. In the best cases, these fishing expeditions reveal information about the potential metabolism and physiology of the uncultivated microbes *(45)*.

A study of this type was conducted on *C. symbiosum* by completely sequencing overlapping 40-Kb fosmid clones. Schleper et al. cloned, sequenced, and assembled three fosmid clones from the microbial community of the sponge *(46)*. The assembled fosmid clones show an unexpected level of diversity in that some regions align perfectly, but others show evidence of genome rearrangement. Interestingly, 16S rRNA sequence diversity of the community is less than 3%. This study also revealed differences in promotor regions for this closely related archaeal symbiont community *(46)*. This study was the first of its kind comparing large insert genomic clones of closely related uncultivated symbionts and clearly exposed our embryonic understanding of genomic diversity within even closely related microbial communities.

Deep Sea Hydrothermal Vent Microbial Symbionts

One of the specialized communities found at hydrothermal vent sites on the East Pacific Rise is dominated by polychaetes from the family Alvinellidae. Alvinellids make complex chitin tube colonies inhabited by several hundred animals on the flanks of black smoker chimneys. These animals live merely centimeters away from chimneys that spew superheated (up to 450°C) water. The 2.5-cm long, filament-covered alvinellids live under the highest known temperatures for metazoans, frequently experiencing 80°C temperatures in their tubes, and they have been observed transiently exposed to a temperature of 105°C *(47)*. They share the community with about 40–50 strains of ε-proteobacteria. The worms and the microbes coexist, with the worms wearing a coat of microbes on their filament-covered backs. It is unclear what specific advantage the microbes confer to the hosts, but speculation has centered around thermotolerance, resistance to toxic heavy metals, and use of the microbes as a direct food source.

The alvinellid microbial community is the subject of a current project to determine the metagenome of this community consortia. This study will sequence a large number of environmental clones and preliminarily identify them by comparisons to databases. The different genes and variants of genes will be arrayed on microarray slides, taken back to the environment, and used as targets for probing with message libraries made from new environmental extractions. The study then becomes one of functional environmental genomics, and it is hoped it will lead to physiological assignment of genes and metabolism for the community.

Deep Sea Tubeworm Chemoautotrophic Symbionts

Another specialized deep sea microbial–invertebrate symbiosis occurs on the oases of East Pacific Rise vent communities at depths of 2500 m. The giant tubeworm *Riftia pachyptila* grows up to 2.5 m long and is the longest lived of known marine invertebrates. Adult *Riftia* lack a mouth, gut, and anus and are completely dependent on endosymbiotic bacteria for nutrition *(48)*. The worms have highly developed circulatory and nervous systems and use a specialized oxygen–sulfide reversible-binding hemoglobin to transport CO_2, O_2, and HS to their symbionts, which are housed in a specialized organ called a *trophosome*. In the trophosome, the symbionts chemoautotrophically fix CO_2 to malate using the Calvin–Benson cycle, providing four- and five-chain sugars back to the host tissues.

The symbionts are members of the γ-proteobacteria related to *E. coli*. They are picked up by the hosts from the environment by horizontal transmission; thus, they have a free-living state. The symbionts show little phylogeographic signal and low levels of genetic diversity; they are phylogenetically noncongruent with their hosts *(1,49)*. The symbionts have a free-living phase in their life cycle and possess flagellin genes *(50)*. They also have histidine kinase signaling systems *(51)*. It is unclear what the recognition mechanism is between juvenile tubeworms and the symbionts, but this is one of the foci of a genome project currently under way.

HUMAN PATHOGENS

The era of genomics was ushered in with completion of the first genome from the human pathogen *Haemophilus influenzae* *(52)*. In addition to being a remarkable technical achievement, the genome provides several insights into *H. influenzae* biology. One of these is the finding that the nonpathogenic *H. influenzae* Rd strain varies from the pathogenic type b strain for factors involved in infectivity *(52)*. Specifically, an adhesion gene cluster (which allows bacterial attachment to host cells) encoding eight fimbrial genes is completely absent in the Rd strain. This type of finding pointed out immediately the utility of complete genome analysis for inferring and providing for new hypotheses regarding pathogen biology.

Helicobacter pylori was shown in 1983 to be the causative agent of human peptic ulcers *(53)*. *H. pylori* lives in the low pH (range 2.0 to 3.0) intragastric environment, where it uses molecular hydrogen as an energy source *(54)*. Infecting almost one-half the world's population, *H. pylori* is probably the most common human pathogen. The *H. pylori* genome revealed several interesting features that correlate with pathogenic lifestyle in an extreme environment. It has a well-developed restriction modification system, at least five different adhesins to attach to host gastric epithelial cells, and a positive inside membrane potential *(55)*.

The bug has several methods for increasing antigen variability. *Helicobacter pylori* has a high number of dinucleotide repeats, allowing for slipped-strand mispairing, which can increase genotypic and phenotypic variation *(55)*. It also has a human blood group antigen mimicry system in its surface lipopolysaccharide molecules, making it less immunogenic than other enteric bacteria *(55)*. Phylogenetic trees, including genes from *H. pylori*, often have "strange" (i.e., unlike 16S rRNA) topologies. This might indicate that *H. pylori* branched off early in the evolution of the Proteobacteria, that *H. pylori* genes undergo rapid evolution, and that there were high levels of lateral gene transfer. *Helicobacter pylori* is commonly passed from mothers to infants in a pseudovertical transmission pattern, making some *H. pylori* genes suitable as tracers of human migration patterns *(56,57)*.

Mycoplasmas are low G+C Gram-positive bacteria that completely lack a cell wall and are obligate human pathogens. In humans, *Mycoplasma pneumoniae* causes atypical pneumonia, and *Mycoplasma genitalium* causes nongonococcal urethritis. These two related *Mycoplasma* genomes were among the first ones completed, allowing their comparison *(58,59)*. Both genomes are extremely small (*M. pneumonia* is 816 Kb, and *M. genitalium* is 580 Kb) and have correspondingly small gene content. *Mycoplasma genitalium* has the smallest gene content (517) and is the smallest genome for any known free-living organism. Mycoplasmas have highly reduced metabolic gene systems, which are compensated by transport of nutrients from the host *(60)*.

Treponema pallidum is a spirochete that causes the venereal disease syphilis in humans. Syphilis has a long and checkered coevolutionary history with humans and repeatedly decimated native peoples after first contact with Europeans. The organism is an obligate pathogen that cannot exist outside its human host. The genome of *T. pallidum* was compared with that of another spirochete, the causative agent of Lyme disease, *Borrelia burgdorferi (61)*. The *T. pallidum* genome is small (1,138,006 bp), is 52.8% G+C, and encodes 1041 predicted ORFs *(61)*. The *T. pallidum* genome contains a complement of DNA replication genes similar to that of *B. burgdorferi* and *M. genitalium (61)*. Like *B. burgdorferi*, *T. pallidum* contains a variety of transport proteins and lacks respiratory electron transport chain genes. Both genomes also contain conserved motility and chemotaxis genes. The outer membrane of *T. pallidum* contains few membrane proteins, presumably allowing for host recognition avoidance *(61)*.

Chlamydiae are intracellular pathogens that cause a wide range of diseases. Two genomes of *Chlamydia* (*Chlamydia trachomatis* MoPn and *Chlamydia pneumoniae* AR39, infective of mice and humans, respectively) were published in a comparative article *(62)*. Both genomes are small (*C. trachomatis* is 1,042,519 bp, encoding 894 ORFs; *C. pneumoniae* is 1,230,230 bp, encoding 1052 ORFs) and show striking synteny, presumably because of their ecological isolation as obligate intracellular pathogens *(62)*. Host species tropism might be determined by only a few genes, perhaps those involved in nucleotide salvaging and metabolism. The comparative genome study also revealed presence of a *C. pneumoniae* bacteriophage that may play a role in pathogenesis *(62)*.

Neisseria meningitidis is a β-proteobacterium that causes meningitis and septicemia. The genome of *N. meningitidis* is 2,272,351 bp long, encoding 2158 ORFs *(63)*. *Neisseria meningitidis* uses phase variation to control gene expression and evade host immune systems. The genome reveals 49 new genes (in addition to 16 previously known) that are potentially phase variable, including genes involved in outer membrane proteins, lipopolysaccharides, pilus, restriction modification, capsule formation, and iron acquisition *(63)*.

Campylobacter jejuni is a foodborne pathogen phylogenetically classified in the Δ-proteobacteria; it is the world's leading cause of bacterial food poisoning, especially from poultry and water *(64)*. The genome is 1,641,481 bp, encoding 1654 proteins, making it one of the densest bacterial genomes *(65)*. The genome has hypervariable sections of short homopolymer runs in genes encoding for biosynthesis and modification of surface structures *(65)*. The genomes of *C. jejuni* and *H. pylori* have little in common (mainly housekeeping functions) in spite of their close relationship, both phylogenetically and physiologically. This may be because of strong selection pressures that have driven these two pathogens to different niches *(65)*.

Complete genomes have been determined for a number of low G+C Gram-positive streptococci including *Streptococcus pyogenes* and *Streptococcus pneumoniae (66,67)*. *Streptococcus pyogenes* is an obligate human pathogen and causes a wide variety of diseases, including sore throat, scarlet fever, impetigo, erysipelas, cellulitis, septicemia, toxic shock syndrome, flesh-eating disease, rheumatic fever, and acute glomerulonephritis *(66)*. *Streptococcus pneumoniae* causes over 1 million deaths each year globally and is one of the top 10 causes of death in the United States *(67)*. Both genomes show a wide variety of virulence-associated genes, many encoding factors localized to the cell surface or targeted for secretion and high levels of horizontal gene transfer from Gram-negative bacteria that share the same niche with the streptococci. *Streptococcus pyogenes* contains four different prophages encoding superantigen proteins.

Vibrios are γ-proteobacteria free living ubiquitously in marine and freshwater systems. They also occur as symbionts of fish *(68)* and cephalopods *(69)*. *Vibrio cholerae* is an important human pathogen in developing regions, where cholera epidemics cause widespread death by a severe form of diarrhea. The etiology of the transition of *V. cholera* from a benign member of the aquatic community to one that causes devastating epidemics may be heavily underpinned by environmental and climatic factors *(70)*. Genome sequencing of *V. cholerae* reveals two circular chromosomes of 2.96 Mb and 1.07 Mb, which encodes a total of 3885 ORFs *(71)*. Most of the genes on the larger chromosome (chromosome 1) are those essential for normal cellular function and pathogenicity, while chromosome 2 contains mostly hypothetical genes and an integron island gene capture system often found on plasmids *(71)*. Chromosome 1 is probably phylogenetically related to *E. coli*; chromosome 2 probably originated by cellular capture of a megaplasmid *(71)*.

Pseudomonas aeruginosa is a γ-proteobacterium commonly found in coastal marine systems, and it forms biofilms on rocks and soils. *Pseudomonas aeruginosa* is also found in association with plant and animal tissues and is an opportunistic human pathogen highly resistant to antibiotics. *Pseudomonas aeruginosa* causes stubborn and nearly impossible to treat infections of burn patients, patients with urinary tract catheters, patients with hospital-acquired pneumonia on respirators and is the predominant cause of death in patients with cystic fibrosis *(72)*. The *P. aeruginosa* genome is large (6.3 Mb) and encodes 5570 ORFs, including a wide variety of paralogous gene families *(72)*. One of these gene families encodes 150 genes for outer membrane proteins that could be the evolutionary result of strong competition for antibiotic resistance in diverse environments.

Staphylococcus aureus is a low G+C Gram-positive bacterium that is a sister taxon to the bacilli, clostridia, and mycoplasmas. *Staphylococcus aureus* is one of the major causes of hospital-acquired infections and can cause toxic shock syndrome and staphylococcal scarlet fever. It can also cause skin infections, food poisoning, pneumonia, sepsis, osteomyelitis, and infectious endocarditis *(73)*. It has grown increasingly resistant to antibiotics over the last 40 years, and some strains are now virtually resistant to all antibiotic treatment. A genome-sequencing project compared two antibiotic-resistant strains of *S. aureus*, one resistant to methicillin (N315) and the other to vancomycin (Mu50). The genomes are close in size (2,813,641 bp and 2,878,040 bp), percentage total G+C (32.8, 32.9), number of ORFs (2595, 2697), and overall sequence similarity

(96%) *(73)*. The antibiotic resistance genes are carried on several bacteriophages encoding several resistance islands. The genomes also have pathogenicity islands (PAIs). PAIs are large (10–200 kb) stretches of pathogen genomic DNA that encode a variety of virulence factors (see Chapter 4). These can include adhesins, toxins, invasins, protein secretion systems, iron uptake systems, and others *(74)*. PAIs probably accumulate genes and move from taxon to taxon by horizontal gene transfer. *Staphylococcus aureus* also contains several highly diverse exotoxin superantigen protein transposons, insertion sequences, and all the known pathogenic factors plus 70 new ones *(73)*.

THE HUMAN MICROBIOME, UNCULTIVATED COMMENSALS: THE "OTHER" HUMAN GENOME

Completion of the draft sequence of the human genome is a milestone in science akin to landing the first humans on the moon. But this accomplishment should be seen as the beginning of the genomics age and should not place us into, as it has been popularly called, the postgenomics era. The human genome provides a rough scaffolding, a wild-type map, of the normal gene complement. It is a starting point to interpret human evolution and biology. An important and, to date, overlooked component of human evolution and biology no doubt is understanding the composition of human microflora.

Inside every person is a complex microhabitat of 500–1000 species of microbes, at least half of which cannot be cultivated. This collection of microbes has been called the *microbiome* and should be included as part of the human genomic milieu *(75)*. If each of these microbes has an approximate genome size of 5 Mbp and encodes 4000 genes per genome, this means that there are 2–4 million nonhuman genes in humans. This suggests that, within the microbiome, there may be 100 times more nonhuman genes than human *(75)*. Some of the gene products from these microbes are secreted and may have an impact on human cellular physiology. Furthermore, the microbes make metabolic by-products and sometimes produce toxins. This begs the question, What is the role of these microbiomic genes and gene products in normal human physiology?

Microbial ecology underwent a revolution with techniques designed around characterizing 16S rRNA gene diversity. These studies attempted to sequence all members of microbial communities, then identify phylogenetic affiliations by database comparisons against the background of cultivated organisms. These microbial molecular ecology studies have been applied to the human uncultivated microflora with eye-opening results. A study of this type was conducted on the human subgingival crevice and detected over 50 unique phylotypes spanning five divisions of the Bacteria *(76)*. The role of these uncultivated microbes in specific association with disease is also unclear. The best example to date is the linking of *H. pylori* and gastric ulcers *(77)*. But, uncultivated microbes have implications in many diseases, including cardiovascular disease, prostatitis, Crohn's disease, kidney stones, Tourette syndrome, diabetes, multiple sclerosis, tropical sprue, Kawasaki disease, Wegener's granulomatosis, sarcoidosis, rheumatoid arthritis, breast cancer, non-Hodgkin's lymphoma, testicular cancer, and prostate cancer *(77)*.

Characterizing genomics of the uncultivated human microflora may shed light on physiological differences between people, provide tools for detection of environmental exposure, and help account for the large proportion of the nongenetic component as causation of many diseases. To date, most studies have centered around rRNA sequence signatures *(76,78)*, but newer techniques such as microarrays *(79,80)* and transcript-filtering subtraction *(81)* can allow more rapid data accumulation, leading to new drug targets, preventive maintenance, and personalized medicine.

CONCLUSION

As more genomic data accumulate from both cultivated and uncultivated organisms, the relationship between an organism's genome and its physiology becomes clearer. In a broader sense, genomes are increasingly revealed to be remnants, mosaics, and composites formed by virtue of their unique ecological and evolutionary histories. Genomics is just beginning to be appreciated in a community context, without regard to the organismal source of given gene products, and to be thought of in terms of metagenomes interacting physiologically with their environment. To date, genomics has been analagous to X-ray crystallographic studies of proteins, which provide determination of one possible structure of many in vivo. Similarly, genomes provide a starting point for understanding organismal and community physiology, ecology, and evolution.

REFERENCES

1. Feldman RA, Black MB, Cary CS, Lutz RA, Vrijenhoek RC. Molecular phylogenetics of bacterial endosymbionts and their vestimentiferan hosts. Mol Marine Biol Biotechnol 1997; 6:268–277.
2. Peek AS, Feldman RA, Lutz RA, Vrijenhoek RC. Cospeciation of chemoautotrophic bacteria and deep sea clams. Proc Natl Acad Sci USA 1998; 95:9962–9966.
3. Galan JE, Collmer A. Type III secretion machines: bacterial devices for protein delivery into host cells. Science 1999; 284:1322–1328.
4. Dale C, Plague GR, Wang B, Ochman H, Moran NA. Type III secretion systems and the evolution of mutualistic endosymbiosis. Proc Natl Acad Sci USA 2002; 99:12,397–12,402.
5. Fuqua WC, Winans SC, Greenberg EP. Quorum sensing in bacteria: the LuxR-LuxI family of cell density-responsive transcriptional regulators. J Bacteriol 1994; 176:269–275.
6. Surette MG, Miller MB, Bassler BL. Quorum sensing in *Escherichia coli*, *Salmonella typhimurium*, and *Vibrio harveyi*: a new family of genes responsible for autoinducer production. Proc Natl Acad Sci USA 1999; 96:1639–1644.
7. Bainton NJ, Bycroft BW, Chhabra SR, et al. A general role for the lux autoinducer in bacterial cell signalling: control of antibiotic biosynthesis in *Erwinia*. Gene 1992; 116:87–91.
8. Fraser CM, Gocayne JD, White O, et al. The minimal gene complement of *Mycoplasma genitalium*. Science 1995; 270:397–403.
9. Funk DJ, Wernegreen JJ, Moran NA. Intraspecific variation in symbiont genomes: bottlenecks and the aphid-buchnera association. Genetics 2001; 157:477–489.
10. Maruyama T, Kimura M. Genetic variability and effective population size when local extinction and recolonization of subpopulations are frequent. Proc Natl Acad Sci USA 1980; 77:6710–6714.
11. Itoh T, Martin W, Nei M. Acceleration of genomic evolution caused by enhanced mutation rate in endocellular symbionts. Proc Natl Acad Sci USA 2002; 99:12,944–12,948.
12. Read TD, Salzberg SL, Pop M, et al. Comparative genome sequencing for discovery of novel polymorphisms in *Bacillus anthracis*. Science 2002; 296:2028–2033.
13. Margulis L. Symbiosis in Cell Evolution. San Francisco: Freeman, 1981.
14. Palmer JD. A single birth of all plastids? Nature 2000; 405:32–33.
15. Moreira D, Le Guyader H, Philippe H. The origin of red algae and the evolution of chloroplasts. Nature 2000; 405:69–72.

16. Van der Auwera G, Hofmann CJ, De Rijk P, De Wachter R. The origin of red algae and cryptomonad nucleomorphs: a comparative phylogeny based on small and large subunit rRNA sequences of *Palmaria palmata*, *Gracilaria verrucosa*, and the *Guillardia theta* nucleomorph. Mol Phylogenet Evol 1998; 10:333–342.

17. Baldauf SL, Roger AJ, Wenk-Siefert I, Doolittle WF. A kingdom-level phylogeny of eukaryotes based on combined protein data. Science 2000; 290:972–977.

18. Yang D, Oyaizu Y, Oyaizu H, Olsen GJ, Woese CR. Mitochondrial origins. Proc Natl Acad Sci USA 1985; 82:4443–4447.

19. Gray MW, Burger G, Lang BF. Mitochondrial evolution. Science 1999; 283:1476–1481.

20. Lang BF, Burger G, O'Kelly CJ, et al. An ancestral mitochondrial DNA resembling a eubacterial genome in miniature. Nature 1997; 387:493–497.

21. Adams KL, Daley DO, Qiu YL, Whelan J, Palmer JD. Repeated, recent and diverse transfers of a mitochondrial gene to the nucleus in flowering plants. Nature 2000; 408:354–357.

22. Gray MW. Mitochondrial genes on the move. Nature 2000; 408:302–305.

23. Huang CY, Ayliffe MA, Timmis JN. Direct measurement of the transfer rate of chloroplast DNA into the nucleus. Nature 2003; 422:72–76.

24. Cavalier-Smith T. Membrane heredity and early chloroplast evolution. Trends Plant Sci 2000; 5:174–182.

25. Douglas S, Zauner S, Fraunholz M, et al. The highly reduced genome of an enslaved algal nucleus. Nature 2001; 410:1091–1096.

26. Woese CR. Bacterial evolution. Microbiol Rev 1987; 51:221–271.

27. Goodner B, Hinkle G, Gattung S, et al. Genome sequence of the plant pathogen and biotechnology agent *Agrobacterium tumefaciens* C58. Science 2001; 294:2323–2328.

28. Wood DW, Setubal JC, Kaul R, et al. The genome of the natural genetic engineer *Agrobacterium tumefaciens* C58. Science 2001; 294:2317–2323.

29. Downie JA, Young JPW. The abc of symbiosis. Nature 2001; 412:597–598.

30. Galibert F, Finan TM, Long SR, et al. The composite genome of the legume symbiont *Sinorhizobium meliloti*. Science 2001; 293:668–672.

31. Kaneko T, Nakamura Y, Sato S, et al. Complete genome structure of the nitrogen-fixing symbiotic bacterium *Mesorhizobium loti*. DNA Res 2000; 7:331–338.

32. Moran NA. Bacterial menageries inside insects. Proc Natl Acad Sci USA 2001; 98:1338–1340.

33. Clark MA, Baumann L, Thao ML, Moran NA, Baumann P. Degenerative minimalism in the genome of a psyllid endosymbiont. J Bacteriol 2001; 183:1853–1861.

34. Moran NA, Mira A. The process of genome shrinkage in the obligate symbiont *Buchnera aphidicola*. Genome Biol 2001; 2:2–54.

35. Shigenobu S, Watanabe H, Hattori M, Sakaki Y, Ishikawa H. Genome sequence of the endocellular bacterial symbiont of aphids *Buchnera* sp APS. Nature 2000; 407:81–86.

36. Moran NA, Wernegreen JJ. Lifestyle evolution in symbiotic bacteria: insights from genomics. Trends Ecol Evol 2000; 15:321–326.

37. Tamas I, Klasson L, Canback B, et al. Fifty million years of genomics stasis in endosymbiotic bacteria. Science 2002; 296:2376–2379.

38. Gil R, Sabater-Munoz B, Latorre A, Silva FJ, Moya A. Extreme genome reduction in *Buchnera* spp: toward the minimal genome needed for symbiotic life. Proc Natl Acad Sci USA 2002; 99: 4454–4458.

39. Preston CM, Wu KY, Molinski TF, DeLong EF. A psychrophilic crenarchaeon inhabits a marine sponge: *Cenarchaeum symbiosum* gen nov sp nov. Proc Natl Acad Sci USA 1996; 93:6241–6246.

40. Webster NS, Watts JEM, Hill RT. Detection and phylogenetic analysis of novel crenarchaeote and euryarchaeote 16S ribosomal RNA gene sequences from a Great Barrier Reef sponge. Marine Biotechnol 2001; 3:600–608.

41. van Hoek AH, van Alen TA, Sprakel VS, et al. Multiple acquisition of methanogenic archaeal symbionts by anaerobic ciliates. Mol Biol Evol 2000; 17:251–258.
42. Stein JL, Marsh TJ, Wu KY, Shizuya H, Delong EF. Characterization of uncultivated prokaryotes: isolation and analysis of a 40-kilobase-pair genome fragment from a planktonic marine archaeon. J Bacteriol 1996; 178:591–599.
43. Beja O, Suzuki MT, Koonin EV, et al. Construction and analysis of bacterial artificial chromosome libraries from a marine microbial assemblage. Environ Microbiol 2000; 2:516–529.
44. Altschul SF, Gish W, Miller W, Myers EW, Lipman DJ. Basic Local Alignment Search Tool. J Mol Biol 1990; 215:403–410.
45. Beja O, Aravind L, Koonin EV, et al. Bacterial rhodopsin: evidence for a new type of phototrophy in the sea. Science 2000; 289:1902–1906.
46. Schleper C, DeLong EF, Preston CM, Feldman RA, Wu KY, Swanson RV. Genomic analysis reveals chromosomal variation in natural populations of the uncultured psychrophilic archaeon *Cenarchaeum symbiosum*. J Bacteriol 1998; 180:5003–5009.
47. Chevaldonne P, Desbruyeres D, Childress JJ. ... and some even hotter. Nature 1992; 359:593.
48. Cavanaugh CM, Gardiner SL, Jones ML, Jannasch HW, Waterbury JB. Prokaryotic cells in the hydrothermal vent tube worm *Riftia pachyptila* Jones: possible chemoautotrophic symbionts. Science 1981; 213:340–341.
49. Di Meo CA, Wilbur AE, Holben WE, Feldman RA, Vrijenhoek RC, Cary SC. Genetic variation among endosymbionts of widely distributed vestimentiferan tubeworms. Appl Environ Microbiol 2000; 66:651–658.
50. Millikan DS, Felbeck H, Stein JL. Identification and characterization of a flagellin gene from the endosymbiont of the hydrothermal vent tubeworm *Riftia pachyptila*. Appl Environ Microbiol 1999; 65:3129–3133.
51. Hughes DS, Felbeck H, Stein JL. A histidine protein kinase homolog from the endosymbiont of the hydrothermal vent tubeworm *Riftia pachyptila*. Appl Environ Microbiol 1997; 63:3494–3498.
52. Fleischmann RD, Adams MD, White O, et al. Whole-genome random sequencing and assembly of *Haemophilus influenzae* Rd. Science 1995; 269:496–512.
53. Warren JR, Marshall B. Unidentified curved bacilli on gastric epithelium in active chronic gastritis. Lancet 1983; 1:1273–1275.
54. Olson JW, Maier RJ. Molecular hydrogen as an energy source for *Helicobacter pylori*. Science 2002; 298:1788–1790.
55. Tomb JF, White O, Kerlavage AR, et al. The complete genome sequence of the gastric pathogen *Helicobacter pylori*. Nature 1997; 388:539–547.
56. Falush D, Wirth T, Linz B, et al. Traces of human migrations in *Helicobacter pylori* populations. Science 2003; 299:1582–1585.
57. Spratt BG. Microbiology: Stomachs out of Africa. Science 2003; 299:1528–1529.
58. Himmelreich R, Hilbert H, Plagens H, Pirkl E, Li BC, Herrmann R. Complete sequence analysis of the genome of the bacterium *Mycoplasma pneumoniae*. Nucleic Acids Res 1996; 24: 4420–4449.
59. Himmelreich R, Plagens H, Hilbert H, Reiner B, Herrmann R. Comparative analysis of the genomes of the bacteria *Mycoplasma pneumoniae* and *Mycoplasma genitalium*. Nucleic Acids Res 1997; 25:701–712.
60. Hutchison CA, Peterson SN, Gill SR, et al. Global transposon mutagenesis and a minimal *Mycoplasma* genome. Science 1999; 286:2165–2169.
61. Fraser CM, Norris SJ, Weinstock GM, et al. Complete genome sequence of *Treponema pallidum*, the syphilis spirochete. Science 1998; 281:375–388.
62. Read TD, Brunham RC, Shen C, et al. Genome sequences of *Chlamydia trachomatis* MoPn and *Chlamydia pneumoniae* AR39. Nucleic Acids Res 2000; 28:1397–1406.

63. Tettelin H, Saunders NJ, Heidelberg J, et al. Complete genome sequence of *Neisseria mening-itidis* serogroup B strain MC58. Science 2000; 287:1809–1815.

64. Blaser MJ. Epidemiologic and clinical features of *Campylobacter jejuni* infections. J Infect Dis 1997; 176:S103–S105.

65. Parkhill J, Wren BW, Mungall K, et al. The genome sequence of the food-borne pathogen *Campylobacter jejuni* reveals hypervariable sequences. Nature 2000; 403:665–668.

66. Ferretti JJ, McShan WM, Ajdic D, et al. Complete genome sequence of an M1 strain of *Streptococcus pyogenes*. Proc Natl Acad Sci USA 2001; 98:4658–4663.

67. Hoskins J, Alborn WE Jr, Arnold J, et al. Genome of the bacterium *Streptococcus pneumoniae* strain R6. J Bacteriol 2001; 183:5709–5717.

68. Haygood MG, Distel DL. Bioluminescent symbionts of flashlight fishes and deep-sea angler-fishes form unique lineages related to the genus *Vibrio*. Nature 1993; 363:154–156.

69. Nyholm SV, Stabb EV, Ruby EG, McFall-Ngai MJ. Establishment of an animal-bacterial association: recruiting symbiotic vibrios from the environment. Proc Natl Acad Sci USA 2000; 97: 10,231–10,235.

70. Colwell RR. Global climate and infectious disease: the cholera paradigm. Science 1996; 274: 2025–2031.

71. Heidelberg JF, Eisen JA, Nelson WC, et al. DNA sequence of both chromosomes of the cholera pathogen *Vibrio cholerae*. Nature 2000; 406:477–483.

72. Stover CK, Pham XQ, Erwin AL, et al. Complete genome sequence of *Pseudomonas aeruginosa* PA01, an opportunistic pathogen. Nature 2000; 406:959–964.

73. Kuroda M, Ohta T, Uchiyama I, et al. Whole genome sequencing of meticillin-resistant *Staphylococcus aureus*. Lancet 2001; 357:1225–1240.

74. Hacker J, Kaper JB. Pathogenicity islands an the evolution of microbes. Annu Rev Microbiol 2000; 54:641–679.

75. Hooper LV, Gordon JI. Commensal host-bacterial relationships in the gut. Science 2001; 292: 1115–1118.

76. Kroes I, Lepp PW, Relman DA. Bacterial diversity within the human subgingival crevice. Proc Natl Acad Sci USA 1999; 96:14,547–14,552.

77. Relman DA. The human body as a microbial observatory. Nat Genet 2002; 30:131–133.

78. Tanner MA, Shoskes D, Shahed A, Pace NR. Prevalence of corynebacterial 16S rRNA sequences in patients with bacterial and "nonbacterial" prostatitis. J Clin Microbiol 1999; 37:1863–1870.

79. Cummings CA, Relman DA. Using DNA microarrays to study host-microbe interactions. Emerg Infect Dis 2000; 6:513–525.

80. Diehn M, Relman DA. Comparing functional genomic datasets: lessons from DNA microarray analyses of host-pathogen interactions. Curr Opin Microbiol 2001; 4:95–101.

81. Weber G, Shendure J, Tanenbaum DM, Church GM, Meyerson M. Identification of foreign gene sequences by transcript filtering against the human genome. Nat Genet 2002; 30:141–142.

V A SURVEY OF MICROBIAL GENOMES

13

A Survey of Plant Pathogen Genomes

C. Robin Buell

INTRODUCTION

Plants are susceptible to a large range of pathogens, including viruses, bacteria, fungi, and nematodes. Plants arm themselves with preformed defense mechanisms and initiate *de novo* defense reponses to ensure continued existence. These defense mechanisms present a challenge for pathogens, which in turn adapt their virulence and pathogenicity mechanisms to evade the host defense responses and thus secure a nutrient source for growth and reproduction. This delicate balance of host defense vs pathogen virulence has resulted in the coevolution of the pathogen and the host plant.

The central dogma in plant disease resistance is the gene-for-gene hypothesis, which was originally proposed by Flor in the 1940s following a genetic study of disease resistance in flax with the flax rust pathogen *Melampsora lini (1)*. In this model, the interaction between the product of a single dominant gene in the host and the product from a single avirulence gene in the pathogen determine the outcome of the host–pathogen interaction. With the advent of molecular techniques, multiple pathogen avirulence genes and host resistance genes have been cloned, confirming this genetic model at the molecular level (see ref. *2* for a review of the chronology of disease resistance research). Exploitation of the gene-for-gene model has enabled researchers to create transgenic plants with modified pathogen resistance profiles (bacterial, fungal, and viral resistance) using either cloned resistance or avirulence genes *(3–7)*, suggesting that novel control mechanisms for crop plants through bioengineering is feasible. Thus, through the discovery of pathogenicity and virulence mechanisms of plant pathogens, new insights, as well as reagents for plant disease control, can be obtained. As a consequence, the primary focus of all plant pathogen genome projects to date has been the identification of virulence and pathogenicity mechanisms.

For the purposes of this review, the focus is on bacterial, fungal, and nematode species for which genome projects have been completed or are in progress. A complete list of completed and ongoing genome-sequencing projects for bacterial plant pathogens is presented in Table 1, with fungal/oomycete and nematode expressed sequence tag (EST) projects listed in Tables 2 and 3, respectively. Web-based resources are available that report the status of current plant pathogen and plant-associated microbial genome projects: http://www.tigr.org/~vinita/PPwebpage.html and http://www.oardc.ohio-state.edu/Phytophthora/genome_links.htm.

From: *Microbial Genomes*
Edited by: C. M. Fraser, T. D. Read, and K. E. Nelson © Humana Press Inc., Totowa, NJ

Table 1
Genome Projects for Bacterial Plant Pathogens

Bacterium	Genome size (Mb)	Disease	Reference	Web address
Complete genomes				
Agrobacterium tumefaciens C58	5.7	Crown gall on numerous species	*22,23*	www.genome.washington.edu, www.cereon.com
Ralstonia solanacearum GMI1000	5.8	Bacterial wilt on tomato, Arabidopsis	*19*	www.sequence.toulouse.inra.fr
Xanthomonas axonopodis pv. *citri* 306	5.2	Citrus canker	*30*	http://genoma4.iq.usp.br/xanthomonas/
Xanthomonas campestris pv *campestris* ATCC3391	5.1	Black rot on crucifers	*30*	http://genoma4.iq.usp.br/xanthomonas/
Xylella fastidiosa 9a5c	2.7	Citrus variegated chlorosis	*8*	www.watson.fapesp.br
Draft genomes				
Pseudomonas syringae pv *syringae* B728a	~6.0	Bacterial brown spot on bean		www.jgi.doe.gov
Xylella fastidiosa–almond (Dixon)	2.6	Almond leaf scorch		www.jgi.doe.gov
Xylella fastidiosa–oleander (ann1)	2.6	Oleander leaf scorch		www.jgi.doe.gov
Genomes in progress				
Clavibacter michiganensis spp *sepedonicus*	~2.5	Potato Ring Rot		http://www.sanger.ac.uk/Projects/Microbes/
Erwinia carotovora spp *atroseptica*	~5.0	Soft rot and black leg on potato		http://www.sanger.ac.uk/Projects/Microbes/
Erwinia chrysanthemi 3937	3.7	Soft rot of a variety of host species		www.genome.wisc.edu, http://www.tigr.org/tdb/mdb/mdbinprogress.html
Pseudomonas syringae pv *tomato* DC3000	6.5	Bacterial speck on tomato		http://www.tigr.org/tdb/mdb/mdbinprogress.html
Xanthomonas campestris pv *campestris* 8004	5.1	Black rot on crucifers		http://www.chgb.org.cn/en_ke.htm
Xylella fastidiosa–Pierce's disease strain	2.6	Pierce's disease on grapes		http://www.lbi.ic.unicamp.br/world/xf-grape/

Table 2
EST Projects for Fungal and Oomycete Plant Pathogens

Species	Number of ESTs	Disease
Blumeria graminis f sp *hordei*	4908	Powdery mildew on barley
Fusarium sporotrichioides	7625	Maize ear rot
Magnaporthe grisea	14,160	Rice blast
Mycosphaerella graminicola	1158	Wheat septoria leaf blotch pathogen
Phytophthora infestans	2129	Late blight on potato and tomato
Phytophthora sojae	2004	Soybean root and stem rot
Total	31,985	

The number of ESTs per species was determined from the 051002 release of dbEST and only includes species with more than 1000 ESTs. The numbers for *M. grisea* represent ESTs deposited for *M. grisea* and the former name for this species, *Pyricularia grisea*.

Table 3
EST Projects for Nematode Plant Pathogens

Species	Number of ESTs	Disease
Globodera pallida	1832	Potato cyst nematode
Globodera rostochiensis	5934	Potato cyst nematode
Heterodera glycines	4327	Soybean cyst nematode
Meloidogyne arenaria	3334	Root knot nematode
Meloidogyne hapla	6157	Northern root knot nematode
Meloidogyne incognita	10,899	Southern root knot nematode
Meloidogyne javanica	5600	Root knot nematode
Total	38,083	

The number of ESTs per species was determined from the 051002 release of dbEST and only includes species with more than 1000 ESTs. Current data on EST sequencing of plant pathogenic nematodes can be found at http://www.nematode.net/.

PLANT PATHOGEN BACTERIAL GENOMES

In spite of the obvious importance of agriculture in daily life, application of genomics to plant pathogens had been limited in scope. Although over 75 complete bacterial genomes have been reported as of May 2002 (http://www.ncbi.nlm.nih.gov/PMGifs/Genomes/micr.html), only six genomes are from plant pathogenic bacteria. The six complete genome sequences were determined for five plant pathogenic bacterial species (Table 1). The draft sequence for several other bacterial plant pathogen genomes have been completed, and a subset of other bacterial genomes are in the process. To date, a broad survey of the genomes of bacterial plant pathogenic species has begun, and through additional genome projects, data on diversity within a plant pathogenic bacterial species will be generated.

Xylella fastidiosa

The first complete bacterial plant pathogen genome was that of *Xylella fastidiosa*. This was accomplished by the Organization for Nucleotide Sequencing and Analysis

(ONSA) consortium in Brazil and was reported in July 2000 *(8)*. *Xyella fastidosa* is a xylem-limited bacterium that is the causal agent of several economically significant diseases *(9,10)*. One isolate in particular, 9a5c, is the causal agent of citrus variegated chlorosis *(11)*, which causes infected citrus plants to appear wilted and have necrotic lesions on infected leaves *(12)*. The bacterium is transmitted by insects such as the sharp-shooter, and current control approaches rely on pruning infected trees and reduction/elimination of the insect vector through insecticide applications *(12)*. In addition to its vector transmission issues, *X. fastidiosa* is a difficult pathogen to manipulate in the laboratory; consequently, both the pathogen and the disease are poorly understood at the physiological and organismal levels. The completion of the *X. fastidiosa* genome was a major turning point in the approach to plant pathology and is a powerful example of the impact that genomics can have in accelerating the understanding of plant pathology.

The genome of *X. fastidiosa* 9a5c contains a 2.68-Mb chromosome and two plasmids of 51.1 kb and 1.28 kb with guanine and cytosine (G+C) contents of 52.7, 49.6, and 55.6%, respectively *(8)*. The chromosome and two plasmids encode a total of 2848 open reading frames (ORFs). A total of 1314 (46%) of the ORFs could be assigned a putative function based on similarity to known genes in the databases. The metabolic and transport systems could be deduced from the genome annotation data and were consistent with the nutrient-limited environment of the xylem *(8)*.

Insights into the physiology and virulence of this pathogen have been made from the analysis of the 9a5c genome sequence. In a number of Gram-negative bacterial pathogens, it has been demonstrated that a cluster of genes termed the *hypersensitive response* and *pathogenicity* (*hrp*) regulon are essential for growth *in planta*. The *hrp* genes encode the components of the type III secretion system in which molecules are delivered to the host cell via a secretion apparatus spanning the inner and outer membrane and determine the outcome of the plant–microbe interaction (see refs. *13–15* for current reviews on type III secretion systems). The group of molecules delivered through the type III secretion are referred to as effectors and include products from the avirulence genes. Surprisingly, no avirulence genes or genes in the type III secretion system were present in the 9a5c genome, suggesting that host specificity in *X. fastidiosa* is determined through mechanisms other than the classical gene-for-gene model *(8)*. However, other genes with putative roles in virulence/pathogenicity were identified in the 9a5c genome, including a cellulase gene and proteins involved in adherence of the bacterium to the host cell wall, such as extracellular polysaccharide production, type 4 fimbria filaments, and afimbrial adhesions *(8)*. These discoveries provide a platform to begin a systematic study of virulence mechanisms in this species.

As mentioned, the species *X. fastidosa* is the causal agent of a large number of diseases, including citrus variegated chlorosis, Pierce's disease on grape, phony peach disease, and leaf scorch on several hosts including almond, plum, oak, and maple *(9,10)*. Diversity in this species has been examined through classical genotyping techniques such as restriction length fragment polymorphisms and random amplified polymorphic deoxyribonucleic acid (DNA) *(16)*. To address further the diversity and host specificity in this species, the Department of Energy Joint Genome Institute (DOE-JGI) sequenced two additional strains of *X. fastidosa*, one from oleander and one from almond (http://www. jgi.doe.gov/JGI_microbial/html/xylella_oleander/xyle_olean_homepage.html; http:// www.jgi.doe.gov/JGI_microbial/html/xylella_almond/xyle_almnd_homepage.html).

Although closure (finishing) of these genomes has not been proposed, the availability of a sequence for two strains isolated from two additional host species provides a mechanism to understand diversity in this species and potentially provides a resource to identify sequences required for pathogenicity or virulence on specific host species through comparative genomics. In addition to these two draft sequences, a strain of *X. fastidiosa* that causes Pierce's disease, a significant disease on grape, is being sequenced by the ONSA (http://www.lbi.ic.unicamp.br/world/xf-grape/), resulting in a total of four isolates of *X. fastidiosa* for which genome sequence will be available.

Ralstonia solanacearum

Another significant Gram-negative bacterial pathogen of plants is the vascular wilt pathogen *Ralstonia solanacearum*. This bacterium is able to infect the xylem vessels of the vascular tissue. Through extensive bacterial multiplication and production of extracellular polysaccharides, the vessels become plugged, and the plant is unable to translocate water through its vascular tissue, resulting in wilting *(17)*.

The isolate selected for sequencing is the GMI1000 strain, which is able to infect the dicotyledonous vegetable crop tomato as well as the model plant *Arabidopsis thaliana (18)*. Unlike citrus, the host of *X. fastidiosa* 9a5c, both *Arabidopsis* and tomato have tractable, well-developed genetic systems. Thus, access to the *R. solanacearum* GMI1000 genome sequence provides a complementary resource to dissect thoroughly a plant–pathogen interaction. The genome of *R. solanacearum* GMI1000 contains a 3.72-Mb chromosome and a 2.1-Mb megaplasmid *(19)*. The G+C content of the chromosome and megaplasmid are similar (67 and 66.9%, respectively). The two molecules have equivalent coding content per megabase, with the chromosome encoding 3448 proteins and the megaplasmid encoding 1681 proteins. A total of 2261 (44.1%) proteins could be assigned a putative function. A biologically interesting division in the localization of functional proteins can be seen between the chromosome and the megaplasmid. With respect to functional classification of genes, a higher frequency of genes with a putative function are located on the chromosome, whereas the megaplasmid contains a higher frequency of hypothetical genes in comparison to the chromosome. With respect to virulence mechanisms, the megaplasmid encodes a number of key genes involved in virulence, including the *hrp* genes essential for growth *in planta* and for the secretion of effector molecules and a number of genes involved in extracellular polysaccharide production.

Unlike *X. fastidiosa*, *R. solanacearum* has been well studied by plant pathologists, and a number of the virulence mechanisms utilized by this pathogen were known to pathologists prior to the release of the genome sequence *(20)*. However, access to the full genome sequence has provided pathologists the ability to probe fully the mechanism of virulence in this pathogen. From the genome sequence, an additional 195 genes were identified as candidate pathogenicity genes, and these include genes that encode effector molecules, adhesion/attachment factors, degradative enzymes, toxins, and oxidative stress response factors *(19)*.

Agrobacterium tumefaciens

Agrobacterium tumefaciens has been well studied for the last three decades as it has been heavily exploited for its ability to transfer DNA to plant nuclei and in the process create transgenic plants. Through conventional molecular biology research, the virulence mechanism by which *A. tumefaciens* transfers DNA has been dissected and reengineered to serve biotechnology functions *(21)*.

The importance of understanding the biology *A. tumefaciens* more thoroughly is reflected in the fact that two groups independently undertook to sequence the same strain of the bacterium *(22,23)*. The genome of *A. tumefaciens* C58 consists of four replicons: a 2.8-Mb circular chromosome, a 2.1-Mb linear chromosome, a 543-kb plasmid (pAtC58), and a 214-kb plasmid (pTiC58) *(23)*. In total, the genome encodes 5419 proteins, of which a large percentage (64.1%) have been assigned a putative function. The G+C content of the replicons is similar, ranging from 56.7 to 59.4%. However, reflective of the transfer of the transferred DNA (T-DNA) to the eukaryotic plant host nuclei, which have a lower G+C content, the T-DNA portion of the pTi58 plasmid has a G+C content of 46% *(23)*.

A large amount of information on the virulence mechanisms of *A. tumefaciens* was known prior to the genome sequence data, and perhaps the most intriguing result of the *A. tumefaciens* genome sequence was the high degree of similarity between *A. tumefaciens* and the two plant symbiotic bacteria, *Sinorhizobium meliloti (24)* and *Mesorhizobium loti (25)*, with *S. meliloti* being more closely related to *A. tumefaciens* than *M. loti (23)*. Both *S. meliloti* and *M. loti* are symbiotic bacteria that form nodules on plant roots that result in the fixation of nitrogen. Alignment of the *A. tumefaciens* C58 circular chromosome with the *S. meliloti* chromosome revealed a high degree of colinearity *(22,23)*. Thus, although one species is beneficial and the other is pathogenic, both *A. tumefaciens* and *S. meliloti* are soil-borne, plant-associated bacteria, and this commonality is reflected in the conservation of genes and gene order (see Chapter 12).

Xanthomonas *species*

Xanthomonas campestris pathovar (pv) *campestris* is the causal agent of black rot of crucifers *(26)* and, similar to *R. solanacearum*, is a vascular pathogen. Worldwide, black rot is considered a significant disease of crucifers *(26)*. All commercially grown crucifer species are susceptible to this pathogen, and control of this disease is complicated by the seed-borne nature of the pathogen and the lack of strong resistance in commercial crucifer varieties *(26)*. Insight into host resistance mechanisms have been made in the cruiciferous weed *A. thaliana,* as some isolates of *X. campestris* pv *campestris* are pathogenic on this model plant species (for review of *Xanthomonas–Arabidopsis* interactions, see ref. *27*).

Xanthomonas axonopodis pv *citri* is the causal agent of citrus canker, an economically significant disease of citrus throughout the world, with quarantine and removal/destruction of infected trees the only available control measures (for an article on the impact of this disease, see ref. *28*). In contrast to *X. campestris* pv *campestris*, *X. axonopodis* pv *citri* infects nonvascular tissue, resulting in the formation of lesions and cankers on leaves and fruits *(29)*.

In a benchmark paper, the Brazilian ONSA group reported the sequence of two species of *Xanthomonas*, *X. campestris* pv *campestris* strain ATCC33913 and *X. axonopodis* pv *citri* strain 306 *(30)*. The genome sizes of these two pathogens are similar; *X. campestris* pv *campestris* has a 5.1-Mb circular chromosome, and *X. axonopodis* pv *citri* has a 5.2-

Mb circular chromosome and two small plasmids of 34 and 65 kbp. There were 4182 and 4313 ORFs identified in *X. campestris* pv *campestris* and *X. axonodopis* pv *citri*, respectively. A high degree of conservation of gene sequence as well as gene order is present in these two pathogens. Approximately 82% of the ORFs in these two pathogens share 80% or more identity at the amino acid level. With respect to colinearity, about 70% of the nontransposable element genes from each genome could be aligned with a corresponding orthologous gene in the other genome. A rather small fraction of genes was unique to each species, with 646 genes unique to *X. campestris* pv *campestris* and 800 genes unique to *X. axonodopis* pv *citri*.

The virulence and pathogenicity mechanisms of these two *Xathomonas* species are highly conserved, and genes encoding the *hrp* regulon, effector molecules, proteases, and cell wall–degrading enzymes are present in both genomes *(30)*. However, the quantity and representation of specific types of virulence and pathogenicity factors differs between the two pathogens. For example, *X. campestris* pv *campestris* has more genes involved in cell wall degradation than *X. axonodopis* pv *citri*, suggesting it may be the number and the specificity of these additional degradative genes that result in the differential symptom formation in citrus canker vs black rot *(30)*. The type of effector genes present in the two pathogens also differed, with *X. campestris* pv *campestris* encoding a broader set of effector molecules than *X. axonodopis* pv *citri (30)*. Clearly, the identification of unique genes for each species, of which the pathogenicity and virulence genes are a subset, provides an excellent resource to dissect systematically the requirements for growth and development of bacterial pathogens that reside in two diverse ecological niches (the vascular system and the mesophyll tissue) and two diverse disease processes (rot vs canker).

Bacterial Plant Pathogen Genomes In Progress

Several additional bacterial pathogen genome projects are in progress, and with the completion of these genomes, the genome sequence for a broader set of bacterial plant pathogen species will be available. Currently, the genome sequence of two species of *Erwinia*, a close relative of *Escherichia coli*, are in progress (Table 1). In addition, the genome sequence of the first Gram-positive plant pathogen, *Clavibacter michiganensis* spp *sepedonicus*, is in progress at the Sanger Center in the United Kingdom (Table 1). One genome in the final stages of sequencing is that of *Pseudomonas syringae* pv *tomato* DC3000, the causal agent of bacterial speck on tomato. The genome of *P. syringae* pv *tomato* DC3000 is about 6.5 Mb and contains a 6.4-Mb chromosome and two plasmids of about 70 kbp (CR Buell, unpublished). Even without a complete genome, data-mining of the unfinished genome has resulted in an expansion of the currently known repertoire of effector molecules for this pathogen that are involved in pathogenicity *(31,32)*. A second pathovar of *P. syringae* that infects bean, *P. syringae* pv *syringae* isolate B728a, has been sequenced to draft level by the DOE-JGI (Table 1) and will provide a resource for comparative analysis of the *P. syringae* pathovar group.

OOMYCETE AND FUNGAL PLANT GENOME PROJECTS

Due to their small genome size and the limited number of genera that infect plants, bacterial rather than oomycete, fungal, or nematode plant pathogens have received the first and foremost attention in the genomics arena. In contrast, a large number of fungal and oomycete genera that represent diverse taxonomic divisions have been shown to infect plants. Not only is there a wide range of genera and species to select for fungal or oomycete pathogen genome sequencing, the genome size of the average fungal/oomycete species is an order of magnitude higher than the average bacterial genome, thereby requiring substantially more funding than required for a bacterial genome sequencing project.

A large number of fungal species are pathogenic on plants, including species from the Ascomycetes, Basidiomycetes, and the imperfect fungi, which lack a sexual stage and propagate solely through asexual mechanisms. Oomycetes, which had been previously classified in the fungal kingdom, are now considered taxonomically distinct from the true fungi and have been placed in the Stramenopiles kingdom *(33,34)*. In an attempt to narrow the species for genome sequencing, the American Phytopathological Society has generated a list of pathogens they have prioritized for sequencing (http://www.apsnet.org/media/ps/top.asp) based on economic relevance of disease, interest by the scientific community, genetic tractability, and access to other tools/resources to leverage the sequence information. In this list, a total of 26 fungal or oomycete species have been recommended by the American Phytopathological Society for genome sequencing, with genome sizes ranging from 20 Mb in *Ustilago maydis* (corn smut) to 237 Mb in *Phytophthora infestans* (late blight in potato/tomato).

Insights into the coding regions of fungal genomes can be obtained through ESTs, in which single-pass sequence is obtained from cyclic DNA clones. Construction of cDNA libraries from specific developmental or biological time points allows sampling of the transcriptome at discrete stages relevant to pathogenicity, virulence, or reproduction. To date, a number of fungal/oomycete plant pathogen EST projects are under way, with a total of 31,985 ESTs in GenBank (Release 051002). Table 2 presents fungal/oomycete pathogens, their disease, and the number of ESTs for each species. Although there are few ESTs, they represent the first step into genomics for the plant pathogenic fungal and oomycete pathogens.

NEMATODE PLANT PATHOGEN GENOME PROJECTS

Plant parasitic nematodes constitute a significant component of disease loss to pathogens in agriculture (for a review of plant pathogenic nematodes, see ref. *35*). As soilborne pathogens, their presence cannot be abated readily as one component in the control of nematodes involves fumigation of the soil. Two genera of nematodes, *Meloidogyne* (the root knot nematode) and *Heterodera* (the cyst nematode), have received the majority of attention with regard to genomics. As with the fungal/oomycete plant pathogens, the genome size of plant parasitic nematodes is prohibitive to full genome sequencing projects with current funding resources. As a consequence, ESTs have been explored as an entry point into plant parasitic nematode genomes (Table 3).

In addition to the gene discovery provided through the EST projects, it is anticipated that major insights into plant parasitic nematode genomes will be made through comparative analyses between the genomes or ESTs of the plant parasitic species with the model nematode species *Caenorhabditis elegans*, which has been completely sequenced *(36)*. In addition to *C. elegans*, *Caenorhabditis briggsae* has been sequenced, and this

genome sequence will provide a second reference genome for plant parasitic nematodes
(http://www.nematode.net/Species.Summaries/Caenorhabditis.briggsae/).

CONCLUSIONS

Plant pathologists are getting their first taste of genomics through the bacterial plant
pathogen genomes that have been completed or are in progress. Access to this resource
will have a profound impact on the field—the equivalent to the cloning of the first pathogen avirulence genes and the complementary host resistance genes and confirmation of
the 40-year-old gene-for-gene hypothesis. Instead of testing one gene at a time, pathologists will be able to perform genomewide assays and uncover all of the components involved
in pathogenicity and virulence. It will be from these studies that novel control mechanisms will be developed that will allow more efficient and productive agriculture. In addition, comparative genomics, which was implemented early for plant pathogens, provides
a mechanism for the identification of genes involved in host specificity between related
strains, pathovars, species, or genera and is likely to continue to be a powerful tool in
the genomics of plant pathogens.

ACKNOWLEDGMENTS

Work on *P. syringae* pv *tomato* is supported by a grant from the National Science
Foundation Plant Genome Research Program (DBI 0077622). The critical reading by Vinita
Joardar is greatly appreciated, as is the administrative support of Ama Kwamena-Poh.

REFERENCES

1. Flor HH. Current status of the gene-for-gene concept. Annu Rev Phytopathol 1971; 9:275–296.
2. Staskawicz BJ. Genetics of plant-pathogen interactions specifying plant disease resistance. Plant
   Physiol 2001; 125:73–76.
3. Gopalan S, Bauer DW, Alfano JR, Loniello AO, He SY, Collmer A. Expression of the *Pseudomonas syringae* avirulence protein AvrB in plant cells alleviates its dependence on the hypersensitive response and pathogenicity (Hrp) secretion system in eliciting genotype-specific
   hypersensitive cell death. Plant Cell 1996; 8:1095–1105.
4. Thilmony RL, Chen Z, Bressan RA, Martin GB. Expression of tomato *Pto* gene in tobacco
   enhances resistance to *Pseudomonas syringae* pv *tabaci* expressing *avrPto*. Plant Cell 1995; 7:
   1529–1536.
5. Rommens CMT, Salmeron JM, Oldroyd GED, Staskawicz BJ. Intergeneric transfer and functional expression of the tomato disease resistance gene *Pto*. Plant Cell 1995; 7:1537–1544.
6. Whitham S, McCormick S, Baker B. The *N* gene of tobacco confers resistance to tobacco mosaic
   virus in transgenic tomato. Proc Natl Acad Sci USA 1996; 93:8776–8781.
7. Hammond-Kosack KE, Tang S, HarrisonK, Jones JDG. The tomato *Cf-9* disease resistance gene
   functions in tobacco and potato to confer responsiveness to the fungal avirulence gene product
   Avr9. Plant Cell 1998; 10:1251–1266.
8. Simpson AJ, Reinach FC, Arruda P, et al. The genome sequence of the plant pathogen *Xylella
   fastidiosa*. Nature 2000; 406:151–157.
9. Wells JM, Raju BC, Hung H-Y, Weisburg WG, Mandelco-Paul L, Brenner DJ. *Xylella fastidiosa*
   gen nov, sp nov: Gram-negative, xylem-limited, fastidious plant bacteria related to *Xathomonas*
   species. Int J Syst Bacteriol 1987; 37:136–143.

10. Hopkins DL. *Xylella fastidiosa*: xylem-limited bacterial pathogens of plants. Annu Rev Phytopathol 1989; 27:271–290.

11. Li WB, Zreik L, Fernandes NG, et al. A triply cloned strain of *Xylella fastidiosa* multiplies and induces symptoms of citrus variegated chlorosis in sweet orange. Curr Microbiol 1999; 39: 106–108.

12. Derrick KS, Timmer LW. Citrus blight and other diseases of recalcitrant etiology. Annu Rev Phytopathol 2000; 38:181–205.

13. Alfano JR, Collmer A. The type III secretion pathway of plant pathogenic bacteria: trafficking harpins, avr proteins, and death. J Bacteriol 1997; 179:5655–5662.

14. He SY. Type III protein secretion systems in plant and animal pathogenic bacteria. Annu Rev Phytopathol 1998; 36:363–392.

15. Staskawicz B, Mudgett MB, Dangl J, Galan JE. Common and contrasting themes of plant and animal diseases. Science 2001; 292:2285–2289.

16. Hendson M, Purcell AH, Chen D, Smart C, Guilhabert M, Kirkpatrick B. Genetic diversity of Pierce's disease strains and other pathotypes of *Xylella fastidiosa*. Appl Environ Microbiol 2001; 67:895–903.

17. Schell MA, Denny TP, Huang J. Extracellular virulence factors of *Pseudomonas solanacearum*: role in disease and their regulation. In: Kado CI, Crosa JH (eds). Molecular Mechanisms of Bacterial Virulence. Dordrecht, The Netherlands: Kluwer, 1994, pp. 311–324.

18. Deslandes L, Pileur F, Liaubet L, et al. Genetic characterization of *RRS1*, a recessive locus in *Arabidopsis thaliana* that confers resistance to the bacterial soilborne pathogen *Ralstonia solanacearum*. Mol Plant Microbe Interact 1998; 11:659–667.

19. Salanoubat M, Genin S, Artiguenave F, et al. Genome sequence of the plant pathogen *Ralstonia solanacearum*. Nature 2002; 415:497–502.

20. Schell MA. Control of virulence and pathogenicity genes of *Ralstonia solanacearum* by an elaborate sensory network. Annu Rev Phytopathol 2000; 38:263–292.

21. Zupan J, Muth TR, Draper O, Zambryski P. The transfer of DNA from *Agrobacterium tumefaciens* into plants: a feast of fundamental insights. Plant J 2000; 23:11–28.

22. Goodner B, Hinkle G, Gattung S, et al. Genome sequence of the plant pathogen and biotechnology agent *Agrobacterium tumefaciens* C58. Science 2001; 294:2323–2328.

23. Wood DW, Setubal JC, Kaul R, et al. The genome of the natural genetic engineer *Agrobacterium tumefaciens* C58. Science 2001; 294:2317–2323.

24. Galibert F, Finan TM, Long SR, et al. The composite genome of the legume symbiont *Sinorhizobium meliloti*. Science 2001; 293:668–672.

25. Kaneko T, Nakamura Y, Sato S, et al. Complete genome structure of the nitrogen-fixing symbiotic bacterium *Mesorhizobium loti*. DNA Res 2000; 7:331–338.

26. Williams PH. Black rot: a continuing threat to world crucifers. Plant Disease 1980; 64:736–742.

27. Buell CR. Interactions of *Arabidopsis* with *Xanthomonas*. In: Somerville CR, Meyerowitz EM (eds). The Arabidopsis Book. Rockville, MD: American Society of Plant Biologists. 2002; DOI 10.119/tab.0031, http://www.aspb.org/publications/arabidopsis/.

28. Brown K. Florida fights to stop citrus canker. Science 2001; 292:2275–2276.

29. Gottwald TR, Graham JH. In: Timmer LW, Garsney SM, Graham JH (eds). Compendium of Citrus Diseases. St. Paul, MN: American Phytopathological Society, 2002, pp. 5–7.

30. Da Silva AC, Ferro JA, Reinach FC, et al. Comparison of the genomes of two *Xanthomonas* pathogens with differing host specificities. Nature 2002; 417:459–463.

31. Fouts DE, Abramovitch RB, Alfano JR, et al. Genome-wide identification of *Pseudomonas syringae* pv *tomato* DC3000 promoters controlled by the HrpL alternative sigma factor. Proc Natl Acad Sci USA 2002; 99:2275–2280.

32. Petnicki-Ocwieja T, Schneider DJ, Tam VC, et al. Genomewide identification of multiple proteins secreted by the Hrp type III protein secretion system of *Pseudomonas syringae* pv *tomato* DC3000 on the basis of N-terminal export-associated patterns, Hrp promoters, and horizontal transfer indicators. Proc Natl Acad Sci USA 2002; 99:7652–7657.
33. Sogin ML, Morrison HG, Hinkle G, Silberman JD. Ancestral relationships of the major eukaryotic lineages. Microbiologia 1996; 12:17–28.
34. Sogin ML, Silberman JD. Evolution of the protists and protistan parasites from the perspective of molecular systematics. Int J Parasitol 1998; 28:11–20.
35. Bird DM, Kaloshian I. Are roots special? Nematodes have their say. Physiol Mol Plant Pathol, in press.
36. The *C. elegans* Sequencing Consortium. Genome sequence of the nematode *C. elegans*: a platform for investigating biology. Science 1998; 282:2012–2018.

14

Anoxygenic Phototrophic Bacteria

F. Robert Tabita and Thomas E. Hanson

INTRODUCTION

A diverse cadre of prokaryotic organisms is able to obtain energy using photochemical processes. Among these organisms are diverse species of anoxygenic (proteobacteria, green sulfur bacteria, green nonsulfur bacteria) and oxygenic (cyanobacteria, prochlorophytes) phototrophic bacteria that couple photochemical energy production to other extremely important processes that influence the biosphere. In addition, extremophiles such as the halophilic archaea may convert light energy to usable forms of chemical energy, but at a considerably slower rate and substantially lower extent. Thus, the archaea will not be considered in this discussion of the phototrophic prokaryotes. In many respects, however, the amount of information available on aspects of the physiology, biochemistry, molecular biology, and ecology of the phototrophic proteobacteria and cyanobacteria is voluminous. To gain a basic appreciation of the complex life styles and interesting means by which these organisms control aspects of their metabolism at the molecular level, the reader is directed to two highly complete volumes published in the mid-1990s. These books contain contributions by leading investigators on all aspects of the physiology and molecular biology of anoxygenic phototrophic bacteria *(1)* and cyanobacteria *(2)*.

The impressive amount of work published since this time precludes any attempt to consider, in a single chapter, everything that has transpired. However, it is feasible to consider how recent advances have been fueled by the genomics revolution. Some comparisons of phototrophic genomes to those from nonphototrophs is made, but we focus on organisms with which we are most familiar and for which complete or draft sequence is available: the purple nonsulfur bacteria *Rhodopseudomonas palustris*, *Rhodobacter sphaeroides*, *Rhodobacter capsulatus*, and *Rhodospirillum rubrum*, as well as the green sulfur bacterium *Chlorobium tepidum (3)*. Furthermore, we confine the discussion to fundamental processes of carbon dioxide (CO_2) fixation, nitrogen fixation, sulfur oxidation potential, and hydrogen metabolism because these metabolic capabilities are what drive much of the interest in these organisms. We stress how newly available genomic sequences have pointed to new directions and approaches relative to fundamental questions of metabolic control.

The nonsulfur purple bacteria are the most metabolically versatile organisms found on earth and have become model organisms for probing a number of important life

From: *Microbial Genomes*
Edited by: C. M. Fraser, T. D. Read, and K. E. Nelson © Humana Press Inc., Totowa, NJ

processes. These organisms are capable of the five major modes of microbial metabolism, including aerobic and anaerobic chemoheterotrophy, aerobic chemoautotrophy, anaerobic photoautotrophy, and photoheterotrophy. *Rhodopseudomonas palustris* is unique in its ability to catalyze more processes in a single cell than any other nonsulfur purple bacterium and possibly more than any extant organism. *Rhodopseudomonas palustris* is capable of all the growth modes listed above and is remarkably flexible in the organic carbon sources (including aromatic hydrocarbons/lignin monomers) utilized for photoheterotrophic growth and electron donors (reduced sulfur compounds as well as dihydrogen) utilized during photoautotrophic growth when CO_2 is the sole carbon source. Probably as a direct result of this flexibility, *R. palustris* is often the most abundant nonsulfur purple bacterium isolated in enrichments *(4)* and is ubiquitously distributed in soils and waters. The unique ability of *R. palustris* to degrade and recycle monomeric components of woody plant tissue, the most abundant polymer on earth, under diverse conditions undoubtedly contributes to its wide distribution.

GENERAL COMPARISONS
OF PHOTOTROPH VS NONPHOTOTROPH GENOMES

A list of completed or in progress phototrophic genome sequencing projects is shown in Table 1. It was compiled from databases at the National Center for Biotechnology Information *(5)*, the Genomes OnLine Database *(6)*, and the Department of Energy Joint Genome Institute (DOE-JGI) (jgi.doe.gov/JGI_microbial/html/). The first observation is the predominance of cyanobacterial and prochlorophyte genome projects compared to a relative paucity of genome projects for all groups of anoxygenic phototrophic bacteria. This disparity needs to be addressed because many unique and biologically significant discoveries await in relatively neglected groups of physiologically diverse anoxygenic phototrophic bacteria.

Using publicly available complete and draft sequence annotation, a global comparison of genomes from phototrophic bacteria (9 genomes, 5 draft, 4 complete) was made with genomes from representative groups of nonphototrophic bacteria (22 genomes) and archaea (9 genomes). Sequences from organisms with extremely reduced genomes such as *Mycoplasma* spp and relatives *(7–9)* or obligate intracellular pathogens were omitted from this comparison. One obvious trend from this comparison is that the genomes of phototrophic bacteria tend to be larger than the mean size for either bacterial or archaeal genomes sequenced to date (Fig. 1). This likely stems from a bias toward sequencing bacterial pathogens and archaeal extremophiles, which tend to be limited in physiological diversity and thus are likely to require a smaller genome to encode required metabolic pathways. This general trend is emphasized in extremely fastidious or obligatorily symbiotic or parasitic bacteria, which tend to have very small genome sizes *(10)*.

Comparing the percentage of total open reading frames (ORFs) from each genome that are related to a defined family in the Clusters of Orthologous Groups (COGs) database *(11,12)* (Fig. 2) yielded interesting findings. Orthologous sequences are likely descendants from a common ancestral sequence or members of the same protein superfamily. The genomes of phototrophic bacteria appeared to contain a larger fraction of ORFs not represented in the COGs database (31–52% of all ORFs in genome) compared to the mean values for bacteria (mean 25%) or archaea (mean 23%). The underrepresentation

Table 1
Completed and In Progress Genome Sequencing Projects for Phototrophic Bacteria

Organism	Size (Mb)	Source	Status
Unicellular cyanobacteria			
Synechocystis sp PCC 680	33.57	Kazusa DNA Research Institute	Published *(50,51)*
Synechococcus sp PCC 6301	2.69	Nagoya University	Sequencing in progress
Synechococcus sp WH8102	2.72	DOE-JGI	Annotation in progress
Synechococcus sp PCC 7942	2.5	Texas A&M University	Sequencing in progress
Synechococcus sp PCC 7002	3.2	Beijing University	Sequencing in progress
Thermosynechococcus elongatus BP1	2.6	Kazusa DNA Research Institute	Published *(52)*
Filamentous cyanobacteria			
Anabaena sp PCC 7120	6.4 + 0.8	Kazusa DNA Research Institute	Published *(53)*
Nostoc punctiforme ATCC 29133	9.76	DOE-JGI	Sequencing in progress
Microcystis aeruginosa	3.15	Institut Pasteur	Sequencing in progress
Prochlorophytes			
Prochlorococcus marinus MED4	1.66	DOE-JGI	Annotation in progress
Prochlorococcus marinus MIT9313	2.4	DOE-JGI	Annotation in progress
Prochlorococcus marinus SS120	1.8	Genoscope	Sequencing in progress
Purple Nonsulfur			
Rhodobacter capsulatus SB1003	3.7	Integrated Genomics	Completed
Rhodopseudomonas palustris CGA009	5.47	DOE-JGI	Publication pending
Rhodobacter sphaeroides 2.4.1	4.4	DOE-JGI	Sequencing in progress
Rhodospirillum rubrum	3.4	DOE-JGI	Annotation in progress
Purple sulfur			
Thermochromatium tepidum	3.30	Integrated Genomics	Annotation in progress
Heliobacteria			
Heliobacillus mobilis	4.2	Integrated Genomics	Completed
Green Nonsulfur			
Chloroflexus aurantiacus J-10-fl	3	DOE-JGI	Sequencing in progress
Green sulfur			
Chlorobium tepidum	2.2	TIGR	Published *(3)*

of phototrophs is noticeable in all COG categories and is, in part, because of the structural genes that encode the photochemical light-harvesting complexes and reaction centers, which are not represented in the COG database and are an obvious phototroph-specific group of gene products. The underrepresentation is, however, not because of

Genome Size

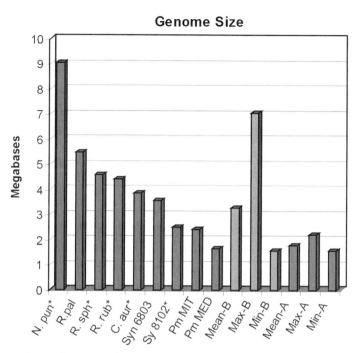

Fig. 1. Comparison of genome size between individual phototrophic bacterial genomes and mean, maximum, and minimum values of groups of 22 complete nonphototrophic bacterial genome sequences (Mean-B, Max-B, and Min-B) and 9 complete archaeal sequences (Mean-A, Max-A, and Min-A). Phototrophic genomes marked with an asterisk denote draft quality sequence (more than one contig); the other sequences are closed as a single contiguous sequence. *N. pun, Nostoc punctiforme*; *R. pal, R. palustris*; *R. rub, R. rubrum*; *C. aur, Chloroflexus aurantiacus*; *Syn* 6803, *Synechocystis* sp strain PCC 6803; *Sy* 8102, *Synechococcus* sp strain WH8102; *Pm* MIT, *Prochlorococcus marinus* MIT9313; *Pm* MED, *P. marinus* strain MED4. Nonphototroph bacterial genomes: *Aquifex aeolicus (54), Bacillus halodurans (55), Bacillus subtilis (56), Campylobacter jejuni (57), Caulobacter crescentus (58), Deinococcus radiodurans (59), Escherichia coli stra*ins K12 *(60)* and O157:H7 *(61), Haemophilus influenzae (62), Helicobacter pylori (63), Lactococcus lactis (64), Mesorhizobium loti (39), Mycobacterium tuberculosis (38), Mycobacterium leprae (65), Neisseria meningiditis* strains MC58 *(66)* and Z2491 *(67), Pasteurella multocida (68), Pseudomonas aeruginosa (69), Streptococcus pyogenes (70), Thermotoga maritima (30), Vibrio cholerae (71),* and *Xylella fastidiosa (72).* Archaeal genomes: *Aeropyrum pernix (73), Archaeoglobus fulgidus (74), Halobacterium* sp strain NRC-1 *(75), Methanococcus jannaschii (76), Methanobacterium thermoautotrophicum (77), Pyrococcus horikoshii (78), Pyrococcus abyssi, Thermoplasma acidophilum (79),* and *T. volcanii (80).*

ORFs predicted to be involved in photopigment biosynthesis (chlorophyll, bacterio-chlorophyll, carotenoids), which are clearly orthologous to proteins found in genomes of nonphototrophs. The results of this comparison indicate that genomes of phototrophic bacteria contain a genetic pool of as yet undescribed physiological potential. Functional genomics and large-scale integrative studies, such as the Microbial Cell Project studies on *R. palustris* and *R. sphaeroides*, sponsored by the Department of Energy will be instrumental in revealing the consequences of this enhanced physiological potential and how it may be exploited and regulated.

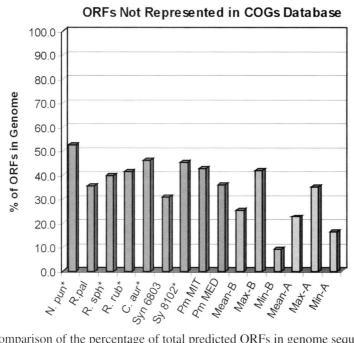

Fig. 2. A comparison of the percentage of total predicted ORFs in genome sequences that are not represented in the Clusters of Orthologous Groups database (www.ncbi.nlm.nih.gov/COG/). Genomes, abbreviations are as in Fig. 1.

CARBON DIOXIDE ASSIMILATION AND ITS CONTROL

Carbon Dioxide Fixation in Nonsulfur Purple Bacteria

All the nonsulfur purple bacteria use the Calvin–Benson–Bassham (CBB) pathway to assimilate CO_2; the green sulfur bacteria employ the reductive tricarboxylic acid cycle (TCA) *(13–15)*. In *Rhodobacter* and *Rhodopseudomonas*, the CBB genes are organized in two main clusters or operons (cbb_I and cbb_{II}), each of which contain genes that encode either form I or form II ribulose 1,5-bisphosphate carboxylase/oxygenase (RubisCO), the enzyme that catalyzes the actual CO_2 fixation reaction (Fig. 3). In *R. palustris* and *R. sphaeroides*, there is a single divergently transcribed *cbbR* gene upstream from the first gene of the cbb_I operon that encodes for a transcriptional regulator (CbbR) that activates transcription of both *cbb* operons *(16)*. The situation is somewhat more complicated in *R. capsulatus* in that there is a divergently transcribed *cbbR* regulatory gene upstream from each operon that is specific for its cognate cbb_I or cbb_{II} operon *(17)*.

Deduced amino acid sequences of the CbbRI and CbbRII proteins of *R. capsulatus* are considerably less related to each other than is *R. capsulatus* CbbRII to the CbbR proteins from the other organisms. In fact, comparative sequence analysis of the *cbbR* and cbb_I genes of *R. capsulatus* indicated that this entire cluster of genes probably arose from some chemoautotrophic bacterium via a lateral gene transfer event *(18)*, undoubtedly accounting for the differences between the CbbR and form I RubisCO proteins of *R. capsulatus* and those of *R. sphaeroides* and *R. palustris*.

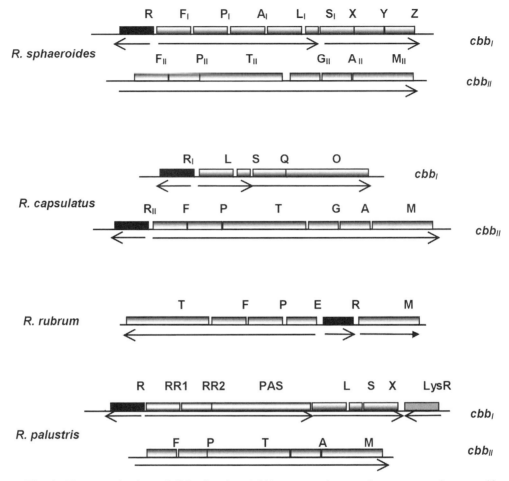

Fig. 3. The organization of CO$_2$ fixation (*cbb*) structural genes into operons in nonsulfur purple bacteria. In green sulfur photosynthetic bacteria (*C. tepidum*), the structural genes of the reductive TCA cycle are not organized in clusters.

In *R. rubrum*, there is one major CBB gene cluster (Fig. 3). Of note are three interesting ORFs juxtaposed between the *cbbR* and *cbbLS* genes of *R. palustris*. These genes encode (5' from *cbbL* and 3' from *cbbR*, in order) two putative response regulators (RR1 and RR2) and a large protein (Pas) that contains several interesting motifs, including 2 PAS domains, one PAC domain, and both phospho-receiver and phospho-donor domains. Recent studies indicated that the Pas protein regulates autotrophic CO$_2$-dependent growth and the transcription of both the *cbb$_I$* and *cbb$_{II}$* operons of *R. palustris* (C.-S. Oh and F. R. Tabita, unpublished manuscript). This unique gene arrangement and unprecedented involvement of the large Pas protein has sparked additional studies in our laboratory of its involvement, along with RR1 and RR2, in mediating CO$_2$-dependent growth as well as CbbR-dependent control.

In all nonsulfur purple bacteria, the same two-component regulatory system (encoding the RegA/B or PrrA/B response regulator–sensor kinase pair) that controls transcription of genes important for photosystem biosynthesis *(19–21)* also is involved in regulating CO_2 fixation *(19)*. Indeed, this last finding led to the notion that the Reg(Prr) system is a global regulator that controls and helps integrate several important processes in these and related bacteria *(22)* (see below). There is also ample evidence for other regulators specific for CO_2 fixation, and it is apparent that the *cbb* operons are differentially controlled in *R. sphaeroides* and *R. capsulatus (23)*.

Carbon Dioxide Fixation in Green Sulfur Bacteria

The metabolic versatility of the green sulfur bacteria is somewhat limited compared to the nonsulfur purple bacteria. As such, CO_2 fixation capability and the synthesis of the key reductive TCA enzymes vary comparatively little under photoautotrophic and photoheterotrophic growth conditions *(24)*. Indeed, unlike the nonsulfur bacteria, it is clear that the CO_2 fixation machinery is indispensable in these organisms, and there is nothing in the genome of *C. tepidum* that would indicate the potential for an alternative lifestyle. The pyruvate ferredoxin oxidoreductase/pyruvate synthase and α-ketoglutarate ferredoxin oxidoreductase/α-ketoglutarate synthase enzymes catalyze reversible reactions in which pyruvate or α-ketoglutarate, respectively, is either oxidized or reductively synthesized. Reductive carboxylation reactions mediated by reduced electron carriers such as ferredoxin are crucial for CO_2-dependent growth, and anything that influences the reduction potential of these electron carries will influence whether the organism degrades or synthesizes pyruvate or α-ketoglutarate via CO_2 fixation *(25,26)*.

NITROGEN FIXATION CAPABILITIES

Many phototrophic bacteria have the capability to reduce dinitrogen to ammonia and grow in the absence of fixed nitrogen compounds. This trait obviously imbues these organisms with a selective advantage in the environment if inorganic or organic sources of nitrogen are scarce. The normal nitrogenase enzyme complex is composed of two major components. Dinitrogenase, or component I, contains the molybdo cofactor, comprised of two different polypeptides, NifD and NifK. Dinitrogenase reductase, component II or the iron protein, is comprised of the NifH polypeptide. The *nifHDK* genes are typically cotranscribed and regulated by the product of the *nifA* gene in the nonsulfur purple bacteria. There is also a complex nitrogen regulatory cascade that responds to the nitrogen and carbon status of the cell; it involves several components, including NtrA, NtrB, and NtrC and the products of the *glnB* and *glnD* genes *(27)*.

Classically, the *nif* genes are upregulated under growth conditions for which N_2 is the nitrogen source or in the presence of a poor organic nitrogen source such as glutamate in nonsulfur purple bacteria. It was originally discovered in *Azotobacter vinelandii* that there are two additional nitrogenase enzyme complexes in which the molybdo cofactor of component I or dinitrogenase is replaced by either vanadium or iron, the so-called vanadium or iron nitrogenases, respectively *(28)*. These are encoded by the *vnfHDK* or *anfHDK* genes. *Rhodopseudomonas palustris*, like *A. vinelandii*, is one of the few organisms that contains all three nitrogenases, and there are specific regulators for each, encoded by the *nifA*, *vnfA*, and *anfA* genes.

The genes of each system are organized in a unique and interesting fashion, and it will be important to elucidate the contribution and role of each nitrogenase system to overall

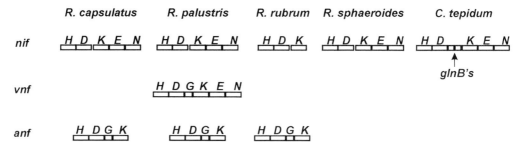

Fig. 4. Clusters of nitrogen fixation genes in purple nonsulfur bacteria and *C. tepidum*, including the molybdenum, vanadium, and iron containing dinitrogenases and their associated genes.

metabolism. By way of comparison, *R. capsulatus* contains the *nif* and *anf* systems, while *R. sphaeroides* and *C. tepidum* possess only the *nif* system (Fig. 4). *R. rubrum* possesses genes for both the *nif* and *anf* systems *(29)*, but there is no current evidence that the *anf* gene products are functional. The *C. tepidum nif* operon is very similar in structure to that found in the methanogenic archaeon *Methanobacterium thermoautotrophicum (3)* suggesting a cross-kingdom lateral transfer of genetic information occurred at some point in the past. Analysis of other genes from *C. tepidum* and other genomes *(30)* also indicated that lateral transfers of substantial amounts of genetic information have occurred more frequently than once suspected.

AROMATIC HYDROCARBON DEGRADATION

There are few phototrophic organisms that have the innate capability to metabolize aromatic hydrocarbons. Prominent among these organisms is *R. palustris*, which can metabolize benzoate derivatives under both aerobic and anaerobic phototrophic conditions *(31)*. *Rhodopseudomonas palustris* has dedicated a large number of genes to encoding the enzymes necessary for this task. For aerobic degradation, *R. palustris* encodes four distinct ring cleavage pathways with specificity for protocatechuate, homoprotocatechuate, homogentisate, and phenylacetate. Each pathway includes a specific dioxygenase. An additional 15 predicted monooxygenases, dioxygenases, or P450s are found in the genome, suggesting an even wider variety of compounds may be transformed than is currently understood. One of the monooxygenases is likely involved in sulfur acquisition as it is located in an operon very similar to that characterized for metabolism of aliphatic sulfonates from *Pseudomonas putida* S313 *(32)*.

Under anaerobic conditions, aromatic hydrocarbons are directed to a well-studied benzoyl–coenzyme A (CoA) degradation pathway that utilizes a novel reductive enzyme *(33)* and catalysts similar to β-oxidation pathways *(34)*. Analysis of the *R. palustris* genome also suggests that the organism may utilize a broader substrate range anaerobically than is currently appreciated because it encodes 42 CoA ligases and eight clusters of β-oxidation genes in addition to those known to be involved in benzoyl–CoA degradation. Presumably, this genetic diversity allows *R. palustris* to utilize the broad range of plant-derived aromatic compounds it is likely to encounter in the soil, contributing to the ubiquitous distribution of this organism in this environment.

INTEGRATIVE CONTROL OF CARBON DIOXIDE FIXATION, NITROGEN FIXATION, AND HYDROGEN METABOLISM

The regulation of CO_2 fixation, N_2 fixation, and H_2 metabolism is linked in the non-sulfur purple bacteria *(22,35)*. This was discovered after knocking out genes that encode key and unique enzymes of the CBB CO_2 fixation pathway. As discussed here, the structural genes of the CBB pathway are found in two major operons, controlled in some cases by a single *cbbR* gene (Fig. 3). Knocking out the two RubisCO genes or the PRK gene(s) *(cbbP)* will cause some interesting adaptations if the organism is to grow at all under phototrophic conditions. Certainly, photoautotrophic growth is not feasible if the CBB cycle is compromised.

Normally, in wild-type strains, the CBB scheme allows nonsulfur purple bacteria to use CO_2 as the electron acceptor when organic carbon is present and oxidized. However, for a CBB pathway knockout strain to grow photoheterotrophically in the absence of an exogenously added electron acceptor, the organism must develop some alternative means to dissipate reducing equivalents produced from the oxidation of the organic carbon source. We have found that these bacteria can develop alternative redox balancing mechanisms using different (in some cases, undefined) ways to accomplish this.

One particular adaptation is of special interest in that many different nonsulfur purple bacteria derepress synthesis of the dinitrogenase enzyme complex in the presence of ammonia, then use the inherent hydrogenase activity of this enzyme system to reduce protons to evolve copious quantities of H_2 gas *(22,35)*. This allows the organism to "vent" excess reducing equivalents produced from organic carbon oxidation; these reducing equivalents normally would accumulate and prevent growth in a strain crippled in its ability to use CO_2 as the preferred electron acceptor. The cell thus goes to a lot of trouble to synthesize the energetically expensive dinitrogenase complex and turn on the *nif* system under conditions (in the presence of ammonia) when it is normally repressed.

These results also establish another important point, that normal CO_2, N_2, and H_2 metabolism is exquisitely regulated, and control of all three processes is highly integrated (see Fig. 5). At this point, it is not clear how the organism senses when to turn on and derepress the *nif* system in the absence of a functional CBB pathway, but the global two-component RegA/B (PrrA/B) regulatory system is somehow involved *(22)*. Studies are currently directed at using modern functional genomics-based methods to elucidate the molecular basis for derepression and enhanced gas evolution.

COMPARATIVE ANALYSIS OF SULFUR METABOLISM IN *RHODOPSEUDOMONAS PALUSTRIS* AND *CHLOROBIUM TEPIDUM*

Potential sulfur oxidation metabolic pathways were analyzed in the genomes of *C. tepidum*, a green sulfur bacterium, and *R. palustris*, one of the few nonsulfur purple phototrophic bacteria that can oxidize reduced sulfur compounds and use the energy from these oxidations to support photoautotrophic CO_2-dependent growth. The genome sequence of *C. tepidum (3)* was accessed at the Institute for Genomic Research (TIGR) (www.tigr.org), and the *R. palustris* sequence was accessed via the Joint Genome Institute (www.jgi.doe.gov).

Chlorobium tepidum and *R. palustris* are quite distinct organisms physiologically. *Chlorobium tepidum* is an obligate phototroph, a sulfur oxidizer, and an anaerobe with

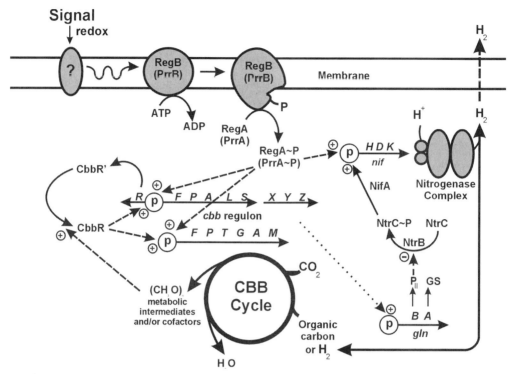

Fig. 5. Conceptual model for the interplay of various factors involved in signal transduction and regulation of *cbb* gene expression in nonsulfur purple photosynthetic bacteria (NSP PS) bacteria. The link between the CO₂ (*cbb*) and nitrogen (*nif*) regulatory system is shown. Primary signals are received at the cytoplasmic membrane. This is thought to affect the redox potential of some key component (?) influencing RegB/PrrB autophosphorylation and the subsequent formation of RegA~P (PrrA~P). Positive regulation is thus conferred by both the CbbR protein and RegA~P(PrrA~P). CbbR' is converted to CbbR (the transcriptionally active form of this molecule), presumably by virtue of binding a coinducer molecule produced under CO₂ fixation conditions or other growth conditions that favor *cbb* transcription. The expression of *glnB* is affected by the *cbb* system *(81)*, with *glnB* influencing *nif* derepression through the Ntr system and NifA. Blockage of the CBB pathway results in H₂ evolution by virtue of the hydrogenase activity of the derepressed nitrogenase complex *(22)*. The nitrogenase complex and its inherent hydrogenase activity thus serve to remove excess reducing equivalents not dissipated in strains unable to use CO₂ as an electron acceptor. The p refers to promoter–operator regions activated in a positive manner (+). (From ref. *82* with permission.)

a limited repertoire for assimilating carbon, nitrogen, and sulfur sources *(36)*; *R. palustris* is a facultative phototroph, sulfur oxidizer, and anaerobe with a considerably wider array of options for assimilating diverse compounds into cellular material. As would be expected from these differences, the 5.45-Mb genome of *R. palustris* is over twice the size of the 2.21-Mb *C. tepidum* genome. Aside from these differences, both organisms share the capacity to utilize reduced sulfur compounds for phototrophic growth.

The availability of complete genome sequences and genetic tools for both organisms will make them extremely useful model systems for understanding phototrophic sulfur metabolism. Predicted pathways for sulfur metabolism from analysis of the *R. palustris* and *C. tepidum* genome sequences are presented in Fig. 6. It is clear that there are many

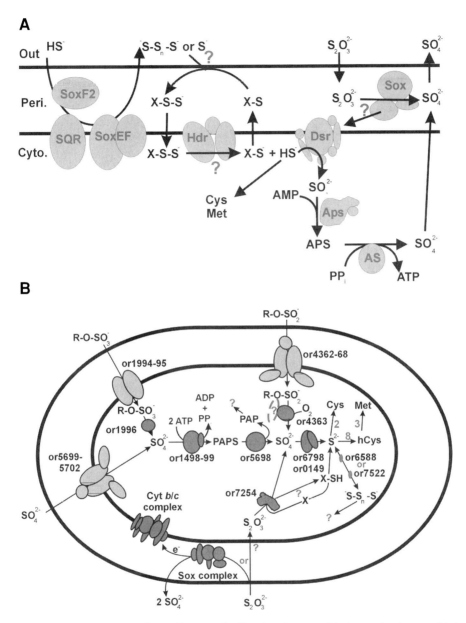

Fig. 6. Proposed pathways for sulfur metabolism in phototrophic bacteria. Areas of interest requiring further experimentation are indicated by red question marks. A) *Chlorobium tepidum*. A similar pathway was proposed in the published analysis of the *C. tepidum* genome sequence (*3*). HS^-, sulfide; ^-S-S_n-S^-, polysulfide; S^o, elemental sulfur; $S_2O_3^{2-}$, thiosulfate; SO_3^{2-}, sulfite; SO_4^{2-}, sulfate; X-S^-, low molecular weight thiol; SQR, sulfide quinone oxidoreductase; Sox, thiosulfate oxidation complex; Hdr, possible heterodisulfide reductase; Dsr, dissimilatory sulfite reductase homologs; Aps, adenosine phosphosulfate reductase; AS, ATP-sulfurylase. B) *Rhodopseudomonas palustris*. Specific ORFs are designated by "or" numbers (i.e., or1994-95) assigned during annotation. Numbers in red by Cys (cysteine) and Met (methionine) production routes indicate the number of possible orthologous enzymes encoded by the genome for each step. Abbreviations: inorganic sulfur compounds as in Fig. 6A; R-O-SO_3^-, alkyl/aryl sulfate; R-O-SO_2^-, alkyl/aryl sulfonate; PAPS, phosphoadenosine phosphosulfate; PAP, phosphoadenosine phosphate; hCys, homocysteine.

open questions, some of which are discussed in this section. As only scant biochemical information is available on sulfur oxidation in these organisms, it is apparent that their genomes provide a rich area for further investigation, particularly the potential integration of sulfur oxidation control to carbon assimilation. Clearly, complete genome sequences will guide and stimulate hypothesis formulation and future experimentation relative to sulfur metabolism.

Duplication of Putative Sulfur Metabolism Genes

Considerable genetic potential for assimilating a wide variety of oxidized sulfur compounds is indicated by the *R. palustris* genome. The *C. tepidum* genome, in contrast, encodes a very limited capacity for the assimilation of oxidized sources of sulfur, consistent with *C. tepidum*'s habitat of sulfidic hot springs *(36)*. Reductive sulfate assimilation is likely to be a maintenance or survival strategy for *C. tepidum* because reduced sulfur compounds are required for phototrophic growth *(36,37)*. The genomes of both organisms contain examples of genetic redundancy with regard to enzymes predicted to be involved in sulfur metabolism.

In the case of *R. palustris*, the redundancy occurs at the step of sulfide fixation. Sulfide fixation is carried out by *O*-acetylserine sulfhydrylase for cysteine biosynthesis and *O*-acetylhomoserine sulfhydrylase (OAHS) to produce homocysteine, which is methylated to yield methionine. *Chlorobium tepidum* encodes a single homolog of each activity; *R. palustris* encodes two potential OAS orthologs and eight potential OAHS orthologs. Four of the OAHS homologs are more closely related to one another than other OAHS protein sequences in the database and are closely associated with *nifE, fixLJ* and *fixK2*, or *vnfNE* genes on the *R. palustris* genome (T. E. Hanson and F. R. Tabita, unpublished results). The significance of this association is unknown, but the association with genes predicted to be involved in nitrogenase cofactor biosynthesis (*nifE* and *vnfNE*) and regulation (*fixLJ* and *fixK2*) is intriguing and may suggest integrative links between sulfur and nitrogen metabolism similar to those described above. The other four homologs are more closely related to functionally characterized OAHS proteins. This large number of potential OAHS activities is not found in the genomes of relatively close phylogenetic relatives, such as *Sinorhizobium meliloti* and *Mesorhizobium loti*, which each only encode two potential OAHS activities *(38,39)*. This observation may suggest that *R. palustris* possesses more diverse pathways for sulfur metabolism or low molecular weight thiol metabolism than is currently appreciated.

The redundancy in *C. tepidum* occurs for proteins predicted to be involved in the interconversion of sulfide and sulfite. During reductive assimilation, sulfite is reduced to sulfide by an assimilatory sulfite reductase prior to incorporation into either methionine or cysteine. *Rhodopseudomonas palustris* encodes homologs of CysI and CysJ proteins, which encode the α- and β-subunits of the assimilatory sulfite reductase. The CysI protein is most similar to that recently characterized in *Pseudomonas aeruginosa (40)*. *Chlorobium tepidum* does not contain homologs of CysI or CysJ, as might be expected for an organism that is not expected to assimilate sulfite reductively. However, the *C. tepidum* genome does encode two copies of *dsrAB* and associated genes from the purple sulfur photosynthetic bacterium *Allochromatium vinosum (41)*, encoding a putative dissimilatory sulfite reductase similar to the enzyme found in sulfate reducers. Genetic studies suggested that the single copy of the *dsr* operon in *A. vinosum* was required for the oxidation of stored elemental sulfur *(41)*. The duplication in *C. tepidum* is 4.6 kb long at greater than 99% deoxyribonucleic acid (DNA) sequence identity covering the genes *dsrCABL*. Intriguingly, the genes encoding adenosine triphosphate (ATP)-sulfurylase, adenosine phosphosulfate (APS) reductase, and a heterodisulfide reductase homolog are closely linked in the *C. tepidum* genome to one copy of the *dsr* operon (Fig. 6) *(3)*. The physiological consequences of the duplication are unknown. It is also unknown whether both copies of the *dsrCABL* genes are expressed or required for growth, although the copy that is not linked with ATP–sulfurylase and associated genes does contain a 2-bp deletion in the *dsrB* gene, causing a frameshift mutation *(3)*.

The RubisCO-Like Protein and Sulfur Metabolism

Initially discovered after parts of the draft sequence of the *C. tepidum* genome became available at the Web site, a protein that clearly resembles RubisCO and shares many of the important active site residues of *bona fide* RubisCO was subsequently isolated and characterized *(42)*. This protein was termed the RubisCO-like protein (RLP) because it was shown not to catalyze CO_2 fixation.

As other genomic sequences became available, several other organisms were shown to contain these curious and interesting RLPs, including archaea, nonsulfur purple bacteria *R. palustris* and *R. rubrum*, and the purple sulfur bacterium *A. vinosum* (Fig. 7). Indeed, there appear to be at least two, and possibly three, subclasses of RLP, which will undoubtedly become clearer as more genomic sequences are completed. A *C. tepidum* mutant strain (Ω::RLP), lacking RLP, does not oxidize elemental sulfur as well as the wild type under phototrophic conditions *(42)*. The phenotype can be physiologically complemented by the addition of cysteine to the growth medium, which suggests that the defect may lie in low molecular weight thiol metabolism.

Low molecular weight thiols have been postulated to be involved in elemental sulfur oxidation in phototrophic bacteria *(43)*. Specifically, glutathione amide appears to cycle between thiol and perthiol forms when *A. vinosum* is oxidizing stored elemental sulfur globules *(44)*. *Chlorobium limicola* and *C. tepidum* appear to contain potentially structurally novel low molecular weight thiols *(45)* (T. E. Hanson and F. R. Tabita, unpublished). We hypothesize that the RubisCO-like protein is involved in synthesizing or cycling these compounds during the oxidation of reduced sulfur compounds. The necessity for this function may arise from the observation that the *C. tepidum* genome does not encode a complete system for the oxidation of thiosulfate.

Both the *R. palustris* and *C. tepidum* genomes contain an operon that encodes a homolog of the well-studied sulfur-oxidizing (*sox*) system of *Paracoccus pantotrophus* GB17, which encodes thiosulfate:cytochrome c oxidoreductase. A review described the components of the system, their functions, and the distribution of *sox* genes among various bacteria *(46)*. The *R. palustris sox* operon is most similar to that found in *P. pantotrophus* for any available genome sequence; the *C. tepidum* operon lacks various genes, including those encoding the SoxCD component. SoxCD is a periplasmic, molybdenum-containing sulfur dehydrogenase that is required for the full oxidation of thiosulfate in vitro *(47)*. The substrate of SoxCD is a protein-bound cysteine-*S*-sulfide (*S*-thiocysteine) bound on SoxY in the SoxYZ subunit of the *sox* complex *(48)*. The products of the reaction are sulfate and the regenerated cysteine in SoxY. The obvious question is, How does *C. tepidum* overcome the lack of this activity and still oxidize thiosulfate completely?

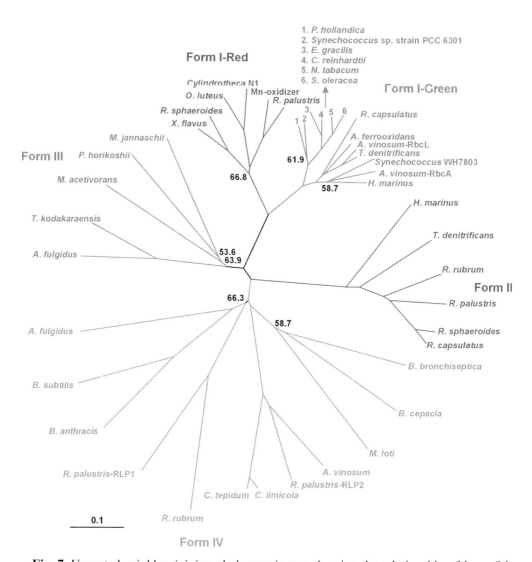

Fig. 7. Unrooted neighbor joining phylogenetic tree showing the relationship of *bona fide* RubisCO sequences (form I red and green, form II purple, form III bronze) and the RubisCO-like protein sequences (form IV cyan). Bootstrap values at the nodes indicate the percentage of times that particular node was present in 1000 trials. Only nodes with less than 70% bootstrap support are indicated. The scale bar represents 0.1 substitutions per site. Complete names for organisms not mentioned elsewhere in the text: *A. ferroxidans, Acidithiobacillus ferroxidans, B. anthracis, Baccilus anthracis, B. cepacia, Burkholderia cepacia, B. subtilis, Bacillus subtilis, B. bronchiseptica, Bordetella bronchiseptica, C. reinhardtii, Chlamydomonas reinhardtii, E. gracilis, Euglena gracilis, H. marinus, Hydrogenovibrio marinus, M. acetivorans, Methanosarcina acetivorans, N. tabacum, Nicotiana tabacum, O. luteus, Olisthodiscus luteus, P. hollandica, Prochlorothrix hollandica, S. oleracea, Spinacia oleracae, T. kodakaraenis, Thermococcus kodakaraensis, T. denitrifcans, Thiobacillus denitrificans, X. flavus, Xanthobacter flavus.*

Preliminary results indicate that the Ω::RLP mutant strain of *C. tepidum* is defective in thiosulfate oxidation in addition to its defects in elemental sulfur oxidation (T. E. Hanson and F. R. Tabita, unpublished). Sulfide oxidation is not affected in strain Ω::RLP. We speculate that there must be an acceptor for the cysteine-*S*-sulfide formed on SoxY during thiosulfate oxidation. One likely candidate for this acceptor is a low molecular weight thiol. Similar to the sulfur oxidation defect in strain Ω::RLP, we hypothesize a defect in low molecular weight thiol metabolism caused by the loss of the RLP may explain the thiosulfate oxidation defect in strain Ω::RLP. The product of the transfer from the *S*-thiocysteine of SoxY would be a low molecular weight perthiol, which is structurally analogous to heterodisulfides produced during methanogenesis and archaeal C1 metabolism. The *C. tepidum* genome encodes a homolog of an archaeal heterodi-sulfide oxidoreductase, with the closest functionally characterized homolog that of *Methanosarcina mazei* Göl *(49)*. These genes are localized in the *C. tepidum* genome along with genes encoding a copy of the *A. vinosum dsr* system, previously implicated in elemental sulfur oxidation *(41)*, as well as genes encoding ATP–sulfurylase and APS reductase. This association leads to a proposed pathway for the oxidation of elemental sulfur globules and the S-thiocysteine of SoxY via a low molecular weight thiol interme-diate through sulfite and APS, leading to sulfate. An attractive feature of this admittedly speculative model is the conservation of one molecule of ATP by reversal of the ATP–sulfurylase reaction.

Rhodopseudomonas palustris also contains two RubisCO-like proteins, RLP1 and RLP2. RLP2 is closely related (66% amino acid identity) to the *C. tepidum* RLP; RLP1 is more distantly related to the *C. tepidum* RLP (31% amino acid identity). *Chlorobium tepidum* RLP and *R. palustris* RLP2 form a distinct cluster within the RLPs along with the RLP sequences from *C. limicola* and *A. vinosum* (Fig. 3). All of these organisms are phototrophic sulfide and thiosulfate oxidizers. The specific role of RLPs in these orga-nisms is still in doubt, and no concrete model is obvious from the comparative analysis of sulfur metabolism genes in *R. palustris* and *C. tepidum*. However, many candidates for future genetic and biochemical analyses of phototrophic sulfur metabolism have been identified.

CONCLUSIONS

Phototrophic bacteria include representative organisms that may be considered the most metabolically versatile organisms on the planet. This ability to grow under so many conditions and to catalyze the many fundamentally important bioprocesses considered here imbues these organisms with the ability to employ sophisticated means to regulate metabolism. Newly sequenced genomes provide the means to probe control further, and as a deeper understanding of the mechanisms by which metabolism is integrated devel-ops, potential applications to use these organisms in carbon sequestration and alternative energy production will be the norm.

ACKNOWLEDGMENTS

Work on the biochemistry and regulation of carbon dioxide assimilation and its inte-gration with the control of nitrogen fixation, hydrogen metabolism, and sulfur oxidation in the Tabita laboratory is supported by grants from the Department of Energy (DE-FG02-01ER63241 and DE-FG02-91ER20033) and the National Institutes of Health (GM45404 and GM24497).

REFERENCES

1. Blankenship RE, Madigan MT, Bauer CE, eds. Anoxygenic Photosynthetic Bacteria. Dordrecht, The Netherlands: Kluwer Academic Publishers,1995.
2. Bryant DA, ed. The Molecular Biology of Cyanobacteria. Dordrecht, The Netherlands: Kluwer, 1994.
3. Eisen JA, Nelson KE, Paulsen IT, et al. The complete genome sequence of *Chlorobium tepidum* TLS, a photosynthetic, anaerobic, green-sulfur bacterium. Proc Natl Acad Sci USA 2002; 99: 9509–9514.
4. Gest H, Favinger JL. Enrichment of purple photosynthetic bacteria from earthworms. FEMS Microbiol Lett 1992; 91:265–270.
5. Wheeler DL, Church DM, Lash AE, et al. Database resources of the National Center for Biotechnology Information: 2002 update. Nucleic Acids Res 2002; 30:13–16.
6. Bernal A, Ear U, Kyrpides N. Genomes OnLine Database (GOLD): a monitor of genome projects world-wide. Nucleic Acids Res 2001; 29:126–127.
7. Glass JI, Lefkowitz EJ, Glass JS, Heiner CR, Chen EY, Cassell GH. The complete sequence of the mucosal pathogen *Ureaplasma urealyticum*. Nature 2000; 407:757–762.
8. Chambaud I, Heilig R, Ferris S, et al. The complete genome sequence of the murine respiratory pathogen *Mycoplasma pulmonis*. Nucleic Acids Res 2001; 29:2145–2153.
9. Himmelreich R, Hilbert H, Plagens H, Pirkl E, Li BC, Herrmann R. Complete sequence analysis of the genome of the bacterium *Mycoplasma pneumoniae*. Nucleic Acids Res 1996; 24: 4420–4449.
10. Gil R, Sabater-Munoz B, Latorre A, Silva FJ, Moya A. Extreme genome reduction in *Buchnera* spp: toward the minimal genome needed for symbiotic life. Proc Natl Acad Sci USA 2002; 99: 4454–4458.
11. Tatusov RL, Natale DA, Garkavtsev IV, et al. The COG database: new developments in phylogenetic classification of proteins from complete genomes. Nucleic Acids Res 2001; 29:22–28.
12. Tatusov RL, Koonin EV, Lipman DJ. A genomic perspective on protein families. Science 1997; 278:631–637.
13. Yoon KS, Hanson TE, Gibson JL, Tabita FR. Autotrophic CO_2 metabolism. In: Lederburg J (ed). Encyclopedia of Microbiology, 2nd ed. San Diego, CA: Academic, 2000, pp. 349–358.
14. Tabita FR. Molecular and cellular regulation of autotrophic carbon dioxide fixation in microorganisms. Microbiol Rev 1988; 52:155–189.
15. Buchanan BB, Arnon DI. A reverse KREBS cycle in photosynthesis: consensus at last. Photosynth Res 1990; 24:47–53.
16. Gibson JL, Tabita FR. The molecular regulation of the reductive pentose phosphate pathway in Proteobacteria and Cyanobacteria. Arch Microbiol 1996; 166:141–150.
17. Vichivanives P, Bird TH, Bauer CE, Tabita FR. Multiple regulators and their interactions in vivo and in vitro with the *cbb* regulons of *Rhodobacter capsulatus*. J Mol Biol 2000; 300:1079–1099.
18. Paoli GC, Soyer F, Shively J, Tabita FR. *Rhodobacter capsulatus* genes encoding form I ribulose-1,5-bisphosphate carboxylase/oxygenase (*cbbLS*) and neighbouring genes were acquired by a horizontal gene transfer. Microbiology 1998; 144:219–227.
19. Qian Y, Tabita FR. A global signal transduction system regulates aerobic and anaerobic CO_2 fixation in *Rhodobacter sphaeroides*. J Bacteriol 1996; 178:12–18.
20. Eraso JM, Kaplan S. *prrA*, a putative response regulator involved in oxygen regulation of photosynthesis gene expression in *Rhodobacter sphaeroides*. J Bacteriol 1994; 176:32–43.
21. Mosley CS, Suzuki JY, Bauer CE. Identification and molecular genetic characterization of a sensor kinase responsible for coordinately regulating light harvesting and reaction center gene expression in response to anaerobiosis. J Bacteriol 1995; 177:3359.

22. Joshi HM, Tabita FR. A global two component signal transduction system that integrates the control of photosynthesis, carbon dioxide assimilation, and nitrogen fixation. Proc Natl Acad Sci USA 1996; 93:14,515–14,520.

23. Gibson JL, Dubbs JM, Tabita FR. Differential expression of the CO_2 fixation operons of *Rhodobacter sphaeroides* by the Prr/Reg two-component system during chemoautotrophic growth. J Bacteriol 2002; 184:6654–6664.

24. Wahlund TM, Tabita FR. The reductive tricarboxylic acid cycle of carbon dioxide assimilation: initial studies and purification of ATP-citrate lyase from the green sulfur bacterium *Chlorobium tepidum*. J Bacteriol 1997; 179:4859–4867.

25. Yoon KS, Hille R, Hemann C, Tabita FR. Rubredoxin from the green sulfur bacterium *Chlorobium tepidum* functions as an electron acceptor for pyruvate ferredoxin oxidoreductase. J Biol Chem 1999; 274:29,772–29,778.

26. Yoon KS, Bobst C, Hemann CF, Hille R, Tabita FR. Spectroscopic and functional properties of novel 2[4Fe-4S] cluster-containing ferredoxins from the green sulfur bacterium *Chlorobium tepidum*. J Biol Chem 2001; 276:44,027–44,036.

27. Kranz RG, Cullen PJ. Regulation of nitrogen fixation. In: Blankenship RE, Madigan MT, Bauer CE (eds). Anoxygenic Photosynthetic Bacteria. Dordrecht, The Netherlands: Kluwer, 1995, pp. 1181–1208.

28. Bishop PE, Premakumar R, Joerger RD, et al. Alternative nitrogen fixation systems in *Azotobacter vinelandii*. In: Bothe H, De Bruijn FJ, Newton WE (eds). Nitrogen Fixation: Hundred Years After. Stuttgart, Germany: Gustav Fisher, 1988, pp. 71–79.

29. Loveless TM, Bishop PE. Identification of genes unique to Mo-independent nitrogenase systems in diverse diazotrophs. Can J Microbiol 1999; 45:312–317.

30. Nelson KE, Clayton RA, Gill SR, et al. Evidence for lateral gene transfer between Archaea and Bacteria from genome sequence of *Thermotoga maritima*. Nature 1999; 399:323–329.

31. Harwood CS, Gibson J. Anaerobic and aerobic metabolism of diverse aromatic compounds by the photosynthetic bacterium *Rhodopseudomonas palustris*. Appl Environ Microbiol 1988; 54:712–717.

32. Kahnert A, Vermeij P, Wietek C, James P, Leisinger T, Kertesz MA. The ssu locus plays a key role in organosulfur metabolism in *Pseudomonas putida* S-313. J Bacteriol 2000; 182:2869–2878.

33. Gibson J, Dispensa M, Harwood CS. 4-Hydroxybenzoyl coenzyme A reductase (dehydroxylating) is required for anaerobic degradation of 4-hydroxybenzoate by *Rhodopseudomonas palustris* and shares features with molybdenum-containing hydroxylases. J Bacteriol 1997; 179:634–642.

34. Egland PG, Pelletier DA, Dispensa M, Gibson J, Harwood CS. A cluster of bacterial genes for anaerobic benzene ring biodegradation. Proc Natl Acad Sci USA 1997; 94:6484–6489.

35. Tichi MA, Tabita FR. Maintenance and control of redox poise in *Rhodobacter capsulatus* strains deficient in the Calvin–Benson–Bassham pathway. Arch Microbiol 2000; 174:322–333.

36. Wahlund TM, Woese CR, Castenholz RW, Madigan MT. A thermophilic green sulfur bacterium from New Zealand hot springs, *Chlorobium tepidum* Sp-Nov. Arch Microbiol 1991; 156: 81–90.

37. Mukhopadhyay B, Johnson E, Ascano M. Conditions for vigorous growth on sulfide and reactor-scale cultivation protocols for the thermophilic green sulfur bacterium *Chlorobium tepidum*. Appl Environ Microbiol 1999; 301–306.

38. Galibert F, Finan TM, Long SR, et al. The composite genome of the legume symbiont *Sinorhizobium meliloti*. Science 2001; 293:668–672.

39. Kaneko T, Nakamura Y, Sato S, et al. Complete genome structure of the nitrogen-fixing symbiotic bacterium *Mesorhizobium loti*. DNA Res 2000; 7:331–338.

40. Hummerjohann J, Kuttel E, Quadroni M, Ragaller J, Leisinger T, Kertesz MA. Regulation of the sulfate starvation response in *Pseudomonas aeruginosa*: role of cysteine biosynthetic intermediates. Microbiology 1998; 144:1375–1386.

41. Pott AS, Dahl C. Sirohaem sulfite reductase and other proteins encoded by genes at the *dsr* locus of *Chromatium vinosum* are involved in the oxidation of intracellular sulfur. Microbiology 1998; 144:1881–1894.

42. Hanson TE, Tabita FR. A ribulose-1,5-bisphosphate carboxylase/oxygenase (RubisCO)-like protein from *Chlorobium tepidum* that is involved with sulfur metabolism and the response to oxidative stress. Proc Natl Acad Sci USA 2001; 98:4397–4402.

43. Brune D. Sulfur oxidation by phototrophic bacteria. Biochim Biophys Acta 1989; 975:189–221.

44. Bartsch R, Newton G, Sherril C, Fahey R. Glutathione amide and its perthiol in anaerobic sulfur bacteria. J Bacteriol 1996; 178:4742–4746.

45. Fahey RC, Buschbacher RM, Newton GL. The evolution of glutathione metabolism in phototrophic microorganisms. J Mol Evol 1987; 25:81–88.

46. Friedrich CG, Rother D, Bardischewsky F, Quentmeier A, Fischer J. Oxidation of reduced inorganic sulfur compounds by bacteria: emergence of a common mechanism? Appl Environ Microbiol 2001; 67:2873–2882.

47. Friedrich CG, Quentmeier A, Bardischewsky F, et al. Novel genes coding for lithotrophic sulfur oxidation of *Paracoccus pantotrophus* GB17. J Bacteriol 2000; 182:4677–4687.

48. Quentmeier A, Friedrich CG. The cysteine residue of the SoxY protein as the active site of protein-bound sulfur oxidation of *Paracoccus pantotrophus* GB17. FEBS Lett 2001; 503:168–172.

49. Ide T, Baumer S, Deppenmeier U. Energy conservation by the H_2:heterodisulfide oxidoreductase from *Methanosarcina mazei* Go1: identification of two proton-translocating segments. J Bacteriol 1999; 181:4076–4080.

50. Kaneko T, Sato S, Kotani H, et al. Sequence analysis of the genome of the unicellular cyanobacterium *Synechocystis* sp strain PCC6803. II. Sequence determination of the entire genome and assignment of potential protein-coding regions [supplement]. DNA Res 1996; 3:185–209.

51. Kaneko T, Sato S, Kotani H, et al. Sequence analysis of the genome of the unicellular cyanobacterium *Synechocystis* sp strain PCC6803. II. Sequence determination of the entire genome and assignment of potential protein-coding regions. DNA Res 1996; 3:109–136.

52. Nakamura Y, Kaneko T, Sato S, et al. Complete genome structure of the thermophilic cyanobacterium *Thermosynechococcus elongatus* BP-1. DNA Res 2002; 9:123–130.

53. Kaneko T, Nakamura Y, Wolk CP, et al. Complete genomic sequence of the filamentous nitrogen-fixing cyanobacterium *Anabaena* sp strain PCC 7120. DNA Res 2001; 8:205–213; 227–253.

54. Deckert G, Warren PV, Gaasterland T, et al. The complete genome of the hyperthermophilic bacterium *Aquifex aeolicus*. Nature 1998; 392:353–358.

55. Takami H, Nakasone K, Takaki Y, et al. Complete genome sequence of the alkaliphilic bacterium *Bacillus halodurans* and genomic sequence comparison with *Bacillus subtilis*. Nucleic Acids Res 2000; 28:4317–4331.

56. Kunst F, Ogasawara N, Moszer I, et al. The complete genome sequence of the Gram-positive bacterium *Bacillus subtilis*. Nature 1997; 390:249–256.

57. Parkhill J, Wren BW, Mungall K, et al. The genome sequence of the food-borne pathogen *Campylobacter jejuni* reveals hypervariable sequences. Nature 2000; 403:665–668.

58. Nierman WC, Feldblyum TV, Laub MT, et al. Complete genome sequence of *Caulobacter crescentus*. Proc Natl Acad Sci USA 2001; 98:4136–4141.

59. White O, Eisen JA, Heidelberg JF, et al. Genome sequence of the radioresistant bacterium *Deinococcus radiodurans* R1. Science 1999; 286:1571–1577.

60. Blattner FR, Plunkett G 3rd, Bloch CA, et al. The complete genome sequence of *Escherichia coli* K-12. Science 1997; 277:1453–1474.

61. Perna NT, Plunkett G 3rd, Burland GV, et al. Genome sequence of enterohaemorrhagic *Escherichia coli* O157:H7. Nature 2001; 409:529–533.

62. Fleischmann RD, Adams MD, White O, et al. Whole-genome random sequencing and assembly of *Haemophilus influenzae* Rd. Science 1995; 269:496–512.

63. Tomb JF, White O, Kerlavage AR, et al. The complete genome sequence of the gastric pathogen *Helicobacter pylori*. Nature 1997; 388:539–547.

64. Bolotin A, Wincker P, Mauger S, et al. The complete genome sequence of the lactic acid bacterium *Lactococcus lactis* sp *lactis* IL1403. Genome Res 2001; 11:731–753.

65. Cole ST, Eiglmeier K, Parkhill J, et al. Massive gene decay in the leprosy bacillus. Nature 2001; 409:1007–1011.

66. Tettelin H, Saunders NJ, Heidelberg J, et al. Complete genome sequence of *Neisseria meningitidis* serogroup B strain MC58. Science 2000; 287:1809–1815.

67. Parkhill J, Achtman M, James KD, et al. Complete DNA sequence of a serogroup A strain of *Neisseria meningitidis* Z2491. Nature 2000; 404:502–506.

68. May BJ, Zhang Q, Li LL, Paustian ML, Whittam TS, Kapur V. Complete genomic sequence of *Pasteurella multocida*, Pm70. Proc Natl Acad Sci USA 2001; 98:3460–3465.

69. Stover CK, Pham XQ, Erwin AL, et al. Complete genome sequence of *Pseudomonas aeruginosa* PA01, an opportunistic pathogen. Nature 2000; 406:959–964.

70. Ferretti JJ, McShan WM, Ajdic D, et al. Complete genome sequence of an M1 strain of *Streptococcus pyogenes*. Proc Natl Acad Sci USA 2001; 98:4658–4663.

71. Heidelberg JF, Eisen JA, Nelson WC, et al. DNA sequence of both chromosomes of the cholera pathogen *Vibrio cholerae*. Nature 2000; 406:477–483.

72. Simpson AJ, Reinach FC, Arruda P, et al. The genome sequence of the plant pathogen *Xylella fastidiosa*. The *Xylella fastidiosa* Consortium of the Organization for Nucleotide Sequencing and Analysis. Nature 2000; 406:151–157.

73. Kawarabayasi Y, Hino Y, Horikawa H, et al. Complete genome sequence of an aerobic hyperthermophilic crenarchaeon, *Aeropyrum pernix* K1. DNA Res 1999; 6:83–101, 145–152.

74. Klenk HP, Clayton RA, Tomb JF, et al. The complete genome sequence of the hyperthermophilic, sulphate-reducing archaeon *Archaeoglobus fulgidus*. Nature 1997; 390:364–370.

75. Ng WV, Kennedy SP, Mahairas GG, et al. Genome sequence of *Halobacterium* species NRC-1. Proc Natl Acad Sci USA 2000; 97:12,176–12,181.

76. Bult CJ, White O, Olsen GJ, et al. Complete genome sequence of the methanogenic archaeon, *Methanococcus jannaschii*. Science 1996; 273:1058–1073.

77. Smith DR, Doucette-Stamm LA, Deloughery C, et al. Complete genome sequence of *Methanobacterium thermoautotrophicum* deltaH: functional analysis and comparative genomics. J Bacteriol 1997; 179:7135–7155.

78. Kawarabayasi Y, Sawada M, Horikawa H, et al. Complete sequence and gene organization of the genome of a hyper-thermophilic archaebacterium, *Pyrococcus horikoshii* OT3. DNA Res 1998; 5:55–76.

79. Ruepp A, Graml W, Santos-Martinez ML, et al. The genome sequence of the thermoacidophilic scavenger *Thermoplasma acidophilum*. Nature 2000; 407:508–513.

80. Kawashima T, Amano N, Koike H, et al. Archaeal adaptation to higher temperatures revealed by genomic sequence of *Thermoplasma volcanium*. Proc Natl Acad Sci USA 2000; 97:14,257–14,262.

81. Qian Y, Tabita FR. Expression of *glnB* and a *glnB*-like gene (*glnK*) in a ribulose bisphosphate carboxylase/oxygenase-deficient mutant of *Rhodobacter sphaeroides*. J Bacteriol 1998; 180: 4644–4649.

82. Tabita FR. Microbial ribulose-1,5-bisphosphate carboxylase/oxygenase: a different perspective. Photosynth Res 1999; 60:1–28.

15

Genomics of Thermophiles

Frank T. Robb

INTRODUCTION: LIFE *IN EXTREMIS*

The concept of normality for living systems is centered on 37°C, pH 7.0, and 50 m*M* NaCl and on atmospheric gases at normal pressure. The life-limiting conditions for higher eukaryotes are conceptual harnesses; the microbial world has very different ideas of normalcy. Microbial life has adapted to extend well beyond the familiar mesophilic envelope and thrives in almost every terrestrial environment. Table 1 provides a list of microorganisms that live on the fringes of the known biosphere. For example, the limiting extremes of temperature for microbial growth is at or above 113°C or below 0°C; pressures may be at least 50 mPa; the pH may be approaching pH 0; and salt conditions may be up to saturation (equivalent to 5*M* NaCl) *(1)*. In many cases, more than one extreme condition is present; for example, many of the barophilic prokaryotes are also adapted to grow at temperatures above the normal boiling point of water. *Pyrolobus fumarii*, the current record holder for high-temperature growth, is a case in point *(2)*. This strain, isolated from a Mid-Atlantic vent at great depth (3650 m), is able to survive exposure to autoclave conditions for up to 1 h and is able to grow at 113°C at a pressure of 25 mPa *(2)*.

Thermoacidophiles such as *Sulfolobus* spp and *Thermoplasma* spp are found in solfataric environments and other low pH habitats, such as hot coal waste. These heated habitats are also the exclusive territory of prokaryotes, free from competition from eukaryotes. The most thermophilic eukaryotes on record are ciliates that cannot grow above 52°C from geothermal hot springs *(3)*. It is remarkable that the wide temperature range of thermophiles is sustained by basic genomic organization and coding conventions that are comparable with "normal" microbial cells. Consequently, the simple and powerful approach of whole genome sequencing provides the most direct method to explore the molecular basis for survival and growth at high temperature.

This chapter focuses on the 15 thermophiles with genomes that were completely sequenced by May 2003 (Table 2). These completed sequencing projects are presented in chronological order, with the optimal growth temperatures of the strains and their dominant mode of energy conservation.

It is notable that, following the historic sequencing of the first microbial genome, the pathogenic bacterium *Haemophilus influenzae* in 1995 *(4)*, the next whole genome sequence described was that of the deep-sea barophilic hyperthermophile *Methanococcus*

From: *Microbial Genomes*
Edited by: C. M. Fraser, T. D. Read, and K. E. Nelson © Humana Press Inc., Totowa, NJ

Table 1
Microbial Life in Extreme Temperature Conditions

Class of extreme	Environment	Organism	Defining growth condition	Reference
High-temperature growth (hyperthermophile)	Submarine vents	*Pyrolobus fumarii* (A) *Methanopyrus kandleri* (A)	T_{max} 113°C T_{max} 110°C	Blochl et al. (1997), Huber et al. (1989), Slesarev et al. (2002)
High-temperature survival	Soil, growth media contaminant	*Moorella thermoaceticum* (spore) (B)	2 h, 121°C, 15 psi	Bryer et al. (2000)
Cold temperature (psychrophile)	Snow, lake water, sediment, ice	Numerous, such as *Vibrio, Arthrobacter, Pseudomonas* (B), and *Methanogenium* (A) spp	–17°C	Carpenter et al. (2000); Cavicchioli and Thomas (2000)
High temperature, acidic	Dry solfataric soil	*Picrophilus oshimae/torridus* (A), *Sulfolobus solfataricus* (A), *Thermoplasma* spp (A)	pH_{opt} 0.7 (1.2 M H_2SO_4)	Schleper et al. (1995); She et al. (2001) Johnson (1998)

A, Archaea; B, Bacteria.

Table 2
Sequenced Thermophile Genomes

Species	Genome size	Completion	T_{opt}	Energy conservation
Methanococcus jannaschii	1.66	1996 (5)	80	Methanogenesis
Archaeoglobus fulgidus VC16	2.18	1997 (6)	83	Sulfate reduction
Methanobacterium thermoautotrophicum ΔH	1.75	1997 (16)	65	Methanogenesis
Pyrococcus horikoshii OT3	1.73	1998 (18)	98	Heterotrophic/S_0 reduction
Aeropyrum pernix K1	1.80	1999 (25)	95	Heterotrophic
Aquifex aeolicus (B)	1.80	1999 (9)	85	H_2 oxidation
Thermotoga maritima MSB8 *(B)*	1.86	1999 (8)	80	Heterotrophic, S_0 reduction
Pyrococcus abyssi	1.75	2000 (19)	95	Heterotrophic, S_0 reduction
Pyrococcus furiosus	1.91	2000 (17)	100	Heterotrophic, S_0 reduction
Sulfolobus tokodaii 7	2.69	2000 (13)	80	Heterotrophic, S oxidation
Sulfolobus solfataricus P2	2.99	2001 (12)	85	Heterotrophic, S oxidation
Thermoplasma volcanium GSS1	1.58	2001 (21)	60	Heterotrophic, S oxidation
Thermoplasma acidophilum	1.58	2001 (20)	63	Heterotrophic, S oxidation
Methanopyrum kandleri AV19	1.69	2001 (22)	103	Methanogenesis
Pyrobaculum aerophilum IM2	2.22	2002 (24)	100	Nitrate reduction
Thermoanaerobacter tencongensis (B)	2.69	2002 (10)	75	Fermentation, S reduction

T_{opt}, optimal growth temperature; B, bacteria.

jannaschii (5). Since then, steady progress has been made in sequencing the genomes of several thermophiles each year.

THERMOPHILES, EXTREME THERMOPHILES, AND HYPERTHERMOPHILES

Thermophiles are defined as microorganisms that require elevated temperatures for growth. This distinguishes them from thermoduric microorganisms such as endospore formers that have specialized adaptive strategies that allow them to survive, but not grow, at elevated temperatures. The following hierarchy has been adopted to distinguish different degrees of adaptation within the thermophiles. Strains able to grow between 50 and 75°C are generically classified as thermophiles, whereas strains that grow from 75 to 90°C are extreme thermophiles, and strains that can grow above 90°C are considered hyperthermophiles *(6).* The hyperthermophiles described to date are quite diverse, with 23 genera recognized as having hyperthermophilic members *(7).* Several genera (e.g., *Pyrolobus, Pyrodictium, Pyrobaculum, Methanopyrus, Pyrococcus, Sulfolobus,* and *Archaeoglobus*) are exclusively hyperthermophilic. Although several Bacteria are also considered hyperthermophiles, isolates that grow optimally at 90°C and above are exclusively Archaea *(6,7).* Hyperthermophilic Bacteria with completed genome projects are *Thermotoga maritima (8), Aquifex aeolicus (9)* and *Thermoanaerobacter tencongensis (10).*

Many of the commonly used genetic tools and methods of genetic analysis are impossible to apply to study of hyperthermophiles and many extreme thermophiles because of their high growth temperatures. Techniques for selection, such as antibiotic resistance, are intractable at high temperatures because most antibiotics are unstable or the proteins that mediate resistance are heat sensitive. Although several methods can be applied to grow extreme thermophiles on solid media using temperature-resistant gelling agents such as Gellan gum or silica gel in place of agar *(11),* in some cases, colony formation is either impossible or too slow to be useful for genetic studies. Genomic sequencing of hyperthermophiles thus provides, in these cases, the only accessible form of genetic analysis. Notable exceptions to this rule are the *Sulfolobus* spp, represented so far by full genome sequences of two species, *Sulfolobus solfataricus (12)* and *Sulfolobus tokodaii (13).* These organisms are aerobic, relatively easy to cultivate, and readily form colonies at 80°C. It has been possible to construct vectors for recombinant expression and to use the thermostable antibiotic hygromycin for selection *(14).* At the other end of the spectrum, the smallest archaeal cell is the obligate symbiont *Nanoarcheum equitans,* so named because it rides as an obligate symbiont on its thermophilic host, an *Ignicoccus* species *(15).* This strain has a genome of less than 500 kbp, and its 165 ribosomal ribonucleic acid (SrRNA) sequence suggests that it may be a new phylum, the *Nanoarcheota.*

COMPARATIVE GENOMIC STUDIES ON ADAPTATIONS REQUIRED FOR THERMOPHILY

Multiple genome sequences are available for several closely related thermophiles. For example, there are now sequences for three thermophilic methanogens, *M. jannaschii (5), Methanobacterium thermoautotrophicum (16),* and *Methanopyrus kandleri (17);* two *Sulfolobus* spp, *S. tokodaii (12)* and *S. solfataricus* P1 *(13);* three *Pyrococcus* spp

(18–20); and two *Thermoplasma* spp. *(21,22)*. As a result, comparative genomic studies have become a very significant factor in understanding adaptive features that enable microbial survival and propagation in hot conditions. Table 2 also references the recently completed sequencing projects on the high-temperature hyperthermophiles *M. kandleri (17)*, a methanogen that grows at temperatures up to 110°C *(23)*, and *Pyrobaculum aerophilum*, a microaerophilic, nitrate reducing Crenarcheote that grows optimally at 100°C by nitrate reduction *(24)*. Finally, an obligate aerobe that is able to grow at up to 100°C, *Aeropyrum pernix,* was sequenced in 1999 *(25)*.

Because of their small size, the intracellular compartments of all classes of thermophiles are exposed to ambient temperatures during growth. It follows that, for biochemical processes to continue to function normally at elevated temperatures, all cellular components must have significant intrinsic thermostability. In many cases, this is because of adaptations that provide intrinsic stability at high temperature; for example, proteins from hyperthermophiles almost without exception have high intrinsic stability. However, components such as deoxyribonucleic acid (DNA) and RNA, and many intermediates in biochemical pathways, are quite unstable in purified form at the optimal growth temperature of the source organisms. In some cases, when a metabolite is extremely labile, such as carbamoyl phosphate, adaptive enzyme structures enable the pathway to proceed by channeling the intermediates internally in an enzyme complex so that degradation in the cytoplasm is avoided *(26)*. Intracellular conditions and adaptive strategies must ensure durability or replacement when turnover is rapid. In terms of cellular survival, hyperthermophilic cells are more thermostable than individual proteins and protein–nucleic acid complexes in vitro, in some cases by a 20°C margin. The following sections cover adaptive stabilization or repair of cellular components in hyperthermophiles.

DNA Repair and Replication

One of the enigmatic features of hyperthermophiles is their ability to grow at temperatures far exceeding the nominal denaturation temperatures of their DNA and RNA. Whereas thermophilic Bacteria frequently have a higher genomic guanine and cytosine (G+C) content that elevates the melting temperature (T_m) of DNA, the G+C content of genomic DNA in hyperthermophiles is spread over a wide range, and most hyperthermophiles have genomes relatively low in G+C content. The major exception to this rule is *M. kandleri,* with a 65% G+C content *(23)*.

Two solutions to this enigma have emerged. Normally, the melting temperature of DNA or RNA is measured under standard solvent conditions such as standard saline citrate. The melting temperature increases at higher ionic strengths and in the presence of some organic solutes. In all cases studied, compatible solutes are present at high concentrations in the cytoplasm of hyperthermophiles. For example, *Pyrococcus furiosus* has an intracellular potassium glutamate concentration of 200–600 mM *(18)*. The counterions for K+ are either di-myo-inositol-1,1'-phosphate or glutamate. *Methanopyrum kandleri*, a strain that grows up to 110°C and cannot grow below 90°C, has a cytoplasmic concentration of up to 2.5M cyclic diphosphoglycerate *(17,23)*. In fact, the hyperthermophiles can be viewed as "closet halophiles" because their cytoplasmic compartments are maintained at highly elevated solute concentrations compared with their growth milieus. These solutes have been critical in many studies for achieving full thermostability in enzymes from hyperthermophiles and to lead to stabilization of DNA duplexes *(27)*. High salt has also prevented DNA breakage in vitro, and covalently closed circular DNA is highly resistant to DNA breakage compared to nicked or linear DNA *(27)*.

Variable temperature *per se* will alter the superhelical density of a replicon as a function of temperature because of a change in the pitch of the DNA helix *(27)*. Therefore, DNA topoisomerase activity has been studied in several hyperthermophiles. Many thermophiles have homologs of the *rgy* gene encoding the enzyme reverse gyrase, which is found in all archaeal and bacterial hyperthermophiles *(28–31)*. Recently, Forterre *(31)* used a comparative genomic approach to find clusters of orthologous groups (COGs) that represent genes found only in hyperthermophiles. This interesting study identified only reverse gyrase, which has an apparent role in maintaining the superhelical density of chromosomes and plasmids in neutral or positively supercoiled condition *(29)*. The enzymes are uniquely bifunctional, consisting of a helicase domain and a topoisomerase I domain *(30)*. The supercoiling is performed by generating positive turns ahead of the mobile helicase as the topoisomerase releases negative turns in the rear, resulting in the net generation of positive supercoils in the replicon *(30)*. Mechanisms of regulation of superhelical density are still unclear. The reverse gyrase from *M. kandleri* is unique in its high activity and the division into two subunits *(32)*, whereas all other reverse gyrases examined are composed of a single large protein with two domains. Another topoisomerase V from *M. kandleri* is apparently another dual-functional enzyme with an important function to carry out backbone scission during base excision repair at high growth temperatures *(33)*.

DNA-binding proteins in Euryarcheota include archaeal histones, which form tetrameric nucleosomes that maintain DNA in positive supercoils in high-salt conditions *(34)*, stabilizing and compacting DNA in vitro. The histones of *M. jannaschii* are of great interest as they have C-terminal extensions, which do not interfere with nucleosome formation, but do provide a great measure of stability *(35,36)*. In the Crenarcheota, DNA-binding proteins, including a novel chromatin-forming protein named Alba *(37)*, contribute to DNA stability by raising the melting temperature of nucleoprotein complexes. The binding properties of Alba are modulated by acetylation in a manner analogous to that of eukaryotic histones *(38)*. Mechanisms of prevention of hydrolytic DNA breakage at high temperature *(39)*, as well as rapid and precise double-strand break repair *(40)*, are dependent on nucleoprotein complexes, which must alternate between being tight enough to protect the DNA, yet able to release according to the demands of transcription and translation. Interestingly, the genome sequences of *Thermoplasma acidophilum (21)* and *Thermoplasma volcanium (22)* do not contain either Crenarcheote or Euryarcheote variants of archaeal histones, but instead contain copies of the bacterial basic, DNA-binding protein HU. Presumably, this was the result of lateral gene transfer; however, it raises the interesting question as to whether the *Thermoplasma* spp represent upward or downward mobility in terms of their temperature ranges.

Most of the genes required for DNA replication have been identified in archaeal genomes *(41)*. Components of replication universally found in Archaea are the proliferating cell nuclear antigen (PCNA) family, which forms the annular clamp at the replication fork, and the minichromosome maintenance (MCM) class of helicases that propagate strand separation of the replication fork *(42)*. Judging by the close similarity to components of eukaryotic replication forks, the archaeal process is similar to that in eukaryotes, whereas DNA replication in eubacterial hyperthermophiles utilizes familiar bacterial replication components. The PCNA protein, which promotes processivity of the replication fork, exists as a single copy in Euryarcheota, but has three closely related isologs in Crenarcheota *(42)*. A variety of replicative DNA polymerases have been identified. In the Euryarcheota, Cann et al. *(43)* identified a unique class of dimeric replicative DNA polymerases with no obvious sequence similarity to either bacterial or eukaryotic polymerase. The additional copies of polymerase appear to be conserved in other sequenced genomes of the Euryarcheota *(44)*. Additional B class DNA polymerases are also found in the Euryarcheota, functioning in lagging strand synthesis and DNA repair. In Crenarcheota, the B polymerases function in both repair and replication *(41)*, often in multiple paralogous genes.

The ligases in Archaea are analogous and similar in sequence to typical eukaryotic enzymes dependent on adenosine triphosphate (ATP) *(45)*, similar to yeast ligases, whereas bacterial thermophiles contain ligases that are typically enzyme dependent on nicotinamide adenine dinucleotide (NAD), although with high intrinsic stability *(46)*.

The extraordinary DNA break repair capabilities of hyperthermophiles imply that exceptionally active reciprocal recombination systems operate during normal growth *(40)*. The homologous recombination system in hyperthermophiles, which is a key to their double-strand break repair capabilities, is emerging from studies of the Holliday junction resolving enzyme in *S. solfataricus (47)*. The Hjc enzyme is a branch-dependent nuclease that dissects the Holliday junction and allows the recombined strands to go their separate ways. The ability to purify and characterize thermostable variants of DNA-modifying enzymes enables biochemical studies that would be difficult or impossible to carry out with the mesophilic homologs of these proteins.

Heat and Cold Shock

The lifestyles of thermophiles in geothermal or hydrothermal habitats exposes them to frequent thermal cycles, both above and below their optimal growth temperatures. Thus, conditions approaching a continuous alternation between both heat and cold shock may be normal for these microorganisms. It is therefore somewhat surprising to find relatively few heat and cold shock genes in the genomes of thermophiles. One of the cold-adaptive proteins, superoxide reductase in the hyperthermophile *P. furiosus (48)*, is active at temperatures well below the 80°C minimal growth temperature *(49)*. This is probably an adaptive response to combat oxidative damage in the event that these oxygen-sensitive cells are flushed out of the vent environment into cold, aerated seawater.

Thermophilic bacteria have the normal complement of groE/groEL and HSP70 molecular chaperones. Interestingly, these genes are also found in the genomes of several Euryarcheota, such as methanogens and halophiles (see ref. *50*, for a review). However, the hyperthermophiles do not appear to encode these heat shock systems. Instead, a heat shock response consisting of the barrel-shaped HSP60 ATP-dependent chaperone, often referred to as the *thermosome*, and a small heat shock protein that is only formed in response to exposures to superoptimal temperatures are employed *(51)*. The chaperonin, an α-crystallin homolog with similarity to the central domain of the vertebrate eye-lens protein, has been crystallized from *M. jannaschii (52)*, and when the *P. furiosus* version is expressed in *Escherichia coli,* it creates exogenous heat resistance that enables the host cells to survive long exposure to 50°C *(53)*.

Protein Thermostability

The discovery of hyperthermophilic organisms living at temperatures near or beyond 100°C has initiated studies of the molecular determinants of protein stability. There has been vigorous debate concerning whether general rules can be discerned that apply to all hyperstable proteins and, if so, whether the rules can be followed during rational design of more thermostable variants of useful enzymes. There are certainly several properties common to thermostable proteins, such as more compact structures with less internal voids, which are achieved by smaller loops and shorter N- and C-termini. In a significant proportion of cases, thermostable enzymes occur in larger oligomeric assemblies than found in mesophiles. When disrupted by site-directed mutagenesis, heat resistance declines. Increased oligomer size can also permit channeling of substrates beween subunits, a strategy that may be common in hyperthermophiles *(54)*. In many cases, the thermostable variants have smaller subunits than homologous proteins from mesophiles *(5; see* refs. *55* and *56* for review).

Methanococcus jannaschii is a barophile isolated from deep sea hydrothermal vents and was the first hyperthermophile to be sequenced completely *(5)*. Its maximal growth rate and the upper temperature limits of growth are only achieved when it is grown under a pressure of 50 mPa *(57; see* Table 2). The genomic information has provided the high-temperature benchmark for comparative proteomic studies with mesophilic species. The adaptation of proteins at very high temperatures under pressure has been studied. Work has shown that moderate pressures ($\leq$100 mPa) below those normally needed for pressure-induced denaturation can dramatically stabilize proteins against thermoinactivation *(58,59)*. Particularly large effects have been observed for thermophilic enzymes at very high temperatures. This phenomenon has important implications for the adaptation of thermophilic proteins *in extremis*.

The contributions of ion pair interactions to the stability of hyperstable proteins has been the subject of debate for several years. In theory, electrostatic interactions are extremely suitable for internal bonding of protein structures because of their relatively long range (up to 4 A) compared with van der Waals forces and their insensitivity to the weakening of water structure near or above 100°C, which causes the hydrophobic effect to weaken. Comparative studies with the deduced proteomes from mesophilic and thermophilic members of related taxa have suggested that the higher growth temperatures are correlated with increased proportions of charged amino acid residues *(60,61)*. Experimental and theoretical studies have found ion pair interactions are mostly stabilizing *(62–64)*.

Crystal structures from an increasing number of hyperthermophilic proteins have been determined and compared to their less-thermostable counterparts *(65,66)*. In some cases, it has been possible to place ion pairs into less-stable proteins by rational design and to improve stability *(67,68)*. Several structural studies that have compared mesophilic proteins with their thermostable homologs have confirmed that ion pair interactions are much more abundant in the hyperthermophilic homologs. Comparative genomic studies have supported this conclusion.

The completion of two genome sequences of *Thermoplasma* spp is very valuable as these strains have relatively low growth temperatures, around 60°C, and therefore their proteomes provide a low-temperature baseline for comparison with high-temperature strains *(21,22)*. The analysis of the amino acid composition of the *T. volcanium*–deduced proteome confirmed the conclusions of two comparative studies *(60,61)*. The general conclusion was that a shift occurs toward a higher abundance of charged amino acid residues, such as Glu, Asp, Arg and Lys, with increasing growth temperatures, at the expense of polar residues such as Ser, Gln, Tyr, and Asn. While it is clear that charged residues can engage in ion pair formation, it has also been argued that intracellular solubility of proteins in hyperthermophiles may depend on increased surface charge in high-salt conditions, accounting for the increase in overall charge of the proteome *(60)*. This is the reason that the proteome of *Halobacterium* strain NRC1 is extremely rich in charged amino acids (see Chapter 21).

A structural genomic approach will contribute significantly to solving this question. In theory, access to the deduced proteome allows access to all of the proteins using recombinant expression techniques. In practice, this applies only to those proteins that can be expressed in an appropriate host, are correctly folded, and undergo appropriate posttranslational modification. Membrane proteins in general are not easily accessible unless they can be expressed in a phylogenetically related host system.

Despite these caveats, high-throughput expression and crystal structure determination projects for *T. maritima (69)*, *M. jannaschii (70–74)*, and *P. furiosus (74)* are under way. When a sufficient number of crystal structures of hyperstable proteins have been solved, the answer to the question as to whether their excess charges are buried or solvent exposed will be clarified by comparing the structures of mesophilic and thermostable proteins. In the case of *T. maritima* and *P. furiosus*, many protein structures have already been solved because these organisms are readily grown to high cell densities, allowing highly expressed native proteins to be purified from cell-free extracts using conventional protocols. In this way, more than 100 native proteins from *P. furiosus* have been purified and characterized. Access to metalloproteins or glycoproteins with prosthetic groups already in place is the major advantage of this method. It has been suggested that correct folding of hyperstable proteins may require exposure to similar solute concentrations and temperatures *(74–76)*, implying that novel protocols and methods for expression may be needed for recombinant expression of many hyperstable proteins.

BIOSYNTHESIS AND METABOLIC SYSTEMS

Thermophiles have a variety of autotrophic and heterotrophic energy conservation systems; in general, the autotrophic strains have comprehensive biosynthetic capacity for nucleotides, amino acids, and vitamins. Among heterotrophic thermophiles, the capacity for biosynthesis is variable, and many strains require vitamins and amino acids for growth *(11)*. Therefore, this section addresses the relationship between energy conservation and biosynthetic capacity.

Biosynthesis

It has been argued that the rigors encountered near the upper temperature limit for living systems dictate a cytoplasmic membrane structure that includes so-called archaeal lipids *(2,23)*. The Archaea have unique membrane lipids with terpenoid or phytanyl chains that have ether linkages instead of ester bonds between the aliphatic chains and the glycerol-derived polar head groups. Diether and tetraether lipids with ether linkages at both ends of the chains are common, especially in lipids of hyperthermophiles *(6)*.

Membranes with high diether and tetraether lipid content, having a monolayer rather than the familiar bilayer cross section, have unmatched stability to temperature extremes and oxidation/reduction, and remain impermeable to protons at extremes of pH and temperature *(11)*.

Lipid biosynthesis is of great interest because the ether lipids of the Archaea necessitate a pathway for isoprenoid biosynthesis in which 3-hydroxy-3-methylglutaryl–coenzyme A reductase (HMG-CoA reductase, E.C. 1.1.1.34) is a key enzyme. In the Eukarya, HMG-CoA reductase is a rate-limiting step in cholesterol biosynthesis and is the target for many cholesterol-lowering pharmaceuticals. In Bacteria, the enzyme is often part of nonessential pathways to produce terpenoid antibiotics *(78)*. Interestingly, the genomes of *Archaeglobus fulgidus* and *Thermoplasma* spp contain a bacterial-type HMG-CoA reductase, not the eukaryal/archaeal version in all other archaeal genomes *(79)*. Presumably, a bacterial gene has supplanted the archaeal version, an example of gene loss affecting an essential function, presumably after the bacterial transferred gene was functional *(80)*.

The heterotrophic thermophiles are also distinguished by apparent concerted gene loss and gain. For example, the *Pyrococcus* genomes are significantly polymorphic with respect to amino acid biosynthetic pathways and several pathways for the uptake and catabolism of the carbohydrates cellobiose, maltose, trehalose, laminarin, and chitin. The *P. furiosus* genome is also significantly larger, at 1.95 Mbp, than those of than *Pyrococcus abyssi* (1.75 Mbp) and *Pyrococcus horikoshii* (1.73 Mbp). The *P. furiosus* and *P. abyssi* genomes encode *trp, aro, arg,* and *ile/val* operons, unlike the genome of *P. horikoshii (81)*. The genomes of *P. abyssi* and *P. horikoshii* are similar in lacking biosynthetic pathways for histidine, cobalamin, a portion of the tricarboxylic acid cycle, and the fermentation capacity (uptake and degradation) for starch, maltose, trehalose, and cellobiose. *Pyrococcus horikoshii*, with the smallest genome of the pyrococci, is also auxotrophic for *Val, Leu*, and *Ile*, and presumably the other aromatic amino acids because it lacks the enzymes of the common aromatic pathway. *Pyrococcus furiosus* has the capacity to synthesize all nucleotides as well as vitamins B_{12} and B_6, which is confirmed by its ability to grow in a defined medium in the absence of these nutrients *(82)*. These properties are reflected in the gene complements for these biosynthetic functions *(82)*. Ettema et al. *(83)* suggested that a mechanism of concerted, *en bloc* insertion and deletion is operating in the genomes of hyperthermophiles, in a similar manner as described for bacteria.

This may have the following implications. Assuming that the environments of the abyssal species of *Pyrococcus* rarely contain available sugars, whereas the conditions of the onshore isolation spots of *P. furiosus* are more often characterized by high-sugar, low-peptide conditions. Because *P. furiosus* can use many carbohydrates efficiently, its growth on sugar substrates in the absence of available amino acids may have led to the selection for the import of the amino acid biosynthetic pathways absent in *P. horikoshii* and *P. abyssi*.

The interesting question that is posed by these three very similar genomes is whether the abyssal *Pyrococcus* spp, with their reduced genome size and decreased metabolic versatility, are undergoing genome reduction following their descent to the deep-sea vents or whether *P. furiosus*, having ascended from the depths, is acquiring new metabolic capacity. The available evidence supports the latter premise.

Although the mechanisms of genome plasticity remain to be discovered, it was reported that the maltose region of *P. furiosus* is apparently a recently imported section of the genome that is not found in the other *Pyrococcus* spp *(84)*. The 19-kb section containing a maltose operon similar in gene order and overall sequence to the mal operon of *E. coli* is flanked by 2 of the 23 insertion sequence (IS) elements in the *P. furiosus* genome. By comparison with the mal region of a related archaeon, *Thermococcus litoralis*, it was discovered that the entire region was very similar in both strains. Only 196 mutations have occurred since the presumed divergence of the *P. furiosus* and *T. litoralis* versions of this region. The analysis of the terminal direct repeats of the transposon indicated that the region moved as a composite transposon *(84)*.

The distribution of IS elements in *Pyrococcus* spp is extremely variable. *Pyrococcus furiosus* has at least 29 full-length IS elements *(18)*. A tandem IS insertion element consisting of 2 intact IS elements was reported. Transfer is plausible because, although *T. litoralis* has a lower growth temperature range than *Pyrococcus* spp, their habitats overlap, and *T. litoralis* was isolated from the same locale (the Italian shoreline) as *P. furiosus (49)*.

Central Metabolism

An unusual feature of the intermediary metabolism of the hyperthermophiles is the frequent use of adenosine 5'-diphosphate (ADP) in reactions in which ATP is normally used. For example, the Embden–Meyerhof–Parnas pathway appeared to be absent or dysfunctional in *Pyrococcus* spp because ATP-dependent phosphofructokinases and glucokinases could not be detected. Unique, ADP-dependent enzymes for these were eventually discovered by astute biochemical analysis *(85)*, and these enzymes were subsequently purified, cloned, and expressed *(86,87)*. As genome sequences were reanalyzed, ADP-dependent kinases have been reported in both mesophilic and thermophilic methanogens *(88,89)*.

This narrative might tempt the reader to conclude that hyperthermophiles are constrained to use ADP in place of ATP in this step in the Embden–Meyerhof–Parnas pathway for reasons of thermostability. However, because the aerobic hyperthermophile *A. pernix* has a "normal" Family B ATP-dependent phosphofructokinase *(90)*, this is not necessarily the case. In any event, the problem of generating the excess of ADP required during normal metabolic flux remains for the hyperthermophiles, and the descriptions of ADP-forming acetyl-CoA-synthetase in *P. furiosus (91)*, *M. jannaschii*, and *A. fulgidus (92)* may point the way to new classes of ADP-generating steps in hyperthermophiles and the possible assignment of as-yet-unidentified genes to unusual metabolic functions.

The metabolism of hydrogen is an extremely well-characterized and important facet of metabolism in thermophiles. Thermophilic hydrogen-forming Archaea and Bacteria encode membrane-bound and cytoplasmic hydrogenases (for review, see ref. *93*). The membrane bound hydrogenases comprise 10- to 14-subunit proteins with the capability of converting intracellular protons into hydrogen that is accumulated extracellularly *(94)*. In the case of *P. furiosus*, the genome contains two putative operons with strong similarity to hydrogenase operons from bacteria such as *Rhodospirillum rubrum* (see Chapter 12). However, a highly unusual putative binding motif for the nickel–iron cluster in the large hydrogenase subunit is characteristic of the *P. furiosus* hydrogenase complexes. Kinetic studies of membrane-bound hydrogenase, soluble hydrogenase, and sulfide dehydrogenase activities allow the formulation of a comprehensive working hypothesis of H_2 metabolism in *P. furiosus* in terms of three pools of reducing equivalents (ferredoxin, nicotinamide adenine dinucleotide phosphate, H_2) connected by devices for transduction, transfer, recovery, and safety valving of energy *(93,95)*.

MEMBRANES AND TRANSPORT SYSTEMS

Thermophiles face two challenges in the operation of selectively permeable transport systems. The first is the control of membrane integrity, which as mentioned in the section on biosynthesis, is achieved by so-called membrane-spanning lipids in archaeal hyperthermophiles *(96)*. The content of diether and tetraether lipids is upregulated during heat stress, and this results in modulation of proton permeability, which is not found in the Bacteria *(96)*. This may be the explanation for the exclusion of bacteria from the upper echelons of the hyperthermophiles. The release of the genome sequences of two *Sulfolobus* spp *(12,13)* has allowed classification of the signal sequences in these thermoacidophiles into four types: secretory signals, twin-Arg signal peptides, lipoprotein signal peptides, and type IV pilinlike signal peptides *(97)*. The deduced proteome of *S. solfataricus* has a total of 4.2% of the proteins that have signal peptides, a much higher level than the proteome of *M. jannaschii* (2%).

The destinations of secreted proteins are controlled by which signal peptide is present, and the majority of transporters appear to be ATP-binding cassette (ABC)-type binding protein–dependent transporters *(98)*. These important uptake components appear to be divided into two families; one has a single linked ATPase encoding gene and is primarily directed toward monosaccharide uptake, and the other has two ATPases and broader specificity toward di- and oligosaccharides *(99)*. Analogous systems have been reported in the Euryarchaeota, such as the maltose uptake system of *T. litoralis*, which is identical to the system in *P. furiosus (84)*. Many of the uptake systems in Archaea are of the tripartite ATP-independent periplasmic type, a relatively widespread type of binding protein transport system that, however, is not found in Eukaryotes. The energy for solute movement is not derived from ATP, but rather from an electrochemical ion gradient *(100)*.

A striking feature of the uptake systems from hyperthermophilic Archaea is their extremely high substrate affinity relative to many mesophilic systems *(98)*. High temperatures are highly destructive for sugars, particularly monosaccharides, and the uptake process must therefore be extremely efficient to capture all of the monomeric sugars. Oligomeric uptake of sugars and peptides is also extremely efficient and mediated primarily by ABC transporters or tripartite ATP-independent periplasmic systems. The latter are driven by either sodium or proton gradients. Some isolates will only grow and take up oligopeptides; for example, Gonzalez et al. described *Thermococcus peptonophilus*, a deep-sea heterotroph that has the fastest growth rate of all *Thermococcus* spp and does not grow on individual amino acids *(101)*.

GENOME PLASTICITY AND PHYLOGENY

Lateral Gene Transfer

Bacterial and archaeal thermophiles often share the same habitats, and there is abundant evidence from genomic analysis that lateral gene transfer (LGT) is common in the group. For example, the *T. maritima* genome has been estimated to have approx 25% of genes that have primary homology to hyperthermophilic Archaea, principally *Pyrococcus* spp *(8)*. Through periodicity analysis, the bacterial signatures that predominate in the *T. maritima* could be distinguished from the archaeal signatures that mark the genes resembling those from Archaea *(102)*.

The observation was made that hyperthermophilic Archaea had extreme codon bias in relation to Bacteria, for example, the exclusive use of AGG and AGA codons for Arg coding in hyperthermophilic Archaea and the absence of UAU encoding Ile *(103)*. This has resulted in the use of "codon usage-adjusting" plasmids such as pSJS to carry out heterologous expression of transfer RNA (tRNAs) to recognize the rare codons in Bacteria. This enables the efficient overexpression of recombinant proteins from hyperthermopiles in *E. coli* *(53*; S. Sandler, personal communication). The genes encoding carbon monoxide dehydrogenase in the extremely thermophilic bacterium *Carboxydothermus hydrogenoformans* were cloned and found to have primary homology to carbon monoxide dehydrogenase from Archaea and a typical archaeal codon bias *(104)*.

These observations beg the question of whether LGT is pervasive in all organisms sharing a niche. If it is, then the attempts to construct phylogenetic trees using individual genes appear to be futile because there is no assurance that the gene has been resident in the genome since the divergence of the lineage from its ancestor. One approach to this takes advantage of the rapidly mounting whole genome count to construct phylogenetic lineages by polling and comparing the gene composition of genomes *(105)*. A tree was constructed using the SHOT (Shared Ortholog and Gene Order Tree Construction Tool) Web facility *(106*; see http://www.bork.embl-heidelberg.de/SHOT).

Figure 1 was constructed with the available thermophile genomes and using the *Halobacterium* NRC1 genome as an outgroup. Overall, however, because the thermophiles have similar genome sizes, this would appear to be a practical method for examining their phylogeny. Although it can be argued that genome content is subject to variable admixtures of genes from unrelated thermophiles because of LGT, these trees at the least provide a basis for examining strain similarity.

In Fig. 1, the acidophiles *Thermoplasma* spp and *Sulfolobus* spp are clustered, as are the Methanogen–Archaeoglobus group. The aerobic Crenarcheote forms a deeper branch with the other aerobic Archaea. These associations are well supported by the major physiological traits of these strains.

The flaw in this study is the position of *Halobacterium* NRC1, which occupies a very deep branch and is on the "eubacterial" end of the Archaea. This is in contrast to 16S rRNA and r-protein phylogenies, which place the haloarchaea at the crown of the euryarchaeal tree. The 16S rRNA phylogeny suggests that *Halobacterium* NRC1 is most closely related to *A. fulgidus* and *M. jannaschii*. However, many genes in the *Halobacterium* NRC1 genome sequence have their highest similarity to the Gram-positive bacterium *Bacillus subtilis* and the radiation-resistant bacterium *Deinococcus radiodurans*, suggesting that strain NRC1 may have acquired a substantial number of genes through LGT (see Chapter 21).

Modifications of the whole genome approach using contextual analysis are also flawed. Gene order is poorly conserved, especially in autotrophic hyperthermophiles such as *Aquifex aeolicus, M. jannaschii*, and *M. kandleri (5,9,17)*.

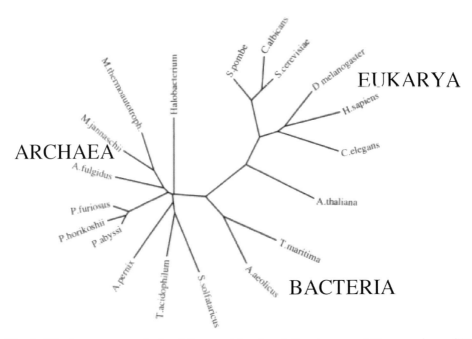

Fig. 1. Whole genome analysis of thermophiles in relation to mesophilic organisms, SHOT method *(107)*.

A major anomaly within the phylogeny of Archaea based on 16S rRNA sequences has been the placement of *M. kandleri,* a methanogen, as a basal root on the common archaeal branch. The remaining methanogens occupy one of the crowns of the Euryarcheota. *Methanopyrum kandleri* is capable of growth under pressure at temperatures up to 110°C, making it the most thermophilic archaeon sequenced to date. It has the highest genomic G+C content of the methanogens. Interestingly, phylogenetic analysis using alignments of concatenated sequences of ribosomal proteins strongly supported a methanogen/ *Pyrococcus* clade *(17).*

It can be argued that this is a reliable approach to organismal phylogeny based on the surmise that genes encoding ribosomal proteins are relatively less likely to be transferred between unrelated strains. The *M. kandleri* placement in an extremely deep branch on the 16S rRNA appears to be an artifact because of the extremely high G+C content of the rRNA (75%) at the base of the archaeal stem. The genomic sequence resolves this anomaly. The genes for methanogenesis are similar in amino acid sequence to those of other methanogens, and the conservation of gene order between the existing methanogen genomes and *M. kandleri* in key loci concerned with energy conservation argues that it is a typical methanogen and confirms that the methanogens are probably a monophyletic group.

It follows that, in this case, extremely high G+C content is likely to be an adaptation that allows growth at up to 110°C. The analysis of the *T. tencongensis* genome indicates that the G+C contents of the genomes bear no relation to the rRNA and tRNA G+C contents *(10).* Collectively, the growth temperature is very significantly correlated with the G+C content of stable RNAs, which rises to a maximum of 65–70% in hyperthermophiles. For example, the genomic G+C content of *P. abyssi* is 42%, whereas the rRNA is 68%. This can lead to an artifactual rRNA-based phylogeny of hyperthermophiles, as noted by Galtier et al. *(105)*, and argues that candidates for common ancestor status among the hyperthermophiles should be viewed with suspicion.

The general prevalence of LGT between Bacteria and Archaea has also been established by the report that the genome of the hyperthermophilic bacterium *T. maritima* has an extraordinarily high proportion of open reading frames (24%) that are most similar to homologs in Archaea *(8)*. Of these shared genes, almost half are most similar to *Pyrococcus* spp homologs. *Thermotoga* and *Pyrococcus* spp share the same marine hydrothermal niches, and their physiology is very similar (i.e., they are both heterotrophic, sulfur reducing, and able to grow above 90°C). It is therefore tempting to speculate that, in both intradomain LGTs such as the *Thermococcus–Pyrococcus* exchanges and in interdomain transfers such as *Thermotoga–Pyrococcus* gene transfers, thermophiles that are able to grow under similar conditions are more likely to exchange genes.

The compact genome sequence of *M. kandleri* has the unusual feature of being essentially free of LGT events *(17)*. The LGT events, or lack of them in the case of the *M. kandleri* genome, provide support for the argument that a common habitat may be conducive to LGT. It is tempting to speculate that, physiologically, *M. kandleri* is a "lone ranger" at the top end of the temperature scale, with fewer opportunities for contact with live cells during active growth.

Another consideration is the unusual physiology of *M. kandleri*, which has an extremely high concentration of phosphate and cyclic diphosphoglycerate in its cytoplasm. One of the requirements for the eventual functioning of a protein imported into a new cell is whether the cytoplasmic composition will permit it to function. The proteins of *M. kandleri* are highly unusual in that they have a requirement for molar concentrations of phosphate for optimal function. Enzyme stability, activity, and subunit assembly were all adversely affected by low phosphate concentration, referred to as a lyotropic condition *(107–109)*. It is very likely that the adaptive mutations required to enable functioning of proteins that were not lyotropic would be a significant barrier toward incorporation of new genes by LGT.

Insertion Elements and Introns

The overall lack of conserved gene order in many thermophilic genomes begs the question as to the mechanisms that allow rapid rearrangement. The IS elements are an obvious source of nonreciprocal recombination and, as repeated sequences, can be important in providing local regions of homology that promote insertion–deletion and inversion events by reciprocal recombination. IS elements have been widely reported in thermophile genomes, and there is experimental evidence for their ongoing activity. For example, in the genome of *S. sulfataricus*, which is rich in several families of IS elements (201 total), the frequent and specific insertion into ura genes could be observed under positive selection by fluoroorotic acid *(110)*. The occurrence of insertion elements of several types has been correlated to genome rearrangements during culture of thermophiles (R. Garrett, personal communication, 2002).

The known IS elements fall into two main types, autonomous IS elements and nonautonomous miniature inverted repeat element-like elements *(111)*. The latter are transmitted adventitiously using the transposases from full-size IS elements and have appropriate direct repeats flanking variable-size "passenger" sequences. In addition, the integration of DNA viruses, plasmids, and DNA fragments from distant regions of the genomes can occur by an integrase mechanism, such as happened with *Sulfolobus* virus SSV1 *(112)* and also in the chromosomes of *A. pernix* and *P. horikoshii*, in which integronlike segments have been reported to border inserted DNA elements such as the plasmid pXQ1.

An important discovery of three protein-encoding genes containing introns was reported *(113)*. A homolog of the eukaryotic centromere-binding factor 5 and a subunit of a small nucleolar ribonucleoprotein in three Archaea (*A. pernix, S. solfataricus*, and *S. tokodaii*) apparently contain introns that are removed at the RNA level by a typical archaeal helix–bulge–helix excision mechanism. This suggests that the possibility of an archaeal origin of introns should be reexamined.

FUTURE TRENDS

The availability of deduced proteomes for hyperthermophiles provides the opportunity for structural genomics studies. High-throughput recombinant gene expression studies are under way for *M. jannaschii, T. maritima*, and *P. furiosus*. It could be argued that a structural genomic study of one of the lower-temperature Archaea, such as the *Thermoplasma* or *Thermoanaerobacter* spp, would serve the ends of determining the basis for protein thermostability by facilitating a comparative study together with the hyperthermophiles.

The primary event that causes cell death at lethal temperatures is very likely the loss of membrane integrity. The study of the adaptive basis for membrane structures that retain integrity above 100°C is becoming clear in biochemical terms. Archaeal lipids have been shown to undergo modification affecting the number of cyclopentane rings, which increases with increasing growth temperature *(114)*. Through this ring temperature compensation mechanism, the plasma membrane of thermoacidophilic Archaea is able to maintain a proton-impermeable and rigid structure and, consequently, a constant and steep proton gradient between the extracellular milieu (often lower than pH 2.5) and intracellular compartment (pH 6.5) over a wide range of growth temperatures. The basis for this temperature-dependent modification is still unclear in terms of the unique biosynthetic pathway for archaeal membrane biosynthesis. The mechanisms will probably be identified by studying the temperature-dependent increases in expression of genes common to hyperthermophiles.

This is one area among many that will benefit from the application of microarray-based studies, already under way using the *P. furiosus* system *(115,116)*; these will be critically important to examine global stress regulation. The discovery of a repressor that mediates the heat shock response in *P. furiosus* *(117)* (Fig. 2) and ongoing work with the bacterial-like Lrp regulator *(118)* promise to lead to rapid gains in information on gene regulation. In addition, although many of the details of the DNA replication fork are emerging from a study of the extreme thermophiles such as *S. solfataricus* or *Mtb. thermoautotrophicum*, the details of the functions of the origin of replication remain elusive. Presumably, the solution to this problem will also provide valuable information on the mechanisms of regulation of the cell cycle in the Archaea.

The mesophilic Archaea have recently been shown to encode the 22nd amino acid, pyrrolysine, using the stop codon UAG and a specialized UAG–tRNA *(119,120)*. Although

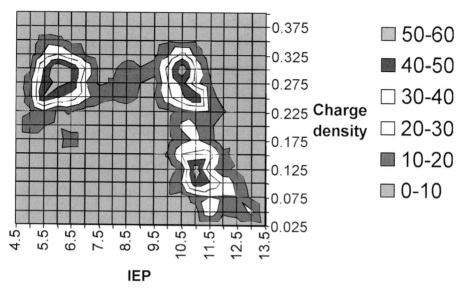

Fig. 2. Mapped charge distribution of the deduced proteome of the hyperthermophile *Pyrococcus furiosus*. Map density, number of proteins per shaded area; IEP, isoelectric point.

not demonstrated for the thermophilic Archaea yet, it seems likely that a reexamination of existing genomes may reveal UAG codons with the appropriate context for pyrrolysine incorporation.

The thermophiles have the potential for advancing both basic and applied science. For instance, the energy bioconversions in these cells are relevant to future production of nonfossil fuels because many of the hyperthermophiles are anaerobes that produce hydrogen or methane as gaseous end products. Further studies of the growth physiology and molecular biology of model organisms such as hyperthermophiles and halophiles will be necessary to determine their potential for the production of gas fuels and the potential application of their extremely thermostable enzymes in biotechnology.

ACKNOWLEDGMENT

This chapter is contribution no. 04-612 from the Center of Maine Biotechnology. The author acknowledges funding support NSF grant MCB02383387 and the Knut and Alice Wallenberg Foundation.

REFERENCES

1. Madigan MT, Oren A. Thermophilic and halophilic extremophiles. Curr Opin Microbiol 1999; 2:265–269.
2. Blochl E, Rachel R, Burggraf S, Hafenbradl D, Jannasch HW, Stetter KO. *Pyrolobus fumarii*, gen and sp nov, represents a novel group of archaea, extending the upper temperature limit for life to 113°C. Extremophiles 1997; 1:14–21.
3. Baumgartner M, Stetter KO, Foissner W. Morphological, small subunit rRNA, and physiological characterization of *Trimyema minutum* (Kahl, 1931), an anaerobic ciliate from submarine hydrothermal vents growing from 28C to 52C. J Eukaryot Microbiol 2002; 49:227–238.

4. Fleischmann RD, Adams MD, White O, et al. Whole-genome random sequencing and assembly of *Haemophilus influenzae* Rd. Science 1995; 269:496–512.

5. Bult CJ, White O, Olsen GJ, et al. Complete genome sequence of the methanogenic archaeon, *Methanococcus jannaschii*. Science 1996; 273:1058–1073.

6. Stetter KO. Extremophiles and their adaptation to hot environments. FEBS Lett 1999; 452:22–25.

7. Huber R, Huber H, Stetter KO. Towards the ecology of hyperthermophiles: biotopes, new isolation strategies and novel metabolic properties. FEMS Microbiol Rev 2000; 24:615–623.

8. Nelson KE, Clayton RA, Gill SR, et al. Evidence for lateral gene transfer between Archaea and bacteria from genome sequence of *Thermotoga maritima*. Nature 1999; 399:323–329.

9. Deckert G, Warren PV, Gaasterland T, et al. The complete genome of the hyperthermophilic bacterium *Aquifex aeolicus*. Nature 1998; 392:353–358.

10. Bao Q, Tian Y, Li W, et al. A complete sequence of the *Thermoanaerobacter tengcongensis* genome. Genome Res 2002; 12:689–700.

11. Robb FT, DasSarma S, Place AR, Sowers KR, Schreier HJ, Fleischmann EM. (eds) Archaea: A Laboratory Manual. Cold Spring Harbor, NY: Cold Spring Harbor Laboratory Press, 1995.

12. She Q, Singh RK, Confalonieri F, et al. The complete genome of the crenarchaeon *Sulfolobus solfataricus* P2. Proc Natl Acad Sci USA 2001; 98:7835–7840.

13. Kawarabayasi Y, Hino Y, Horikawa H, et al. Complete genome sequence of an aerobic thermoacidophilic crenarchaeon, *Sulfolobus tokodaii* strain7. DNA Res 2001; 8:123–140.

14. Cannio R, Contursi P, Rossi M, Bartolucci S. Thermoadaptation of a mesophilic hygromycin B phosphotransferase by directed evolution in hyperthermophilic Archaea: selection of a stable genetic marker for DNA transfer into *Sulfolobus solfataricus*. Extremophiles 2001; 5: 153–159.

15. Huber H, Hohn MJ, Rachel R, Fuchs T, Wimmer VC, Stetter KO. A new phylum of Archaea represented by a nanosized hyperthermophilic symbiont. Nature 2002; 417:63–67.

16. Smith DR, Doucette-Stamm LA, Deloughery C, et al. Complete genome sequence of *Methanobacterium thermoautotrophicum* deltaH: functional analysis and comparative genomics. J Bacteriol 1997; 179:7135–7155.

17. Slesarev AI, Mezhevaya KV, Makarova KS, et al. The complete genome of hyperthermophile *Methanopyrus kandleri* AV19 and monophyly of archaeal methanogens. Proc Natl Acad Sci USA 2002; 99:4644–4649.

18. Robb FT, Maeder DL, Brown JR, et al. Genomic sequence of the hyperthermophile, *Pyrococcus furiosus:* implications for physiology and enzymology. Methods Enzymol 2001; 330: 134–157.

19. Kawarabayasi Y, Sawada M, Horikawa H, et al. Complete sequence and gene organization of the genome of a hyper-thermophilic archaebacterium, *Pyrococcus horikoshii* OT3. DNA Res 1998; 5:147–155.

20. Cohen GN, Barbe V, Flament D, et al. An integrated analysis of the genome of the hyperthermophilic archaeon *Pyrococcus abyssi*. Mol Microbiol 2003; 47:1495–1512.

21. Ruepp A, Graml W, Santos-Martinez ML, et al. The genome sequence of the thermoacidophilic scavenger *Thermoplasma acidophilum*. Nature 2000; 407:508–513.

22. Kawashima T, Amano N, Koike H, et al. Archaeal adaptation to higher temperatures revealed by genomic sequence of *Thermoplasma volcanium*. Proc Natl Acad Sci USA 2000; 97:14,257–14,262.

23. Burggraf S, Stetter KO, Rouviere P, Woese CR. *Methanopyrus kandleri*: an archaeal methanogen unrelated to all other known methanogens. Syst Appl Microbiol 1991; 14:346–351.

24. Fitz-Gibbon ST, Ladner H, Kim UJ, Stetter KO, Simon MI, Miller JH. Genome sequence of the hyperthermophilic crenarchaeon *Pyrobaculum aerophilum*. Proc Natl Acad Sci USA 2002; 99:984–989.

25. Kawarabayasi Y, Hino Y, Horikawa H, et al. Complete genome sequence of an aerobic hyper-thermophilic crenarchaeon, *Aeropyrum pernix* K1. DNA Res 1999; 6:83–101, 145–152.

26. Massant J, Verstreken P, Durbecq V, et al. Metabolic channeling of carbamoyl phosphate, a thermolabile intermediate: evidence for physical interaction between carbamate kinase-like carbamoyl-phosphate synthetase and ornithine carbamoyltransferase from the hyperthermophile *Pyrococcus furiosus*. J Biol Chem 2002; 277:18,517–18,522.

27. Marguet E, Forterre P. Stability and manipulation of DNA at extreme temperatures. Methods Enzymol 2001; 334:205–215.

28. Confalonieri F, Elie C, Nadal M, de La Tour C, Forterre P, Duguet M. Reverse gyrase: a helic-ase-like domain and a type I topoisomerase in the same polypeptide. Proc Natl Acad Sci USA 1993; 90:4753–4757.

29. Bouthier de la Tour C, Portemer C, Kaltoum H, Duguet M. Reverse gyrase from the hyperther-mophilic bacterium *Thermotoga maritima*: properties and gene structure. J Bacteriol 1998;180: 274–281.

30. Borges KM, Bergerat A, Bogert AM, DiRuggiero J, Forterre P, Robb FT. Characterization of the reverse gyrase from the hyperthermophilic archaeon *Pyrococcus furiosus*. J Bacteriol 1997; 179:1721–1726.

31. Forterre P. A hot story from comparative genomics: reverse gyrase is the only hyperthermo-phile-specific protein. Trends Genet 2002; 18:236–237.

32. Kozyavkin SA, Krah R, Gellert M, Stetter KO, Lake JA, Slesarev AI. A reverse gyrase with an unusual structure. A type I DNA topoisomerase from the hyperthermophile *Methanopyrus kand-leri* is a two-subunit protein. J Biol Chem 1994; 269:11,081–11,089.

33. Belova GI, Prasad R, Kozyavkin SA, Lake JA, Wilson SH, Slesarev AI. A type IB topoisom-erase with DNA repair activities. Proc Natl Acad Sci USA 2001; 98:6015–6020.

34. Marc F, Sandman K, Lurz R, Reeve JN. Archaeal histone tetramerization determines DNA affinity and the direction of DNA supercoiling. J Biol Chem 2002; 277:30,879–30,886.

35. Sandman K, Bailey KA, Pereira SL, Soares D, Li WT, Reeve JN. Archaeal histones and nucleo-somes. Methods Enzymol 2001; 334:116–129.

36. Li WT, Sandman K, Pereira SL, Reeve JN. MJ1647, an open reading frame in the genome of the hyperthermophile *Methanococcus jannaschii*, encodes a very thermostable archaeal histone with a C-terminal extension. Extremophiles 2000; 2:115–122.

37. Bell SD, Botting CH, Wardleworth BN, Jackson SP, White MF. The interaction of Alba, a conserved archaeal chromatin protein, with Sir2 and its regulation by acetylation. Science 2002; 296:148–151.

38. Wardleworth BN, Russell RJ, Bell SD, Taylor GL, White MF. Structure of Alba: an archaeal chromatin protein modulated by acetylation. EMBO J 2002; 21:4654–4662.

39. Peak MJ, Robb FT, Peak JG. Extreme resistance to thermally induced DNA backbone breaks in the hyperthermophilic archaeon *Pyrococcus furiosus*. J Bacteriol 1995; 177:6316–6318.

40. DiRuggiero J, Santangelo N, Nackerdien Z, Ravel J, Robb FT. Repair of extensive ionizing-radi-ation DNA damage at 95°C in the hyperthermophilic archaeon *Pyrococcus furiosus*. J Bacteriol 1997; 179:4643–4645.

41. Bohlke K, Pisani FM, Rossi M, Antranikian G. Archaeal DNA replication: spotlight on a rapidly moving field. Extremophiles 2002; 6:1–14.

42. Daimon K, Kawarabayasi Y, Kikuchi H, Sako Y, Ishino Y. Three proliferating cell nuclear antigen-like proteins found in the hyperthermophilic archaeon *Aeropyrum pernix*: interactions with the two DNA polymerases. J Bacteriol 2002; 184:687–689.

43. Cann IK, Komori K, Toh H, Kanai S, Ishino Y. A heterodimeric DNA polymerase: evidence that members of Euryarchaeota possess a distinct DNA polymerase. Proc Natl Acad Sci USA 1998; 95:14,250–14,255.

44. Gueguen Y, Rolland JL, Lecompte O, et al. Characterization of two DNA polymerases from the hyperthermophilic euryarchaeon *Pyrococcus abyssi*. Eur J Biochem 2001; 268:5961–5969.

45. Gunther S, Montes M, de DA, del VM, Atencia EA, Sillero A. Thermostable *Pyrococcus furiosus* DNA ligase catalyzes the synthesis of (di)nucleoside polyphosphates. Extremophiles 2002;6: 45–50.

46. Tong J, Barany F, Cao W. Ligation reaction specificities of an NAD(+)-dependent DNA ligase from the hyperthermophile *Aquifex aeolicus*. Nucleic Acids Res 2000; 28:1447–1454.

47. Kvaratskhelia M, Wardleworth BN, Norman DG, White MF. A conserved nuclease domain in the archaeal Holliday junction resolving enzyme Hjc. J Biol Chem 2000; 275:25,540–25,546.

48. Yeh AP, Hu Y, Jenney FE Jr, Adams MW, Rees DC. Structures of the superoxide reductase from *Pyrococcus furiosus* in the oxidized and reduced states. Biochemistry 2000; 39:2499–2508.

49. Fiala G, Stetter KO. *Pyrococcus furiosus* sp nov represents a novel genus of marine heterotrophic archaebacteria growing optimally at 100°C. Arch. Microbiol 1985; 145:56–61.

50. Hickey AJ, Conway de Macario E, Macario AJ. Transcription in the archaea: basal factors, regulation, and stress gene expression. Crit Rev Biochem Mol Biol 2002; 37:199–258.

51. Laksanalamai P, Maeder DL, Robb FT. Regulation and mechanism of action of the small heat shock protein from the hyperthermophilic archaeon *Pyrococcus furiosus*. J Bacteriol 2001; 183: 5198–5202.

52. Kim KK, Kim R, Kim SH. Crystal structure of a small heat-shock protein. Nature 1998; 394: 595–599.

53. Laksanalamai P, Maeder DL, Jiemjit A, Bue Z, Robb FT. Oligomeric structures are necessary for cellular thermoadaption in the archaeal small heat shock protein. Extremophiles 2003; 7:79–83.

54. Jaenicke R, Bohm G. The stability of proteins in extreme environments. Curr Opin Struct Biol 1998; 8:738–748.

55. Robb FT, Maeder DL. Novel evolutionary histories and adaptive features of proteins from hyperthermophiles. Curr Opin Biotechnol 1998; 9:288–291.

56. Robb FT, Clark DS. Adaptation of proteins from hyperthermophiles to high pressure and high temperature. J Mol Microbiol Biotechnol 1999; 1:101–105.

57. Kaneshiro SM, Clark DS. Pressure effects on the composition and thermal behavior of lipids from the deep-sea thermophile *Methanococcus jannaschii*. J Bacteriol 1995; 177:3668–3672.

58. Sun MM, Tolliday N, Vetriani C, Robb FT, Clark DS. Pressure-induced thermostabilization of glutamate dehydrogenase from the hyperthermophile *Pyrococcus furiosus*. Protein Sci 1999; 8:1056–1063.

59. Sun MM, Caillot R, Mak G, Robb FT, Clark DS. Mechanism of pressure-induced thermostabilization of proteins: studies of glutamate dehydrogenases from the hyperthermophile *Thermococcus litoralis*. Protein Sci 2001; 10:1750–1757.

60. Cambillau C, Claverie JM. Structural and genomic correlates of hyperthermostability. J Biol Chem 2000; 275:32,383–32,386.

61. Haney PJ, Badger JH, Buldak GL, Reich CI, Woese CR, Olsen GJ. Thermal adaptation analyzed by comparison of protein sequences from mesophilic and extremely thermophilic *Methanococcus* species. Proc Natl Acad Sci USA 1999; 96:3578–3583.

62. Anderson DE, Becktel WJ, Dahlquist FW. pH-induced denaturation of proteins: a single salt bridge contributes 3-5 kcal/mol to the free energy of folding of T4 lysozyme. Biochemistry 1990; 29:2403–2408.

63. Kumar S, Nussinov R. Salt bridge stability in monomeric proteins. J Mol Biol 1999; 293:1241–1255.

64. Lounnas V, Wade RC. Exceptionally stable salt bridges in cytochrome P450cam have functional roles. Biochemistry 1997; 36:5402–5417.

65. Elcock AH. The stability of salt bridges at high temperatures: implications for hyperthermophilic proteins. J Mol Biol 1998; 284:489–502.

66. Yip KSP, Stillman TJ, Britton KL, et al. The structure of *Pyrococcus furiosus* glutamate dehydrogenase reveals a key role for ion-pair networks in maintaining enzyme stability at extreme temperatures. Structure 1995; 3:1147–1115.

67. Vetriani C, Maeder DL, Tolliday N, et al. Protein thermostability above 100°C: a key role for ionic interactions. Proc Natl Acad Sci USA 1998; 95:12,300–12,305.

68. Consalvi V, Chiaraluce R, Giangiacomo L, et al. Thermal unfolding and conformational stability of the recombinant domain II of glutamate dehydrogenase from the hyperthermophile *Thermotoga maritima.* Protein Eng 2000; 13:501–507.

69. Lesley SA, Kuhn P, Godzik A, et al. Structural genomics of the *Thermotoga maritima* proteome implemented in a high-throughput structure determination pipeline. Proc Natl Acad Sci USA 2002; 99:11,664–11,669.

70. Wang W, Kim R, Jancarik J, Yokota H, Kim SH. Crystal structure of phosphoserine phosphatase from *Methanococcus jannaschii,* a hyperthermophile, at 1.8 A resolution. Structure (Camb) 2001; 10:9, 65–71.

71. Lee BI, Chang C, Cho SJ, et al. Crystal structure of the MJ0490 gene product of the hyperthermophilic archaebacterium *Methanococcus jannaschii,* a novel member of the lactate/malate family of dehydrogenases. J Mol Biol 2001; 307:1351–1362.

72. Lee DY, Ahn BY, Kim KS. A thioredoxin from the hyperthermophilic archaeon *Methanococcus jannaschii* has a glutaredoxin-like fold but thioredoxin-like activities. Biochemistry 2000; 39:6652–6659.

73. Wang H, Boisvert D, Kim KK, Kim R, Kim SH. Crystal structure of a fibrillarin homologue from *Methanococcus jannaschii,* a hyperthermophile, at 1.6 A resolution. EMBO J 2000; 19: 317–323.

74. Roy R, Adams MW. Tungsten-dependent aldehyde oxidoreductase: a new family of enzymes containing the pterin cofactor. Met Ions Biol Syst 2002; 39:673–697.

75. DiRuggiero J, Robb FT, Jagus R, et al. Cloning, characterization and in vitro expression of an extremely thermostable glutamate dehydrogenase from a novel hyperthermophilic Archeon, ES4. J Biol Chem 1993; 268:17,767–17,744.

76. Frankenberg RJ, Hsu TS, Yakota H, Kim R, Clark DS. Chemical denaturation and elevated folding temperatures are required for wild-type activity and stability of recombinant *Methanococcus jannaschii* 20S proteasome. Protein Sci 2001; 10:1887–1896.

77. Grottesi A, Ceruso MA, Colosimo A, Di Nola A. Molecular dynamics study of a hyperthermophilic and a mesophilic rubredoxin. Proteins 2002; 46:287–294.

78. Dairi T, Motohira Y, Kuzuyama T, Takahashi S, Itoh N, Seto H. Cloning of the gene encoding 3-hydroxy-3-methylglutaryl coenzyme A reductase from terpenoid antibiotic-producing *Streptomyces* strains. Mol Gen Genet 2000; 262:957–964.

79. Boucher Y, Huber H, L'Haridon S, Stetter KO, Doolittle WF. Bacterial origin for the isoprenoid biosynthesis enzyme HMG-CoA reductase of the archaeal orders Thermoplasmatales and Archaeoglobales. Mol Biol Evol 2001; 18:1378–1388.

80. Bochar DA, Stauffacher CV, Rodwell VW. Sequence comparisons reveal two classes of 3-hydroxy-3-methylglutaryl coenzyme A reductase. Mol Genet Metab 1999; 66:122–127.

81. Lecompte O, Ripp R, Puzos-Barbe V, et al. Genome evolution at the genus level: comparison of three complete genomes of hyperthermophilic archaea. Genome Res 2001; 11:981–993.

82. Maeder DL, Weiss RB, Dunn DM, et al. Divergence of the hyperthermophilic archaea *Pyrococcus furiosus* and *P. horikoshii* inferred from complete genomic sequences. Genetics 1999; 152:1299–1305.

83. Ettema T, van der Oost J, Huynen M. Modularity in the gain and loss of genes: applications for function prediction. Trends Genet 2001; 17:485–487.

84. DiRuggiero J, Dunn D, Maeder DL, et al. Evidence of recent lateral gene transfer among hyperthermophilic archaea. Mol Microbiol 2000; 38:684–693.

85. Kengen SW, de Bok FA, van Loo ND, Dijkema C, Stams AJ, de Vos WM. Evidence for the operation of a novel Embden–Meyerhof pathway that involves ADP-dependent kinases during sugar fermentation by *Pyrococcus furiosus*. J Biol Chem 1994; 269:17,537–17,541.

86. Tuininga JE, Verhees CH, van der Oost J, Kengen SW, Stams AJ, de Vos WM. Molecular and biochemical characterization of the ADP-dependent phosphofructokinase from the hyperthermophilic archaeon *Pyrococcus furiosus*. J Biol Chem 1999; 274:21,023–21,028.

87. Kengen SW, Tuininga JE, de Bok FA, Stams AJ, de Vos WM. Purification and characterization of a novel ADP-dependent glucokinase from the hyperthermophilic archaeon *Pyrococcus furiosus*. J Biol Chem 1995; 270:30,453–30,457.

88. Verhees CH, Tuininga JE, Kengen SW, Stams AJ, van der Oost J, de Vos WM. ADP-dependent phosphofructokinases in mesophilic and thermophilic methanogenic archaea. J Bacteriol 2001; 183:7145–7153.

89. Sakuraba H, Yoshioka I, Koga S, et al. ADP-dependent glucokinase/phosphofructokinase, a novel bifunctional enzyme from the hyperthermophilic archaeon *Methanococcus jannaschii*. J Biol Chem 2002; 277:12,495–12,498.

90. Ronimus RS, Kawarabayasi Y, Kikuchi H, Morgan HW. Cloning, expression and characterisation of a Family B ATP-dependent phosphofructokinase activity from the hyperthermophilic crenarchaeon *Aeropyrum pernix*. FEMS Microbiol Lett 2001; 202:85–90.

91. Mai X, Adams MW. Purification and characterization of two reversible and ADP-dependent acetyl coenzyme A synthetases from the hyperthermophilic archaeon *Pyrococcus furiosus*. J Bacteriol 1996; 178:5897–5903.

92. Musfeldt M, Schonheit P. Novel type of ADP-forming acetyl coenzyme A synthetase in hyperthermophilic archaea: heterologous expression and characterization of isoenzymes from the sulfate reducer *Archaeoglobus fulgidus* and the methanogen *Methanococcus jannaschii*. J Bacteriol 2002; 184:636–644.

93. Silva PJ, van den Ban EC, Wassink H, et al. Enzymes of hydrogen metabolism in *Pyrococcus furiosus*. Eur J Biochem 2000; 267:6541–6551.

94. Kelly RM, Adams MW. Metabolism in hyperthermophilic microorganisms. Antonie Van Leeuwenhoek 1994; 66:247–270.

95. Adams MW, Holden JF, Menon AL, et al. Key role for sulfur in peptide metabolism and in regulation of three hydrogenases in the hyperthermophilic archaeon *Pyrococcus furiosus*. J Bacteriol 2001; 183:716–724.

96. Albers SV, van de Vossenberg JL, Driessen AJ, Konings WN. Adaptations of the archaeal cell membrane to heat stress. Front Biosci 2000; 5:D813–D820.

97. Albers SV, Driessen AM. Signal peptides of secreted proteins of the archaeon *Sulfolobus solfataricus*: a genomic survey. Arch Microbiol 2002; 177:209–216.

98. Koning SM, Albers SV, Konings WN, Driessen AJ. Sugar transport in (hyper)thermophilic archaea. Res Microbiol 2002; 153:61–67.

99. Kelly DJ, Thomas GH. The tripartite ATP-independent periplasmic (TRAP) transporters of bacteria and archaea. FEMS Microbiol Rev 2001; 25:405–424.

100. Albers SV, Van de Vossenberg JL, Driessen AJ, Konings WN. Bioenergetics and solute uptake under extreme conditions. Extremophiles 2001; 5:285–294.

101. Gonzalez JM, Kato C, Horikoshi K. *Thermococcus peptonophilus* sp nov, a fast-growing, extremely thermophilic archaebacterium isolated from deep-sea hydrothermal vents. Arch Microbiol 1995; 164:159–164.

102. Worning P, Jensen LJ, Nelson KE, Brunak S, Ussery DW. Structural analysis of DNA sequence: evidence for lateral gene transfer in *Thermotoga maritima.* Nucleic Acids Res 2000; 28:706–709.

103. Borges KM, Brummct S, Davis MF, et al. Survey of the genome of the hyperthermophile, *Pyrococcus furiosus.* Genome Sci Technol 1996; 1:37–46.

104. Gonzalez JM, Robb FT. Genetic analysis of the *Carboxydothermus hydrogenoformans* carbon monoxide dehydrogenase genes cooF and cooS(1). FEMS Microbiol Lett 2000; 191:243–247.

105. Galtier N, Tourasse N, Gouy M. A nonhyperthermophilic common ancestor to extant life forms. Science 1999; 283:220–221.

106. Korbel JO, Snel B, Huynen MA, Bork P. SHOT: a Web server for the construction of genome phylogenies. Trends Genet 2002; 18:158–162.

107. Breitung J, Borner G, Scholz S, Linder D, Stetter KO, Thauer RK. Salt dependence, kinetic properties and catalytic mechanism of *N*-formylmethanofuran:tetrahydromethanopterin formyl-transferase from the extreme thermophile *Methanopyrus kandleri.* Eur J Biochem 1992; 210: 971–981.

108. Ermler U, Merckel M, Thauer R, Shima S. Formylmethanofuran:tetrahydromethanopterin formyltransferase from *Methanopyrus kandleri*—new insights into salt-dependence and thermo-stability. Structure 1997; 5:635–646.

109. Shima S, Herault DA, Berkessel A, Thauer RK. Activation and thermostabilization effects of cyclic 2, 3-diphosphoglycerate on enzymes from the hyperthermophilic *Methanopyrus kandleri.* Arch Microbiol 1998; 170:469–472.

110. Martusewitsch E, Sensen CW, Schleper C. High spontaneous mutation rate in the hyperthermophilic archaeon *Sulfolobus solfataricus* is mediated by transposable elements. J Bacteriol 2000; 182:2574–2581.

111. Brugger K, Redder P, She Q, Confalonieri F, Zivanovic Y, Garrett RA. Mobile elements in archaeal genomes. FEMS Microbiol Lett 2002; 206:131–141.

112. She Q, Brugger K, Chen L. Archaeal integrative genetic elements and their impact on genome evolution. Res Microbiol 2002; 153:325–332.

113. Watanabe Y, Yokobori S, Inaba T, et al. Introns in protein-coding genes in Archaea. FEBS Lett 2002; 510:27–30.

114. Gabriel JL, Chong PL. Molecular modeling of archaebacterial bipolar tetraether lipid membranes. Chem Phys Lipids 2000; 105:193–200.

115. Schut GJ, Zhou J, Adams MW. DNA microarray analysis of the hyperthermophilic archaeon *Pyrococcus furiosus:* evidence for a new type of sulfur-reducing enzyme complex. J Bacteriol 2001; 183:7027–7036.

116. Shockley KR, Ward DE, Chhabra SR, Conners SB, Montero CI, Kelly RM. Heat shock response by the hyperthermophilic archaeon *Pyrococcus furiosus.* Appl Environ Microbiol 2003; 69: 2365–2371.

117. Vierke G, Engelmann A, Hebbeln C, Thomm M. A novel archaeal transcriptional regulator of heat shock response. J Biol Chem 2002; October 14; Epub.

118. Dahlke I, Thomm M. A *Pyrococcus* homolog of the leucine-responsive regulatory protein, LrpA, inhibits transcription by abrogating RNA polymerase recruitment. Nucleic Acids Res 2002; 30:701–710.

119. Srinivasan G, James CM, Krzycki JA. Pyrrolysine encoded by UAG in Archaea: charging of a UAG-decoding specialized tRNA. Science 2002; 296:1459–1462.

120. Ibba M, Soll D. Genetic code: introducing pyrrolysine. Curr Biol 2002; 12:R464–R466.

16

The Genomes of Pathogenic Enterobacteria

Julian Parkhill and Nicholas R. Thomson

INTRODUCTION

The Enterobacteriaceae (enterics) form a diverse group of bacteria that take their name from the fact that they generally, but not exclusively, inhabit the gastrointestinal tract of animals. The group includes many of the most important pathogens of humans and domestic livestock and has furnished microbiologists with one of their most enduring laboratory tools, *Escherichia coli* K12. For these reasons, the biology of enteric bacteria has been intensively studied over many decades *(1)*, and they have, unsurprisingly, been subject to the close scrutiny of the upstart science of genomics.

The core of the enteric family is formed by the two genera, *Escherichia* and *Salmonella*; this review focuses mainly on these organisms. To date, the complete sequences of five *E. coli* strains and two salmonellae have been described, and many more are to come. However, the wider family includes other human pathogens, such as *Yersinia pestis*, *Yersinia enterocolitica*, and *Klebsiella pneumonia*; plant pathogens such as the *Erwiniaceae*; insect pathogens like *Photorhabdus*; and insect symbionts like *Buchnera* and *Wigglesworthia*. Many of these pathogens have been subject to genome sequencing or are under study (Table 1). Although most of these are not directly discussed here, many of the results described, and lessons learned, from *Escherichia* and *Salmonella* are directly relevant to these organisms.

Comparisons among these organisms have shown that they have a core set of genes and functions along a generally colinear genomic backbone with many regions and points of difference occurring along it. The differences between the genomes fall into several independent, but overlapping, types; these are discussed in the sections below. These include large insertions and deletions (including pathogenicity islands), integrated bacteriophages, small insertions and deletions, point mutations, and chromosomal rearrangements.

LARGE INSERTIONS AND DELETIONS
AND THE CONCEPT OF PATHOGENICITY ISLANDS

The genome sequence of the nonpathogenic *E. coli* K12 reference strain MG1655 was published in 1997 *(2)*; since then, it has acted as a backdrop against which to compare all subsequent enterobacterial genome sequences. To date, two other genotypes of *E. coli* have been sequenced to completion and the information published (although, as shown

From: *Microbial Genomes*
Edited by: C. M. Fraser, T. D. Read, and K. E. Nelson © Humana Press Inc., Totowa, NJ

Table 1
Enterobacterial Genomes Complete and In Progress

	Size (Mb)	%G+C	Predicted genes	Reference
Complete				
Buchnera aphidicola SG	0.641	25.3	545	*(97)*
Buchnera aphidicola BP	0.616	25.3	504	*(98)*
Buchnera aphidicola APS	0.641	26.3	564	*(99)*
Escherichia coli K12	4.639	50.8	4289	*(2)*
Escherichia coli O157:H7 (EPEC) EDL933	5.529	50.4	5349	*(3)*
Escherichia coli O157:H7 (EPEC) RIMD (Sakai)	5.498	50.5	5361	*(4)*
Escherichia coli CFT073 (UPEC)	5.231	50.5	5379	*(5)*
Salmonella enterica serovar Typhi CT18	4.809	52.1	4599	*(21)*
Salmonella enterica serovar Typhimurium LT2	4.857	52.2	4489	*(27)*
Shigella flexneri 2a	4.607	50.9	4434	*(6)*
Wigglesworthia brevipalpis	0.698	22.5	611	*(100)*
Yersinia pestis CO-92	4.654	47.6	4012	*(67)*
Yersinia pestis KIM	4.601	47.6	4198	*(94)*

Incomplete	Reference (http://)
Erwinia carotovora	www.sanger.ac.uk/Projects/Microbes/
Erwnia chrysanthmi	www.ahabs.wisc.edu/~pernalab/
Escherichia coli 042	www.sanger.ac.uk/Projects/Microbes/
Esherichia coli DH10B	www.hgsc.bcm.tmc.edu/microbial/
Escherichia coli E238/69	www.sanger.ac.uk/Projects/Microbes/
Escherichia coli K1	www.genome.wisc.edu/sequencing.htm
Klebsiella pneumoniae	genome.wustl.edu/projects/bacterial/
Photorhabdus luminescens	www.pasteur.fr/recherche/unites/gmp/
Photorhabdus asymbiotica	www.sanger.ac.uk/Projects/Microbes/
Salmonella bongori	www.sanger.ac.uk/Projects/Microbes/
Salmonella enterica serovar Arizonae	genome.wustl.edu/projects/bacterial/
Salmonella enterica serovar Choleraesuis	www.salmonella.org/
Salmonella enterica serovar Diarizonae	genome.wustl.edu/projects/bacterial/
Salmonella enterica serovar Dublin	www.salmonella.org/
Salmonella enterica serovar Enteritidis LK5	www.salmonella.org/
Salmonella enterica serovar Enteritidis PT4	www.sanger.ac.uk/Projects/Microbes/
Salmonella enterica serovar Gallinarum	www.sanger.ac.uk/Projects/Microbes/
Salmonella enterica serovar Paratyphi A	genome.wustl.edu/projects/bacterial/
Salmonella enterica serovar Pullorum	www.salmonella.org/
Salmonella enterica serovar Typhi Ty2	www.genome.wisc.edu/sequencing.htm
Salmonella enterica serovar Typhimurium DT104	www.sanger.ac.uk/Projects/Microbes/
Salmonella enterica serovar Typhimurium SL1344	www.sanger.ac.uk/Projects/Microbes/
Serratia marcescens	www.sanger.ac.uk/Projects/Microbes/
Shigella dysenteriae	www.sanger.ac.uk/Projects/Microbes/
Shigella flexneri 2a	www.genome.wisc.edu/sequencing.htm
Shigella sonnei	www.sanger.ac.uk/Projects/Microbes/
Yersinia enterocolitica	www.sanger.ac.uk/Projects/Microbes/
Yersinia pseudotuberculosis	www.bbrp.llnl.gov/bbrp/html/ microbe.html

Information taken from GOLD (*2a*; http://wit.integratedgenomics.com/GOLD/) and the Institute for Genomic Research (http://www.tigr.org/tdb/mdb/mdbinprogress.html)

in Table 1, investigations of many more are in progress: the enterohemorrhagic *E. coli* (EHEC) O157:H7 *(3,4)* and the uropathogenic *E. coli* (UPEC) *(5)*. In addition, information on the genome of *Shigella flexneri* strain 2a has been published *(6)*. Current opinion holds that the shigellae are part of the *E. coli* species complex *(7)*, and they are treated as such in this chapter.

Escherichia coli O157:H7, first isolated in 1982 *(8)*, is recognized as the major cause of diarrhea, hemorrhagic colitis, and sporadically hemolytic uremic syndrome *(8,9)*. Two separate groups have published a genome sequence of an EHEC strain. Perna et al. *(3)* published the sequence of strain EDL933, which was linked to the originally recorded outbreak. The published sequence of this strain contained two physical gaps (4 kb and 54 kb), which correspond to prophage-related regions. Hayashi et al. *(4)* published the complete sequence of *E. coli* O157:H7 strain RIMD 0509952, referred to as Sakai, which was taken from a patient in the 1996 Japanese outbreak.

The UPEC sequenced was strain CFT073 and is a member of the varied group of extra-intestinal *E. coli*. UPEC are attributed as the cause of neonatal meningitis/sepsis and the majority of urinary tract infections.

A comparison of all the pathogenic *E. coli* with the archetypal nonpathogen *E. coli* K12 revealed that the genomes are essentially colinear, displaying both conservation in sequence and gene order (Fig. 1). The conserved sequence, with respect to *E. coli* K12, amounts to 4.1 Mb for EHEC and 3.9 Mb for the UPEC. The genes that are predicted to be encoded within this conserved sequence display more than 95% sequence identity and have been termed the *core genes*. Similar observations were also made for the 4.6 Mb *S. flexneri* genome, which also shares 3.9 Mb of common sequence with *E. coli* K12 *(6)*.

A comparison of all the *E. coli* genomes revealed that genes shared by all three amounted to 2996 *(5)* from a total of 4288, about 5400, and about 5500 predicted protein-coding sequences for *E. coli* K12, EHEC, and UPEC, respectively. The region encoding these core genes is known as the *backbone sequence*.

It was also apparent from these comparisons that interdispersed throughout this backbone sequence were large regions unique to the different genotypes. Moreover, several studies had shown that some of these unique loci were present in clinical disease-causing isolates, but apparently absent from their comparatively benign relatives *(10)*.

One such well-characterized region is the locus of enterocyte effacement (LEE; Fig. 2) *(11)*, which was first discovered in enteropathogenic *E. coli* (EPEC). An EPEC infection results in effacement of the intestinal microvilli and the intimate adherence of bacterial cells to enterocytes. EPEC also subvert the structural integrity of the cell and force the polymerization of actin, which accumulates below the adhered EPEC cells, forming cuplike pedestals *(12)*. This is called an attachment and effacing (AE) lesion. The ability to cause an AE lesion was localized to a 35-kb region, LEE, which displayed an anomalous guanine and cytosine (G+C) content. LEE was subsequently found in all of the bacteria known to be able to elicit an AE lesion, including clinical isolates of EHEC, *Hafnia alvei* and *Citrobacter freundii (11–16)*.

Detailed analysis of the LEE has shown that it is located adjacent to the *selC* transfer ribonucleic acid (tRNA) gene in EPEC and constitutes an insertion with respect to the analogous locus in *E. coli* K12 (Fig. 2). Sequence analysis of this region revealed that LEE encodes an adhesin (intimin), a type III secretion system, as well as secreted protein effectors *(11,17)*, which have been shown to be essential for pathogenicity. Mutations

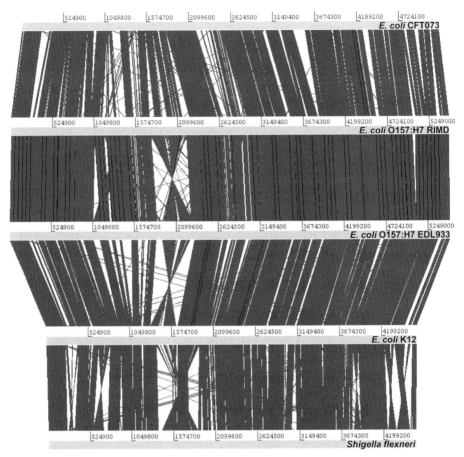

Fig. 1. Global comparison between five *E. coli* strains. DNA:DNA matches computed using BLASTN and displayed using the Artemis Comparison Tool (ACT; http://www.sanger.ac.uk/Software/ACT) between the five published *E. coli* genomes. The genomes are, from the top: *E. coli* CFT073 (UPEC), *E. coli* O157:H7 RIMD, *E. coli* O157:H7 EDL933, *E. coli* K12, and *S. flexneri*. The gray bars between the genomes represent individual BLASTN matches. Some of the shorter and weaker BLASTN matches have been removed to show the overall structure of the comparison.

within LEE attenuate virulence and remove the ability to cause the AE lesion *(11,13)*. In addition, the introduction of LEE from EPEC into *E.coli* K12 renders this relatively benign strain capable of eliciting an AE lesion *(18)*.

Many similar regions linked to virulence have been characterized in both Gram-negative and Gram-positive bacteria (for a review, see ref. *19*). This led to the concept of pathogenicity islands (PAIs) and the formulation of a definition to describe their features. Characteristically, PAIs are inserted adjacent to stable RNA genes and have an atypical G+C content. In addition to virulence-related functions, these regions often carry genes encoding transposase or integraselike proteins and are unstable and self-mobilizable *(19,20)*. As mentioned here, PAIs have also been found to be limited in their phylogenetic distribution when comparing pathogenic and nonpathogenic isolates. It was also noted that PAIs carry a high proportion of gene fragments or disrupted genes when compared to the backbone regions *(21)*.

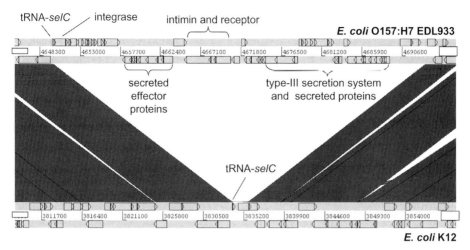

Fig. 2. The LEE pathogenicity island in *E. coli* O157:H7 EDL933. DNA:DNA matches, computed using BLASTN and displayed using ACT, between *E. coli* O157:H7 EDL933 (top) and *E. coli* K12. Important genes in the LEE island are labeled. The overall size of the inserted DNA is about 45 kb and includes the bacteriophage 933L (genes to the right of the integrase).

As with many generic descriptors, the term *pathogenicity island* has been used widely to describe many genomic loci. Some of these loci, although sharing some of the features of PAIs, are unlikely to impinge on pathogenicity. To take account of this, the concept of PAIs has been extended to include islands or strain specific loops, which represent discrete genetic loci that are lineage specific but are as yet not known to be involved in virulence *(3,4)*.

E. coli *Unique Regions/Islands*

In addition to the backbone sequence, both EHEC and UPEC carry an additional approx 1.3–1.4 Mb of unique sequence. Comparing *E. coli* K12 and EHEC O157:H7 strain EDL933, the unique regions were divided into K islands and O islands, referring to regions unique to *E. coli* K12 or *E. coli* O157:H7, respectively *(3)*. These K islands and O islands correspond to the K loops (K12 specific) and S loops (O157:H7 strain specific) used to describe the same features of the EHEC O157:H7 strain Sakai vs *E. coli* K12 comparison *(4)*.

Using the notation of Hayashi et al. *(4)*, *E. coli* O157:H7 possessed 296 S loops (>19 kb), compared to 325 K loops identified in *E. coli* K12. Approximately half of the genes carried on the S loops were phage related (48.2%). Of the remaining genes, 33% could not be ascribed a function, and 15% were related to virulence. Among the larger S loops, 4 were virulence related, encoding fimbrial proteins unique to O157:H7. A further 5 fimbrial clusters were also identified that were partially conserved in *E. coli* K12, leaving open the possibility that they could also facilitate binding to strain-specific target sites. Along with the fimbrae, 14 adhesins/invasins were identified, including the previously characterized intimin carried on the EHEC LEE PAI and the Iha adhesin *(22)*.

In addition to the type III secretion system known to be located on LEE, a novel type III secretion system was also discovered. Interestingly, this novel type III secretion system is more similar to the *Salmonella enterica* subspecies *enterica* serovar Typhimurium

(*S. typhimurium*) secretion system carried on another PAI, *Salmonella* pathogenicity island 1 (SPI-1) (see next section). In addition to the previously characterized phage-encoded Shiga-like toxins (Stx) *(23,24)* and enterohemolysin *(25)*, two other toxins were identified. One of these toxin genes was predicted to encode a large RTX-family (*repeats in structural toxin*) protein (5292 aa), which was carried on an S loop, alongside genes that may facilitate its secretion and activation.

Welch et al. *(5)* used a similar notation to describe the UPEC-specific islands (UIs). Genes encoded within UIs included 12 distinct fimbrial systems, such as the two *pap* operons, which were known to be uropathogen specific. In addition to these, several other fimbrial systems were identified in UPEC, such as the *yad* chaperone–usher system, but these were also known to be present in the other two sequenced genotypes, K12 and EHEC. However, even these ubiquitous fimbrial systems displayed a high level of sequence variation, which suggested that they interact with different target sites.

UIs were also found to carry 7 autotransporters, a novel RTX family toxin, and 5 *fimE* and *fimB* recombinase systems, all of which are associated with aspects of host interaction or, in the case of the recombinases, phase switching (rapid random phenotypic variation). In total, the S loops of EHEC account for 25.3% (1.393 Mb) of the total genome. The analogous figure for UIs was 24.9% (1.303 Mb). Comparing the coding sequences carried on S loops and UIs with the genes of *E. coli* K12, Welch et al. *(5)* showed that, apart from the core 2996 genes shared by all three genotypes, there were 1827 genes present in UPEC but absent from *E. coli* K12. The equivalent figure for EHEC was 1387, and even *E. coli* K12 possessed 585 genes that were not present in the other two *E. coli*. Interestingly, only a small percentage (11%) of the 1827 genes unique to UPEC were also found in EHEC.

Analysis of the nucleotide composition and codon usage of the S loops and UIs also showed that they display an atypical G+C content (47–48%; this figure excludes phage-related genes for EHEC) with respect to the backbone sequence (50.05%). In addition, there was a preponderance of rare codons within coding sequences in these regions *(3–5)*.

These observations are consistent with the hypothesis that pathogenic *E. coli* genotypes have evolved from a much smaller nonpathogenic relative by the acquisition of foreign deoxyribonucleic acid (DNA). This laterally acquired DNA has been attributed with conferring on the different genotypes the ability to colonize alternative niches in the host and the ability to cause a range of different disease outcomes.

Comparison of E. coli with the Salmonellae

Currently there are over 2300 *Salmonella* serovars in two species, *S. enterica* and *Salmonella bongori (26)*. All the salmonellae are highly related, sharing a median DNA identity for the reciprocal best match of between 85 and 90% *(27,28)*. Despite this homogeneity, there are significant differences in the pathogenesis and host range of the different serovars. For example *S. enterica* subspecies *enterica* serovar Typhi (*S. typhi*) is only able to infect humans and causes severe typhoid fever. Whereas *S. typhimurium* causes gastroenteritis in humans, a systemic infection in mice, and has a broad host range *(27)*, *S. enterica* serovar Pullorum primarily infects chickens, causing fowl typhoid *(29)*.

Like *E. coli,* the salmonellae are known to possess PAIs (*Salmonella* pathogenicity islands, SPIs), which are thought to have been acquired laterally. The gene products encoded by pathogenicity island SPI-1 *(30,31)* and SPI-2 *(32,33)* have been shown to be important for different stages of the infection process. Both of these islands carry type III secretion systems and their associated secreted protein effectors. SPI-1 confers all salmonellae with the ability to invade epithelial cells. SPI-2 is important for various aspects of the systemic infection, allowing *Salmonella* to spread from the intestinal tissue into the blood and eventually to infect, and survive within, macrophages of the liver and spleen (reviewed in ref. *34*).

SPI-3 (17 kb), like LEE and PAI-1 of UPEC, is inserted alongside the *selC* tRNA gene *(11,20)*. SPI-3 carries the gene *mgtC*, which is required for intramacrophage survival and growth in the low-magnesium environment thought to be encountered in the phagosome *(35)*.

Other *Salmonella* SPIs encode type III–secreted effector proteins, chaperone–usher fimbrial operons, Vi antigen biosynthetic genes, a type IVB pilus operon, and many other determinants associated with enteropathogenicity *(21,36–39)*. The majority of the *S. typhi* SPIs are also present in *S. typhimurium*, the exceptions are SPI-7 (Vi pathogenicity island), SPI-8, and SPI-10.

There are several features of SPIs that make them unique from the *E. coli* PAIs. SPI-1 is present in all members of the genus, including *S. bongori. Salmonella enterica* and *S. bongori* are thought to have split from a common ancestor during the early stages of *Salmonella* evolution. Therefore, SPI-1 is thought to have been acquired by *Salmonella* soon after the preceding split from *E. coli*. SPI-2 is present in all subspecies of *S. enterica*, but absent from *S. bongori (40)* and so represents a more recent acquisition than SPI-1. Thus, unlike many of the *E.coli* PAIs, SPIs are lineage specific rather than strain specific. However, it is worth noting that both SPI-1 and SPI-2 were not detected in any other enterobacterial pathogens *(41)* and so are restricted in distribution like all other true PAIs.

Although the mobile nature of PAIs is mentioned frequently in the literature, there is little direct experimental evidence to support these observations. One explanation for this is that, on integration, the PAI mobility genes subsequently become degraded, thereby fixing their position. Certainly, there is evidence to support such an idea because many proposed PAIs carry integrase or transposase pseudogenes or remnants. An excellent example of this is the high-pathogenicity island (HPI) first characterized in the yersinias *(42)*. The *Yersinia* HPIs can be split into two lineages based on the integrity of the phage integrase gene (*int*) carried in the island: (1) *Y. enterocolitica* biotype 1B and (2) *Y. pestis* and *Y. pseudotuberculosis (43)*. The *Y. enterocolitica* HPI *int* gene carries a point mutation, whereas the analogous gene is intact in the *Y. pestis* and *Y. pseudotuberculosis* HPIs.

The HPI is a 35 to 43-kb island that carries genes for the production and uptake of the siderophore yersiniabactin, as well as genes, such as *int,* proposed to be involved in the mobility of this island. HPI is flanked by 17-bp direct repeats also thought to be essential for mobility. HPI-like elements are widely distributed in enterobacteria, including *E. coli, Klebsiella, Enterobacter*, and *Citrobacter* spp *(4,44,45)*, like many prophages, HPIs are found adjacent to *asn*–tRNA genes. tRNA genes are common sites for bacteriophage integration into the genome *(46)*. The reason for this is thought to be that tRNA genes are highly conserved between the genera and found in multiple copies. Integration at these sites commonly involves site-specific recombination between short stretches of identical DNA located on the phage (*attP*) and at the integration site on bacterial genomes (*attB*).

The *Yersinia* HPI, as well as most other PAIs, is thought to move in a manner similar to bacteriophages. To address this issue, Rakin et al. *(47)* constructed a mini-HPI–based plasmid that carried the *int* gene, a selectable marker, and a reformed *attP* site (based on the sequence of the 17-bp direct repeats that flank the HPI). This construct was able to integrate freely in *E. coli recA* strains. Moreover, this construct could integrate at multiple *asn*–tRNA loci *(47,48)*.

The tRNA genes represent common sites for the integration for many other PAIs and bacteriophages. Perhaps the most heavily utilized integration site in the enterics is the *selC* tRNA locus. This site is occupied by the LEE PAI in Shiga toxin–producing *E. coli* and EPEC and by PAI-I in UPEC *(5)*. In enterotoxigenic *E. coli* (ETEC) the *selC* site is occupied by a novel island encoding Tia (an adhesin) *(49)*, in *Shigella flexneri* SHI-2 (encoding aerobactin biosynthesis) is found here *(50)*, and this site is occupied by SPI-3 in *S. typhimurium* and *S. typhi (21,35)*. It has also been shown that, in some of Stx-producing *E. coli* strains that naturally lack the LEE island, the *selC* tRNA locus is occupied by a PAI, denoted the locus of proteolysis activity, reflecting the presence of a serine protease gene within the island *(51)*. Also, the same PAI can be integrated not only at paralogous tRNAs (as shown with HPI; see discussion above), but also at unrelated tRNA loci. LEE is found at both the *selC* tRNA locus and alongside the *pheU* tRNA in *E. coli (52,53)*.

Welch et al. *(5)* published a comprehensive comparison of UPEC and EHEC; it showed that 13 of the UIs, identified in the genomic sequence, were integrated alongside tRNA genes. This compares to 10 of the EHEC-specific islands adjacent to tRNA genes. Moreover, 9 of these islands are located at the same tRNA gene in both genotypes. In addition, 10 other common loci shared by UPEC and EHEC are occupied with different organism-specific islands in both genomes. At least one of these other regions is also integrated into a stable RNA gene, *ssrA (4,5)*.

However, the picture is still more complicated; many unique islands that appear as insertions in whole genome comparisons are not simply the result of a single insertion and may be the product of successive events *(54)*. A well-characterized example of this is SPI-2 of *S. typhimurium*. SPI-2 is thought to be a composite of at least two independent insertion events. Starting at the integration site, the *valV*–tRNA gene, the first 25.3 kb of SPI-2 carries genes encoding the type III secretory apparatus and protein effectors. This section of SPI-2 is only present in *S. enterica* and is absent from *S. bongori*. The next section of SPI-2 (14.5 kb), which is present in *S. bongori* as well as *S. enterica,* carries genes for the respiration of tetrathionate. This region is proposed to be part of a more ancient insertion *(54)*. A comparison of uropathogen-specific islands also showed a high degree of diversity when comparing the same UI sequenced from different UPEC strains *(5)*.

What is clear is that PAIs have been fundamental in the evolution of many pathogenic members of the *Enterobacteriaceae*. This is also evident in the natural populations, in which the genomes of enteric bacteria have been shown to vary markedly. For example, the genomes of *E. coli* have been found to vary in size by as much as 1 Mb, from between 4.5 and 5.5 Mb *(55,56)*. Looking for the presence of a selection of PAIs, a correlation was observed by Boyd and Hartl, who saw that the isolates carrying PAIs tended to have larger genomes *(57)*, indicating that PAIs accounted for a significant portion of this genetic diversity. One explanation for the capacity of *E. coli* and *Salmonella* to acquire large sections of foreign DNA is that some strains have a defective mismatch repair mechanism *(58)*, a side effect of which is an enhanced ability to recombine and acquire DNA. Of course, looking at the data from more recent genome projects, it is now known that bacteriophages contribute significantly to this apparent fluidity in genome size.

BACTERIOPHAGES

Integrated bacteriophages, or prophages, are also commonly found in bacterial genome sequences. The impact of bacteriophages on bacterial evolution should not be underestimated. As mentioned in the section on *E. coli* unique regions, of the S loops reported for EHEC O157:H7 EDL933, nearly 50% were phage related (see Chapter 5). Looking at all the enteric genomes with information published thus far, the sheer number and diversity of prophages contained in these bacterial genome sequences has been a revelation and has had a dramatic effect on the understanding of bacterial evolution. In addition to the 18 prophage sequences detected in the genome of EHEC strain Sakai *(4)*, the genomes of *E. coli* K12, UPEC, and *S. flexneri* have all been shown to carry multiple prophage or prophage-like elements *(2,3,5,6)*. This diversity is not restricted to comparisons between different species or genera; a comparison of the genome sequences of EHEC O157:H7 strain EDL933 and strain Sakai revealed that the complement and integration sites of prophages varied markedly, as did internal regions within highly related phage *(4,59)*.

The number and diversity of prophages represent a rich source of additional genetic variation. In addition to genes essential for their own replication, phages often carry genes that, for example, prevent superinfection by other bacteriophages, such as *old* and *tin (60,61)*. However, other genes carried in prophages appear to be of nonphage origin and can encode determinants that enhance the virulence of the bacterial host by a process called *lysogenic conversion* (reviewed in ref. *61a*).

In addition to the presence of the LEE PAI and the ability to elicit an AE lesion, one of the other defining characteristics of EHEC is the production of Stx. Shiga toxins are a family of potent cytotoxins that, on entry into the eukaryotic cell, act as glycosylases and cleave the 28S ribosomal RNA (rRNA), thereby inactivating the ribosome and consequently preventing protein synthesis *(62)*. EHEC are known to produce two variants of Stx, both encoded on a different prophage: CP933V (Sp15) and BP933W (Sp5) *(3,4, 63)*. The *stx* genes are located within the lambda-like phage such that their expression is linked to the induction of the prophage following the derepression of the phage late genes. This has strong implications for the use of chemotherapeutic agents such as mitomycin C and antibiotics commonly used to treat diarrheal diseases, which may induce the production of these toxins *(64)*.

Other determinants carried by prophages in the EHEC genome include an enterohemolysin *hly2 (65)*, an array of tRNA genes *(4)*, and genes involved in serum resistance, such as *lom* (for a more detailed treatise, see ref. *66*).

Other enterics such as *S. typhi*, *S. typhimurium*, and *Y. pestis* are also known to harbor significant numbers of prophages *(21,27,67)*. The principal virulence determinants of the salmonellae are the type III secretion systems, carried on SPI-1 and SPI-2, and their associated protein effectors *(68,69)*. A significant number of these type III secreted effector proteins are carried in the genomes of prophages. These include *sseI* and *gogB* (a leucine-rich YopM family effector protein), which are carried by the λ-like prophage Gifsy 1 and Gifsy 2 in *S. typhimurium*, respectively *(70,71)*. Like the EHEC prophages, these *Salmonella* prophages also have a dramatic influence on the ability of their bacterial hosts to cause disease. For example, curing *S. typhimurium* strains of Gifsy 2 results in a more than 100-fold attenuation of virulence in a mouse model of disease *(71,72)*.

Both *S. typhi* and *S. typhimurium* carry the P2-like prophage SopE *(73)*. As its name suggests, the SopE phage carries the *sopE* gene, the product of which is secreted by the SPI-1 type III secretion system and promotes efficient entry of the bacterium into cells by activating RhoGTPases *(74)*. Like the other effector proteins carried in Gifsy 1 and 2 phages, *sopE* lies in the variable tail fiber region of the SopE phage. Detailed analysis of this region has shown that the *sopE*-containing region is a 3-kb cassette, termed the sopE moron *(73)*, thought to be transmissible between related and unrelated bacteriophages. This mechanism is postulated to facilitate horizontal gene transfer between phages and subsequently different bacterial hosts, thus circumventing some of the barriers for horizontal gene transfer brought about by bacteriophage superinfection immunity *(73)*.

SMALL INSERTIONS AND DELETIONS

Although it is undeniable that the large PAIs discussed above are fundamental to any explanation of the differing phenotypes of these strains, many other differences exist and must be considered when looking at the overall genomic picture of this group of organisms.

The comparisons between *E. coli* K12 and *E. coli* O157:H7 *(3)* and between *S. typhi* and *S. typhimurium* *(21)* indicated that many small differences existed, aside from the large islands discussed above. In fact, analysis of the differences between *S. typhi* and *S. typhimurium* and of those between *S. typhi* and *E. coli* K12 clearly shows that the majority of insertions or deletions between these pairs of organisms are small. The number of separate insertion or deletion events showed that there are 145 events of 10 genes or fewer compared with 12 events of 20 genes or more for the *S. typhi* and *S. typhimurium* comparison. The *S. typhi* vs *E. coli* comparison revealed 504 events of 10 genes or fewer compared with just 25 events of 20 genes or more (Fig. 3A). Even taking into account the fact that the larger islands contain many more genes per insertion or deletion event, it is clear that nearly equivalent numbers of species-specific genes are attributable to insertion or deletion events involving 10 genes or fewer as are due to events involving 20 genes or more (Fig. 3B). Insertion or deletion events involving 10 genes or fewer are responsible for 377 gene differences for the *S. typhi* and *S. typhimurium* comparison, whereas events involving 20 genes or more are responsible for 631. Similarly, events involving 10 genes or fewer are responsible for 1287 gene differences for the *S. typhi* and *E. coli* comparison, whereas events involving 20 genes or more are responsible for 1019 (Fig. 3B).

It should be clear from this discussion that the acquisition and exchange of small islands is likely to be important to the overall phenotype of the organisms; however, a few examples underline the point. SspH2 is a leucine-rich protein secreted by the *Salmonella* type III secretion system encoded by SPI-2. In *S. typhi*, SspH2 is encoded by a unique

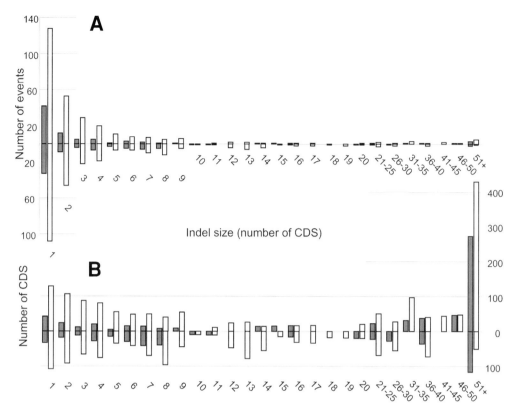

Fig. 3. Graph showing insertion and deletion events between *S. typhi* and *S. typhimurium/E. coli*. The *x*-axis shows the size of the insertion/deletion events expressed as number of coding sequences. Bars above the axis show insertions in *S. typhi* relative to *S. typhimurium* (gray) and *E. coli* (white). Bars below the axis show deletions in *S. typhi* relative to *S. typhimurium* (gray) and *E. coli* (white). The *y*-axis in plot A shows the number of insertion/deletion events; in plot B, it shows the total number of CDSs in all the events of that size.

region relative to *E. coli* K12, which contains just one other intact gene and three pseudo-genes of phage origin. The *S. typhimurium* equivalent of this island contains additional genes of phage origin, suggesting the entire region is a prophage remnant. Elsewhere, the single *S. typhimurium* gene *envF* (encoding a putative lipopotein) is replaced in *S. typhi* by a five-gene block encoding distant homologs of the *Campylobacter jejuni* toxin sub-unit CdtB and the *Bordetella pertussis* toxin subunits PtxA and PtxB. Many other such examples exist for other organisms and in other comparisons.

In the majority of these cases, no evidence exists for genes that might allow these islands to be self-mobile. It is theoretically possible that these are examples of lineage-specific gene loss, and that these groups of genes existed in the common ancestor of both strains or species. However, for this to be the case, an ancestral chromosome con-siderably larger than that in any extant member of the group would have to be postulated, and this seems very unlikely. It is far more likely that small islands of this type are indeed exchanged between members of a species and constitute part of the species gene pool. It is easy to see that, once acquired by one member of the species, they can be easily exchanged by generalized transduction mechanisms *(1)*, followed by homologous recombination between the near-identical flanking genes to allow integration into the chromosome.

This sort of mechanism of genetic exchange would also allow nonorthologous gene replacement, involving the exchange of related genes at identical regions in the backbone. A specific example of this can be seen in the capsular switching of *Neisseria meningitidis (75)* and *Streptococcus pneumoniae (76,77)*, for which different sets of genes responsible for the biosynthesis of different capsular polysaccharides are found at identical regions in the chromosome and are flanked by conserved genes. The implied mechanism for capsular switching involves replacement of the polysaccharide-specific gene sites by homologous recombination between the chromosome and exogenous DNA in the flanking genes.

This type of gene exchange may well be more general, and good candidates for this occur among the numerous chaperone–usher fimbrial systems of *E. coli* and *Salmonella*. These systems are variable in number and occur in both *E. coli (3)* and *S. enterica (78)*. Perna et al. *(3)* noted that these fimbrial operons were among the most variable sequences in both *E. coli* K12 and *E. coli* O157:H7, and that the most variable gene apparently orthologous between the two genomes was *yadC*, which encodes a fimbrial subunit in one of these operons. This observation also extends to the comparison between *S. typhi* and *E. coli* K12. Figure 4 shows the comparison between the operon containing *yadC* in *E. coli* and the apparently orthologous *sta* fimbrial operon in *S. typhi*. It can be seen that they reside in the same chromosomal context, and the flanking genes show very high levels of similarity. This contrasts with the fimbrial genes themselves, which are only weakly conserved. It is possible that this may represent rapid sequence divergence under selective pressure (possibly from the host immune system). Alternatively, it may be that similar, but not orthologous genes have been exchanged at the same chromosomal location by homologous recombination between the conserved flanking genes and exogenously acquired DNA. Interestingly, this operon is not present in *S. typhimurium*, although another chaperone–usher system is present just 4 kb downstream (Fig. 4), further illustrating the exchange of small functional islands.

POINT MUTATIONS AND PSEUDOGENES

One of the most surprising observations to come from some of the enterobacterial genome projects was that certain species/strains appear to contain a large number of *pseudogenes*, that is, genes that appear to be untranslatable because of the presence of stop codons, frameshifts, internal deletions, or insertion sequence (IS) element insertions. This has been the cause of much debate, especially against the background of the common assumption that the bacterial genome is a highly "streamlined" system that does not carry "junk DNA."

It is certainly the case that there exist specific mechanisms to allow read-through of stop codons and correction of frameshift mutations through programmed ribosomal frameshifting *(1)*. However, in many cases, it has been possible to identify specific phenotypic correlates of these genomic mutations. For others, the increase in comparative genomics has allowed more confidence to be placed in predictions of nonfunctionality. Given that one close relative has a mutation and the other does not, how likely is it that

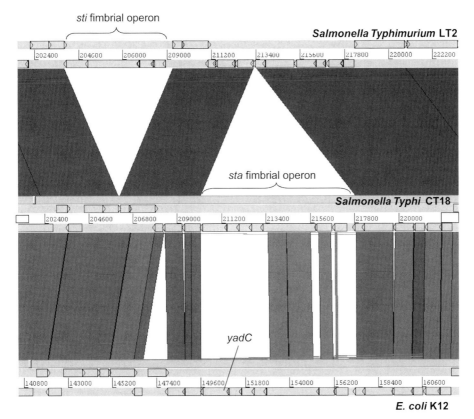

Fig. 4. Chaperone–usher fimbrial operons in *Salmonella* and *E. coli*. DNA:DNA matches, computed using BLASTN and displayed using ACT, between *S. typhimurium* LT2, *S. typhi* CT18, and *E. coli* K12. The gray bars between the genomes represent individual BLASTN matches. The *sti* and *sta* fimbrial operons in *S. typhi* and *S. typhimurium* are labeled, as is the *yadC* gene, which is part of the equivalent operon in *E. coli*.

the first strain will have acquired both the mutation and the specific mechanism to suppress it in such a short period of evolutionary time? Assuming that these predictions of inactive genes are correct, what might be inferred about the biology, and evolution, of the strains that contain them?

As described in the section comparing *E. coli* with the salmonellae, *S. typhi*, the causative agent of typhoid fever, is host restricted and appears only capable of infection of a human host, whereas *S. typhimurium*, which causes a milder disease in humans, has a much broader host range. In the analysis of the *S. typhi* genome, Parkhill et al. *(21)* were able to identify over 200 pseudogenes, and *S. typhimurium* was predicted to contain only around 39 *(27)*.

It is clear that the pseudogenes in *S. typhi* are not randomly spread throughout the genome: They are overrepresented in genes that are unique to *S. typhi* when compared to *E. coli* (59% of the pseudogenes lie in the unique regions compared to 33% of all *S. typhi* genes in unique regions), and many of the pseudogenes in *S. typhi* have intact counterparts in *S. typhimurium* that have been shown to be involved in aspects of virulence and host interaction. Specific examples of this include the leucine-rich repeat protein *slrP* [involved in host range specificity in *S. typhimurium (79)* and secreted through a type III system]; other type III–secreted effector proteins, including *sseJ (80)*, *sopE2 (81)*, and *sopA (82,83)*; and the genes *shdA*, *ratA*, and *sivH*, which are present in an island unique to salmonellae infecting warm-blooded vertebrates *(84)*. Many other inactivated genes may also have been involved in virulence or host interaction, including components of 7 of the 12 chaperone–usher fimbrial systems.

Given this distribution of pseudogenes, it was suggested that the host specificity of *S. typhi* might be because of the loss of an ability to interact with a broader host range caused by functional inactivation of the necessary genes *(21)*. In contrast to other organisms containing multiple pseudogenes, such as *Mycobacterium leprae (85)*, most of the pseudogenes in *S. typhi* are caused by a single mutation, suggesting that they have been inactivated relatively recently. This ties in well with the fact that, worldwide, *S. typhi* is seen to be clonal *(86)*, and the serovar may be only a few tens of thousands of years old *(86a)*.

These observations together suggest an evolutionary scenario in which the recent ancestor of *S. typhi* changed its niche in a human host, evolving from an organism (similar to *S. typhimurium*) limited to localized infection and invasion around the gut epithelium into one capable of invading the deeper tissues of the human host. This change of niche was likely to have involved a small population, leading to an evolutionary bottleneck: conditions that are thought to increase the fixation of mutations by genetic drift *(87)*.

A similar evolutionary scenario has been suggested for another recently evolved enteric bacterium, *Y. pestis*. This organism has also recently changed from a gut bacterium (*Y. pseudotuberculosis*), transmitted via the fecal–oral route, to an organism capable of utilizing a flea vector for systemic infection *(88,89)*. Again, this change in niche was accompanied by pseudogene formation, and genes involved in virulence and host interaction are overrepresented in the set of genes inactivated *(67)*.

A further example is *S. flexneri* 2a, a member of the species *E. coli* (which is predicted to have over 250 pseudogenes), and is again restricted to a human host *(6)*.

These organisms clearly demonstrate that enteric evolution is a process that involves gene loss as well as gene gain, and that the remnants of the genes lost in this process of evolution can be readily detected.

REARRANGEMENTS: GENOME INTEGRITY AND FLUIDITY

As soon as the first chromosomal maps of *Escherichia* and *Salmonella* were produced, it was observed that many of the markers were in similar positions on the two chromosomes, and that there was an overall colinearity in the two genomes *(90,91)*. This original observation has been amply supported by the complete genome sequences available. Figure 1 shows that the genomes of the four sequenced *E. coli* strains are almost entirely colinear except for a 440-kb inversion around the replication terminus in *E. coli* O157:H7 EDL933 *(3)*. Given that these are all members of the same (admittedly broad) species, this is not surprising.

What is more surprising is that this colinearity extends to the genomes of *Salmonella* (Fig. 5) *(21,27)*. There is just one inversion around the terminus that differentiates the

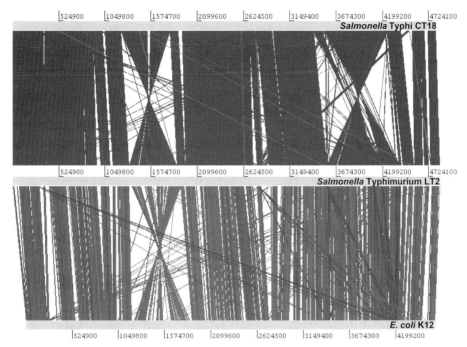

Fig. 5. Global comparison between *Salmonella* and *E. coli*. DNA:DNA matches, computed using BLASTN and displayed using ACT, between *S. typhi* CT18, *S. typhimurium* LT2, and *E. coli* K12. The gray bars between the genomes represent individual BLASTN matches. Some of the shorter and weaker BLASTN matches have been removed to show the overall structure of the comparison.

global gene order of *S. typhimurium* from *E. coli* K12. Such reciprocal inversions around the origin and terminus are the most common form of chromosomal rearrangement between related bacteria *(92,93)* and have been hypothesized as caused by either direct recombination between replication forks *(92)* or the fact that only those regions of DNA close to the replication forks are unpackaged and available for recombination *(94)*.

Two possible explanations for this overall conservation of gene order are that the genomes are stable because very little recombination occurs, or that the specific gene order in *Escherichia* and *Salmonella* is a product of, and is maintained by, selection. The explanation invoking a lack of recombination seems unlikely because the organisms are fully recombination proficient, and there is ample evidence of genome rearrangements between these organisms and more distant relatives such as *Yersinia (67,94)*. In addition, it is clear from the finer details of the genomic comparisons that chromosomal recombinations have occurred in these organisms.

If, as seems likely, the explanation is that this conserved gene order is maintained by selection in the face of recombination, then it becomes necessary to identify the basis of this selection. There are several possibilities. First, the enteric bacteria generally rely heavily for coregulation of functionally related genes by clustering them in cotranscribed operons; this is likely to place constraints on successful intraoperon recombination events.

Second, in rapidly growing bacteria, chromosomal replication will often occur substantially in advance of cell division, leading to genes closest to the origin being present in more copies per cell than genes close to the terminus. It is likely that this gene–dosage effect will have an effect on specific expression levels in the cell, and the organism may well have adapted to utilize it.

Third, genes are predominantly transcribed in the same direction as the movement of the replication fork, probably to avoid collisions between the replication and the transcription machineries; again, recombination may well disrupt this process.

Other possibilities exist, including selection and mutational pressures on the position and orientation of genes *(95,96)*, and it is probable that any, or indeed all, of these may be acting together. It is interesting to note that this conservation of gene order does not hold in the case of the comparison of *E. coli* vs *Shigella*. It is evident that *S. flexneri* has many inversions and translocations that are not reciprocal around the origin or terminus of replication *(6)* (Fig. 1). All of these are caused by recombination between the numerous perfectly duplicated IS elements that litter the genome of *S. flexneri* (which contains over 300 IS elements; more than seven times as many as the other *E. coli* strains).

It appears that, whatever the selection pressures that are maintaining the gene order and orientation in most enterics, they can be overcome by the widespread recombinations that must follow from this level of IS element expansion within the genome. This is again supported by observations in another branch of the entrics. *Yersinia pestis* has undergone a similar level of IS element expansion, and comparisons between highly related *Y. pestis* strains *(94)* indicate that recent recombination has occurred; indeed, some recombinational events were occurring during the clonal growth of a single strain of *Y. pestis (67)*.

CONCLUSIONS

It can be seen that the enteric bacteria, as exemplified by the *Escherichia* and *Salmonella* families, have managed to square the circle in that they have genomes that are both remarkably stable and extremely fluid at the same time. They share a conserved set of core functional genes encoded on a stable genomic backbone. However, overlaid on this backbone are a number of mechanisms for specific genome variation, including large-scale gene acquisition (PAIs and bacteriophages) and small-scale gene acquisition, balanced by gene loss through excision, deletion, and mutation. These mechanisms have enabled this family to become extremely successful in terms of the range of environmental and pathogenic niches in which it is able to survive and propagate and in its ability to evolve rapidly to fill new niches.

ACKNOWLEDGMENTS

We would like to thank Stephen Bentley and Matt Holden for critical reading of the manuscript.

REFERENCES

1. Neidhardt FC, Curtiss R. *Escherichia coli* and *Salmonella*: Cellular and Molecular Biology. 2nd ed. Washington, DC: ASM Press, 1996.
2. Blattner FR, Plunkett G, Bloch CA, et al. The complete genome sequence of *Escherichia coli* K-12. Science 1997; 277:1453–1474.

2a. Bernal A, Ear U, Kyrpides N. Genome Online Databases (GOLD): a monitor of genome projects worldwide. Nuc Acids Res 2001; 29:126–127.

 3. Perna NT, Plunkett G 3rd, Burland V, et al. Genome sequence of enterohaemorrhagic *Escherichia* coli O157:H7. Nature 2001; 409:529–533.

 4. Hayashi T, Makino K, Ohnishi M, et al. Complete genome sequence of enterohemorrhagic *Escherichia coli* O157:H7 and genomic comparison with a laboratory strain K-12. DNA Res 2001; 8:11–22.

 5. Welch RA, Burland V, Plunkett G 3rd, et al. Extensive mosaic structure revealed by the complete genome sequence of uropathogenic *Escherichia coli*. Proc Natl Acad Sci USA 2002; 99: 17,020–17,024.

 6. Jin Q, Yuan Z, Xu J, et al. Genome sequence of *Shigella flexneri* 2a: insights into pathogenicity through comparison with genomes of *Escherichia coli* K12 and O157. Nucleic Acids Res 2002; 30:4432–4441.

 7. Pupo GM, Lan R, Reeves PR. Multiple independent origins of *Shigella* clones of *Escherichia coli* and convergent evolution of many of their characteristics. Proc Natl Acad Sci USA 2000; 97:10,567–10,572.

 8. Riley LW, Remis RS, Helgerson SD, et al. Hemorrhagic colitis associated with a rare *Escherichia coli* serotype. N Engl J Med 1983; 308:681–685.

 9. Karmali MA, Petric M, Lim C, Fleming PC, Steele BT. *Escherichia coli* cytotoxin, haemolytic-uraemic syndrome, and haemorrhagic colitis. Lancet 1983; 2:1299–1300.

10. Knapp S, Hacker J, Jarchau T, Goebel W. Large, unstable inserts in the chromosome affect virulence properties of uropathogenic *Escherichia coli* O6 strain 536. J Bacteriol 1986; 168: 22–30.

11. McDaniel TK, Jarvis KG, Donnenberg MS, Kaper JB. A genetic locus of enterocyte effacement conserved among diverse enterobacterial pathogens. Proc Natl Acad Sci USA 1995; 92: 1664–1668.

12. Levine MM. *Escherichia coli* that cause diarrhea: enterotoxigenic, enteropathogenic, enteroinvasive, enterohemorrhagic, and enteroadherent. J Infect Dis 1987; 155:377–389.

13. Jarvis KG, Giron JA, Jerse AE, McDaniel TK, Donnenberg MS, Kaper JB. Enteropathogenic *Escherichia coli* contains a putative type III secretion system necessary for the export of proteins involved in attaching and effacing lesion formation. Proc Natl Acad Sci USA 1995; 92: 7996–8000.

14. Frankel G, Candy DC, Everest P, Dougan G. Characterization of the C-terminal domains of intimin-like proteins of enteropathogenic and enterohemorrhagic *Escherichia coli*, *Citrobacter freundii*, and *Hafnia alvei*. Infect Immun 1994; 62:1835–1842.

15. Donnenberg MS, Yu J, Kaper JB. A second chromosomal gene necessary for intimate attachment of enteropathogenic *Escherichia coli* to epithelial cells. J Bacteriol 1993; 175:4670–4680.

16. Schauer DB, Falkow S. Attaching and effacing locus of a *Citrobacter freundii* biotype that causes transmissible murine colonic hyperplasia. Infect Immun 1993; 61:2486–2492.

17. Kenny B, Finlay BB. Protein secretion by enteropathogenic *Escherichia coli* is essential for transducing signals to epithelial cells. Proc Natl Acad Sci USA 1995; 92:7991–7995.

18. McDaniel TK, Kaper JB. A cloned pathogenicity island from enteropathogenic *Escherichia coli* confers the attaching and effacing phenotype on *E. coli* K-12. Mol Microbiol 1997; 23: 399–407.

19. Hacker J, Blum-Oehler G, Muhldorfer I, Tschape H. Pathogenicity islands of virulent bacteria: structure, function and impact on microbial evolution. Mol Microbiol 1997; 23:1089–1097.

20. Blum G, Ott M, Lischewski A, et al. Excision of large DNA regions termed pathogenicity islands from tRNA- specific loci in the chromosome of an *Escherichia coli* wild-type pathogen. Infect Immun 1994; 62:606–614.

21. Parkhill J, Dougan G, James KD, et al. Complete genome sequence of a multiple drug resistant *Salmonella enterica* serovar Typhi CT18. Nature 2001; 413:848–852.
22. Tarr PI, Bilge SS, Vary JC Jr, et al. Iha: a novel *Escherichia coli* O157:H7 adherence-conferring molecule encoded on a recently acquired chromosomal island of conserved structure. Infect Immun 2000; 68:1400–1407.
23. O'Brien AD, Marques LR, Kerry CF, Newland JW, Holmes RK. Shiga-like toxin converting phage of enterohemorrhagic *Escherichia coli* strain 933. Microb Pathog 1989; 6:381–390.
24. O'Brien AD, Newland JW, Miller SF, Holmes RK, Smith HW, Formal SB. Shiga-like toxin-converting phages from *Escherichia coli* strains that cause hemorrhagic colitis or infantile diarrhea. Science 1984; 226:694–696.
25. Schmidt H, Beutin L, Karch H. Molecular analysis of the plasmid-encoded hemolysin of *Escherichia coli* O157:H7 strain EDL 933. Infect Immun 1995; 63:1055–1061.
26. Boyd EF, Wang FS, Whittam TS, Selander RK. Molecular genetic relationships of the salmonellae. Appl Environ Microbiol 1996; 62:804–808.
27. McClelland M, Sanderson KE, Spieth J, et al. Complete genome sequence of *Salmonella enterica* serovar Typhimurium LT2. Nature 2001; 413:852–856.
28. Reeves P, Stevenson G. Cloning and nucleotide sequence of the *Salmonella typhimurium* LT2 gnd gene and its homology with the corresponding sequence of *Escherichia coli* K12. Mol Gen Genet 1989; 217:182–184.
29. Shivaprasad HL. Fowl typhoid and pullorum disease. Rev Sci Tech 2000; 19:405–424.
30. Mills DM, Bajaj V, Lee CA. A 40 kb chromosomal fragment encoding *Salmonella typhimurium* invasion genes is absent from the corresponding region of the *Escherichia coli* K-12 chromosome. Mol Microbiol 1995; 15:749–759.
31. Galan JE. Molecular genetic bases of *Salmonella* entry into host cells. Mol Microbiol 1996; 20:263–271.
32. Shea JE, Hensel M, Gleeson C, Holden DW. Identification of a virulence locus encoding a second type III secretion system in *Salmonella typhimurium*. Proc Natl Acad Sci USA 1996; 93:2593–2597.
33. Ochman H, Soncini FC, Solomon F, Groisman E. A. Identification of a pathogenicity island required for *Salmonella* survival in host cells. Proc Natl Acad Sci USA 1996; 93:7800–7804.
34. Kingsley RA, Baumler AJ. Pathogenicity islands and host adaptation of *Salmonella* serovars. Curr Top Microbiol Immunol 2002; 264:67–87.
35. Blanc-Potard AB, Groisman EA. The *Salmonella selC* locus contains a pathogenicity island mediating intramacrophage survival. EMBO J 1997; 16:5376–5385.
36. Wood MW, Jones MA, Watson PR, Hedges S, Wallis TS, Galyov EE. Identification of a pathogenicity island required for *Salmonella* enteropathogenicity. Mol Microbiol 1998; 29:883–891.
37. Galyov EE, Wood MW, Rosqvist R, et al. A secreted effector protein of *Salmonella dublin* is translocated into eukaryotic cells and mediates inflammation and fluid secretion in infected ileal mucosa. Mol Microbiol 1997; 25:903–912.
38. Hashimoto Y, Li N, Yokoyama H, Ezaki T. Complete nucleotide sequence and molecular characterization of ViaB region encoding Vi antigen in *Salmonella typhi*. J Bacteriol 1993; 175:4456–4465.
39. Zhang XL, Tsui IS, Yip CM, et al. *Salmonella enterica* serovar typhi uses type IVB pili to enter human intestinal epithelial cells. Infect Immun 2000; 68:3067–3073.
40. Hensel M, Shea JE, Baumler AJ, Gleeson C, Blattner F, Holden DW. Analysis of the boundaries of *Salmonella* pathogenicity island 2 and the corresponding chromosomal region of *Escherichia coli* K-12. J Bacteriol 1997; 179:1105–1111.
41. Lee CA. Pathogenicity islands and the evolution of bacterial pathogens. Infect Agents Dis 1996; 5:1–7.

42. Buchrieser C, Prentice M, Carniel E. The 102-kilobase unstable region of *Yersinia pestis* comprises a high-pathogenicity island linked to a pigmentation segment which undergoes internal rearrangement. J Bacteriol 1998; 180:2321–2329.

43. Rakin A, Urbitsch P, Heesemann J. Evidence for two evolutionary lineages of highly pathogenic *Yersinia* species. J Bacteriol 1995, 177:2292–2298.

44. Schubert S, Cuenca S, Fischer D, Heesemann J. High-pathogenicity island of *Yersinia pestis* in enterobacteriaceae isolated from blood cultures and urine samples: prevalence and functional expression. J Infect Dis 2000; 182:1268–1271.

45. Bach S, de Almeida A, Carniel E. The *Yersinia* high-pathogenicity island is present in different members of the family Enterobacteriaceae. FEMS Microbiol Lett 2000; 183:289–294.

46. Reiter WD, Palm P, Yeats S. Transfer RNA genes frequently serve as integration sites for prokaryotic genetic elements. Nucleic Acids Res 1989; 17:1907–1914.

47. Rakin A, Noelting C, Schropp P, Heesemann J. Integrative module of the high-pathogenicity island of *Yersinia*. Mol Microbiol 2001; 39:407–415.

48. Hare JM, Wagner AK, McDonough KA. Independent acquisition and insertion into different chromosomal locations of the same pathogenicity island in *Yersinia pestis* and *Yersinia pseudotuberculosis*. Mol Microbiol 1999; 31:291–303.

49. Fleckenstein JM, Lindler LE, Elsinghorst EA, Dale JB. Identification of a gene within a pathogenicity island of enterotoxigenic *Escherichia coli* H10407 required for maximal secretion of the heat-labile enterotoxin. Infect Immun 2000; 68:2766–2774.

50. Moss JE, Cardozo TJ, Zychlinsky A, Groisman EA. The selC-associated SHI-2 pathogenicity island of *Shigella flexneri*. Mol Microbiol 1999; 33:74–83.

51. Schmidt H, Zhang WL, Hemmrich U, et al. Identification and characterization of a novel genomic island integrated at selC in locus of enterocyte effacement-negative, Shiga toxin-producing *Escherichia coli*. Infect Immun 2001; 69:6863–6873.

52. Wieler LH, McDaniel TK, Whittam TS, Kaper JB. Insertion site of the locus of enterocyte effacement in enteropathogenic and enterohemorrhagic *Escherichia coli* differs in relation to the clonal phylogeny of the strains. FEMS Microbiol Lett 1997; 156:49–53.

53. Sperandio V, Kaper JB, Bortolini MR, Neves BC, Keller R, Trabulsi LR. Characterization of the locus of enterocyte effacement (LEE) in different enteropathogenic *Escherichia coli* (EPEC) and Shiga-toxin producing *Escherichia coli* (STEC) serotypes. FEMS Microbiol Lett 1998; 164:133–139.

54. Hensel M, Nikolaus T, Egelseer C. Molecular and functional analysis indicates a mosaic structure of *Salmonella* pathogenicity island 2. Mol Microbiol 1999; 31:489–498.

55. Bergthorsson U, Ochman H. Distribution of chromosome length variation in natural isolates of *Escherichia coli*. Mol Biol Evol 1998; 15:6–16.

56. Ochman H, Bergthorsson U. Rates and patterns of chromosome evolution in enteric bacteria. Curr Opin Microbiol 1998; 1:580–583.

57. Boyd EF, Hartl DL. Chromosomal regions specific to pathogenic isolates of *Escherichia coli* have a phylogenetically clustered distribution. J Bacteriol 1998; 180:1159–1165.

58. LeClerc JE, Li B, Payne WL, Cebula TA. High mutation frequencies among *Escherichia coli* and *Salmonella* pathogens. Science 1996; 274:1208–1211.

59. Makino K, Yokoyama K, Kubota Y, et al. Complete nucleotide sequence of the prophage VT2-Sakai carrying the verotoxin 2 genes of the enterohemorrhagic *Escherichia coli* O157:H7 derived from the Sakai outbreak. Genes Genet Syst 1999; 74:227–239.

60. Mosig G, Yu S, Myung H, et al. A novel mechanism of virus-virus interactions: bacteriophage P2 Tin protein inhibits phage T4 DNA synthesis by poisoning the T4 single-stranded DNA binding protein, gp32. Virology 1997; 230:72–81.

61. Myung H, Calendar R. The *old* exonuclease of bacteriophage P2. J Bacteriol 1995; 177:497–501.

61a. Davis BM, Waldor MK. Filamentous phages linked to virulence of *Vibrio cholerae*. Curr Opin Micribiol 2003; 6:35–42.

62. Donohue-Rolfe A, Acheson DW, Keusch GT. Shiga toxin: purification, structure, and function. Rev Infect Dis 1991; 13(Suppl 4):S293–S297.

63. Plunkett G 3rd, Rose DJ, Durfee TJ, Blattner FR. Sequence of Shiga toxin 2 phage 933W from *Escherichia coli* O157:H7: Shiga toxin as a phage late-gene product. J Bacteriol 1999; 181:1767–1778.

64. Wagner PL, Neely MN, Zhang X, et al. Role for a phage promoter in Shiga toxin 2 expression from a pathogenic *Escherichia coli* strain. J Bacteriol 2001; 183:2081–2085.

65. Beutin L, Stroeher UH, Manning PA. Isolation of enterohemolysin (Ehly2)-associated sequences encoded on temperate phages of *Escherichia coli*. Gene 1993; 132:95–99.

66. Boyd EF, Brussow H. Common themes among bacteriophage-encoded virulence factors and diversity among the bacteriophages involved. Trends Microbiol 2002; 10:521–529.

67. Parkhill J, Wren BW, Thomson NR, et al. Genome sequence of *Yersinia pestis*, the causative agent of plague. Nature 2001; 413:523–527.

68. Hansen-Wester I, Hensel M. *Salmonella* pathogenicity islands encoding type III secretion systems. Microbes Infect 2001; 3:549–559.

69. Lostroh CP, Lee CA. The *Salmonella* pathogenicity island-1 type III secretion system. Microbes Infect 2001; 3:1281–1291.

70. Figueroa-Bossi N, Uzzau S, Maloriol D, Bossi L. Variable assortment of prophages provides a transferable repertoire of pathogenic determinants in *Salmonella*. Mol Microbiol 2001; 39:260–271.

71. Figueroa-Bossi N, Bossi L. Inducible prophages contribute to *Salmonella* virulence in mice. Mol Microbiol 1999; 33:167–176.

72. Figueroa-Bossi N, Coissac E, Netter P, Bossi L. Unsuspected prophage-like elements in *Salmonella typhimurium*. Mol Microbiol 1997; 25:161–173.

73. Mirold S, Rabsch W, Tschape H, Hardt WD. Transfer of the *Salmonella* type III effector *sopE* between unrelated phage families. J Mol Biol 2001; 312:7–16.

74. Hardt WD, Chen LM, Schuebel KE, Bustelo XR, Galan JE. *S. typhimurium* encodes an activator of Rho GTPases that induces membrane ruffling and nuclear responses in host cells. Cell 1998; 93:815–826.

75. Swartley JS, Marfin AA, Edupuganti S, et al. Capsule switching of *Neisseria meningitidis*. Proc Natl Acad Sci USA 1997; 94:271–276.

76. Dillard JP, Caimano M, Kelly T, Yother J. Capsules and cassettes: genetic organization of the capsule locus of *Streptococcus pneumoniae*. Dev Biol Stand 1995; 85:261–265.

77. Dillard JP, Yother J. Genetic and molecular characterization of capsular polysaccharide biosynthesis in *Streptococcus pneumoniae* type 3. Mol Microbiol 1994; 12:959–972.

78. Townsend SM, Kramer NE, Edwards R, et al. *Salmonella enterica* serovar Typhi possesses a unique repertoire of fimbrial gene sequences. Infect Immun 2001; 69:2894–2901.

79. Tsolis RM, Townsend SM, Miao EA, et al. Identification of a putative *Salmonella enterica* serotype typhimurium host range factor with homology to IpaH and YopM by signature-tagged mutagenesis. Infect Immun 1999; 67:6385–6493.

80. Miao EA, Miller SI. A conserved amino acid sequence directing intracellular type III secretion by *Salmonella typhimurium*. Proc Natl Acad Sci USA 2000; 97:7539–7544.

81. Bakshi CS, Singh VP, Wood MW, Jones PW, Wallis TS, Galyov EE. Identification of SopE2, a *Salmonella* secreted protein which is highly homologous to SopE and involved in bacterial invasion of epithelial cells. J Bacteriol 2000; 182:2341–2344.

82. Wood MW, Jones MA, Watson PR, et al. The secreted effector protein of *Salmonella dublin*, SopA, is translocated into eukaryotic cells and influences the induction of enteritis. Cell Microbiol 2000; 2:293–303.

83. Zhang S, Santos RL, Tsolis RM, et al. The *Salmonella enterica* serotype typhimurium effector proteins SipA, SopA, SopB, SopD, and SopE2 act in concert to induce diarrhea in calves. Infect Immun 2002; 70:3843–3855.

84. Kingsley RA, Baumler AJ. Host adaptation and the emergence of infectious disease: the *Salmonella* paradigm. Mol Microbiol 2000; 36:1006–1014.

85. Cole ST, Eiglmeier K, Parkhill J, et al. Massive gene decay in the leprosy bacillus. Nature 2001; 409:1007–1011.

86. Reeves MW, Evins GM, Heiba AA, Plikaytis BD, Farmer JJ 3rd. Clonal nature of *Salmonella typhi* and its genetic relatedness to other salmonellae as shown by multilocus enzyme electrophoresis, and proposal of *Salmonella bongori* comb nov. J Clin Microbiol 1989; 27:313–320.

86a. Kidgell C, Reichard U, Wain J, et al. *Salmonella typhi,* the causative agent of typhoid fever, is approximately 50,000 years old. Infect Genet Evol 2002; 2:39–45.

87. Andersson DI, Hughes D. Muller's ratchet decreases fitness of a DNA-based microbe. Proc Natl Acad Sci USA 1996; 93:906–907.

88. Perry RD, Fetherston JD. *Yersinia pestis*—etiologic agent of plague. Clin Microbiol Rev 1997; 10:35–66.

89. Achtman M, Zurth K, Morelli G, Torrea G, Guiyoule A, Carniel E. *Yersinia pestis*, the cause of plague, is a recently emerged clone of *Yersinia pseudotuberculosis*. Proc Natl Acad Sci USA 1999; 96:14,043–14,048.

90. Sanderson KE, Hessel A, Liu S, Rudd KE. The genetic map of *Salmonella typhimurium*, edition VIII. In: Neidhardt FC, Curtiss R (eds). *Escherichia coli* and *Salmonella*: Cellular and Molecular Biology. 2nd ed. Washington, DC: ASM Press, 1996, pp. 1903–1999.

91. Berlyn MKB, Brooks Low K, Rudd KE. Linkage map of *Escherichia coli* K12, Edition 9. In: Neidhardt FC, Curtiss R (eds). *Escherichia coli* and *Salmonella*: Cellular and Molecular Biology. 2nd ed. Washington, DC: ASM Press, 1996, pp. 1715–1902.

92. Tillier ER, Collins RA. Genome rearrangement by replication-directed translocation. Nat Genet 2000; 26:195–197.

93. Eisen JA, Heidelberg JF, White O, Salzberg SL. Evidence for symmetric chromosomal inversions around the replication origin in bacteria. Genome Biol 2000; 1:RESEARCH0011.

94. Deng W, Burland V, Plunkett G 3rd, et al. Genome sequence of *Yersinia pestis* KIM. J Bacteriol 2002; 184:4601–4611.

95. Roth JR, Benson N, Galitski T, Haack K, Lawrence JG, Miesel L. Rearrangements of the bacterial chromosome: Formation and applications. In: Neidhardt FC, Curtiss R (eds). *Escherichia coli* and *Salmonella*: Cellular and Molecular Biology. 2nd ed. Washington, DC: ASM Press, 1996, pp. 2256–2276.

96. Mackiewicz P, Mackiewicz D, Gierlik A, et al. The differential killing of genes by inversions in prokaryotic genomes. J Mol Evol 2001; 53:615–621.

97. Tamas I, Klasson L, Canback B, et al. Fifty million years of genomic stasis in endosymbiotic bacteria. Science 2002; 296:2376–2379.

98. van Ham RC, Kamerbeek J, Palacios C, et al. Reductive genome evolution in *Buchnera aphidicola*. Proc Natl Acad Sci USA 2003; 100:581–586.

99. Shigenobu S, Watanabe H, Hattori M, Sakaki Y, Ishikawa H. Genome sequence of the endocellular bacterial symbiont of aphids *Buchnera* sp APS. Nature 2000; 407:81–86.

100. Akman L, Yamashita A, Watanabe H, et al. Genome sequence of the endocellular obligate symbiont of tsetse flies, *Wigglesworthia glossinidia*. Nat Genet 2002; 32:402–407.

17

Obligate Intracellular Pathogens

Siv G. E. Andersson

INTRODUCTION

The eukaryotic cell represents an extremely attractive growth habitat for bacteria starving for nutrients and protection. However, to survive and reproduce in the hostile interior of another organism, the invading bacteria must be able to enter the host, multiply inside a selected set of cells, and finally exit and reestablish this cycle in another individual *(1)*. During all stages of this process, the bacterium has to avoid being killed by the host immune system *(1)*. Thus, making the transition to an intracellular growth habitat is not a trivial process. *Obligate intracellular parasites* are defined as intracellular bacteria that multiply in a strict host-associated manner, whereas *facultative intracellular parasites* have explored the intracellular milieu and retained the ability to grow outside their chosen hosts *(1)*. The dichotomy between facultative vs obligate intracellular parasitism is mainly a reflection of the ability vs inability to culture these organisms in vitro *(1)*. This difference in lifestyle is as yet not defined at the molecular level.

It can be speculated that the inability of obligate host-associated bacteria to grow in a free-living mode is because of the absence of genes encoding essential enzymes or compounds freely available in the intracellular environment. Indeed, a number of such dependencies have been demonstrated. However, despite numerous attempts to cultivate obligate intracellular bacteria on artificial growth media, it has not yet been possible to restore growth in vitro in the laboratory. Another possibility is that replication and cell division is regulated by cues provided by the host cell, although such potential regulatory molecules have yet to be identified. Whatever the basis for obligate host dependence, the interior of the eukaryotic cell has been an attractive target for invasion by bacteria because a variety of unrelated species have explored this growth habitat independent of each other. This may not be surprising because metabolites are present in abundance in the eukaryotic cytoplasm, and there is little or no competition from other bacteria. However, the cost for gaining access to this luxurious environment is paid directly in nucleotides, which in the long term leads to complete host dependence and severe genome degeneration *(2)*.

RICKETTSIAE AND CHLAMYDIAE

Two model systems have been particularly important for the understanding of obligate intracellular parasitism: chlamydiae and rickettsiae. Historically, the name *rickettsiae*

From: *Microbial Genomes*
Edited by: C. M. Fraser, T. D. Read, and K. E. Nelson © Humana Press Inc., Totowa, NJ

was used for a variety of small, rod-shaped bacteria that could not be cultivated *(3)*. However, molecular sequence analysis of selected genes, such as the ribosomal ribonucleic acid (rRNA) genes, revealed that they consisted of an assemblage of phylogenetically unrelated bacteria, including genera such as the *Rickettsia*, *Coxiella*, and *Ehrlichia* *(4)*. With the development of cultivation methods for rickettsiae in the yolk sacs of chicken embryos in the 1940s, it became possible to recover enough material for more detailed biochemical studies. However, despite these improvements, progress on the biology, epidemiology, and phylogeny of these organisms was still very slow, mainly because of the fastidious growth requirements and the highly pathogenic nature of these bacteria. In the beginning of the 1990s, fewer than a few dozen genes from each species had been characterized at the sequence level.

In the mid-1990s, automated deoxyribonucleic acid (DNA) sequencing brought about a revolution in microbiology through the provision of genome sequence data from bacteria, archaea, and fungi. The first complete genome sequences of two obligate intracellular pathogens, *Rickettsia prowazekii* and *Chlamydia trachomatis*, were published in 1998 *(5,6)*, and a few related strains and species have since then been characterized at the genomic level *(7–10)*. Genetic manipulation of a rickettsial gene was also accomplished in 1998 by the successful transformation of a rifampicin resistance gene into the *R. prowazekii* genome *(11)*. This represented the first step toward the development of a genetic system for *Rickettsia*.

The future challenge is to learn how to exploit the genome sequence data and improve transformation capabilities to understand the in vivo behavior of *Rickettsia*. The aim of this chapter is to provide an overview of how the acquisition of whole genome sequence data from *Rickettsia* species has already advanced the knowledge of their physiology and evolution. The specific focus is on the genomes of *R. prowazekii* and *Rickettsia conorii*. However, general concepts of obligate intracellular parasitism are discussed from a broader perspective that also includes discussion of other obligate intracellular parasites.

EPIDEMIOLOGY, PHYLOGENY, AND DISEASE MANIFESTATIONS

The Latin name for the typhus pathogen, *Rickettsia prowazekii*, was given to honor the two scientists who first discovered it, H. T. Ricketts *(12)* and S. J. M. Prowazek. Sadly, both scientists contracted typhus and died in the course of trying to elucidate the secrets of the typhus pathogen. In a global perspective, the typhus pathogen has been a source of human disasters for hundreds of years, causing the deaths of tens of millions of people. The clinical syndromes of epidemic typhus were first described during the 16th century in the Mediterranean area. From there, the disease followed the armies as they moved through Europe, and it spread to civilians via the transmission of infected lice, as discovered by Nicolle in 1909 *(13,14)*. His finding that epidemic typhus is transmitted by the human body louse *Pediculus humanus corporis* was rewarded with the Nobel Prize in 1928. This discovery was of the greatest practical importance during World War I and II when hygenic measures such as shaving and washing and burning of clothes were taken to minimize the spread of the disease. Although epidemic outbreaks of typhus are currently very rare, the disease is still a major problem in some African countries according to the World Health Organization. Indeed, because of bad

Fig. 1. The human body louse *Pediculus humanus*, host and vector of the obligate intracellular parasite *Rickettsia prowazekii*.

sanitation there was an outbreak of louse-borne epidemic typhus in Burundi as recently as 1995, when the disease swept across most of the country *(15)*.

Humans, Vectors, and Reservoirs

Humans are the main host and currently the only known natural reservoir for *R. prowazekii*, which uses the human body louse as its vector of transmission *(16)*. Lice are strict blood-sucking insects (Fig. 1), and millions of bacterial cells are excreted with their feces near the skin bite lesion. These cells may induce the deadly disease if rubbed or scratched into the skin. Surprisingly, infections with the pathogen are also fatal to the louse transmitting the disease. This is because *R. prowazekii* multiplies intensively inside the midgut epithelial cells of the louse, inducing cell lysis. Depending on the amount of bacteria in the gut, the louse may be killed within 1 to 2 weeks, as compared to a normal life span of about a month.

Most other *Rickettsia* species have natural life cycles that include animal hosts and arthropod vectors such as fleas, mites, and ticks (Table 1) *(16,17)*. Several *Rickettsia* species have established symbiotic relationships with their arthropod vectors, in which they are maintained through transovarial transmission (i.e., from infected females to infected ova) *(18)*. Although individual fleas have been infected with both *Rickettsia typhi* and *Rickettsia felis* in the laboratory, natural coinfections of two or more species have not been observed. A tick infected by *Rickettsia peacockii* was resistant to infection as well as to transovarial transmission of *Rickettsia rickettsii*, suggesting that *Rickettsia* spp may induce cytoplasmic incompatibility. However, there is no phylogenetic support for long-term interactions and coevolution of ticks and *Rickettsia (19)*, as is observed for aphids and their endosymbionts *(20)*.

Table 1
Rickettsia **Species, Vectors, and Human Diseases**

Group	Species	Disease	Vector
TG	*R. prowazekii*	Epidemic typhus	Human body lice
TG	*R. typhi*	Murine tuphus	Rat fleas
SFG	*R. conorii*	Mediterranean spotted fever	Ticks
SFG	*R. sibirica*	Siberian tick typhus	Ticks
SFG	*R. rickettsii*	Rocky Mountain spotted fever	Ticks
SFG	*R. montana*	Unknown	Ticks
SFG	*R. rhipicephali*	Unknown	Ticks
SFG	*R. australis*	Queensland tick typhus	Ticks
SFG	*R. akari*	Rickettsia pox	Mites
—	*R. canada*	Unknown	Ticks
—	*R. bellii*	Unknown	Ticks
—	AB bacterium	Unknown	Beetle

TG, typhus group; SFG, spotted fever group; —, not classified.

Rickettsia *in a Phylogenetic Context*

The phylogenetic position of *Rickettsia*, as inferred from rRNA sequence data, is within the α-proteobacteria *(4)*. The genus *Rickettsia* is composed of two main groups; the typhus group (TG) *Rickettsia* and the spotted fever group (SFG) *Rickettsia* (Table 1) *(19)*. A few species, such as *Rickettsia bellii* and *Rickettsia canada*, have been shown to be phylogenetically close to, but distinct from the two main groups *(21,22)*. A more distantly related species has recently been reclassified from *Rickettsia tsutsugamushi* to *Orientia tsutsugamushi* to emphasize its early divergence from the other *Rickettsia* species *(23,24)*.

The Typhus Group Rickettsia

There are only two members, *R. prowazekii* and *R. typhi*, in the TG *Rickettsia*; both are pathogenic for humans. The incubation period for epidemic typhus caused by *R. prowazekii* is typically 10 to 14 days. The clinical symptoms at the onset of the disease are high fever, headache, and development of a rash on trunks, limbs, and axillar areas after 5 to 7 days. The central nervous system is invaded, and neurological disorders are common. Patients may fall into a coma, during which the body temperature is high, and the blood pressure low. The disease is fatal in 10–30% of cases.

Humans who contract typhus may retain *Rickettsia* in a chronic form, known as Brill–Zinsser disease, that may be activated under stressful condition *(19)*. A single case of Brill–Zinsser disease may initiate a new outbreak of epidemic typhus if lice infestations are high in the population. Fortunately, the disease can still be cured by use of antibiotics, and natural antibiotic resistances have not yet been observed in rickettsial populations. Because epidemic outbreaks are strictly dependent on lice infestations, improved hygenic measures have severely limited the spread of the disease during the last century. A milder form of typhus in humans, murine typhus, is caused by *R. typhi*, the other member of the TG. Rats are the main reservoirs for this species, which is transferred to humans via rats or rat fleas. For the purpose of this review, we use *R. prowazekii* as a representative of the TG *Rickettsia* because complete genome sequence data are available for this species *(5)*.

The Spotted Fever Group Rickettsia

There are currently more than 20 different members of the SFG *Rickettsia*. These include *R. conorii*, which the causative agent of Mediterranean spotted fever. This obligate intracellular pathogen is transmitted by the dog brown tick to humans. Disease symptoms include high fever, headache, myalgia, and arthralgia. Another human pathogen is *R. rickettsii*, the causative agent of Rocky Mountain spotted fever. Like *R. conorii*, this pathogen is transferred to humans via the bite of an infected tick. The disease causes fever as well as skin eruptions on all parts of the body *(25)*. It occurs in North, South, and Central America and can be lethal if left untreated.

Additional diseases caused by the pathogenic members of the SFG are African tick typhus and rickettsial pox. It is interesting to note that the pathogenic species *R. rickettsii*, *Rickettsia parkeri*, and *Rickettsia sibirica* are phylogenetically distinct from the nonpathogenic species *Rickettsia rhipicephali* and *Rickettsia montana*. Other species, such as *R. felis* and *Rickettsia helvetica*, are early diverging within the SFG. The difference at the molecular level between the pathogenic and nonpathogenic species has not yet been completely elucidated. Complete genome sequence data are available for one species in this group, *R. conorii*, which in this review is used as a representative of the SFG *Rickettsia (7)*.

Attachment, Entry, Multiplication, and Exit

Rickettsia may enter the host cells by induced phagocytosis, with phagocytic vesicles formed simultaneously with the binding of *Rickettsia* to receptors of the host cell. Although *Rickettsia* can enter various nucleated cells in vitro, the primary target in vivo is the epithelium, which it enters by an actin-dependent process *(26)*. Once inside the cell, the bacterium induces lysis of the phagosomal membrane prior to phagosome–lysosome fusion to gain access to the cytoplasm. *Rickettsia* is unique among obligate intracellular pathogens in that it multiplies directly in the cell cytoplasm without any surrounding host-derived membrane. Some species, such as *R. canada*, may even replicate inside the cell nucleus. During the multiplication process, the pathogen causes only modest injuries to the host cells. Cell death is ultimately caused by physical disruption of the host cells when there are too many internal parasites to fit into the cytoplasmic space.

Some species, such as *R. conorii* and *R. rickettsii*, are able to polymerize actin into comet-tail structures *(27,28)*. They have been observed to move within the cytosol of infected cells, similar to the actin-based motility of *Listeria* or *Shigella*. This process may facilitate cell-to-cell spread from the initial site of cell invasion, which may occur without any further extracellular stage *(29)*. Indeed, endothelial cell culture plaques induced by *R. rickettsii* have been reported, indicative of cell-to-cell transmission in association with cytophatic effects. However, the infection of endothelial cells has also been shown to induce antiapoptotic effects essential for host cell survival. Thus, *Rickettsia* seems to be able to modulate the apoptotic response of the host cell to its own advantage by allowing the host cell to remain as a site of infection *(30)*.

Table 2
Comparison of Genome Features
for *Rickettsia prowazekii* (5) and *Rickettsia conorii* (7)

Feature	R. prowazekii	R. conorii
Genome size (bp)	1,111,523	1,268,755
Genic G+C content (%)	29.0	32.4
Protein coding genes (no.)	834	1374
Pseudogenes (no.)	>12	>2
Noncoding content (%)	<76	<81

THE ARCHITECTURES OF THE *RICKETTSIA* GENOMES

The genomes of *R. prowazekii* and *R. conorii* are very small, only 1.11 and 1.29 Mb, respectively (Table 2) (5,7). The overall architecture of the two genomes is essentially the same, with the exception of a few rearrangements near the terminus of replication (Fig. 2). Symmetric deoxyribonucleic acid (DNA) inversions at the regions surrounding the origins of replication and termination have been observed in *Chlamydia (31)*. The symmetric nature of these rearrangements is thought to be the outcome of recombination events at the open replication forks. Such translocation and inversion events have since been identified in a variety of genomes, suggesting that the replicating DNA at the open replication fork is particularly vulnerable to recombination events.

The difference in genome size between the two *Rickettsia* genomes is reflected in their gene content; *R. prowazekii* contains 834 annotated genes (5), whereas *R. conorii* contains as many as 1372 annotated genes (7). The *R. conorii* genome contains 804 of the 834 genes previously identified in *R. prowazekii*. The remaining 552 open reading frames (ORFs) are uniquely present in *R. conorii (7)*. However, a closer inspection of the *R. prowazekii* genome has revealed short gene remnants for 229 of the 552 unique genes. This suggests that *R. prowazekii* has eliminated more than 200 genes since it diverged from *R. conorii*, and that another 200 genes or more have been extensively degraded.

Thus, as much as one third of the coding capacity of the ancestral *R. prowazekii* genome has been lost since its divergence from *R. conorii*. From the human perspective, it is remarkable that the genome of the highly pathogenic species *R. prowazekii is* essentially only a degraded version of that of its close relative *R. conorii* which is a less-pathogenic species.

NUTRITIONAL DEPENDENCE, METABOLISM, AND REGULATION

Compared to most natural growth environments encountered by free-living bacteria, the eukaryotic cytoplasm is exceptionally rich in nutrients. The lifestyle of obligate intracellular bacteria is also different from that of their free-living relatives in that they only multiply in a single growth habitat, namely, the eukaryotic host cell. In contrast, free-living bacteria constantly switch between excessive nutrient consumption and the harshest of starvation conditions. They must therefore be equipped with a battery of

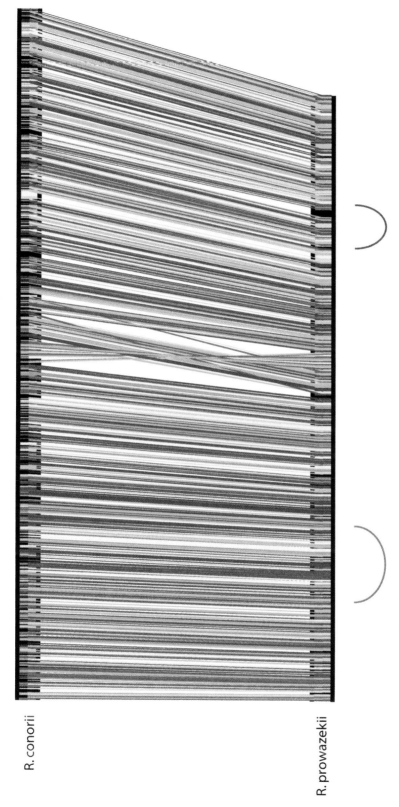

Fig. 2. Schematic representation of the *Rickettsia prowazekii* and *Rickettsia conorii* genomes. Lines between the two genomes show regions of gene synteny. Crossed lines in the central region show rearrangements near the terminus of replication. Arched lines indicate the distribution of repeated sequences longer than 100 nucleotides.

survival strategies, including regulatory genes that read changes in the external environment and trigger appropriate responses. Obligate intracellular pathogens, on the other hand, only need to respond to a few external signals. Indeed, one of the most dramatic consequences of adaptations to intracellular lifestyles is the extreme loss of genes involved in regulatory processes. Nevertheless, gene regulation has been demonstrated in Rickettsia. For example, several genes involved in adenosine triphosphate (ATP) biosynthesis and transport, such as citrate synthase and ATP/ADP (adenosine 5'-diphosphate) translocase, have been regulated in response to the availability of ATP, as discussed in more detail in a separate section.

Loss of Small-Molecule Biosynthetic Genes

Given the availability of compounds such as amino acids and nucleoside monophosphates in the eukaryotic cytoplasm, it is not surprising that genes coding for proteins involved in the biosynthesis of amino acids and nucleotides are absent from the *Rickettsia* genomes *(5,7)*. As expected, this is explained by the identification of transport systems for nucleoside monophosphates in *R. prowazekii*. Another seemingly nonfunctional biosynthetic pathway in several *Rickettsia* species is that for *S*-adenosylmethionine synthatese, which is required for the biosynthesis of *S*-adenosylmethionine (SAM) *(32)*. Also, in this case, a transport system for SAM has been identified in *R. prowazekii (32a)*. There are also no genes for glycolysis, suggesting that pyruvate is imported from the cytoplasm. This lack of biosynthetic genes shows that *Rickettsia* is completely dependent on its host for supply of a variety of small molecules.

Exploitation of Cytosolic ATP

A particularly interesting feature of obligate intracellular pathogens is their ability to import ATP from the cytoplasm. This transport function is uniquely present in bacteria (*Rickettsia* and *Chlamydia*; *5–10*) and eukaryotes (*Encephalitozoon cuniculi*; *33*) that are the obligate intracellular parasites. A similar type of ATP/ADP transport system has also been identified in the inner membrane of plastids *(34–38)*. Mitochondria also have membrane proteins that mediate ATP/ADP transport functions, but these have evolved from a more general type of transporter and are not phylogenetically related to the ATP/ADP transport system in obligate intracellular parasites and plastids *(39)*. No free-living bacteria have been identified that are capable of ATP/ADP transport. This is not surprising because the direction of transfer (ATP in exchange for ADP or vice versa) is dependent on the concentration of ATP and ADP across the membrane. With the exception of the interior of the eukaryotic cells, no naturally occurring growth habitats are rich in ATP, which means that ATP would be exported to the external environment if such transporters were present in free-living bacteria.

The *R. prowazekii* and *R. conorii* genomes contain as many as five genes (*tlc*) for ATP/ADP translocases *(5,7)*; one, *tlc*1, has been shown experimentally to catalyze ATP and ADP transport *(40)*. The transcription level of the *tlc*1 gene in *R. prowazekii* is sensitive to the ATP pool of the host cell cytoplasm *(40)*. The expression level is highest at an early stage when the cytoplasm is still rich in ATP, but decreases in cells that are heavily infected with *R. prowazekii*. This transcription pattern contrasts with that of genes required for the production of ATP (e.g., the citrate synthase gene), which are most highly expressed when the ATP pool of the host cell cytoplasm is low *(40)*. Thus, the early production of ATP/ADP translocases enables *R. prowazekii* to exploit the ATP of the host cell efficiently, whereas the subsequent reduction in expression avoids ATP produced internally by aerobic respiration leaking into the cytoplasm.

Two paralogous genes coding for nucleotide transport proteins have also been identified in *C. trachomatis*, Npt1 and Npt2, that exhibit 68% and 61% similarity, respectively, to the ATP/ADP transporter from *R. prowazekii* at the amino acid level. *Npt1* catalyzes ATP and ADP transport in an exchange mode, providing chlamydia with energy; *npt2* catalyzes the net uptake of ribonucleoside triphosphates required for anabolic reactions *(41)*. *Rickettsia, Chlamydia*, and *E. cuniculi* have evolved independently from phylogenetically distinct microbial ancestors. Nevertheless, the shared acquisition of ATP/ADP transporters is probably a major reason for the successful establishment of these parasitic relationships.

GENOME DETERIORATION IN *RICKETTSIA*

The *R. prowazekii* genome is remarkable in that as much as 24% of the genome consists of noncoding DNA *(5)*. The corresponding estimate in the annotated *R. conorii* genome is 19% *(7)*. Another obligate intracellular pathogen, *Mycobacterium leprae*, has an even higher fraction of noncoding DNA, as much as 56% *(42)*. As initially suggested *(5)*, it has now been convincingly demonstrated that the intergenic sequences in *R. prowazekii* consist of decayed genes that are no longer active, but have not yet been completely eliminated *(32,43,44)*. For example, the lack of coding potential in these regions was first verified by comparative analyses of closely related strains and species, which revealed sequence similarities between genes in one species and pseudogenes or noncoding DNA in another species *(32,43,44)*. A systematic comparison of the complete *R. prowazekii* and *R. conorii* genomes identified gene remnants in *R. prowazekii* for 229 of the 552 ORFs uniquely present in *R. conorii* *(7)*. Many of these have no sequence similarity to genes in other species (orphans), and these tend to be shorter than genes for which homologous genes are present in other species. This suggests that a subset of the orphans may represent gene fragments rather than complete genes.

The acquisition and evolution of junk DNA in bacterial genomes has been studied most extensively in *Rickettsia*. Such studies are facilitated by the high alanine and threonine (A+T) content of these genomes. The overall guanine and cytosine (G+C) content values range from 29 to 30% in members of the TG and from 32 to 33% in members of the SFG. The mutation bias toward A+T is particularly well pronounced in noncoding areas of the genome and at third codon positions, where G+C content values may be as low as 10–15%. There is no correlation between codon usage patterns and putative expression levels, suggesting that natural selection for codon usage has not been effective in these species *(45)*. The lack of variation in nucleotide frequency statistics for genic regions greatly simplifies judgments about the coding status of any particular sequence. Thus, sequences coding for essential gene functions are expected to have higher G+C content values and lower substitution rates than sequences that represent noncoding DNA. Next, we discuss the gradual process, as inferred from comparative sequence analyses, which whereby genomic segments and gene sequences may have been discarded from the *Rickettsia* genomes.

Loss of Multicopy Gene Sequences

Because the loss of redundant genes may not necessarily be a lethal event, multicopy genes are expected to be the first targets for sequence elimination during adaptations to intracellular growth environments. Indeed, the small genomes of obligate intracellular bacteria have low contents of repeated sequences and no or only a few genetic parasites *(46)*. In contrast, free-living bacteria tend to have large genomes with high contents of repeated sequences and self-propagating DNA, such as transposons and bacteriophages *(46)*. A global survey revealed a correlation among genome size, repeat content, and life-style, with obligate host–associated bacteria such as *Rickettsia, Chlamydia*, and *Buchnera* occupying one extreme of the spectrum *(47)*.

Prime targets for elimination of redundant gene sequences are genes coding for rRNAs (*rrs, rrl*, and *rrf*) and elongation factor Tu (*tuf*). These genes are present in two or more copies in most bacterial genomes *(48)*. In contrast, only one copy of the rRNA and the *tuf* genes are present in *Rickettsia* species *(49–51)*. A comparative study of several *Rickettsia* species showed that the loss of duplicate copies of the rRNA and the *tuf* genes occurred prior to the divergence of species within the genus *Rickettsia (52)*. Duplicated genes like the rRNA and the *tuf* genes are targets not only for sequence elimination, but also for genome rearrangements. Indeed, it has been suggested that the inversion at the single *tuf* gene in *Rickettsia* resulted from an intrachromosomal recombination event at the two ancestral *tuf* genes, followed by a deletion of one copy *(50)*. Recombination events at multicopy genes may have induced rearrangements as well as deletions of large segments of DNA at an early stage of evolution of the ancestral *Rickettsia* genomes, leading to rapid sequence elimination at an early stage of the adaptation to the intracellular growth habitat.

Loss of Short Repeated Sequences

A fundamental difference between the two *Rickettsia* genomes is that the *R. conorii* genome contains a much higher density of short repetitive DNA sequences than *R. prowazekii* (Fig. 2) *(7)*. As many as 10 repeat families containing a total of 656 repeated elements were identified in the *R. conorii* genome. These vary in size between 19 and 172 bp and constitute 3.2% of the entire genome. The repeats are very G+C rich (approx 40%), which partially explains the difference in G+C content between the two genomes; 29% in *R. prowazekii* versus 32% in *R. conorii* (Table 2).

A comparative study of intergenic regions in *Rickettsia* species showed that short repeated sequences in one or more species often flank short deletions in the other species *(32,44,51,52)*. As expected, if these deletions were mediated by the short repetitive sequences, only one copy of the repeat is present in the strains in which the intervening segment has been deleted. The repetitive elements thus identified range in size from 7 to 30 bp and are dispersed throughout the intergenic regions. For example, a short sequence repeat of 7 bp is located at the 5' end of the *fmt–rrl* spacer region in *R. rickettsii* and flanks a 28-bp sequence that is uniquely present in *R. rickettsii (51)*. Likewise, a short repetitive element of 7 bp flanks an intervening sequence that has been lost twice independently, in *R. montana* and *R. felis (52)*. This suggests that short repeated elements act as targets for recombination events, causing loss of the intervening segments as well as of one copy of the repeated sequence. Given that the *R. prowazekii* genome is smaller and has a much lower fraction of repeats than *R. conorii*, it may be speculated that short repeated elements have been consumed much faster in *R. prowazekii* than in *R. conorii* because of a higher rate of recombination.

Deterioration of Rickettsia *Palindromic Elements*

A particularly interesting type of repeated sequences is the *Rickettsia* palindromic element (RPE), which is reported to be present in 45 copies in the *R. conorii* genome, 19 of which are located inside genes *(53,54)*. Because repeated elements are normally found in noncoding areas, the presence of these repeats inside genes is highly remarkable. When the *R. conorii* RPEs were used to search for similar sequences in the *R. prowazekii* genome, 10 highly divergent RPEs were identified in protein-coding genes also in this species *(53)*. A more exhaustive search for repetitive elements in the *R. conorii* genome *(7)* identified a total of 656 interspersed repeated sequences, which were classified into 10 distinct families *(55)*. Many of these full-length and partial repeat sequences are located within ORFs annotated as genes in the *R. conorii* genome *(55)*.

The RPEs were first suggested to represent recent intragenomic proliferation *(53, 54)*. However, the finding of identical insertion sites for RPEs in a variety of *Rickettsia* species, including members from both the TG and the SFG, argues that the RPEs were acquired prior to the divergence of species within *Rickettsia (52,55)*. Many of the RPEs are partial, and a comparative study of two RPEs in a wide variety of *Rickettsia* species suggests that these elements are deteriorating in several species *(52)*. If so, some of the ancestral RPEs may already have been lost, or they may no longer be recognizable by sequence similarity searches. Indeed, highly fragmented RPE sequences in *R. prowazekii* and *R. typhi* were identified only because of homologous locations to RPE insertion sites in other species *(52)*. Thus, the simplest explanation for genome-specific differences in the host proteins targeted by RPEs is loss in some lineages rather than recent proliferation in other lineages *(52,55)*.

The RPEs are suggested to have contributed to protein evolution by inserting into RNA and protein-coding genes in *R. conorii (53,54)*. The location of the in-frame insertions seems to be compatible with the three-dimensional fold and function. Two classes of RPEs are predicted to induce α-helices and one β-strand *(55)*; however, it cannot be excluded that these conformations are induced in response to the surrounding structural environment at the site of insertion *(55)*. Indeed, the lack of conservation across *Rickettsia* species suggests that the RPEs are neutral genomic passengers and may not necessarily contribute any functional role *(52)*.

If all of the identified RPEs in *R. conorii* originated prior to speciation in *Rickettsia*, the decay of RPEs must have occurred more rapidly in the TG than in the SFG. This is reminiscent of the finding that gene loss also has been much more extensive in *R. prowazekii* than in *R. conorii (7)*. In total, there are more than 500 genes uniquely present in *R. conorii (7)*. Of these, more than 200 are present as highly degraded gene remnants in the *R. prowazekii* genome *(5)*, showing that gene loss occurs more rapidly in this species. Thus, although the *R. conorii* genome is larger and contains many more short repeated sequences than the *R. prowazekii* genome, it is most likely subjected to the same reductive processes as *R. prowazekii*. However, the fixation rate for deletion mutations may differ in the two species, possibly because of differences in generation time or population structures.

Degradation of Single-Gene Sequences

The first information about the detailed pattern of degradation was obtained from the *metK* gene that codes for S-adenosylmethionine synthetase, an enzyme required for the biosynthesis of the essential cofactor SAM *(32)*. This gene contains a single termination codon in the Madrid E strain of *R. prowazekii*, but the reading frame is open in the Breinl strain of *R. prowazekii* as well as in *R. typhi*. In all other *Rickettsia* species, this gene seems to have accumulated mutations in a random manner. It has been shown that there is a transport system for SAM in *Rickettsia* that presumably has rendered this gene nonessential *(32a)*.

The mutations in the *metK* gene, which are supposed to represent neutral mutations, consist mainly of deletions of one or a few bases *(32)*. In a study containing 26 inactivated genes, it was found that as many as 1536 nucleotides had been deleted, as compared to insertions of only 31 nucleotides *(44)*. Even in this case, most deletions were short, and the large number of nucleotides affected was mainly because of two large deletions of 599 bp and 767 bp, respectively. The mean and median sizes of deletions were estimated as 4 and 51 nucleotides, respectively, per event in *Rickettsia (43,44)*. The strong bias for deletion mutations means that a gene inactivated by the accumulation of internal stop codons or frameshift mutations will eventually be eliminated solely as a function of the rate at which mutations are produced by the replication–repair machinery.

Expression of Weakly Fragmented Genes

A small suite of genes in the *R. conorii* genome appears to have been hit by mutations quite recently, just like the *metK gene* in *R. prowazekii (7)*. These were inferred from the identification of a number of short, colocated ORFs in *R. conorii* that are similar to full-length orthologs in other species, including in some cases genes in *R. prowazekii*. In total, 37 ancestral genes seemed to have been degraded into as many as 105 short ORFs *(7)*. Of these genes, 14 are represented by intact orthologs in *R. prowazekii*; the remaining 23 could not be identified in this species. Vice versa, it was found that the *R. prowazekii* genome contains 11 genes that are split into 23 ORFs, all of which are represented by intact orthologs in *R. conorii (7)*.

Curiously, it was found that several, but not all, of these short ORFs produce RNA, raising the possibility that some of these short ORFs may still be functional *(7)*. To search for functional constraints on the expressed gene fragments in *R. conorii*, we analyzed the substitution frequencies for these genes *(56)*. To this end, we compared the frequencies of substitutions of split genes vs full-length genes, of split genes with different expression characteristics, and of expressed vs unexpressed gene fragments. The overall nonsynonymous substitution frequency, which represents the average number of substitutions at sites that cause amino acid replacements, was estimated as 0.07 per site for orthologous, full-length genes *(56)*. The split genes yielded higher substitution frequencies than the orthologous genes, irrespective of whether the ORF was expressed or not, indicating that these short ORFs do not correspond to functional genes *(56)*.

This suggests that promoters may either be created by mutations or be recruited from existing sequences inside the fragmented genes. Because there are only a few regulatory genes in the *Rickettsia* genomes, transcription may be less well controlled, making it difficult to prevent unwanted transcription initiation at A+T-rich regions, especially inside degrading genes. Indeed, bacterial promoters are A+T rich, and potential promoter sequences are very frequent in the A+T-rich genomes of *R. prowazekii* and *R. conorii* *(5,7)*. This could in principle lead to temporary retention of a partial gene function, compensating for the introduction of stop codons and frameshift mutations in these genes. Alternatively, transcription may be initiated randomly, with no functional consequences at the protein level.

Our interpretation is that the split genes in *R. conorii* represent degraded genes in which mutations have started to accumulate *(56)*. The enhanced substitution frequencies at nonsynonymous sites suggest that the split genes are no longer functional, and that the expression driven by some of these fragments is most likely a temporary phenomenon *(56)*. It remains to be determined whether some or a few of the expressed split genes still carry any function.

WILL THESE GENOMES EVENTUALLY BECOME EXTINCT?

All sequence data analyzed to date, including genes, pseudogenes, noncoding DNA, and repeated sequences, suggest that the process of sequence elimination has occurred more rapidly in the TG than in the SFG. Previous studies of sequences that evolve in a neutral manner showed that there is a mutational bias for deletions, and that the sizes of these are normally larger than of insertions in both groups of *Rickettsia (32,44)*. However, genes that evolve within the SFG seem to have a higher survival probability than those that evolve within the TG. Because this difference is observed even for genes that represent inactivated genetic material in the TG and the SFG, the difference probably results from different rates of deletions rather than from different functional constraints on the encoded gene products.

Another major difference between the TG and and SFG *Rickettsia* is the high fraction of short repeated and inverted repeated sequences in the SFG that are not found in the TG. As discussed above *(48,50,51)*, short repeated sequences serve as hot spots for recombinations in which the intervening region is inverted or eliminated. Even very short repeats of fewer than 10 nucleotides seem to be able to mediate these processes *(51,52)*. Even if recombination-mediated deletions occur much less frequently than deletions in which only a few nucleotides are eliminated per event, recombination at repeated sequences is expected to account for most of the difference because of the larger number of nucleotides eliminated per event.

Thus, the difference in size and rate of genome degradation may be related to some intrinsic difference in the mechanisms by which deletion mutations are produced, or it may be related to the different population structures of the various species. For example, there is a striking difference between ticks that transmit *R. conorii* and lice that transmit *R. prowazekii*. Whereas a tick has a life span of 5–6 years, mates once per year, and takes a blood meal only twice in its lifetime, a louse has a life span of only a single month, mates once per day, and takes a blood meal five times per day. Thus, the bacterial population expands much more frequently in lice than in ticks. If deletion mutations arise primarily during the replication phase of the parasites lifestyle, the pathogen utilizing a vector with a shorter life span is expected to go through more generations per year, increasing the probabilities for all types of deletion mutations to accumulate in the population.

Nevertheless, in both cases, it seems likely that the initial genome downsizing occurred by homologous recombination at repeated sequences. As a result, large blocks of DNA were first discarded along with the removal of long repeated sequences. Shorter and shorter repeated sequences were gradually consumed in the recombination processes, until sequence elimination primarily results from the accumulation of short deletions within genic sequences, as in modern *R. prowazekii*. The slow rate at which single, inactivated genes are eliminated by short deletions affecting only a few nucleotides per event leads to a temporary increase in the genomic content of pseudogenes and "junk" DNA.

Taken together, the data obtained so far suggest that *R. conorii* and *R. prowazekii* are exposed to similar evolutionary pressures, but the differences in their lifestyles make these processes operate at different rates. The suggestion is that *R. conorii* is slowly on its way to becoming a copy of the modern *R. prowazekii* genome, which will by then be even smaller in size. Eventually, all repeats will be consumed, and gene degradation will occur by infinitesimally small steps.

Does this suggest that genome degradation eventually will come to a halt? Indeed, a comparative study of pairs of genomes from bacteria with different lifestyles suggests that gene degradation may eventually slow, at least in obligate host–associated bacterial endosymbionts, in which there is a selection for bacterial gene functions acting at the level of the host *(47)*. Thus, two bacterial endosymbionts that diverged 50 million years ago have extremely reduced genomes of only 650 kb, but these have nevertheless remained perfectly stable in architecture and only lost a few genes since their divergence *(47)*. One possible explanation for this extreme stability is that important recombination genes and repeated sites have been discarded from these genomes, thereby reducing the potential for further repeat-mediated deletion events *(47)*. This does not exclude the possibility that these small genomes may one day be replaced by bacteria with larger genomes that invade the same growth niche.

RICKETTSIA AND MITOCHONDRIA

Mitochondria are perhaps the best example that intracellularly replicating bacterial genomes may persist for hundreds of millions of years, albeit in extremely reduced forms, provided there is a selection for the encoded gene functions. It is remarkable that the deadly human typhus pathogen is one of the closest modern relatives of the bacterial ancestor that once established the symbiotic relationship that became the foundation for all higher organisms. The relationship between mitochondria and α-proteobacteria, including *Rickettsia*, is particularly evident in phylogenetic reconstructions based on genes involved in bioenergetic and translation processes *(57,58)*.

However, only a small fraction of the mitochondrial proteins are encoded by the mitochondrial genomes; the majority are encoded from the nuclear genomes. To study the origin and evolution of the nuclear genes coding for mitochondrial proteins, we analyzed a set of more than 400 yeast genes located in the nuclear genome and experimentally shown to code for mitochondrial proteins *(59)*. We found that approximately 50% of these have bacterial homologs and are putatively of bacterial origin. These are predominantly associated with functional categories such as translation, energy, and small-molecule biosynthesis. Phylogenetic reconstructions confirmed a close relationship to α-proteobacteria for a subset of these. The remaining 50% have no bacterial homologs and are putatively of eukaryotic origin. These are mostly involved in membrane, transport, regulation, messenger RNA stability, and splicing.

A study of the identity of messenger RNAs located in polysomes free in the cytoplasm vs those attached to the mitochondria *(60)* revealed an intriguing correlation with our predictions of gene origin. Thus, genes classified as putatively of eukaryotic origin *(59)* were primarily translated on polysomes free in the cytosol *(60)*; those classified as putatively of bacterial origin *(59)* were primarily translated on polysomes attached to the mitochondrion *(60)*. Given this correlation, it is tempting to speculate that genes originally transferred to the nuclear genome utilized a bacterial cotranslational secretion system for import into the mitochondrion. Once alternative import systems were established, proteins encoded by genes that evolved from within the nuclear genome also could be imported back into the mitochondrion.

Taken together, these events suggest that genes coding for key mitochondrial proteins involved in bioenergetic and translation processes were acquired from the ancestral α-proteobacterial endosymbiont. Many of these genes were transferred from the mitochondrial genome into the nuclear genome. However, a majority of the modern mitochondrial proteins seem not to be of α-proteobacterial origin, but recruited from nuclear genes for service in the mitochondrion. Thus, the modern mitochondrial proteome has a dual origin, with a key set of proteins derived from an ancestral α-proteobacterial endosymbiont that also gave rise to modern *Rickettsia (61,62)*.

CONCLUDING REMARKS

One of the major findings of the analysis of the complete genome sequences of the obligate intracellular parasites *R. prowazekii* and *R. conorii* is that these genomes, despite their small size, are still in a process of deterioration. The availability of genome sequence data from *Rickettsia* has enabled study of the process of gene inactivation, degradation, and elimination in great detail. Mechanistically, genome deterioration is explained by a variety of different mutational mechanisms, each of which dominates at a particular time subsequent to the transition to the intracellular milieu.

At an early stage, recombination events in multicopy genes and repeated sequences are likely to cause extensive sequence loss in a few events *(46,47)*. This process is expected to be associated with a reduction of gene copy numbers. Indeed, *Rickettsia*, as well as most other host-associated bacteria, has single copies of genes otherwise present in multiple copies, such as the rRNA genes and genes for elongation factor Tu *(49–52)*. During the next stage, recombination events at short repeated sequences cause the elimination of intervening sequences as well as of one repeated element per deletion event. Such deletion events have been demonstrated in a variety of *Rickettsia* species *(44,51, 52)*. However, as more and more of the repeated sequences are consumed in these processes, the frequency of such events decreases, and the continued elimination of inactivated genes finally occurs by very short intragenic deletions, possibly induced by replication slippage. Because such events eliminate fewer and fewer nucleotides per event, the process of degeneration proceeds at a slower and slower rate.

These reductive evolutionary processes, initially discovered in *Rickettsia*, have since been observed in a variety of other obligate host–associated pathogens. *Mycobacterium leprae* is currently the best example of an obligate intracellular pathogen in which massive gene disintegration occurs *(42)*. In free-living bacterial genomes, the overall flux of DNA is probably much greater because of a much higher frequency of repeated sequences and genetic parasites *(47)*. However, in these organisms, the loss of sequences is compensated by a corresponding influx of sequences, also with the help of phages, plasmids, and other movable genetic elements.

Although the process of gene degradation may gradually slow in obligate host–associated bacteria, at least under selective conditions, many obligate intracellular parasites have presumably become extinct because of extensive deterioration and small populations. Another major threat posed by an extreme degree of host specialization is that the chosen vector or host may become extinct. If, as in the case of *R. prowazekii*, the population size of its vector, the louse, is greatly reduced by improved hygenic measures, there is not much hope for the bacterial pathogen. Thus, it is conceivable that *R. prowazekii* may disappear from the ecosystem long before its genome has been degraded to its smallest possible size.

REFERENCES

1. Moulder JW. Comparative biology of intracellular parasitism. Microbiol Rev 1985; 49:298–337.
2. Andersson SGE, Kurland CG. Reductive evolution of resident genomes. Trends Microbiol 1998; 6:263–278.
3. Winkler HH. Rickettsia species (as organisms). Annu Rev Microbiol 1990; 44:131–153.
4. Olsen GJ, Woese CR, Overbeek R. The winds of (evolutionary) change, breathing new life into microbiology. J Bacteriol 1994; 176:1–6.
5. Andersson SGE, Zomorodipour A, Andersson JO, et al. The genome sequence of *Rickettsia prowazekii* and the origin of mitochondria. Nature 1998; 396:133–140.
6. Stephens RS, Kalman S, Lammel C, et al. Genome sequence of an obligate intracellular pathogen of humans: *Chlamydia trachomatis.* Science 1998; 282:754–759.
7. Ogata H, Audic S, Renesto-Audiffren P, et al. Mechanisms of evolution in *Rickettsia conorii* and *R. prowazekii.* Science 2001; 293:2093–2098.
8. Kalman S, Mitchell W, Marathe R, et al. Comparative genomics of *Chlamydia pneumoniae* and *C. trachomatis.* Nature Genetics 1999; 21:395–389.
9. Read TD, Brunham RC, Shen C, et al. Genome sequence of *Chlamydia trachomatis* MoPn *and Chlamydia pneumoniae* AR39. Nucleic Acids Res 2000; 28:1397–1406.
10. Shirai M, Hirakawa H, Kimoto M, et al. Comparison of whole genome sequences of *Chlamydia pneumoniae* J128 from Japan and CWL029 from USA. Nucleic Acids Res 2000; 28:2311–2314.
11. Rachek LI, Tucker AM, Winkler HH, Wood DO. Transformation of *Rickettsia prowazekii* to rifampicin resistance. J Bacteriol 1998; 180:2118–2124.
12. Ricketts HT. JAMA 1909; 52:379–380.
13. Nicolle C, Comte C, Conseil E. C R Acad Sci 1909; 149:486–189.
14. Gross L. How Charles Nicolle of the Pasteur Institute discovered that epidemic typhus is transmitted by lice: reminiscences from my years at the Pasteur Institute in Paris. Proc Natl Acad Sci USA 1996; 93:10,539–10,540.
15. Raoult D, Ndihokubwayo JB, Tissot-Dupont H, et al. Outbreak of epidemic typhus associated with trench fever in Burundi. Lancet 1998; 352:353–358.
16. Hackstadt T. The biology of Rickettsiae. Inf Agents Dis 1996; 5:127–143.
17. Weiss E, Moulder JW. The rickettsias and chlamydias. Order 1. Rickettsiales Giesszckiewicz 1939, 25. In: Krieg NR, Holt JG (eds). Bergeys Manual of Systematic Bacteriology. Baltimore, MD, Williams and Wilkins, 1984, pp. 687–729.

18. Azad AF, Beard CB. Rickettsial pathogens and their arthropod vectors. Emerg Infect Dis 1998; 4:179–186.

19. Raoult D, Roux V. Rickettsioses as paradigms of new or emerging infectious diseases. Clin Microbiol Rev 19:694–719.

20. Moran NA, Munson MA, Baumann P, Ishikawa H. A molecular clock in endosymbiotic bacteria is calibrated using the insect hosts. Proc R Soc London Ser B 1993; 253:167–171.

21. Roux V, Ridkina E, Eremeeva M, Raoult D. Citrate synthase gene comparison, a new tool for phylogenetic analysis, and its application for the Rickettsiae. Int J Syst Bacteriol 1997; 47:252–261.

22. Andersson SGE, Stothard DR, Fuerst P, Kurland CG. Molecular phylogeny and rearrangement of rRNA genes in *Rickettsia* species. Mol Biol Evol 1999; 16:987–995.

23. Tamura C, Ohashi A, Urakami N, Miyamura S. Classification of *Rickettsia tsutsugamushi* in a new genus, *Orientia* gen nov, as *Orientia tsutsugamushi* comb nov. Int J Syst Bacteriol 1995; 45:589–591.

24. Stothard DR. The Evolutionary History of the Genus *Rickettsia* as Inferred from 16S and 23S Ribosomal RNA Genes and the 17 kilodalton Cell Surface Antigen Gene. Ph.D. thesis. Columbus: Ohio State University, 1995.

25. Walker DH. Rocky Mountain spotted fever: a seasonal alert. Clin Infect Dis 1995; 20:1111–1117.

26. Walker TS. Rickettsial interactions with human endothelial cells in vitro: adherence and entry. Infect Immun 1984; 44:205–210.

27. Gouin E, Gantelet H, Egile C, et al. A comparative study of the actin-based motilities of the pathogenic bacteria *Listeria monocytogenes*, *Shigella flexneri* and *Rickettsia conorii*. J Cell Sci 1999; 112:1697–1708.

28. Heinzen RA, Grieshaber SS, Van Kirk LS, Devin CJ. Dynamics of actin-based movement by *Rickettsia rickettsii* in vero cells. Infect Immun 1999; 67:4201–4207.

29. Walker DH, Firth WT, Edgell CJ. Human endothelial cell culture plaques induced by *Rickettsia rickettsii*. Infect Immun 1982; 37:301–306.

30. Clifton DR, Goss RA, Sahni SK, et al. NF-κB-dependent inhibition of apoptosis is essential for host cell survival during *Rickettsia rickettsii* infection. Proc Natl Acad Sci USA 1998; 95: 4646–4651.

31. Tillier ERM, Collins RA. Genome rearrangement by replication-directed translocation. Nature Genet 2000; 26:195–197.

32. Andersson JO, Andersson SGE. Genome degradation is an ongoping process in *Rickettsia*. Mol Biol Evol 1999; 16:1178–1191.

32a. Tucker A, Winkler HH, Driskell LO, Wood DO. *S*-adenosylmethionine transport in *Rickettsia prowazekii*. J Bacteriol 2003; 185:3031–3035.

33. Katinka MD, Duprat S, Cornillot E, et al. Genome sequence and gene compaction of the eukaryotic parasite *Encephalitozoon cuniculi*. Nature 2001; 414:450–453.

34. Heldt HW. Adenine nucleotide translocation in spinach chloroplasts. FEBS Lett 1969; 5:11–14.

35. Pozueta-Romero J, Frehner M, Viale AM, Akazawa T. Direct transport of ADPglucose by an adenylate translocator is linked to starch biosynthesis in amyloplasts. Proc Natl Acad Sci USA 1991; 88:5769–5773.

36. Schunemann D, Borchert S, Flugge UI, Heldt HW. ADP/ATP translocator from pea root plastids. Comparison with translocators from spinach chloroplasts and pea leaf mitochondria. Plant Physiol (Rock) 1993; 103:131–137.

37. Kampfenkel K, Möhlman T, Batz O, Van Montagu M, Inze D, Neuhaus HE. Molecular characterization of an *Arabidopsis thaliana* cDNA encoding a novel putative adenylate translocator of higher plants. FEBS Lett 1995; 374:351–355.

38. Möhlman T, Tjaden J, Schwöppe C, Winkler HH, Kampfenkel K, Neuhaus H. R. Occurence of two plastidic ADP/ATP transporters in *Arabidopsis thaliana*. Eur J Biochem 1998; 252:353–359.

39. Kuan J, Saier MH. The mitochondrial carrier family of transport proteins: structural, functional, and evolutionary relationships. Crit Rev Biochem Mol Biol 1993; 28:209–233.

40. Cai J, Winkler HH. Transcriptional regulation in the obligate intracytoplasmic bacterium *Rickettsia prowazekii*. J Bacteriol 1996; 178:5543–5545.

41. Tjaden J, Winkler HH, Schwoppe C, Van der Laan MV, Mohlman T, Neuhause HE. Two nucleotide transport proteins in *Chlamydia trachomatis*, one for net nucleoside triphosphate uptake and the other for transport of energy. J Bacteriol 1999; 181:1196–1202.

42. Cole ST, Eiglemeier K, Parkhill J, et al. Massive gene decay in the leprosy bacillus. Nature 2001; 409:1007–1011.

43. Andersson JO, Andersson SGE. Insights into the evolutionary process of genome degradation. Curr Opin Genet Dev 1999; 9:664–671.

44. Andersson JO, Andersson SGE. Pseudogenes, junk DNA and the dynamics of *Rickettsia* genomes. Mol Biol Evol 2001; 18:829–839.

45. Andersson SGE, Sharp PM. Codon usage and base composition in *Rickettsia prowazekii*. J Mol Evol 1996; 42:525–536.

46. Frank AC, Amiri H, Andersson SGE. Genome deterioration: loss of repeated sequences and accumulation of junk DNA. Genetica 2002; 115:1–12.

47. Tamas I, Klasson L, Canback B, et al. Fifty million years of genomic stasis in endosymbiotic bacteria. Science 2002; 28:2376–2379.

48. Andersson SG, Kurland CG. Genomic evolution drives the evolution of the translation system. Biochem Cell Biol 1995; 73:775–787.

49. Andersson SGE, Zomorodipour A, Winkler HH, Kurland CG. Unusual organization of the rRNA genes in *Rickettsia prowazekii*. J Bacteriol 1995; 177:4171–4175.

50. Syvanen AC, Amiri H, Jamal A, Andersson SGE, Kurland CG. A chimeric disposition of the elongation factor genes in *Rickettsia prowazekii*. J Bacteriol 1996; 178:6192–6199.

51. Andersson SGE, Stothard DR, Fuerst P, Kurland CG. Molecular phylogeny and rearrangement of rRNA genes in *Rickettsia* species. Mol Biol Evol 1999; 16:987–995.

52. Amiri H, Alsmark CM, Andersson SGE. Proliferation and deterioration of *Rickettsia* palindromic elements. Mol Biol Evol 2002; 19:1234–1243.

53. Ogata H, Audic S, Barber V, et al. Selfish DNA in protein coding genes. Science 2000; 290:347–350.

54. Ogata H, Audic S, Claverie J-M. Response. Science 2001; 291:299–304.

55. Ogata H, Audic S. Abergel C, Fournier P-E, Claverie J-M. Protein coding palindromes are a unique but recurrent feature in *Rickettsia*. Genome Res 2002; 12:808–816.

56. Davids W, Amiri H, Andersson SGE. Small RNAs in *Rickettsia*: are they functional? Trends Genet 2002; 18:331–334.

57. Sicheritz-Ponten T, Kurland CG, Andersson SGE. A phylogenetic analysis of the cytochrome *b* and cytochrome *c* oxidase I genes supports an origin of mitochondria from within the Rickettsiaceae. Biochem Biophys Acta 1998; 1365:545–551.

58. Gray MW, Burger G, Lang BF. Mitochondrial evolution. Nature 1999; 283:1476–1481.

59. Karlberg O, Canbäck B, Kurland CG, Andersson SGE. The dual origin of the yeast mitochondrial proteome. Yeast 2000; 17:170–187.

60. Marc P, Mageot A, Devaux F, Blugeon C, Corral-Debrinski M, Jacqu C. Genome-wide analysis of mRNAs targeted to yeast mitochondria. EMBO Rep 2002; 3:159–164.

61. Andersson SGE, Kurland CG. Origins of mitochondria and hydrogenosomes. Curr Opin Microbiol 1999; 5:535–541.

62. Kurland CG, Andersson SGE. Origin and evolution of the mitochondrial proteome. Microbiol Mol Biol Rev 2000; 64:786–820.

18

Low G+C Gram-Positive Genomes

Steven R. Gill

INTRODUCTION

The family of low guanine and cytosine (G+C) Gram-positive bacteria is a diverse group that includes nonpathogens, emerging pathogens, and some of the most aggressive pathogens known to humans, several of which are potential agents of biowarfare. The combined impact of infections caused by low G+C Gram-positive pathogens such as Enterococci, Staphylococci, Streptococci, and the Clostridia on human morbidity and mortality is staggering. Emergence of antimicrobial resistance to these pathogens has led to an urgent need for novel therapeutics and methods of control. Likewise, the use of *Bacillus anthracis* as an agent of bioterrorism has invigorated research efforts aimed at understanding their physiology and virulence. At the other end of the disease spectrum are the mycoplasmas, which are recognized more for their utility as a model system to define the minimal or core genome than as chronic pathogens.

Reasons for sequencing these genomes are as diverse as their associated diseases and unique physiologies. Many are sequenced to identify the genes and gene families responsible for their pathogenicity and to identify metabolic pathways that allow them to survive in a broad range of host environments. These sequencing efforts also allow study of genome diversity, which enables pathogens to display adaptive pathogenic phenotypes such as antimicrobial resistance and increasing virulence. Others are sequenced to develop an understanding of their metabolic pathways, which can be exploited for the chemical and dairy industries. In this chapter, information gleaned from genome sequencing efforts in each species of the low G+C Gram-positive bacteria is summarized.

MYCOPLASMAS

Mycoplasmas are members of the class Mollicutes, the smallest known microorganisms capable of autonomous growth *(1)*. They parasitize a wide range of hosts, including humans, animals, insects, and plants. Although mycoplasmas evolved from low G+C Gram-positive ancestors, the rigid Gram-positive cell wall has been replaced by a cytoplasmic membrane containing lipoteichoic acid, lipoglycans, and sterols *(1)*. On the *Mycoplasma* surface is the differentiated terminal structure or attachment organelle responsible for their cytoadherence to host cells *(2)*.

From: *Microbial Genomes*
Edited by: C. M. Fraser, T. D. Read, and K. E. Nelson © Humana Press Inc., Totowa, NJ

The most remarkable features of the mycoplasmas are their reduced genome size (~0.5 to 1.0 Mb) and their preference for reading UGA as a tryptophan codon instead of a stop codon *(1)*. In the early 1980s, several investigators *(3,4)* first proposed the use of mycoplasmas for defining the minimal genetic components required for a self-replicating cell. This led to genome-sequencing efforts on several *Mycoplasma* species *(5–8)* and a project using global transposon mutagenesis to define the minimal *Mycoplasma* genome *(9)*.

The **Mycoplasma genitalium** *and* **Mycoplasma pneumoniae** *Genomes*

Mycoplasma genitalium is found in the urogenital tract, where it is a causative agent of nongonococcal urethritis. *Mycoplasma pneumoniae* is a pathogen of the respiratory tract, where it causes atypical pneumonia in children and young adults. Host tissue specificity is variable; *M. pneumoniae* has been isolated from urogenital clinical specimens *(10)*, and *M. genitalium* has been isolated from the respiratory tract of patients infected with *M. pneumoniae (11)*.

The *M. genitalium* genome is a single circular 580,070 bp chromosome with an average G+C content of 32% and 480 predicted open reading frames (ORFs) *(5)*. In contrast, the *M. pneumoniae* genome is 816,394 bp with an average G+C content of 40% and 677 predicted ORFs *(6,12)*. Homologs of all the proposed *M. genitalium* ORFs are contained within the *M. pneumoniae* genome. The additional 209 ORFs in *M. pneumoniae* not found in *M. genitalium* fall into two primary groups: ORFs with functional similarities to those only present in *M. pneumoniae* and that may underlie the biological differences between the two organisms and ORFs also present in *M. genitalium* but that have been duplicated or amplified in *M. pneumoniae (12)*. Examination of gene order between these two genomes revealed that they could be subdivided into six genomic segments, with the order of genes in these segments conserved. However, the order of these six genomic segments differs in both genomes, likely the result of translocation between segments *(12)*.

Although many genes in essential pathways required for survival, such as pathways for replication and transcription, were confirmed, there are several in key cellular processes that either are absent or cannot be identified *(5,6,12)*. While the inability to identify genes for essential functions may be biologically important, it may also reflect a significant divergence or evolution of genes such that they are unrecognizable by sequence search algorithms *(13)*. For example, two genes thought involved in assembly and processivity of deoxyribonucleic acid (DNA) polymerase III, DNAθ, and DNAδ have not been identified. In addition, the organisms lack any two-component signal transduction system or other regulatory systems found in other bacteria for controlling gene expression *(13)*. Most remarkably, the reduction in genome size among *Mycoplasma* species is associated with a significant reduction in the number and enzymatic components of biosynthetic pathways. The diminished biosynthetic capacity requires them to use metabolic products of their hosts when growing in vivo, and there is necessity for a rich, complex artificial media when grown in vitro *(14)*.

The **Ureaplasma urealyticum** *Genome*

The third *Mycoplasma* genome sequenced, *Ureaplasma urealyticum*, is a member of the related genus *Ureaplasma* and a common commensal of the human urogenital tract.

It has emerged as a significant pathogen during pregnancy, for which it has been associated with inflammation, premature spontaneous delivery, septicemia, meningitis, and pneumonia in newborn infants *(15)*. The genome is a single circular 751,719 bp chromosome with an average G+C content of 25.5% and is predicted to encode 613 ORFs *(7)*. Of these 613 ORFs, only 324 have homologs in the *M. genitalium* and *M. pneumoniae* genomes. This is in contrast to *M. genitalium*, for which all ORFs have homologs in the *M. pneumoniae* genome. The evolutionary divergence of *U. urealyticum* from *M. genitalium* and *M. pneumoniae* is further supported by the lack of synteny between the *U. urealyticum* genome and those of *M. genitalium* and *M. pneumoniae*.

Of the 289 ORFs in *U. urealyticum* but not in *M. genitalium* or *M. pneumoniae*, 76 have been given functional assignments *(7)*. Many of these unique ORFs are involved in iron acquisition and adenosine triphosphate (ATP) production by urea hydrolysis. The most remarkable feature of this microorganism is its ability to generate 95% of its ATP through hydrolysis of urea by urease *(16)*. When comparing ORFs encoded on the *U. urealyticum* genome with the minimal or core set defined by transposon mutagenesis *(9)*, *U. urealyticum* has homologs for 69 of the *M. genitalium* nonessential genes and 255 of the putative essential genes *(7)*.

The Mycoplasma pulmonis *Genome*

Mycoplasma pulmonis is the causative agent of marine respiratory mycoplasmosis and is a useful model for studying mycoplasmal respiratory infections *(17)*. The genome is a single circular 963,879 bp chromosome with an average G+C content of 26.6%, and it is predicted to encode 782 ORFs, making the *M. pulmonis* genome the largest *Mycoplasma* genome sequenced *(8)*. When compared to the other sequenced Mollicutes, a majority of the additional genes in *M. pulmonis* are transport and membrane proteins *(8)*. The most remarkable feature of the *M. pulmonis* genome is the diversity of mechanisms used to generate antigenic diversity. These include phase-variable expression of highly variable surface lipoprotein genes in the *vsa* locus *(18)* and slipped-strand mispairing of homopolymeric repeats upstream of surface proteins *(8)*. Assuming experimental efforts verify these two mechanisms, this is the first indication of two different mechanisms controlling phase variation of surface antigens in a single *Mycoplasma* cell.

The Minimal Genome

Efforts to define a minimal genome have relied on both *in silico* and experimental approaches. *In silico* comparison of the *Haemophilus influenzae* and *M. genitalium* genomes identified a common set of 256 putative core genes required for autonomous replication and life *(19)*. Most recently, global transposon mutagenesis of *M. genitalium* and *M. pneumoniae* has been used to define the minimal *Mycoplasma* genome required for growth in vitro *(9)*. In this experiment, pools of in vitro cultured *M. genitalium* and *M. pneumoniae* with random chromosomal Tn4001 insertions were screened for viable bacteria containing transposon insertions in nonessential genes or intergenic regions. A core set of as few as 265 or as many as 350 *Mycoplasma* ORFs were identified as essential for replication in vitro. These core genes encode proteins in pathways for replication and transcription, transport proteins to derive nutrients from the environment, have metabolic pathways to generate ATP and reduce power, and have components for maintaining cellular homeostasis. The number of essential *Mycoplasma* genes is likely to increase when the organisms are grown in vivo under conditions for which the cells must adhere to host tissue and metabolites such as nucleosides and amino acids are not provided in the in vitro culture media.

Mycoplasma Genome Projects in Progress

There are several additional ongoing *Mycoplasma* genome-sequencing projects, including *Mycoplasma mycoides* subsp *mycoides* SC, *Mycoplasma hyopneumoniae*, *Mycoplasma capricolum*, *Mycoplasma arthritidis*, *Mycoplasma gallisepticum*, *Mycoplasma penetrans*, *Mycoplasma alligatoris*, *Spiroplasma citri*, and *Spiroplasma kunkelii (8)*. The availability of the genome sequence of these *Mycoplasma* species will make this group of organisms good experimental models for genome comparison, studies of microbial evolution, and further characterization of the minimal genome.

LISTERIA

The Listeria monocytogenes *and* Listeria innocua *Genomes*

The *Listeria* are Gram-positive, facultative anaerobic rods that are found in a wide array of environments, including soil, water, foods, and feces of animals and humans *(20)*. The pathogenic *Listeria* include *Listeria monocytogenes*, which infects both humans and other vertebrates, and *Listeria ivanovii*, which rarely infects humans *(21)* but frequently infects sheep and cattle *(20)*, for which it causes abortions, stillbirths, and neonatal infections. The most frequent disease associated with *L. monocytogenes* is listeriosis, a foodborne disease that often leads to serious localized and generalized infections, such as meningitis, meningoencephalitis, abortion, septicemia, gastroenteritis, and perinatal infections *(20)*. Prior to genome sequencing, several *L. monocytogenes* virulence factors were identified as part of a 10-kb virulence locus absent from *Listeria innocua (22)*, a nonpathogenic species of *Listeria* frequently used as a host for heterologous expression of *L. monocytogenes* genes *(23)*. These factors include (1) the surface protein internalin (InlA), which is essential for movement of *L. monocytogenes* across the intestinal barrier; (2) the invasion protein InlB; (3) LLO and PlcA, which facilitate escape from the phagocytic vacuole; and (4) ActA and PlcB, which provide the means for intracellular actin-based motility and cell-to-cell spread *(20,24)*.

The genome of *L. monocytogenes* EGD-e (serovar 1/2a) is 2,944,528 bp, with an average G+C content of 39%; it is predicted to encode 2853 ORFs. The genome of *L. innocua* CLIP 11262 (serovar 6a) is 3,011,209 bp, with an average G+C content of 37%; it is predicted to encode 2973 ORFs. In addition, both genomes contain prophages, and *L. innocua* harbors an 81,905 bp plasmid. The two *Listeria* genomes exhibit a conserved, colinear organization and overall synteny with the genomes of *Bacillus subtilis (25)* and *Staphylococcus aureus (26,27)*. When compared to *L. innocua*, the 270 genes unique to *L. monocytogenes* are dispersed throughout the genome in multiple regions between 1 and 25 kb long. This gene dispersal is similar to that identified in analysis of the *Escherichia coli* O157:H7 and *E.coli* K12 genomes *(28)*, but unlike streptococcal genome diversity, which is generated by prophages *(29)*.

Comparative analysis of the *L. monocytogenes* strain EGD-e (serovar 1/2a) and *L. innocua* strain CLIP 11262 (serovar 6a) genomes facilitated identification of a subset of unique *L. monocytogenes* genes associated with virulence, including the secreted proteins, lipase and chitinase *(22)*. When compared to other sequenced Gram-positive genomes, *L. monocytogenes* encodes far more (41 total) cell surface proteins containing the LPXTG cell wall linkage motif required by the sortase enzyme *(30)* than *Streptococcus pyogenes* (13 total) and *S. aureus* (18 total) *(27,31,32)*. Of the internalin family, 19 and 8 additional members containing the LPXTG motif *(30)* were identified in the *L. monocytogenes* and *L. innocua* genomes, respectively *(22)*.

A nearly identical number of transcriptional regulators were identified in *L. monocytogenes* (209 total) and *L. innocua* (203 total). Similarly, both *L. monocytogenes* and *L. innocua* encode a nearly equal number of two-component regulatory systems *(22)*. This similarity in the *L. monocytogenes* and *L. innocua* regulatory networks is reflective of their need to respond to diverse environmental stimuli, but is surprising considering the relative lack of *L. innocua* virulence factors relative to *L. monocytogenes (20, 33)*. A similar balance of regulatory networks exists between *S. aureus*, an aggressive pathogen, and its less-virulent relative *Staphylococcus epidermidis (27)*. One regulatory factor in *L. monocytogenes* but absent from *L. innocua* is the primarily regulator factor (PrfA), which activates known virulence genes *(20,33)*.

The genomes of two additional *Listeria* are currently being sequenced: *L. ivanovii*, a ruminant pathogen and the causative agent of abortions, stillbirths, and neonatal septicemia in cattle and sheep *(20)*, and *L. monocytogenes* serotype 4b, which has been associated with most major foodborne illness outbreaks *(34,35)*.

ENTEROCOCCI

The Enterococci are Gram-positive, facultative anaerobic cocci that are commensal members of the healthy human and animal intestinal flora and are commonly found in soil, sewage, water, and food. They are also among the leading causes of antibiotic-resistant, hospital-acquired infections in the United States *(36)*. Of the two enterococcal species associated with human infections, *Enterococcus faecalis* and *Enterococcus faecium*, about 80% of all infections are caused by *E. faecalis (37)*. The dominance of *E. faecalis* in enterococcal infections may be a result of its relative abundance over *E. faecium* in the intestinal flora or a consequence of enhanced virulence *(37)*. Emergence of resistance to vancomycin, the antibiotic of last resort for many multiply resistant Gram-positive pathogens, has made treatment of enterococcal infections an increasing challenge *(38,39)*. The genome of *E. faecalis* strain V583 has been completed, and the genome of *E. faecium* has been sequenced to 8× (http://www.jgi.doe.gov/index.html).

The Enterococcus faecalis *Genome*

The genome of *E. faecalis* V583, the first vancomycin-resistant clinical isolate *(38)*, is 3,218,030 bp, with a G+C content of 38% and 3,490 predicted ORFs *(40)*. A comparison with other sequenced genomes indicated that over 85% of *E. faecalis* ORFs have their best match to low G+C Gram-positive organisms. In addition, three circular plasmids, from 17,963 to 66,320 bp long, were identified and sequenced. Two of the plasmids are structurally similar to the pheromone-responsive plasmids pAD1 and pCF10, and the third belongs to the family of pAMβ1 broad host range plasmid *(41–43)*.

Seven regions of the *E. faecalis* genome (10.2% of the entire genome) consist of putative prophages closely related to phages from other low G+C Gram-positive bacteria.

This is similar to the prophage content in the group A Streptococci *(29)* and indicative of the close evolutionary relation of the Enterococci and Streptococci. A large pathogenicity island (~150 kb) identified in the V583 genome *(44)* encodes many of the *E. faecalis* virulence determinants, including genes for bile acid hydrolase and a well-known surface colonization protein (Esp). A nearly identical pathogenicity island was identified in *E. faecalis* strains MMH594 and V586, but with additional insertion sequence (IS) elements and, in MMH594, a 2.8-kb region encoding a cytolysin operon *(44)*. Several cell surface proteins that likely function as adhesins or aggregation factors were identified. Remarkably, many of the putative surface-exposed proteins contain stretches of homopolymeric sequence or iterative nucleotide motifs that may enable phase variation by a slippage-type mechanism. A similar slippage-type mechanism is used by other pathogens, such as *Streptococcus pneumoniae (45)*, to generate phase variation of surface antigens.

Vancomycin resistance in *E. faecalis* V583 is associated with two distinct mobile elements. The first element contains *van*A resistance genes from the *E. faecalis* conjugative transposon *Tn*1546 *(46)* and is responsible for the high level of vancomycin resistance (vancomycin-resistant *Enterococcus*, VRE). A second locus contains homologs of the *S. pneumoniae* VncRS two-component signal transduction system *(47)*, which has been associated with tolerance to vancomycin. The probable conjugal transfer of *van*A genes from VRE to *S. aureus* in two active human infections was responsible for the first clinical occurrence of vancomycin-resistant *S. aureus* (VRSA) *(48)*.

STAPHYLOCOCCI

The Staphylococci are a diverse group of Gram-positive, facultative anaerobes that infect both humans and animals. They cause a variety of diseases, ranging from minor skin infections and food poisoning, to more severe and sometimes fatal infections such as endocarditis, toxic shock syndrome, and meningitis. In spite of large-scale efforts to control their spread, they are a major cause of both hospital- and community-acquired infections worldwide *(49–51)*.

The two major opportunistic pathogens of this genus, *S. aureus* and *S. epidermidis* *(50–52)*, are carried on nasal mucosal cells and throughout the cutaneous ecosystem, in which they generally live in a benign relationship with their host. However, breach of the cutaneous organ system by trauma, inoculation needles, or direct implantation of medical devices enables the staphylococci to gain entry into the host and acquire the role of a pathogen.

Staphylococcus epidermidis is primarily associated with foreign body infections (e.g., infections of implanted medical devices); *S. aureus* is a much more aggressive pathogen that often causes acute and pyogenic infections *(50–52)*. The persistence of methicillin-resistant *S. aureus* and methicillin-resistant *S. epidermidis* (MRSA and MRSE, respectively) and the emergence of *S. aureus* isolates exhibiting intermediate levels of resistance to vancomycin (VISA), have made effective treatment and control of staphylococcal infections increasingly difficult *(48,53)*. Aside from the difficulty of treating hospital-acquired MRSA and VISA (H-MRSA and H-VISA, respectively), perhaps the most formidable problem facing infectious disease clinicians is the emergence of community-acquired MRSA (C-MRSA) and community-acquired methicillin-sensitive *S. aureus* (C-MSSA), some of which are more virulent than the typical H-MRSA *(54–56)*.

Although much is already known about their biology and virulence mechanisms, genome sequencing of *S. aureus* and *S. epidermidis* was undertaken to identify unknown mechanisms of virulence, explore genome diversity, and more clearly define the basis of methicillin and vancomycin resistance. Multiple sequencing projects at different centers worldwide have led to the completed sequences of seven staphylococcal genomes (www.tigr.org):

1. Methicillin-resistant *S. aureus* hospital isolates (H-MRSA) COL *(27)*, N-315 *(32)*, and the UK hospital-acquired strain representative of the epidemic MRSA strain E-MRSA-16 (strain 252) *(55)*
2. Mu50, which is an H-MRSA/H-VISA *(32)*
3. A laboratory strain NCTC8325 *(57)*
4. MRSA community isolate MW-2 (C-MRSA) *(58)*
5. Hypervirulent community-acquired MSSA strain (strain 476) *(55)*
6. An MRSE isolate (strain RP62A) *(27)*

THE *STAPHYLOCOCCUS AUREUS* GENOME

The genomes of the multiple *S. aureus* strains are single circular chromosomes between 2.8 and 2.9 Mb with an average G+C content of 30%; they are predicted to encode approximately 2600 ORFs *(27,32,58)*. Approximately 80% of the ORFs encoded by *S. aureus* have their best match to other low G+C Gram-positive genomes. The *S. aureus* genomes typically also contain multiple IS elements, prophages, and pathogenicity islands *(59)*. The acquisition of virulence and resistance factors by the *S. aureus* prophages, pathogenicity islands, and other mobile elements, followed by transfer between staphylococci is likely responsible for generating genome diversity and the increasing virulence of some strains *(59,60)*. Examples of pathogenicity islands include SaPI1 *(60)*, which contains the toxic shock syndrome gene *tsst* and SaPIn2, which contains the cluster of staphylococcal exotoxinlike (SET) proteins *(32)*. The genomes of the more virulent MW-2 (C-MRSA) and strain 476 (C-MSSA) contain four novel pathogenicity islands not carried by any of the other five sequenced *S. aureus* strains. It is likely that the novel allelic enterotoxin genes sel2 and sec4 carried by one of these islands (vSa3) are two of the factors contributing to their enhanced virulence *(58)*.

Staphylococcal resistance genes are typically located on plasmids or other mobile genetic elements *(61)*. The gene encoding methicillin resistance (*mec*) is carried on a mobile element, the staphylococcal cassette chromosome or SSCmec, which integrates near the genome origin of replication *(62,63)*. Four allelic forms of SSCmec (types I, II, III, and IV), each with a distinct size (between approx 24 and 100 kb) and genetic organization, have been identified in multiple strains *(62,63)*. The mechanism for intermediate levels of resistance to vancomycin in *S. aureus* (VISA with vancomycin minimum inhibitory concentration [MIC] = 8 µg/mL) appears to be complex and involve multiple metabolic pathways.

Although a novel two-component regulator *vraSR* has a demonstrated role *(26)* in VISA, several pathways involved in cell wall synthesis and assembly are likely involved in the resistance mechanism *(64)*. In the first occurrence of clinical *S. aureus* isolates with high levels of vancomycin resistance (VRSA) (vancomycin MIC > 32 µg/mL), two such isolates were identified in 2002 from two patients in US hospitals *(48)*. The mechanism of VRSA resistance differs from VISA in that the *vanA* gene, which is responsible for high levels of vancomycin resistance in the enterococci (VRE) *(39)* was identified in both VRSA isolates and likely was transferred by conjugation from VRE also found in the infection.

The *S. aureus* genome encodes a multitude of virulence factors, including proteases and other degradative enzymes, exotoxins, enterotoxins, hemolysins, and surface proteins, which function as adhesins for binding to host tissue *(49)*. Genes for at least 70 putative new virulence factors have been identified in the *S. aureus* genomes, with many carried on pathogenicity islands. One new virulence factor is a putative (~1 kDa) adhesin (Ebh) identified as an endothelial cell adhesin *(65)* and fibronectin-binding protein *(66)*, predicted to have a role in endocarditis *(32)*. A similar adhesin protein was also identified in the *S. epidermidis* genome *(27)*.

Expression of *S. aureus* virulence factors is regulated by the accessory gene regulator (*agr*) locus *(67)*, additional two-component signal transduction mechanisms related to *agr*, and a family of staphylococcal accessory regulators (*sar* family) *(68,69)*. A total of 16 *agr*-like two-component signal transduction gene pairs were identified in the *S. aureus* genome along with 10 members of the *sar* family *(27,32)*. Overall, the prophages, pathogenicity islands, and other mobile elements in the *S. aureus* genome likely provide a flexibility that enables the bacteria to adapt to multiple host environments and survive the onslaught of antimicrobials.

The Staphylococcus epidermidis *Genome*

The *S. epidermidis* strain RP62A genome is a single circular 2,619,000 bp chromosome, with an average G+C content of 30%, and it is predicted to encode 2,586 ORFs *(27)*. As with *S. aureus*, approximately 80% of the ORFs encoded by *S. epidermidis* have their best match to other low G+C Gram-positive genomes. One prophage, similar to *B. subtilis* SPP1 *(70)*, was identified in *S. epidermidis* RP62A. This strain also contains a 28-kb β-lactamase plasmid that encodes resistance to ampicillin. In contrast to *S. aureus*, no pathogenicity islands were identified in *S. epidermidis*.

Both genomes exhibit a conserved colinear organization that spans a common 2.2-Mb region of high gene conservation. Genome divergence occurring in this 2.2-Mb region is either the result of insertion of small gene clusters or of prophages and other mobile elements *(27)*. The region outside the 2.2-Mb conserved region spans the origin of replication of the two genomes and exhibits significant divergence.

Of the 2586 ORFs encoded by *S. epidermidis*, 2067 have homologs in *S. aureus*. The common genes are those that support the core functions of both staphylococci. Interestingly, the SSCmec mobile element in *S. epidermidis* is nearly identical to the type II SSCmec from *S. aureus* N315. This supports the accumulating data that show movement of SSCmec mobile elements through multiple staphylococcal species *(62,63)*.

Similar to the *Listeria* genomes, comparison of both staphylococcal genomes revealed that the inherent differences in virulence could primarily be attributed to the absence of virulence factors, such as enterotoxins and exotoxins, in *S. epidermidis*. In spite of the lack of virulence factors, the *S. epidermidis* genome encodes a nearly identical complement of *sar*, *agr*, and *agr*-like regulatory genes. These regulatory factors are likely involved in regulation of core function genes in both *S. aureus* and *S. epidermidis* and also are necessary for adaptation to diverse environmental stimuli.

At least two additional staphylococcal genomes are being sequenced: *Staphylococcus carnosis*, an avirulent *Staphylococcus* (F. Götz, personal communication) and a bovine *S. aureus* isolate *(71)*. Comparison of human and bovine *S. aureus* genomes will likely lead to identification of host-specific genes and species-specific virulence factors.

BACILLUS

Members of the genus *Bacillus* are Gram-positive, aerobic and facultative anaerobic, spore-forming, rod-shaped bacteria that are saprophytic inhabitants of the soil. They are a diverse group that demonstrate a broad range of phenotypes and pathological effects *(72)*. Some species are pathogens for humans and animals; others are utilized for applications in industrial microbiology. *Bacillus subtilis* is perhaps the most thoroughly studied and characterized Gram-positive bacterium and is considered a model for conversion of vegetative cells to an endospore *(73)* and for the study of bacterial differentiation and partitioning or segregation of the bacterial chromosome *(74)*. In addition, the ability of *B. subtilis* to secrete large amounts of extracellular proteins efficiently has made it a model for the study of protein secretion and production of heterologous proteins in Gram-positive bacteria *(75)*. *Bacillus anthracis*, a pathogen of humans and animals, is likely the most recognized bacterial pathogen of the early 2000s and has become a model for the application of genomic forensics *(76)*. The first of three genomes *Bacillus* species have been sequenced: *B. subtilis*, *B. halodurans*, and *B. anthracis*.

The Bacillus subtilis Genome

Sequencing of the genome of *B. subtilis* presented technical difficulties not encountered with previously sequenced bacterial genomes. The primary difficulty was the inability to construct plasmid-sequencing libraries in *E. coli* that were representative of the entire *B. subtilis* genome. Many of the approaches used to overcome this difficulty, including alternative low copy number cloning vectors *(77)* and polymerase chain reaction techniques *(78)*, have been applied to subsequent sequencing efforts on other bacterial genomes.

The genome of *B. subtilis* strain 168 is a single circular 4,214,810 bp chromosome with an average G+C content of 43.5%, and it is predicted to encode 4,100 ORFs *(25)*. Comparison of the *B. subtilis* genome with that of *E. coli* revealed that a large fraction of these genomes shared similar functions, with approx 100 putative operons apparently well conserved between these two organisms *(25)*. A comparison of the 450 genes encoded by the *M. genitalium* genome with the *B. subtilis* genome indicated that 300 of the *M. genitalium* ORFs have homologs in the *B. subtilis* genome, many of which represent the minimal *Mycoplasma* genome.

Although the *B. subtilis* genome encodes many functional components needed for gene regulation systems, two-component signal transduction pathways, quorum sensing, protein secretion, and sporulation, several components essential in *E. coli* are missing. Interestingly, MukB, which is a key player in *E. coli* chromosome partitioning *(79)*, is not found in *B. subtilis*. However, the Smc protein in *B. subtilis* is weakly similar to MukB *(79,80)* and likely a functional homolog. This may indicate subtle differences in the chromosome partitioning mechanism between Gram-positive and Gram-negative bacteria. Genes that are unique to *B. subtilis* and likely reflective of its relationship with soil and plants include those involved in the degradation of plant-derived molecules, such as opines. The *B. subtilis* genome contains multiple prophages (including Spβ, PBSX, and skin) or cryptic phages, which have played significant roles in gene transfer and generation of genome diversity.

The **Bacillus halodurans** *Genome*

Bacillus halodurans is an alkaliphilic organism for which optimal growth conditions require a pH greater than 9.5 *(81)*. Interest in *B. halodurans* covers two broad areas: mechanisms utilized by *B. halodurans* for adaptation to alkaline environments and continued development of *B. halodurans* as a producer of commercial products, such as proteases and cellulases. A genome-sequencing project of *B. halodurans* strain C-125 was undertaken to identify specific functions in alkaliphilic *Bacillus* strains and to identify common *Bacillus* functions through a comparison with the *B. subtilis* genome.

The *B. halodurans* genome is a single circular 4,202,353 bp chromosome with an average G+C content of 43.7%, and it is predicted to encode 4066 putative ORFs. Comparison of the *B. subtilis* and *B. halodurans* genomes revealed that about 1500 genes encoding many of the common housekeeping functions are well conserved in a common region of these two genomes *(82)*. Contrary to *B. subtilis* strain 168, which contains at least 3 intact prophages, no intact prophages were identified in the *B. halodurans* genome. At least 15 novel IS elements were identified in the *B. halodurans* genome, in which they primarily insert into noncoding regions *(83)*. Several factors that likely contribute to the alkaliphilic lifestyle were identified, including multiple unique sigma factors that may control transcriptional responses required for life in an alkaline environment.

The **Bacillus anthracis** *Genome*

Bacillus anthracis is the most virulent member of the *Bacillus* family. It is typically an inhabitant of the soil, where it forms spores, which can spread to animals and humans and subsequently cause the often-lethal disease, anthrax. The ease with which stable *B. anthracis* spores can be produced and refined in the laboratory has led to the development of *B. anthracis* as the anthrax bioterror weapon. Very early work with *B. anthracis*, such as Pasteur's work with vaccine protection of sheep *(84)*, was a key factor in the development of the study of bacteriology and immunology.

Bacillus anthracis is a member of the *Bacillus cereus* group of *Bacillus*, which includes *Bacillus thuringiensis* and *B. cereus* and is phylogenetically separate from the other sequenced *Bacillus* species, *B. subtilis (25)* and *B. halodurans (82)*. Known factors that confer virulence to *B. anthracis*, including the anthrax toxin complex and the poly-D-glutamic acid capsule, are carried on two completely sequenced plasmids: pX01 (181 kb) and pX02 (93.5 kb) *(85,86)*. The contribution of unidentified *B. anthracis* genome-encoded factors to virulence is unknown. One purpose of the *B. anthracis* genome-sequencing effort is to identify novel potential virulence factors and possible candidates for protective vaccine development. A second *B. anthracis* genome-sequencing effort was carried out on the virulent Florida anthrax isolate, with the goal of identifying novel sequence polymorphisms to develop new forensic tools for tracking anthrax outbreaks *(76)*.

The first *B. anthracis* genome sequenced was derived from the virulent Ames strain cured of plasmids pX01 and pX02 in Porton Down, UK. This genome is a 5,227,297 bp circular chromosome with an average G+C content of 35.4%, and it is predicted to encode 5,753 ORFs *(87)*. In addition to predicted ORFs, the *B. anthracis* genome contains few predicted IS elements, but harbors 4 prophages, which comprise 2.8% of the chromosome (Fig. 1). The prophages encode no apparent virulence genes, but do encode putative secreted and surface proteins, which may have an impact on interaction with external host environments.

Surprisingly, the majority of the *B. anthracis* potential chromosomal virulence factors have homologs in *B. cereus* 10987 (*B. cereus* genome-sequencing project; www.tigr.org) and are not located on pathogenicity islands. Although the *B. anthracis* genome encodes multiple predicted extracellular proteins, experimental data *(88)* indicate very few of these proteins are actually secreted. This is likely a result of a frameshift in the PlcR positive regulator, which is known to upregulate expression of multiple extracellular proteins in *B. cereus* and *B. thuringiensis (89)*. Searches of the *B. anthracis* genome for candidate surface proteins indicated at least 29 proteins that are potential candidates for vaccine development. The most remarkable outcome of this genome analysis is the lack of novel virulence factors specific for *B. anthracis* encoded on the *B. anthracis* chromosome. Evidence that all chromosome-encoded potential virulence factors are also found on the *B. cereus* 10987 chromosome lends further support to the concept that *B. anthracis*, *B. cereus*, and *B. thuringiensis* are a single species, and that species-specific virulence factors are transferred by exchange of chromosomal deoxyribonucleic acid (DNA) and resident plasmids in their natural environment *(87,90)*.

A much larger *B. anthracis* genome-sequencing effort is currently under way at the Institute for Genomic Research (www.tigr.org) with the goal of sequencing multiple *B. anthracis* isolates, and a comparison of their single-nucleotide polymorphisms to identify families of isolates and develop forensic tools for investigation of anthrax outbreaks. In the first report from this work, Read et al. *(76)* compared the genomes of the virulent Florida anthrax isolate and the Ames strain and identified 60 new markers (including single-nucleotide polymorphisms, inserted or deleted sequences, and tandem repeats). This effort has stimulated the development and refinement of comparative analysis tools, such as MUMmer *(91)*, which will also be useful for the analysis of other genomes.

CLOSTRIDIA

The Clostridia are Gram-positive, spore-forming, rod-shaped anaerobes that include two diverse groups: (1) toxin-producing pathogens such as *Clostridium difficile*, *Clostridium botulinum*, *Clostridium tetani*, and *Clostridium perfringens* and (2) terrestrial species such as *Clostridium acetobutylicum*, which have been exploited by the chemical industry for their ability to produce acetone, butanol, and other organic compounds through fermentation of carbon sources *(92,93)*. The genomes of two Clostridia have been sequenced: *C. acetobutylicum* and *C. perfringens*. *Clostridium acetobutylicum* was used in the early 1900s in an acetone–butanol–ethanol fermentation process for the production of acetone *(94,95)* and interest has recently been renewed for use in the development of energy- and cost-efficient petrochemical processes for production of

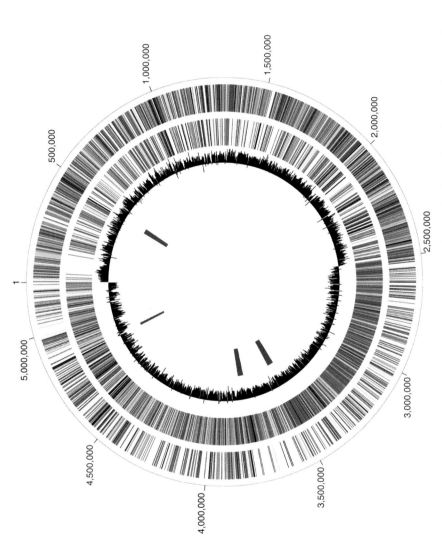

Fig. 1. Circular representation of the *B. anthracis* genome showing the position and orientation of genes (outer and second circle), GC skew (third circle), and location of bacteriophages (fourth circle). The strong bias in gene orientation (74% of the coding sequences are transcribed in the same orientation as movement of the replication fork) is shown by the orientation of genes in the outer and second circles. This type of gene orientation bias is a common feature of low G+C Gram-positive organisms. The four bacteriophages encode multiple surface proteins and are present in other *B. anthracis* genomes. (From Read et al., unpublished data.)

solvents *(95,96)*. Interest in genetic engineering of *C. acetobutylicum* and development of strains with modified solventogenic capacity *(97,98)* led to sequencing of type strain ATCC824. *Clostridium perfringens* is widely distributed in the environment, where it is found in soil and sewage and as part of the normal intestinal microbial flora of ani--mals and humans *(99)*. Entry of *C. perfringens* spores or vegetative cells into the body through an injury or wound initiates a gangrenous infection, causing massive destruction to host tissues that frequently leads to death *(100)*.

The Clostridium acetobutylicum *Genome*

The *C. acetobutylicum* type strain ATCC824 genome is 3,940,880 bp with an average G+C content of 30%, and it encodes 3740 predicted ORFs *(101)*. In addition to the chromosome, strain ATCC824 harbors two cryptic phages and a 200-kb plasmid, pSOL1, which has previously been shown to encode many of the genes involved in solvent production *(102)*. Further analysis of pSOL1 indicated that it encodes 178 putative ORFs involved in both the solventogenesis (production of butanol, ethanol, and acetone) and alcohologenesis (production of butanol and ethanol only) mechanisms carried out by *C. acetobutylicum*. Comparison with other bacterial genomes indicated a close evolutionary relationship with another Gram-positive, spore-forming bacteria, *B. subtilis*. One significant difference between *C. acetobutylicum* and *B. subtilis* is a greater and more diverse number of sporulation genes in *B. subtilis*, including genes involved in regulating sporulation and in spore structure *(101)*.

Horizontal gene transfer from archaea, hyperthermophiles, and eukaryotes to *C. acetobutylicum* appears to have shaped its metabolic abilities *(101)*. The nitrogen fixation operon in *C. acetobutylicum* is conserved with that of another nitrogen fixer, the archaeon *Methanobacterium thermoautotrophicum*. The *C. acetobutylicum* aromatic amino acid biosynthesis operon is also conserved with that of a hyperthermophilic bacterium, *Thermotoga maritima*. Genome analysis also enabled the identification of a complete set of genes involved in degradation of carbohydrate polymers, such as cellulose, xylan, levan, pectin, and starch. Most remarkable is the identification of a unique system involved in the extracellular hydrolysis of organic polymers. Members of this system have unique ChW repeat (clostridial hydrophobic, with a conserved W, tryptophan) domain repeated in ORFs that contain enzymatic domains encoding glycosyl hydrolases or proteases potentially involved in degradation of polysaccharides and proteins. Interestingly, one member of this system contains the leucine-rich (internalin) domain, which is found in *L. monocytogenes*, in which it has been shown to play a critical role in host cell invasion *(103)*.

The Clostridium perfringens *Genome*

The strain selected for sequencing was *C. perfringens* strain 13, a type A strain that is a natural soil isolate and causes gas gangrene in humans *(104)*. The genome is a 3,031,430 bp circular chromosome with an average G+C content of 28.6%, and it is predicted to encode 2660 ORFs *(105)*. Similar to *C. acetobutylicum (101)*, the *C. perfringens* genome contains few mobile genetic elements. A single cryptic prophage and seven intact transposase genes were identified in the *C. perfringens* genome, compared to two prophages and three transposase genes in *C. acetobutylicum*. The two clostridia genomes show little overall synteny, with most of the homologous genes in *C. perfringens* located inversely relative to the *C. acetobutylicum* genome *(105)*. When compared to *B. subtilis (25)*, both *C. perfringens* and *C. acetobutylicum* share a core complement of 61 *Bacillus* genes related to sporulation, but are lacking over 80 additional *Bacillus* sporulation genes, many of which encode spore coat- and germination-related proteins. These differences in sporulation-specific genes indicate a significant difference in the sporulation mechanisms used by these two closely related Gram-positive bacteria. In contrast to *C. acetobutylicum*, which contains a complete set of genes for amino acid biosynthesis, *C. perfringens* lacks many of these genes and is likely dependent on the host environment for provision of needed amino acids *(101)*.

The most significant differences between the two clostridia are those that contribute to the pathogenic lifestyle of *C. perfringens*. Several previously unknown virulence-associated genes were identified in the genome, including multiple hemolysins, proteinases, hyaluronidases, sialidases, an enterotoxin, and the host attachment factors fibronectin-binding protein and collagen adhesin *(105)*. Unlike other Gram-positive pathogens, such as the Staphylococci and Streptococci, none of these virulence-associated genes are contained in pathogenicity islands. Many of the *C. perfringens* virulence factors are regulated by the VirR/VirS and other two-component signal transduction systems. Also, unlike other Gram-positive pathogens, which produce many of their toxins and enzymes during the later stages of growth or establishment of an infection, those of *C. perfringens* are produced early during the growth phase, reflecting a need for *C. perfringens* to gain nutritional factors such as amino acids through degradation of host material. This is similar to *S. pneumoniae* and other Gram-positive pathogens, for which degradative enzymes serve dual roles in both degradation of host tissue and nutrient acquisition for the bacteria *(45)*.

Ongoing genome-sequencing projects on the pathogenic clostridia *C. difficile* and *C. botulinum* will facilitate genomic comparisons with *C. acetobutylicum* and identification of common metabolic and degradative capabilities.

STREPTOCOCCI

The Streptococci are a large and diverse group of Gram-positive cocci that live as either commensal or pathogenic microorganisms in association with animal hosts. They colonize different regions of the body and are notable for their ability to cause severe and frequently fatal infections, including wound infections, pneumonia, septicemia, Endocarditis, and scarlet fever *(106–108)*. There are two major groups of streptococci based on blood hemolysis type. The alpha-hemolytic Streptococci include *S. pneumoniae* and the oral or viridans streptococci such as *Streptococcus mutans*. The beta-hemolytic Streptococci are further classified into groups A through G on the basis of surface carbohydrate antigens. Group A streptococci (GAS) are human pathogens and primarily members of the species *Streptococcus pyogenes*. Group B *Streptococci* (GBS) are primarily members of *Streptococcus agalactiae*, which causes disease in both humans and cattle. Complete genome sequences have been determined for *S. pneumoniae (45)*, *S. pyogenes* (GAS) *(29,31,109)*, *S. agalactiae* (GBS) *(22,110)*, and *S. mutans (111)*.

The **Streptococcus pneumoniae** *Genome*

Streptococcus pneumoniae is one of the major invasive bacterial pathogens of children. It is the most common cause of otitis media and acute respiratory infection and is responsible for approximately 1.1 million deaths annually worldwide *(107)*. The increase in penicillin resistance and emergence of strains resistant to multiple antibiotics are making it more difficult to treat infections caused by *S. pneumoniae*. The genomic sequences of two *S. pneumoniae* strains, TIGR4 (virulent, serotype 4 isolate) *(45)* and strain R6 (an avirulent isolate) *(112)*, have been determined to identify mechanisms of pathogenicity and aid in the search for new therapies.

The TIGR4 isolate has a 2,160,837 bp circular genome with an average G+C content of 39.7%, and it is predicted to encode 2236 ORFs. The R6 isolate has a circular 2,038,615 bp genome with an average G+C content of 40%, and it is predicted to encode 2043 ORFs. One significant difference between TIGR4 and R6 is the presence of an approximately 7-kb deletion within the approximately 18-kb region encoding the capsular biosynthesis genes in isolate R6. Insertion elements are relatively abundant in the *S. pneumoniae* genome, in which they make up 5% of the TIGR4 genome *(45)* and may provide regions of homology for homologous recombination between IS elements and their flanking genes. Interestingly, no prophages were identified in either TIGR4 or R6. This contrasts to GAS, for which prophage DNA makes up 7–12% of their genomes and are responsible for generating genomic diversity and increasing virulence. Three other remarkable features of the *S. pneumoniae* genome are the number of sugar transporters, the dual role of extracellular enzyme systems responsible for degradation of host tissue, and the complete lack of a tricarboxylic acid (TCA) cycle. First, more than 30% of transporters in *S. pneumoniae* are predicted to transport sugar (see Chapter 7). This is likely a reflection of growth within a respiratory tract environment rich in glycoproteins and murein polysaccharides that can be utilized by the bacteria. Second, the *S. pneumoniae* genome encodes enzymes such as hydrolases and neuraminidases that likely participate in degradation of host tissue and polymers, facilitating colonization of host tissues. In their complementary or dual role, the products of these degradative activities are transported back into the bacteria and serve as substrates for synthesis of essential nutrients. Similar dual-role extracellular enzyme systems have been identified in *S. agalactiae (22,110)*, *S. aureus (27)*, and *C. perfringens (105)* and likely will be found in other Gram-positive pathogens. Finally, because an intact TCA cycle is completely absent, *S. pneumoniae* does not have the ability to synthesize the precursors of most amino acids. The TCA cycle is also absent in the other sequenced Streptococci, GAS and GBS.

The Group A Streptococcus *Genome*

The GAS are strict human pathogens and are responsible for a wide variety of diseases, including pharyngitis, scarlet fever, poststreptococcal acute rheumatic fever, impetigo, septicemia, necrotizing fasciitis, and toxic shock syndrome *(113)*. Further classification of GAS strains is based on serological differences in M protein, an antiphagocytic cell surface molecule with specific M serotypes associated with selected infection types. The genomes of three *S. pyogenes* (GAS) serotype strains have been sequenced: M1 (strain SF370), which is a common cause of pharyngitis and invasive infections *(31)*; M3 (strain MGAS315), which is associated with unusually severe infections and a high mortality rate *(29)*; and M18 (strain MGAS8232), which has been associated with acute rheumatic fever outbreaks in the United States *(109)*.

The genomes of the three GAS strains are all approx 1.9 Mb long, have an average G+C content of 38%, and encode between 1752 and 1889 predicted ORFs. The three genomes are colinear in structure over about a 1.7-Mb core region that encodes several known streptococcal virulence factors, such as streptolysin O and the hyaluronic acid capsule *(113)*. Outside of this essential core, the major source of variation is multiple phages, which differ in number, composition, and integration site within the genome. The phages are typically 30–50 kb long, and their number per genome ranges as follows: MGAS315 (M3) with six phages that comprise 12.4% of the M3 genome, MGAS8232 (M18) with five phages which comprise 10.8% of the M18 genome, and SF370 (M1) with four phages that contribute 7.0% of M1 genome. The phages encode multiple virulence factors, such as streptococcal pyrogenic exotoxin (SpeK), that contribute to the unique virulence characteristics of each strain. Because these phages are chimeric in that they share regions of homology and encode similar virulence factors, Beres et al. *(29)* suggested that recombination events between phage-encoded toxin genes provide a mechanism for creation of new, and more virulent, GAS subclones. When considering multiple-genome sequences from related bacterial species, prophages have an exceptional impact on GAS genome diversity.

The Group B Streptococci Genome

Streptococcus agalactiae, a GBS, was first identified as a pathogen in bovine mastitis and is typically a commensal organism that colonizes the genital or gastrointestinal tract of 25–40% of healthy women. As an opportunistic pathogen, it can cause invasive and life-threatening disease in pregnant women, newborn infants, and susceptible adults *(114,115)*. Disease in immunocompromised adults is currently responsible for the majority of serious *S. agalactiae* infections. While *S. agalactiae* and the other pathogenic streptococci share some virulence determinants, the genome of *S. agalactiae* was sequenced to identify unique virulence factors that contribute to its pathogenicty and mechanisms of genome divergence, which may account for the emergence of *S. agalactiae* as a major human pathogen.

The genomes of two GBS isolates have been sequenced: *S. agalactiae* serotype III (strain NEM316) *(22)* and *S. agalactiae* serotype V (strain 2603 V/R) *(110)*. The *S. agalactiae* serotype III isolate has a circular 2,211,485 bp chromosome with a G+C content of 35%, and it is predicted to encode 2082 ORFs. The *S. agalactiae* serotype V isolate has a circular 2,160,267 bp chromosome with a G+C content of 35%, and it encodes 2175 predicted ORFs. A number of identified pathways in *S. agalactiae* reflect an ability to adapt to various host environments. First, similar to *S. pneumoniae*, a large array *(17)* of sugar-specific phosphoenolpyruvate-dependent phosphotransferase system enzyme II complexes were identified in *S. agalactiae*, in which they likely contribute to a broad catabolic capacity. Second, similar to the example of dual metabolism in *S. pneumoniae*, NEM316 encodes four exported peptidases, three oligopeptide-specific ABC transporters, and many intracellular peptidase genes. Working in concert, these genes are able to degrade host proteins and import the resulting oligopeptides back into the cell, providing essential nutrients. Third, the *S. agalactiae* genome encodes more *(21)* two-component regulatory systems than is found in the other streptococcal genomes. This may reflect an increased capacity for *S. agalactiae* to monitor multiple host environments or control expression of virulence factors *(110)*.

Genome diversity exhibited by GBS appears to be controlled by mobile genetic elements, including pathogenicity islands, prophages, or the new type of integrative plasmid identified in *S. agalactiae* NEM316. Many of the 315 genes unique to *S. agalactiae* and the majority of virulence factors are carried on multiple pathogenicty islands dispersed around the genome. The role these mobile elements play in gene acquisition and genome diversity in GBS is likely similar to the role of prophages for generation of genome diversity in GAS.

Streptococcus mutans

Streptococcus mutans is a key member of the oral microbial flora and the primary causative agent of human dental caries. As with other oral Gram-positive Streptococci, such as *Streptococcus gordonii*, *S. mutans* has also been associated with bacterial endocarditis *(116)*. The genome of *S. mutans* UA159, Bratthall serotype c, is a single circular 2,030,936 bp consisting with an average G+C content of 36.8%, and it is predicted to encode 1963 ORFs *(111)*. Similar to *S. pneumoniae* TIGR4 and R6, the genome of *S. mutans* contains no prophages, but has a significant number of IS elements *(45,112)*. In addition, the *S. mutans* genome contains a potential conjugative transposon, *Tn*Smu1, which is similar to *Tn*916 from *E. faecalis*, and a putative 40-kb pathogenicity island *(111)*.

Indicative of its survival in the oral cavity, in which there is an abundance of carbohydrates, *S. mutans* is likely able to metabolize a wider variety of carbohydrates than any other Gram-positive organism sequenced to date *(111)*. As with other Gram-positive bacteria, most sugars are transported by a phosphoenolpyruvate sugar phosphotransferase system. The end products of carbohydrate metabolism lead to acidification of the local oral environment and a shift in the oral bacterial flora that favors *S. mutans* and initiation of dental caries in the host. Multiple virulence factors were identified in the *S. mutans* genome, including adhesins, proteases, extracellular enzymes, and surface proteins. Novel adhesins include a protein similar to streptococcal extracellular matrix-binding protein and a fibronectin-binding protein. Several proteases, one similar to the HtpX protease from *S. gordonii* *(117)*, are primarily involved in degradation of host structural proteins, which are then used for bacterial nutrition. Numerous novel cell surface proteins, including six with C-terminal LPXTG motifs were also identified.

Additional Streptococcal Genomes

There are currently at least 10 additional streptococcal genome projects under way or complete. Those that are underway include *Streptococcus pleomorphus*, *Streptococcus suis*, *Streptococcus uberis*, *Streptococcus equi*, *Streptococcus thermophilus*, and additional oral streptococci (*Streptococcus sanguis*, *S. gordonii*, *Streptococcus mitis*, and *Streptococcus sobrinus*).

LACTOCOCCUS: THE LACTOCOCCUS LACTIS GENOME

The lactic acid bacteria are a diverse group of microorganisms that convert carbohydrates into lactic acid. Members of this group include pathogenic Streptococci and other genuses such as *Lactococcus* and *Lactobacillus*, which produce lactic acid through homolactic fermentation. Homolactic fermentation is the backbone of the dairy industry, in which it is responsible for souring milk used in the production of cheese, yogurt, and other dairy products *(118)*. As the workhorse of the dairy industry, researchers strive to improve growth of *Lactococcus lactis* and optimize its production of lactic acid *(118)*. The genome sequence of *L. lactis* IL1403 was determined *(119)* with the goal of acquiring a more complete understanding of its physiology to develop more efficient fermentative processes.

The genome of *L. lactis* IL1403 is 2,365,589 bp with an average G+C content of 35.4% and 2310 predicted ORFs. A significant proportion of the genome (9.2%) is comprised of IS elements or prophages. This is similar to the contribution of phages to other lactic acid bacteria genomes that have been sequenced, such as *S. pyogenes* (GAS) and *S. agalactiae* (GBS) *(119)*. Comparative analysis by sequencing or comparative genomic hybridization of multiple *Lactococcus* species is needed to measure the contribution of prophages to genome diversity.

Of primary interest is the energy metabolism of *L. lactis*, leading to the production of high amounts of lactic acid by homolactic fermentation *(120,121)*. Anaerobic glycolysis is the primary energy-generating process in *L. lactis*, and all genes for this pathway have been identified. Unexpectedly, enzymes required for aerobic respiration are also encoded on the *L. lactis* genome, indicating possible alternative means of energy generation *(119)*. Mixed acid fermentation by *L. lactis* to form fermentation products other than lactic acid is also used by the dairy industry. Genome analysis has identified a novel gene-encoding pyruvate oxidase (*poxL*), which may be responsible for switching between fermentation modes.

Additional members of the lactic acid bacteria currently being sequenced include *Lactobacillus acidophilus*, which is a dental cariogen and is used for production of dairy products such as acidophilus milk, and *Lactobacillus bulgaricus*, which is also used for production of dairy products such as yogurt and cheese.

SUMMARY

Genome-sequencing efforts on these low G+C Gram-positive bacteria have revealed many novel virulence factors, unexpected metabolic pathways, and mechanisms for generation of genome diversity. Although these factors and pathways reflect adaptation of the bacteria to unique environmental and host tissue environments, there are also common themes shared by several of these species. For example, the contribution of prophages and other mobile elements to generation of genomic diversity in the low G+C Gram-positive bacteria is only now beginning to be explored. Another relationship shared among many of these species is the strong bias for transcription of genes in the same direction as replication (Fig. 1). Many of the pathogens have evolved multiple degradative enzymes and transport systems, which enable them to utilize nutrients obtained from degradation of host tissue. Continued comparative analyses of multiple low G+C Gram-positive genomes, both by genome sequencing and comparative genomic hybridization, will undoubtedly lead to novel insights and the role of horizontal gene transfer in shaping their genomes.

REFERENCES

1. Razin S, Yogev D, Naot Y. Molecular biology and pathogenicity of mycoplasmas. Microbiol Mol Biol Rev 1998; 62:1094–1156.
2. Krause DC. *Mycoplasma pneumoniae* cytadherence: unravelling the tie that binds. Mol Microbiol 1996; 20:247–253.

3. Colman SD, Hu PC, Bott KF. *Mycoplasma pneumoniae* DNA gyrase genes. Mol Microbiol 1990; 4:1129–1134.
4. Su CJ, Baseman JB. Genome size of *Mycoplasma genitalium*. J Bacteriol 1990; 172:4705–4707.
5. Fraser CM, Gocayne JD, White O, et al. The minimal gene complement of *Mycoplasma genitalium*. Science 1995; 270:397–403.
6. Himmelreich R, Hilbert H, Plagens H, Pirkl E, Li BC, Herrmann R. Complete sequence analysis of the genome of the bacterium *Mycoplasma pneumoniae*. Nucleic Acids Res 1996; 24:4420–4449.
7. Glass JI, Lefkowitz EJ, Glass JS, Heiner CR, Chen EY, Cassell GH. The complete sequence of the mucosal pathogen *Ureaplasma urealyticum*. Nature 2000; 407:757–762.
8. Chambaud I, Heilig R, Ferris S, et al. The complete genome sequence of the murine respiratory pathogen *Mycoplasma pulmonis*. Nucleic Acids Res 2001; 29:2145–2153.
9. Hutchison CA, Peterson SN, Gill SR, et al. Global transposon mutagenesis and a minimal *Mycoplasma genome*. Science 1999; 286:2165–2169.
10. Goulet M, Dular R, Tully JG, Billowes G, Kasatiya S. Isolation of *Mycoplasma pneumoniae* from the human urogenital tract. J Clin Microbiol 1995; 33:2823–2825.
11. Baseman JB, Dallo SF, Tully JG, Rose DL. Isolation and characterization of *Mycoplasma genitalium* strains from the human respiratory tract. J Clin Microbiol 1988; 26:2266–2269.
12. Himmelreich R, Plagens H, Hilbert H, Reiner B, Herrmann R. Comparative analysis of the genomes of the bacteria *Mycoplasma pneumoniae* and *Mycoplasma genitalium*. Nucleic Acids Res 1997; 25:701–712.
13. Peterson SN, Fraser CM. The complexity of simplicity. Genome Biol 2001; 2:COMMENT2002.
14. Jensen JS, Hansen HT, Lind K. Isolation of *Mycoplasma genitalium* strains from the male urethra. J Clin Microbiol 1996; 34:286–291.
15. Cassell GH, Waites KB, Watson HL, Crouse DT, Harasawa R. *Ureaplasma urealyticum* intrauterine infection: role in prematurity and disease in newborns. Clin Microbiol Rev 1993; 6:69–87.
16. Smith DG, Russell WC, Ingledew WJ, Thirkell D. Hydrolysis of urea by *Ureaplasma urealyticum* generates a transmembrane potential with resultant ATP synthesis. J Bacteriol 1993; 175:3253–3258.
17. Davidson MK, Lindsey JR, Parker RF, Tully JG, Cassell GH. Differences in virulence for mice among strains of *Mycoplasma pulmonis*. Infect Immun 1988; 56:2156–2162.
18. Shen X, Gumulak J, Yu H, French CT, Zou N, Dylovig K. Gene rearrangements in the *vsa* locus of *Mycoplasma pulmonis*. J Bacteriol 2000; 182:2900–2908.
19. Mushegian AR, Koonin EV. A minimal gene set for cellular life derived by comparison of complete bacterial genomes. Proc Natl Acad Sci USA 1996; 93:10,268–10,273.
20. Vazquez-Boland JA, Kuhn M, Berche P, et al. *Listeria* pathogenesis and molecular virulence determinants. Clin Microbiol Rev 2001; 14:584–640.
21. Cummins AJ, Fielding AK, McLauchlin J. *Listeria ivanovii* infection in a patient with AIDS. J Infect 1994; 28:89–91.
22. Glaser P, Rusniok C, Buchrieser C, et al. Genome sequence of *Streptococcus agalactiae*, a pathogen causing invasive neonatal disease. Mol Microbiol 2002; 45:1499–1513.
23. Gaillard JL, Berche P, Frehel C, Gouin E, Cossart P. Entry of *L. monocytogenes* into cells is mediated by internalin, a repeat protein reminiscent of surface antigens from gram-positive cocci. Cell 1991; 65:1127–1141.
24. Lecuit M, Vandarmael-Pournin S, Lefort J, et al. A transgenic model for listeriosis: role of internalin in crossing the intestinal barrier. Science 2001; 292:1722–1725.
25. Kunst F, Ogasawara N, Moszer I, et al. The complete genome sequence of the Gram-positive bacterium *Bacillus subtilis*. Nature 1997; 390:249–256.
26. Kuroda M, Kuwahara-Arai K, Hiramatsu K. Identification of the up- and downregulated genes in vancomycin-resistant *Staphylococcus aureus* strains Mu3 and Mu50 by cDNA differential hybridization method. Biochem Biophys Res Commun 2000; 269:485–490.

27. Gill SR, et al. Comparative genomics of *Staphylococcus aureus* and *Staphylococcus epidermidis*. 2002; in preparation.

28. Perna NT, Plunkett G 3rd, Burland V, et al. Genome sequence of enterohaemorrhagic *Escherichia coli* O157:H7. Nature 2001; 409:529–533.

29. Beres SB, Sylva GL, Barbian KD, et al. Genome sequence of a serotype M3 strain of group A *Streptococcus*: phage-encoded toxins, the high-virulence phenotype, and clone emergence. Proc Natl Acad Sci USA 2002; 99:10,078–10,083.

30. Navarre WW, Schneewind O. Surface proteins of Gram-positive bacteria and mechanisms of their targeting to the cell wall envelope. Microbiol Mol Biol Rev 1999; 63:174–229.

31. Ferretti JJ, McShan WM, Ajdic D, et al. Complete genome sequence of an M1 strain of *Streptococcus pyogenes*. Proc Natl Acad Sci USA 2001; 98:4658–4663.

32. Kuroda M, Ohta T, Uchiyama I, et al. Whole genome sequencing of meticillin-resistant *Staphylococcus aureus*. Lancet 2001; 357:1225–1240.

33. Glaser P, Frangeul L, Buchrieser C, et al. Comparative genomics of *Listeria* species. Science 2001; 294:849–852.

34. Herd M, Kocks C. Gene fragments distinguishing an epidemic-associated strain from a virulent prototype strain of *Listeria monocytogenes* belong to a distinct functional subset of genes and partially cross-hybridize with other *Listeria* species. Infect Immun 2001; 69: 3972–3979.

35. He W, Luchansky JB. Construction of the temperature-sensitive vectors pLUCH80 and pLUCH88 for delivery of Tn917::NotI/SmaI and use of these vectors to derive a circular map of *Listeria monocytogenes* Scott A, a serotype 4b isolate. Appl Environ Microbiol 1997; 63: 3480–3487.

36. Richards MJ, Edwards JR, Culver DH, Gaynes RP. Nosocomial infections in combined medical-surgical intensive care units in the United States. Infect Control Hosp Epidemiol 2000; 21:510–515.

37. Huycke MM, Sahm DF, Gilmore MS. Multiple-drug resistant enterococci: the nature of the problem and an agenda for the future. Emerg Infect Dis 1998; 4:239–249.

38. Sahm DF, Kissinger J , Gilmore MS, et al. In vitro susceptibility studies of vancomycin-resistant *Enterococcus faecalis*. Antimicrob Agents Chemother 1989; 33:1588–1591.

39. Bonten MJ, Willems R, Weinstein RA. Vancomycin-resistant enterococci: why are they here, and where do they come from? Lancet Infect Dis 2001; 1:314–325.

40. Paulsen I. Role of mobile elements in the evolution of vancomycin resistant *Enterococcus faecalis* V583. Science 2003; 299:2071–2074.

41. Bensing BA, Manias DA, Dunny GM. Pheromone cCF10 and plasmid pCF10-encoded regulatory molecules act post-transcriptionally to activate expression of downstream conjugation functions. Mol Microbiol 1997; 24:285–294.

42. Bruand C, Le Chatelier E, Ehrlich SD, Janniere L. A fourth class of theta-replicating plasmids: the pAM beta 1 family from Gram-positive bacteria. Proc Natl Acad Sci USA 1993; 90:11,668–11,672.

43. de Freire Bastos MC, Tanimoto K, Clewell DB. Regulation of transfer of the *Enterococcus faecalis* pheromone-responding plasmid pAD1: temperature-sensitive transfer mutants and identification of a new regulatory determinant, traD. J Bacteriol 1997; 179:3250–3259.

44. Shankar N, Baghdayan AS, Gilmore MS. Modulation of virulence within a pathogenicity island in vancomycin-resistant *Enterococcus faecalis*. Nature 2002; 417:746–750.

45. Tettelin H, Nelson KE, Paulson IT, et al. Complete genome sequence of a virulent isolate of *Streptococcus pneumoniae*. Science 2001; 293:498–506.

46. Weigel L, Clewell DB, Gill SR, et al. Genetic analysis of a high-level vancomycin-resistant isolate of *Staphylococcus aureus*. Science 2003; 302:1569–1571.

47. Novak R, Henriques B, Charpentier E, Normark S, Tuomanen E. Emergence of vancomycin tolerance in *Streptococcus pneumoniae*. Nature 1999; 399:590–593.

48. CDC. Vancomycin-resistant *Staphylococcus aureus*—Pennsylvania, 2002. Centers for Disease Control and Prevention. MMWR Morb Mortal Wkly Rep 2002; 51:902.

49. Projan SJ, Novick RP. The molecular basis of pathogenicity. In: Crossley KB, Archer GL (eds). The Staphylococci in Human Disease. New York: Churchill Livingstone, 1997, pp. 55–81.

50. Steinberg JP, Clark CC, Hackman BO. Nosocomial and community-acquired *Staphylococcus aureus* bacteremias from 1980 to 1993: impact of intravascular devices and methicillin resistance. Clin Infect Dis 1996; 23:255–259.

51. Thylefors JD, Harbarth S, Pittet D. Increasing bacteremia due to coagulase-negative staphylococci: fiction or reality? Infect Control Hosp Epidemiol 1998; 19:581–589.

52. Rupp ME, Archer GL. Coagulase-negative staphylococci: pathogens associated with medical progress. Clin Infect Dis 1994; 19:231–243; quiz 244–245.

53. Srinivasan A, Dick JD, Perl TM. Vancomycin resistance in staphylococci. Clin Microbiol Rev 2002; 15:430–438.

54. Groom AV, Wolsey DH, Naimi TS, et al. Community-acquired methicillin-resistant *Staphylococcus aureus* in a rural American Indian community. JAMA 2001; 286:1201–1205.

55. Enright MC, Day NP, Davies CE, Peacock SJ, Spratt BG. Multilocus sequence typing for characterization of methicillin-resistant and methicillin-susceptible clones of *Staphylococcus aureus*. J Clin Microbiol 2000; 38:1008–1015.

56. Naimi TS, LeDell KH, Boxrud DJ, et al. Epidemiology and clonality of community-acquired methicillin-resistant *Staphylococcus aureus* in Minnesota, 1996–1998. Clin Infect Dis 2001; 33:990–996.

57. Iandolo JJ, Worrell V, Groicher KH, et al. Comparative analysis of the genomes of the temperate bacteriophages phi 11, phi 12 and phi 13 of *Staphylococcus aureus* 8325. Gene 2002; 289: 109–118.

58. Baba T, Takeuchi F, Kuroda M, et al. Genome and virulence determinants of high virulence community-acquired MRSA. Lancet 2002; 359:1819–1827.

59. Novick RP, Schlievert P, Ruzin A. Pathogenicity and resistance islands of staphylococci. Microbes Infect 2001; 3:585–594.

60. Lindsay JA, Ruzin A, Ross HF, Kurepina N, Novick RP. The gene for toxic shock toxin is carried by a family of mobile pathogenicity islands in *Staphylococcus aureus*. Mol Microbiol 1998; 29:527–543.

61. Firth N, Apisiridej S, Berg T, et al. Replication of staphylococcal multiresistance plasmids. J Bacteriol 2000; 182:2170–2178.

62. Hiramatsu K, Cui L, Kuroda M, Ito T. The emergence and evolution of methicillin-resistant *Staphylococcus aureus*. Trends Microbiol 2001; 9:486–493.

63. Ito T, Katayama Y, Asada K, et al. Structural comparison of three types of staphylococcal cassette chromosome mec integrated in the chromosome in methicillin-resistant *Staphylococcus aureus*. Antimicrob Agents Chemother 2001; 45:1323–1336.

64. Cui L, Murakami H, Kuwahara-Arai F, Hanaki H, Hiramatsu K. Contribution of a thickened cell wall and its glutamine nonamidated component to the vancomycin resistance expressed by *Staphylococcus aureus* Mu50. Antimicrob Agents Chemother 2000; 44:2276–2285.

65. Heilmann C, et al. Characterization of the 113 kDa giant Staphylococcal surface protein (gssp) from *Staphylococcus aureus* involved in adherence to endothelial cells. International Symposium on Staphylococci and Staphylococcal Infections—Abstracts 2002; 2002:151.

66. Clarke SR, et al. Components of *Staphylococcus aureus* expressed during human infection. General Society for Microbiology—2002 Abstracts, 2002:5.

67. Peng HL, Novick RP, Kreiswirth B, Kornblum J, Schlievert P. Cloning, characterization, and sequencing of an accessory gene regulator (*agr*) in *Staphylococcus aureus*. J Bacteriol 1988; 170:4365–4372.

68. Cheung AL, Projan SJ. Cloning and sequencing of sarA of *Staphylococcus aureus*, a gene required for the expression of agr. J Bacteriol 1994; 176:4168–4172.

69. Cheung AL, Yeaman MR, Sullam PM, Witt MD, Bayer AS. Role of the sar locus of *Staphylococcus aureus* in induction of endocarditis in rabbits. Infect Immun 1994; 62:1719–1725.
70. Alonso JC, Luder G, Strege AC, et al. The complete nucleotide sequence and functional organization of *Bacillus subtilis* bacteriophage SPP1. Gene 1997; 204:201–212.
71. Herron LL, Chakravarty R, Dwan C, et al. Genome sequence survey identifies unique sequences and key virulence genes with unusual rates of amino acid substitution in bovine *Staphylococcus aureus*. Infect Immun 2002; 70:3978–3981.
72. Priest FG. Systematics and ecology of *Bacillus*. In: Sonenshein AL, Hoch JA, Losick R (eds). *Bacillus subtilis* and other Gram-Positive Bacteria. Washington, DC: American Society for Microbiology, 1993.
73. Stragier P, Losick R. Molecular genetics of sporulation in *Bacillus subtilis*. Annu Rev Genet 1996; 30:297–241.
74. Shapiro L, Losick R. Dynamic spatial regulation in the bacterial cell. Cell 2000; 100:89–98.
75. Harwood CR. *Bacillus subtilis* and its relatives: molecular biological and industrial workhorses. Trends Biotechnol 1992; 10:247–256.
76. Read TD, Salzberg SL, Pop M, et al. Comparative genome sequencing for discovery of novel polymorphisms in *Bacillus anthracis*. Science 2002; 296:2028–2033.
77. Azevedo V, Alvarez E, Zumstein E, et al. An ordered collection of *Bacillus subtilis* DNA segments cloned in yeast artificial chromosomes. Proc Natl Acad Sci USA 1993; 90:6047–6051.
78. Sorokin A, Lapidus A, Capuano V, et al. A new approach using multiplex long accurate PCR and yeast artificial chromosomes for bacterial chromosome mapping and sequencing. Genome Res 1996; 6:448–453.
79. Soppa J. Prokaryotic structural maintenance of chromosomes (SMC) proteins: distribution, phylogeny, and comparison with MukBs and additional prokaryotic and eukaryotic coiled-coil proteins. Gene 2001; 278:253–264.
80. Hirano M, Hirano T. Hinge-mediated dimerization of SMC protein is essential for its dynamic interaction with DNA. EMBO J 2002; 21:5733–5744.
81. Horikoshi K. Alkaliphiles: some applications of their products for biotechnology. Microbiol Mol Biol Rev 1999; 63:735–750, table of contents.
82. Takami H, Nakasone K, Takaki Y, et al. Complete genome sequence of the alkaliphilic bacterium *Bacillus halodurans* and genomic sequence comparison with *Bacillus subtilis*. Nucleic Acids Res 2000 28:4317–4331.
83. Takami H, Han CG, Takaki Y, Ontsuno E. Identification and distribution of new insertion sequences in the genome of alkaliphilic *Bacillus halodurans* C-125. J Bacteriol 2001; 183:4345–4356.
84. Pasteur L, Chamberland, Roux. Summary report of the experiments conducted at Pouilly-le-Fort, near Melun, on the anthrax vaccination. 1881 [classical article]. Yale J Biol Med 2002; 75:59–62.
85. Okinaka R, Cloud K, Hampton O, et al. Sequence, assembly and analysis of pX01 and pX02. J Appl Microbiol 1999; 87:261–262.
86. Okinaka RT, et al. Sequence and organization of pXO1, the large *Bacillus anthracis* plasmid harboring the anthrax toxin genes. J Bacteriol 1999; 181:6509–6515.
87. Read TD, Peterson SN, Tourasse N, et al. The complete genome sequence of *Bacillus anthracis* Ames and comparison to closely related bacteria. Nature 2003; 423:81–86.
88. Mignot T, Mock M, Robichon D, et al. The incompatibility between the PlcR- and AtxA-controlled regulons may have selected a nonsense mutation in *Bacillus anthracis*. Mol Microbiol 2001; 42:1189–1198.
89. Agaisse H, Gominet M, Okstad OA, Kolsto AB, Lereclus D. PlcR is a pleiotropic regulator of extracellular virulence factor gene expression in *Bacillus thuringiensis*. Mol Microbiol 1999; 32:1043–1053.

90. Helgason E, Okstad OA, Caugant DA, et al. *Bacillus anthracis*, *Bacillus cereus*, and *Bacillus thuringiensis*—one species on the basis of genetic evidence. Appl Environ Microbiol 2000; 66: 2627–2630.

91. Delcher AL, Phillippy A, Carlton J, Salzberg SL. Fast algorithms for large-scale genome alignment and comparison. Nucleic Acids Res 2002; 30:2478–2483.

92. Stackebrandt E, Kramer I, Swiderski J, Heppe H. Phylogenetic basis for a taxonomic dissection of the genus *Clostridium*. FEMS Immunol Med Microbiol 1999; 24:253–258.

93. Keis S, Bennett CF, Ward VK, Jones DT. Taxonomy and phylogeny of industrial solvent-producing clostridia. Int J Syst Bacteriol 1995; 45:693–705.

94. Durre P. New insights and novel developments in clostridial acetone/butanol/isopropanol fermentation. Appl Microbiol Biotechnol 1998; 49:639–648.

95. Woods DR. The genetic engineering of microbial solvent production. Trends Biotechnol 1995; 13:259–264.

96. Mitchell WJ. Physiology of carbohydrate to solvent conversion by clostridia. Adv Microb Physiol 1998; 39:31–130.

97. Ravagnani A, Jennert KC, Steiner E, et al. Spo0A directly controls the switch from acid to solvent production in solvent-forming clostridia. Mol Microbiol 2000; 37:1172–1185.

98. Green EM, Boynton ZL, Harris LM, et al. Genetic manipulation of acid formation pathways by gene inactivation in *Clostridium acetobutylicum* ATCC 824. Microbiology 1996; 142(Pt 8): 2079–2086.

99. Hatheway CL. Toxigenic clostridia. Clin Microbiol Rev 1990; 3:66–98.

100. Rood JI. Virulence genes of *Clostridium perfringens*. Annu Rev Microbiol 1998; 52:333–360.

101. Nolling J, Breton G, Omelchenko MV, et al. Genome sequence and comparative analysis of the solvent-producing bacterium *Clostridium acetobutylicum*. J Bacteriol 2001; 183:4823–4838.

102. Cornillot E, Nair RV, Papoutsakis ET, Soucaille P. The genes for butanol and acetone formation in *Clostridium acetobutylicum* ATCC 824 reside on a large plasmid whose loss leads to degeneration of the strain. J Bacteriol 1997; 179:5442–5447.

103. Marino M, Braun L, Cossart P, Ghosh P. Structure of the InlB leucine-rich repeats, a domain that triggers host cell invasion by the bacterial pathogen *L. monocytogenes*. Mol Cell 1999; 4:1063–1072.

104. Mahony DE, Moore TI. Stable L-forms of *Clostridium perfringens* and their growth on glass surfaces. Can J Microbiol 1976; 22:953–959.

105. Shimizu T, et al. Complete genome sequence of *Clostridium perfringens*, an anaerobic flesh-eater. Proc Natl Acad Sci USA 2002; 99:996–1001.

106. Greenwood B. The epidemiology of pneumococcal infection in children in the developing world. Philos Trans R Soc Lond B Biol Sci 1999; 354:777–785.

107. Klein DL, Eskola J. Development and testing of *Streptococcus pneumoniae* conjugate vaccines. Clin Microbiol Infect 1999; 5(Suppl 4):S17–S28.

108. Musher DM. Infections caused by *Streptococcus pneumoniae*: clinical spectrum, pathogenesis, immunity, and treatment. Clin Infect Dis 1992; 14:801–807.

109. Smoot JC, Barbian KD, Van Gompel JJ, et al. Genome sequence and comparative microarray analysis of serotype M18 group A *Streptococcus* strains associated with acute rheumatic fever outbreaks. Proc Natl Acad Sci USA 2002; 99:4668–4673.

110. Tettelin H, Masignani V, Cieslewicz MJ, et al. Complete genome sequence and comparative genomic analysis of an emerging human pathogen, serotype V *Streptococcus agalactiae*. Proc Natl Acad Sci USA 2002; 99:12,391–12,396.

111. Ajdic D, McShan WM, McLaughlin RE, et al. Genome sequence of *Streptococcus mutans* UA159, a cariogenic dental pathogen. Proc Natl Acad Sci USA 2002; 99:14,434–14,439.

112. Hoskins J, Alborn WE Jr, Arnold J, et al. Genome of the bacterium *Streptococcus pneumoniae* strain R6. J Bacteriol 2001; 183:5709–5717.

113. Cunningham MW. Pathogenesis of group A streptococcal infections. Clin Microbiol Rev 2000; 13:470–511.
114. Schuchat A, Deaver-Robinson K, Plikaytis BD, et al. Multistate case-control study of maternal risk factors for neonatal group B streptococcal disease. The Active Surveillance Study Group. Pediatr Infect Dis J 1994; 13:623–629.
115. Schuchat A, Wenger JD. Epidemiology of group B streptococcal disease. Risk factors, prevention strategies, and vaccine development. Epidemiol Rev 1994; 16:374–402.
116. Herzberg MC. In: Stevens DL, Kaplan EL (eds). Streptococcal Infections. New York: Oxford University Press, 2000, pp. 333–370.
117. Vickerman MM, Mathu NM, Minick PE, Edwards CA. Initial characterization of the *Streptococcus gordonii* htpX gene. Oral Microbiol Immunol 2002; 17:22–31.
118. Kleerebezem M, Boels IC, Groot MN, et al. Metabolic engineering of *Lactococcus lactis*: the impact of genomics and metabolic modelling. J Biotechnol 2002; 98:199–213.
119. Bolotin A, Wincker P, Mauger S, et al. The complete genome sequence of the lactic acid bacterium *Lactococcus lactis* ssp lactis IL1403. Genome Res 2001; 11:731–753.
120. Cocaign-Bousquet M, Even S, Lindley ND, Loubiere P. Anaerobic sugar catabolism in *Lactococcus lactis*: genetic regulation and enzyme control over pathway flux. Appl Microbiol Biotechnol 2002; 60:24–32.
121. Tanaka K, Komiyama A, Sonomoto K, et al. Two different pathways for D-xylose metabolism and the effect of xylose concentration on the yield coefficient of L-lactate in mixed-acid fermentation by the lactic acid bacterium *Lactococcus lactis* IO-1. Appl Microbiol Biotechnol 2002; 60:160–167.

19

Genomics of Actinobacteria, the High G+C Gram-Positive Bacteria

Stephen D. Bentley, Roland Brosch, Stephen V. Gordon, David A. Hopwood, and Stewart T. Cole

INTRODUCTION TO ACTINOBACTERIA

According to Woese and coworkers, all life forms can be classified into one of three domains, the Archaea, Bacteria, or Eucarya *(1)*. The Bacteria fall into 15 classes, one of which is the actinobacteria, the diverse group of Gram-positives with a typically high proportion of guanine and cytosine nucleotides, commonly referred to as the high G+C Gram-positive class. Organisms have been assigned to this class on the basis of their chemotaxonomy (the chemical composition of their peptidoglycans, lipids, etc.), their deoxyribonucleic acid (DNA) base composition (>50% G+C), and their 16S ribosomal ribonucleic acid (rRNA) sequence similarities. Examination by Stackebrandt and colleagues using 16S rRNA sequence comparisons led to this class being termed the *Actinobacteria (2)*. Within the *Actinobacteria* is the order *Actinomycetales*, which is divided into 10 suborders; these in turn contain various families; some of these family members are the subject of this chapter (Fig. 1).

There is enormous biological diversity in the *Actinobacteria* because some members, like the *Streptomyces* species, have complex life cycles and developmental processes that lead to spore formation. Other species, such as certain pathogenic mycobacteria, enter a persistent nonreplicating state after the initial growth phase. A variety of colonial morphologies and pigments are encountered among the *Actinobacteria*, which have lifestyles that vary from saprophytic to obligately pathogenic. This biodiversity is also reflected in the findings of genomic analyses of the leading Actinobacteria studied thus far because of their importance for human and veterinary medicine, biotechnology, and ecology.

Table 1 presents selected features of the seven completely sequenced actinobacterial genomes. Genome sizes range from 2.5 to 8.7 Mb, and this is reflected by variations in gene content from 1600 to over 7800. The number of rRNA operons varies from one in the slow-growing mycobacteria to six in the more rapidly growing corynebacteria and streptomycetes. The G+C compositions also show a broad distribution, from 53.5% in the corynebacteria to 72.1% for the streptomycetes. *Mycobacterium leprae*, which has undergone extensive gene decay *(3)*, has the lowest gene density, and at one gene every 2 kb, this is half that of the other members (Table 1).

From: *Microbial Genomes*
Edited by: C. M. Fraser, T. D. Read, and K. E. Nelson © Humana Press Inc., Totowa, NJ

Families   **Suborders**   **Orders**

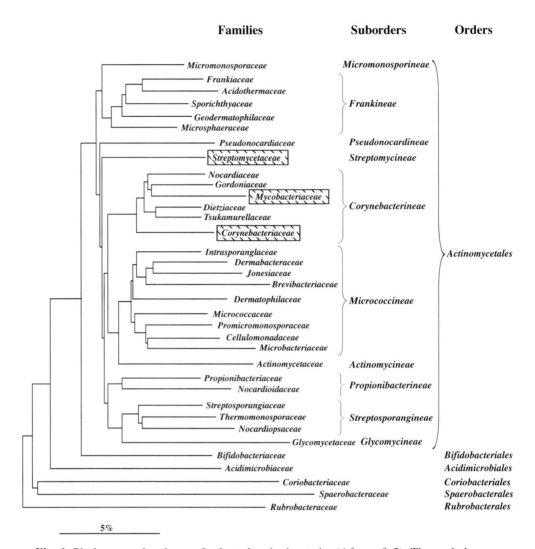

Fig. 1. Phylogeny relatedness of selected actinobacteria. (After ref. *2*). The scale bar represents 5 nucleotide substitutions per 100 nucleotides of the 16S rDNA/rRNA sequence. Families containing members subjected to complete genome sequencing at time of writing are boxed.

The genome sequences of two corynebacteria have recently been completed: *Corynebacterium diphtheriae* and *Corynebacterium glutamicum*. Toxigenic strains of *C. diphtheriae* cause an acute, communicable respiratory disease, and it is anticipated that the genome sequence of this organism will stimulate research into prevention and cure. *Corynebacterium glutamicum*, on the other hand, is an industrial microbe widely used for the fermentative production of amino acids. Here, the emphasis is on strain improvement, with the genome sequence enabling directed construction of high producers. Publications on the analysis of these two corynebacterial genomes are expected. Accordingly, the remainder of this chapter focuses on the published actinobacterial genomes.

Table 1
Comparison of Genome Features

Feature	C. diphtheriae NCTC13139	C. glutamicum ATCC13032	M. leprae TN	M. bovis AF2122/97	M. tuberculosis H37Rv	M. tuberculosis CDC1551	S. coelicolor A3(2)
Genome size (bp)	2,488,635	3,309,401	3,268,203	4,345,492	4,411,532	4,403,836	8,667,507
G+C percentage	53.50	54.72	57.79	65.63	65.61	65.60	72.12
Protein coding (%)	87.9	87.3	49.5	90.8	90.8	92.7	88.9
Protein coding genes	2320	3099	1605	3953	3994	4250	7825
Pseudogenes	45	NA	1116	23	6[a]	NA	55
Gene density (bp/gene)	1073	1067	2037	1108	1104	1036	1107
Average gene length (bp)	962	933	1011	1003	1004	960	991
Average unknown gene length (bp)	NA	NA	338	653	584	NA	NA
rRNA operons	5	6	1	1	1	1	6
tRNA	54	60	45	45	45	45	63
Other stable RNA	NA	2	2	2	2	2	3

NA, not available.
[a]Excluding IS elements.

GENOME SEQUENCES
OF *MYCOBACTERIUM TUBERCULOSIS* STRAINS

Introduction

The first actinobacterial genome to be sequenced was that of the paradigm strain of the agent of human tuberculosis, *Mycobacterium tuberculosis* H37Rv *(4)*, the principal member of the highly related *M. tuberculosis* complex. Other members of this complex include *Mycobacterium africanum*, the cause of human tuberculosis in parts of Africa; *Mycobacterium bovis*, the bovine tuberculosis agent; *M. bovis* BCG, the live tuberculosis vaccine; *Mycobacterium canettii*, a rarely occurring human pathogen of smooth colonial morphology; and *Mycobacterium microti*, a pathogen of voles and shrews that is generally avirulent in humans. The *M. tuberculosis* complex is tightly knit, and early population genetic analysis revealed very limited diversity at the DNA level *(5)*. From these findings, it was inferred that tuberculosis may have emerged in humans as recently as 15,000 years ago *(6)*.

Genomics of **Mycobacterium tuberculosis**

On first examination of the 4.41-Mb genome sequence of *M. tuberculosis* H37Rv, 3,924 protein-coding genes were predicted by bioinformatic analysis. On subsequent reanalysis, using lower cutoffs and newer algorithms, this number rose to nearly 4,000 *(7)*. The genome sequence of a new clinical isolate, CDC1551 *(8)*, has been determined. This appears to be highly transmissible, causing widespread skin test conversion in a rural area of the United States *(9)*. Its genome sequence is nearly identical (99.94%) to that of *M. tuberculosis* H37Rv, but is slightly smaller at 4,403,836 bp (Table 1). Automated annotation methods predict 4,250 genes for proteins in strain CDC1551, but many of these are short (~30 codons) and may, therefore, be overassigned.

Repetitive DNA and Mobile Genetic Elements

Gene duplication and repetitive DNA elements are major players that affect genome architecture. Over 51% of the genes in *M. tuberculosis* H37Rv have arisen as a result of gene duplication or domain-shuffling events *(10)*, and 3.4% of the genome is composed of insertion sequences (ISs) and prophages (phiRv1, phiRv2). There are 56 loci harboring IS elements belonging to the well-known IS*3*, IS*5*, IS*21*, IS*30*, IS*110*, IS*256*, and IS*L3* families, as well as a new IS family, IS*1535*, with six members, that appears to employ a frameshifting mechanism to produce its transposase *(11)*. A novel repeated sequence, the REP13E12 family, is present in seven copies in the *M. tuberculosis* H37Rv chromosome, and this repeat contains an attachment site for phage phiRv1. Interestingly, in *M. tuberculosis* strains Erdman and CDC1551, phiRv1 has integrated in a different copy of REP13E12 than in *M. tuberculosis* H37Rv *(4,8,11,12)*.

IS*6110*, a member of the IS*3* family, is the most abundant IS element and has played an important role in shaping the genome of tubercle bacilli. There are 16 copies of IS*6110* in *M. tuberculosis* H37Rv, but only 4 copies in strain CDC1551. IS*6110* is a useful epidemiological tool as it transposes frequently, thereby generating restriction fragment length polymorphisms *(13)*. While transposition can lead to gene inactivation through insertion, it has also contributed significantly to gene loss as a result of deletion of genetic material between near-adjacent copies of the element *(14)*. Four such examples, RvD2–RvD5, have been described in *M. tuberculosis* H37Rv, and these have removed more than 9 genes *(15)*. One of these loci, RvD2, appears to be a hot spot for diversity and displays great variability between strains *(16)*.

Genomics and the Biology of Mycobacterium tuberculosis

The information deduced from the genome sequence provided new and valuable insight into the biology of the tubercle bacilli and highlighted the importance of lipid metabolism to its lifestyle as at least 8% of the genome is dedicated to this activity. While *M. tuberculosis* was long known to contain a remarkable array of lipids, glycolipids, lipoglycans, and polyketides in its "waxy coat" *(17)* and the genome sequence revealed many of the genes required for their production, it was surprising to note the presence of numerous genes and proteins that could confer lipolytic functions. Estimates of the concentrations of potential substrates available to a pathogen in host tissues suggest that lipids and sterols are more abundant than carbohydrates. While *M. tuberculosis* has the prototype β-oxidation cycle required for lipid catabolism, catalyzed by the multifunctional fatty acid degradation (Fad)A/FadB proteins, it also appears to have approx 100 enzymes potentially involved in alternative lipid oxidation pathways in which exogenous lipids could be metabolized following the degradation of host cell membranes *(4)*. The acetyl–coenzyme A thus derived could then be used for the synthesis of mycobacterial cell wall components or fed into the Krebs cycle, glyoxylate shunt, or other central metabolic pathways.

The genome sequence also revealed numerous metabolic facets of *M. tuberculosis* H37Rv that were unknown or obscure. For instance, although the finding of complete sets of genes required for anabolic functions such as amino acid or vitamin biosynthesis is fully consistent with the ability to culture the tubercle bacillus in defined medium, this contrasts with the situation in several other intracellular pathogens in which these genes tend to be absent, or streamlined, because the corresponding metabolic products are scavenged from the host. Although this finding is compatible with the hypothesis that *M. tuberculosis* has only recently become a human pathogen *(6)*, it most likely indicates that the availability of these metabolites within the phagosomal vacuole is limiting, thereby imposing a selective pressure for maintaining the genes. A similar trend is seen with the leprosy bacillus, discussed in a separate section *(3)*.

In addition to lipolysis, energy can also be generated by the catabolism of a range of carbohydrates, alcohols, ketones, and hydrocarbons. The genome predicts the existence of the glycolytic and pentosephosphate pathways, as well as more than 200 oxidoreductases, oxygenases, and dehydrogenases that might allow the metabolism of other carbon sources. Especially noteworthy is a 20-member family of cytochrome P450–containing monooxygenases that may catalyze the introduction of oxygen groups into organic molecules as part of their synthesis or degradation *(18,19)*. Such enzymes are commonly found in soil microorganisms, in which they degrade organic matter; their presence in *M. tuberculosis* indicates that, like most other mycobacteria, the tubercle bacillus may have occupied this niche before becoming an obligate pathogen. Furthermore, like many fungi, *M. tuberculosis* has a P450 enzyme system that effects steps in sterol biosynthesis *(20,21)*.

Adenosine triphosphate (ATP) is most likely to be produced by oxidative phosphory-lation under aerobic growth conditions in which electron donors such as reduced nico-tinamide adenine dinucleotide (NADH) are oxidized by NADH oxidase and the electrons transferred to oxygen via a chain involving a quinone cytochrome-*b* reductase complex and cytochrome-*c* oxidase. However, analysis of the genome suggests that *M. tuberculo-sis* can also respire anaerobically using NADH, or similar donors, if alternative electron acceptors are available. These include nitrate, nitrite, and fumarate *(4)*, and the important role of nitrate reductase in mycobacterial virulence has been delineated in a mutational study *(22)*. It is probable that such enzyme systems contribute significantly to growth during infection as there is low availability of oxygen in abscesses and granulomas.

One of the major findings of the *M. tuberculosis* genome project was the identification of large gene families that were either unknown previously or poorly understood. Fore-most among these were the novel PE (ProGlu) and PPE (ProProGlu) families com-prised of at least 100 and 67 members, respectively *(4,23)*. Each family has a conserved N-terminal domain of approx 110 and 180 amino acid residues, with the characteristic motifs PE or PPE at positions 8–9 or 8–10, respectively. The N-terminal domain is gen-erally followed by a C-terminal extension that is often of highly repetitive sequence.

Of particular interest are the PE proteins belonging to the PGRS (polymorphic G+C-rich sequence) subclass as nearly half of their amino acid content is composed of gly-cine, and this occurs in tandem repetitions of the motif AsnGlyGlyAlaGlyGlyAla, or variants thereof, and the PPE proteins belonging to the MPTR (major polymorphic tandem repeat) subclass, which contain multiple repetitions of the motif AsnXGlyXGly AsnXGly. There has been recent progress in the characterization of these proteins, and some have been localized in the cell envelope, where they are surface exposed *(24–27)*. Their genes represent a major source of diversity both within and between the species, and as a direct consequence, the corresponding proteins vary extensively because of the expansion or contraction of the repetitive units. There is some evidence that these pro-teins could be variable surface antigens *(24,27,28)*, but their precise biological role remains obscure. Several of them appear to be involved in pathogenesis *(29,30)*; others act as adhesions that influence phagocytosis. The identification of the hitherto unknown PE and PPE proteins is important testimony to the power of genomics as a tool for discovery.

GENOMICS OF *MYCOBACTERIUM BOVIS*

Introduction

Mycobacterium bovis shares over 99.9% identity at the DNA level with the other members of the *M. tuberculosis* complex. However, it can be differentiated on the basis of host range, physiological characteristics, and virulence. For example, *M. tuberculosis* rarely causes disease in domestic or wild animals, but *M. bovis* is a consummate patho-gen of diverse animal species. Primary isolates of *M. bovis* show a requirement for pyru-vate in media in which glycerol is the sole carbon source; *M. tuberculosis* strains show no such requirement. Cell wall lipid profiles between the strains also show variation, with *M. bovis* containing phenolic glycolipids in its cell wall that are absent from *M. tuberculosis*. Clearly, the reasons for these phenotypic differences are encoded in the genome. Through knowledge of the genome, and by focusing on regions that show vari-

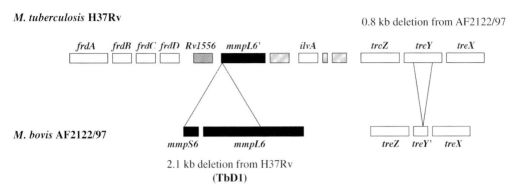

Fig. 2. Examples of deletions in the genomes of tubercle bacilli. *Mycobacterium tuberculosis* has lost most of the *mmpS6* gene and part of *mmpL6* as the result of a deletion, whereas these genes are intact in *M. bovis (42)*. In contrast, the *treY* gene of *M. bovis* has incurred a deletion relative to its ortholog in *M. tuberculosis*.

ation between the species, a genetic description of the biological properties of *M. bovis* should be derived.

Comparison of Mycobacterium bovis *and* Mycobacterium tuberculosis

Prior to the availability of the *M. bovis* genome sequence, comparative genomics of the *M. tuberculosis* complex was performed using hybridization-based methods with micro- and macroarrays. These experiments revealed several regions of difference (RD), because of deletions (Fig. 2), from the genome of *M. bovis*, ranging in size from about 1 to 12.7 kb *(31,32)*. The deletions affect a range of metabolic functions and putative virulence factors, although the role they played in the evolution of the bacterium is unclear. Although they may represent host-adaptive mutations, it is also possible that they represent the fixation of deleterious mutations or the removal of redundancy.

Loss of the RD5 locus removed the genes for three phospholipase C enzymes from the genome, a known virulence factor in *Listeria* and *Clostridium* species *(33)*. However, a fourth phospholipase gene, *plcD*, is intact in *M. bovis* and may compensate for the loss of the other genes. The RD7 locus encompasses one of the *mce* operons, originally described by Arruda and colleagues as coding for a putative mycobacterial invasin *(34)*. The genome sequence revealed that there are in fact four *mce* operons in *M. tuberculosis*, encoding a family of 24 proteins *(4)*. It is therefore possible that loss of one *mce* operon may be compensated by the remaining loci.

At a gross level, it is clear that gene deletion has been a major force in shaping the *M. bovis* genome. However, single-nucleotide polymorphisms (SNPs) could prove to be equally powerful in driving the biology of the organism. SNPs have previously been shown to be responsible for the inherent pyrazinamide resistance of *M. bovis* through mutation of the *pncA* gene *(35)*. A single base change in the principal sigma factor was also sufficient to attenuate *M. bovis (36)*. In light of this last work, it is striking to note that the *M. bovis sigM* homolog shows a frameshift relative to the *M. tuberculosis* gene. It is probable that this alteration affects the transcription of a gene regulon under defined conditions and may shed light on host adaptive mutations.

Antigenic Variation

An obvious difference between *M. bovis* and other tubercle bacilli is the elevated expression of two serodominant antigens, Mpb70 and Mpb83, in the bovine bacillus, in which Mpb70 can account for 10% of culture filtrate proteins *(37)*. Modification of the antigen repertoire presented by the bacterium may be a strategy for immune modulation, or it is possible that these proteins play a direct role in virulence and pathogenesis. A further group of antigens affected by deletions from *M. bovis* is the ESAT-6 family. The ESAT-6 protein was originally described as a potent T-cell antigen secreted by *M. tuberculosis (38)*. *In silico* analysis of the *M. tuberculosis* proteome revealed that ESAT-6 was one of a family of 22 proteins that contains other T-cell antigens, such as CFP-10 and CFP-7 *(10)*. The demonstration of an interaction between ESAT-6 and CFP-10 suggests that other members of the family may also act in pairs, possibly in a mix-and-match arrangement *(39)*. However, 4 members of this family, encoded by the RD5 and RD8 loci, are missing from *M. bovis*. The effects of the loss of these proteins are difficult to predict, although their absence could affect the function of other members if they act in combination.

Cell Wall Variation

Cell wall components of pathogenic bacteria are known to show variation in protein sequences, reflecting selective pressures on these structures *(40,41)*. Alteration of surface-exposed proteins has an impact on antigenic variation, ligand–receptor interactions, and host–bacteria communication. Therefore, among the variable genes identified across the *M. tuberculosis* complex would be those encoding membrane, secreted, or transport proteins.

Perhaps one of the most striking examples of this is sequence variation in the PE-PGRS and PPE protein families. Although initially of unknown function, there is now a considerable body of evidence to suggest that at least some of these proteins are surface exposed and play a role in immune modulation *(24,26,27)*. Comparison of the *M. bovis* and *M. tuberculosis* genomes reveals the genes encoding these proteins are by far the most variable loci between the species. Two examples of this are shown in Fig. 3. The PPE-MPTR Rv1917c protein shows extensive variation between the bacilli, with the *M. bovis* homolog containing three discrete insertions (Fig. 3A). In fact, the Rv1917c locus was originally described as a variable region upstream from *katG* and has been shown to have application in minisatellite typing. Similarly, the *M. bovis* homolog of the PE-PGRS protein Rv3508 shows blocks of unique sequence relative to *M. tuberculosis* (Fig. 3B). This indicates that these proteins can support extensive sequence polymorphism without loss of function, providing an ideal substrate for action of selective pressures.

Analysis of the *M. bovis* sequence revealed that only one region was present in *M. bovis* but absent from the majority of *M. tuberculosis* strains *(42)*. This locus, subsequently termed TbD1, contains the gene *mmpS6* and the 5' region of *mmpL6* (Fig. 2). The Mmp proteins are a family of membrane-spanning proteins, with similarity to RND transporters *(43)*, which have been shown to be involved in the export of cell wall lipids. Hence, the TbD1 locus in *M. bovis* may be involved in the transport of lipids, such as phenolic glycolipid, that are lacking in *M. tuberculosis* strains. Conversely, inspection has revealed frameshifts in the *M. bovis mmpL1* and *pks6* homologs. Pks6 likely plays a role in syn-

A Rv1917c

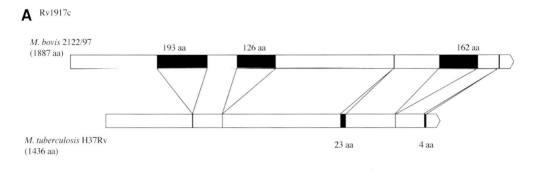

B Rv3508

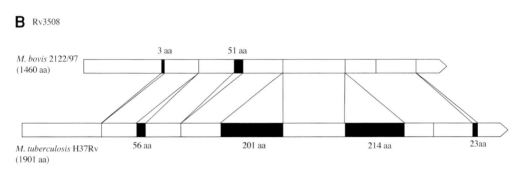

Fig. 3. Variation in the PPE genes. Orthologous genes in *M. tuberculosis* and *M. bovis* are shown schematically, and the positions of insertions and their sizes are indicated.

thesis of complex lipids that would be exported via the MmpSL1 transporter. Hence, loss of these functions would again be predicted to affect the cell wall of the bacillus.

Metabolic Insights

Among the systems showing deleterious mutations in *M. bovis* is the metabolism of inorganic sulfate. The atsA encodes arylsulfatase, which catalyzes the hydrolysis of sulfate esters to release inorganic sulfate. However, *atsA* is truncated in *M. bovis*. This builds on work that showed that inactivation of CysTWA, the sole inorganic sulfate transporter in *M. bovis* BCG, did not affect the organism's ability to persist in vivo *(44)*. It also reflects the situation in *M. leprae*, in which *cysTWA* are pseudogenes *(3)*. Hence, acquisition of inorganic sulfate does not appear to be required for *in vivo* survival. Conversely, it shows that, for *M. bovis*, there is a dependency on organic sulfur sources in vivo, perhaps transported through amino acid permeases.

An 808-bp deletion has been identified that truncates the *treY* gene, encoding a malto-oligosyltrehalose (MOT) synthase (Fig. 2). This enzyme catalyzes the conversion of a terminal α-(1,4) glucose linkage to an α-(1,1) linkage. The terminal MOT disaccharide unit is then cleaved by the action of the MOT trehalohydratase (encoded by *treZ*), releasing free trehalose. It is not clear which phenotypic effects the *treY* deletion would have on *M. bovis* as two other pathways for the synthesis of trehalose are intact, and their expression may be coordinated. Furthermore, it is intriguing to note that this deletion is proximal to the TbD1 deletion (Fig. 2). Polymerase chain reaction (PCR) analysis across diverse *M. bovis* strains revealed that, although the *treY* gene was disrupted in the majority of strains, it was intact in *M. bovis* BCG. This suggests that the deletion may be useful as a marker for deep phylogeny.

GENOME SEQUENCE OF *MYCOBACTERIUM LEPRAE*

Introduction

In some respects, *M. leprae* may be considered an outlying member of the genus *Mycobacterium*, especially in terms of its G+C content, which, at 57%, is considerably lower than that of all other mycobacteria, and its genome size, which is at least 1.1 Mb smaller than those of all other mycobacterial species examined (Table 1). The genomic approach was without doubt the most powerful means of characterizing this otherwise untractable bacterium. The leprosy bacillus cannot be cultivated in laboratory medium and displays an exceptionally long generation time of about 14 days in infected mice. Analysis of the genome sequence provided clear explanations for some of these properties because a large number of mutationally inactivated genes or pseudogenes were uncovered that appear devoid of function *(3)*.

Genome Sequence of **Mycobacterium leprae**

On examination of the 3.27-Mb genome sequence of an armadillo-derived Indian isolate of the leprosy bacillus, 1605 genes were identified that encode proteins, as were 50 genes that code for stable RNA species (Table 1). On comparison with the genome sequence of *M. tuberculosis*, an extreme case of reductive evolution was revealed because less than half of the genome (49.5%) contains functional genes, although pseudogenes were highly abundant, accounting for at least 27% of the sequence. The pseudogenes appear to have lost their function as a result of one or more mutations, including in-frame stop codons, frameshifts, deletions, and less commonly, insertions. Over 1116 pseudogenes were identified with functional counterparts in the tubercle bacilli, and the true number is certainly greater than this as some genes may have mutated beyond all recognition or may never have had an ortholog in *M. tuberculosis*. Such pseudogenes may occur in the remaining 23.5% of the genome, which appears to be completely noncoding or "empty."

The distribution of the 1116 recognizable pseudogenes was essentially random throughout the genome, whereas the 1605 potentially active genes tended to occur in clusters, which were often flanked by long stretches of seemingly noncoding DNA. This process of gene decay appears to have been accompanied by a reduction in G+C content. Interestingly, the G+C content of the putative functional genes at 60.1% is higher than that of the identifiable pseudogenes (56.5%), which in turn is greater than that of the empty region of the genome (54.5%), which likely underwent the most extensive decay. These observations suggest that the relatively high G+C content of *M. leprae*, and by extension the other mycobacteria, is imposed by the codon preference of the active genes; random mutation within the noncoding regions leads to a more neutral G+C content that is closer to that of the host. Deamination of cytosine residues in the DNA to thymidine is a possible mechanism to account for this trend and would account for the leprosy bacillus having the lowest G+C content of all mycobacteria. It is noteworthy that the genomes of microorganisms that have undergone reductive evolution are generally richer in adenine (A) + thymine (T) *(45)*.

Reductive Evolution in **Mycobacterium leprae**

Reductive evolution has been documented in a number of important human pathogens, including the obligate intracellular parasites *Rickettsia* and *Chlamydia* spp (see Chapter 17) *(46)*. It is thought that genes become inactivated once their functions are no longer required in highly specialized niches that may correspond to evolutionary dead ends. In some endosymbionts found in aphids, like the *Buchnera* spp related to the enteric bacteria, reductive evolution has proceeded so extensively that the genome is believed to have shrunk from approx 4.5 to 0.64 Mb *(47)*. Here, deletion appears to have been the dominant means of genome downsizing because *Buchnera* spp contain few pseudogenes. Until the genome sequence of *M. leprae* became available, the most extensive genome degradation reported in a pathogen was in *Rickettsia prowazekii*, the typhus agent, in which only 76% of the potential coding capacity was used *(48)*. In comparison with *M. leprae*, the level of gene loss detected in *R. prowazekii* was modest, and it is notable that elimination of pseudogenes by deletion lags far behind gene inactivation in both pathogens, in contrast to *Buchnera (47)*.

An elegant hypothesis, known as Muller's ratchet, has been proposed to explain reductive evolution. This involves the stochastic loss of genetic material or gene function and often results in decreased fitness. In part, this is because of the inability of organisms with no sex cycle to acquire DNA and hence to repair genetic lesions through acquisition of new genes or by recombination. A major consequence is scant genetic variability. Clearly, as a consequence of its highly specialized niche, the organism with which *M. leprae* could most likely exchange DNA is the human or another ancestral mammalian host. There is some evidence that horizontal gene transfer may have taken place, and this is illustrated by *proS*, encoding the prolyl–tRNA (transfer RNA) synthetase. Of all the aminoacyl–tRNA synthetase genes in *M. leprae, proS* is the only one with no strict counterpart in *M. tuberculosis*. Surprisingly, the domain structure of ProS is eukaryotic-like because it is more similar to the enzymes of *Drosophila*, humans, and yeast and to *Borrelia burgdorferi*. It has been proposed that horizontal transfer of tRNA synthetase genes occurs frequently, and that the pathogen *B. burgdorferi* may have acquired *proS* from its host *(49)*. Further support for this hypothesis was provided by comparison of the genetic neighborhood as the *M. leprae proS* is both displaced and inverted with respect to the *M. tuberculosis* genome *(3)*, consistent with recent acquisition. The domain structure of the *M. tuberculosis* enzyme (and all other sequenced mycobacteria) is typically prokaryotic; taken together, these findings suggest that *M. leprae* may at one time have had a prokaryotic *proS* gene that was replaced by that of a mammalian host.

Repetitive DNA in **Mycobacterium leprae**

About 2% of the *M. leprae* genome is composed of repetitive DNA that is believed to have mediated genomic rearrangements *(50)*. None of the IS elements, of which there are more than 26 different examples, appears to be functional. There are, however, 4 families of dispersed repeats present in 5 copies or more, RLEP (37 copies), REPLEP (15 copies), LEPREP (8 copies), and LEPRPT (5 copies). None of these repetitive elements contains open reading frames, but they all display some features reminiscent of transposable elements. Notably, LEPREP, which is 2383 bp long, contains a 54-bp palindromic inverted repeat and has a 6-bp inverted repeat (5'-CTAGTG) at its ends.

BLASTX (Basic Alignment Search Tool X) searches revealed extensive sequence similarity to transposases from *Pseudomonas putida* and *Agrobacterium tumefaciens* and to putative group II intron maturase-related proteins from fungi. Although there is no sequence similarity to known transposable elements, RLEP occurs predominantly at the 3' end of genes and, in several cases, within pseudogenes, suggesting that it was once capable of transposition. REPLEP, 881 bp, is bounded in most cases by an 8-bp inverted repeat; LEPRPT is 1254 bp long *(50)*. Importantly, on comparison of the genome sequences of *M. leprae* and *M. tuberculosis*, many of these repetitive sequences were found at sites of discontinuity in gene order, and evidence has been presented that convincingly shows that loss of synteny, inversion, and genome downsizing resulted from recombination between dispersed copies of these repetitive elements *(50)*.

Reductive Evolution and the Biology of Mycobacterium leprae

In both the *M. tuberculosis* complex and *Streptomyces coelicolor*, extensive gene duplication has occurred, leading to the existence of large protein families and much functional redundancy *(4,10,51)*. The level of gene duplication in *M. leprae* was estimated at about 34% *(52)*, and on classification of the proteins into families, the largest functional groups were found to be involved in the metabolism and modification of fatty acids and polyketides, the transport of metabolites, cell envelope synthesis, and gene regulation. The same trend was observed in *M. tuberculosis*, for which 52% of the genes are believed to have arisen as a result of duplication or domain shuffling *(10)*. When the pseudogenes are included in the analysis, it becomes clear that there was once more extensive redundancy in the leprosy bacillus, but that reductive evolution led to the loss of many genes and selective retention of certain functions.

As in the tubercle bacilli, the largest protein family *(52)* contains enzymes involved in polyketide synthesis and fatty acid metabolism; this again underscores the importance of these functions to slow-growing mycobacterial pathogens. However, this enzyme repertoire is much less extensive than in the tubercle bacillus, which has a cell envelope that has greater diversity of lipids, glycolipids, and carbohydrates *(17)*. The second- and third-largest protein families in *M. tuberculosis* H37Rv are those for the 167 PE and PPE proteins; only 12 of these proteins are predicted in *M. leprae*, although about 30 pseudogenes also exist. None of the putative PE or PPE proteins produced by *M. leprae*, contain the multiple C-terminal repetitions that are suspected of involvment in antigenic variation *(4,24,25)*. Retraction of these gene families, which are exceptionally G+C rich, may have also contributed to the smaller genome size and G+C content of *M. leprae*.

Gene decay and genome downsizing have eliminated entire metabolic pathways, together with their regulatory circuits and accessory functions, particularly those involved in catabolism. When the genes that have been lost or inactivated during the reductive evolutionary process are examined closely, clear trends emerge that are consistent with Darwinian evolutionary theory. For instance, comparative genomic analysis indicated that both *M. tuberculosis* and *S. coelicolor* are capable of anaerobic respiration using nitrate as the terminal electron acceptor.

In this reaction, nitrate reductase harnesses electrons initially obtained from formate through the action of formate dehydrogenase-N from the quinone pool to generate a proton-motive force *(4,51)*. Nitrate reductase has three subunits encoded by *narGHI*, whereas formate dehydrogenase-N contains three subunits, and formate is generated from pyruvate via acetyl–coenzyme A *(53)*. Both enzymes use molybdopterin, a complex cofactor containing molybdenum, which is synthesized by at least nine *moe/moa* genes. Molybdate is taken up from the extracellular medium by the ATP-binding cassette (ABC) transporter encoded by *modABC (4)*. *Mycobacterium leprae* has pseudogenes corresponding to both nitrate reductase and formate dehydrogenase-N and for most of the proteins required for molybdate transport and insertion into the pterin ring. Apparently, once the ability to use the formate–nitrate pathway was lost, the genes for the entire system acquired mutations and decayed as none of their functions was required. It is also interesting that the aerobic respiratory chain of *M. leprae* is truncated as only the 3' end of the NADH oxidase operon, *nuoA-N,* remains. Curiously, although there is an intact *nuoA-N* operon in *S. coelicolor*, a truncated copy was also found *(51)*.

In contrast to the catabolic and energy-generating functions, the anabolic pathways of *M. leprae* appear to be essentially intact. This suggests that the inability to culture the leprosy bacillus is not because of the lack of a particular amino acid, vitamin, or other cofactor, but rather because of the choice of carbon and energy source. This might imply that *M. leprae* is limited to growth on very few carbon sources, or even a limited and rather specialized combination, on which it can maintain balanced carbon metabolism under quite selective redox conditions.

COMPARATIVE MYCOBACTERIAL GENOMICS AND EVOLUTIONARY CLUES

Introduction

For many years, genetic studies of the mycobacteria lagged behind those of the streptomycetes because of the lack of adequate tools. However, the last decade has seen the development of various genetic systems based on mycobacterial phages, plasmids, and transposons *(54–58)*, as well as the conception of efficient vectors that allow gene knockouts to be made *(59,60)*. In parallel, the knowledge of the genetic organization of these organisms has been greatly enriched thanks to genomics (Tables 1 and 2). In contrast to the situation in the 1970s and 1980s, when molecular genetics was predominantly restricted to model organisms, the mycobacteria are becoming one of the best-characterized groups genetically, creating a strong background for comparative genomic analyses.

Evolution of the Mycobacterium tuberculosis Complex

As mentioned in the introductory section to discussion of *M. tuberculosis* strains, this complex comprises highly related bacteria, most of which cause tuberculosis in mammals, but differ in their host range and pathogenesis in humans. Whereas *M. tuberculosis*, *M. canettii*, and *M. africanum* are exclusively human pathogens, *M. microti* is primarily a rodent pathogen, and *M. bovis* displays a very broad spectrum, infecting most mammalian species, including humans. The organism that is best known for its avirulence in humans is the BCG vaccine strain, obtained by Calmette and Guérin after 230 serial passages of a virulent *M. bovis* isolate. In spite of its extensive use, the reason BCG is attenuated is still unknown. This remarkably safe live vaccine has never regained virulence, suggesting that irreversible genomic changes, such as deletions, occurred dur-

Table 2
Useful Web Sites for Actinobacterial Genomes

Corynebacterium diphtheria
   http://www.sanger.ac.uk/Projects/C_diphtheriae/
Mycobacterium bovis
   http://www.sanger.ac.uk/Projects/M_bovis/
Mycobacterium leprae
   http://genolist.pasteur.fr/Leproma/
   http://www.sanger.ac.uk/Projects/M_leprae/
Mycobacterium marinum
   http://www.sanger.ac.uk/Projects/M_marinum/
Mycobacterium paratuberculosis
   http://www.cbc.umn.edu/ResearchProjects/AGAC/Mptb/Mptbhome.html
Mycobacterium smegmatis, Mycobacterium avium
   http://www.tigr.org/tdb/mdb/mdbinprogress.html
   http://tigrblast.tigr.org/ufmg/
Mycobacterium tuberculosis
   Strain H37Rv
      http://genolist.pasteur.fr/TubercuList/
      http://www.sanger.ac.uk/Projects/M_tuberculosis/
   Strain CDC1551
      http://www.tigr.org/tigr-scripts/CMR2/
GenomePage3.spl?database=gmt
   Strain 210 (Beijing type)
      http://tigrblast.tigr.org/ufmg/
Mycobacterium ulcerans
   http://www.pasteur.fr/recherche/unites/Lgmb/
Streptomyces coelicolor
   http://www.sanger.ac.uk/Projects/S_coelicolor/
   http://jiio16.jic.bbsrc.ac.uk/S.coelicolor/

ing the attenuation process. The search for the basis of this attenuation has inspired a variety of comparative studies *(31,32,61,62)*. The combined findings showed that there are at least 18 RDs, ranging from 0.3 to 12.7 kb and representing 120 genes, that are present in *M. tuberculosis* H37Rv, but absent from BCG Pasteur. Some of these may account for the phenotypic differences between the vaccine and *M. tuberculosis*.

Specific PCR analysis of the RDs showed that most of the regions absent from BCG were also missing from other strains of *M. bovis*, indicating that some of them reflect the evolutionary divergence of *M. tuberculosis* and *M. bovis*, as well as genomic modifications that were introduced during the attenuation process of BCG. As an example, the absence of RD7–RD10 from *M. microti*, *M. bovis*, and BCG most probably reflects the result of such events. From close inspection of the flanking sequences, it is apparent that deletions occurred, in genes that are still intact in *M. tuberculosis* and *M. canettii*, at exactly the same site in *M. microti*, *M. bovis*, and *M. bovis* BCG. This observation argues strongly against the possibility of the RDs resulting from insertion of genes into *M. tuberculosis (63)*.

Interestingly, *M. africanum* strains lack RD9 as well, so some of these deletions must have occurred in a common ancestor of an evolutionary lineage that now comprises *M. africanum*, *M. microti*, and *M. bovis*, but has diverged from the lineage of *M. tuberculosis* isolates. Assuming that horizontal gene transfer between different lineages of the tubercle bacilli is rare, the observation that certain deletions are conserved among several members of the complex allowed a pathway for the evolution of the members of the *M. tuberculosis* complex to be proposed *(42)*. In this scheme (Fig. 4), *M. bovis* and *M. bovis* BCG are the final members of a separate lineage that has accumulated the most deletions relative to *M. tuberculosis* strains.

This scheme is also in good agreement with the distribution of informative SNPs, such as the one in the *pncA* gene that causes resistance to pyrazinamide in *M. bovis (35)*, and the finding that the *M. bovis* AF2122/97 genome sequence is approx 66 kb smaller than that of *M. tuberculosis* H37Rv (Table 1). This is an important observation because it contradicts the often-proposed hypothesis that the human tubercle bacillus *M. tuberculosis* was derived from *M. bovis* because the bovine bacillus crossed the species barrier at the time of the domestication of cattle *(64,65)*. According to the distribution of RDs and SNPs, *M. tuberculosis* strains appear to be more closely related to the common ancestor of the *M. tuberculosis* complex than do *M. bovis* strains. It seems plausible that a separate linage represented by *M. africanum* (RD9), *M. microti* (RD7, RD8–RD10), and *M. bovis* (RD4, RD5, RD7–RD10, RD12, RD13) evolved from the progenitor of today's *M. tuberculosis* isolates and adapted to new hosts. Although further confirmation by paleopathologists is needed, the finding that the human pathogens *M. canettii*, *M. tuberculosis*, and *M. africanum* carry several genes that have been truncated or deleted in the predominantly animal pathogens *M. microti* and *M. bovis* suggests that the common ancestor of the tubercle bacilli could have been a human pathogen already.

As shown in Fig. 2, comparative genomics also identified the TbD1 region, which was specifically absent from almost all *M. tuberculosis* isolates, but present in all other members of the complex. A PCR assay for this deletion can now be used as an efficient means for the rapid and unambiguous identification of *M. tuberculosis* strains.

Evolution of the Mycobacteria in General

To obtain a more global evolutionary picture, comparative genomics involving other mycobacterial species with genome sequences that are presently being determined (Table 2) will certainly be valuable for understanding the transition of harmless soil bacteria into obligate intracellular pathogens. Phylogenetically, the closest relatives of the *M. tuberculosis* complex are the environmental species *Mycobacterium marinum* and *Mycobacterium ulcerans*. Whereas *M. marinum* is principally an ectothermic pathogen that rarely causes human infections, *M. ulcerans* is responsible for a debilitating cutaneous disease that is reaching epidemic proportions in parts of West Africa. The *Mycobacterium avium* complex, another group of slow-growing mycobacterial subspecies, includes both veterinary and opportunistic human pathogens. Genome-sequencing projects of *M. avium* and *Mycobacterium paratuberculosis* are under way (Table 2). In addition, the genome sequence of *Mycobacterium smegmatis* mc^{2}155, a fast-growing species, is being determined (Table 2). One of the most astonishing preliminary findings concerns the size of the *M. smegmatis* genome, which, at about 7 Mb, is much larger than those

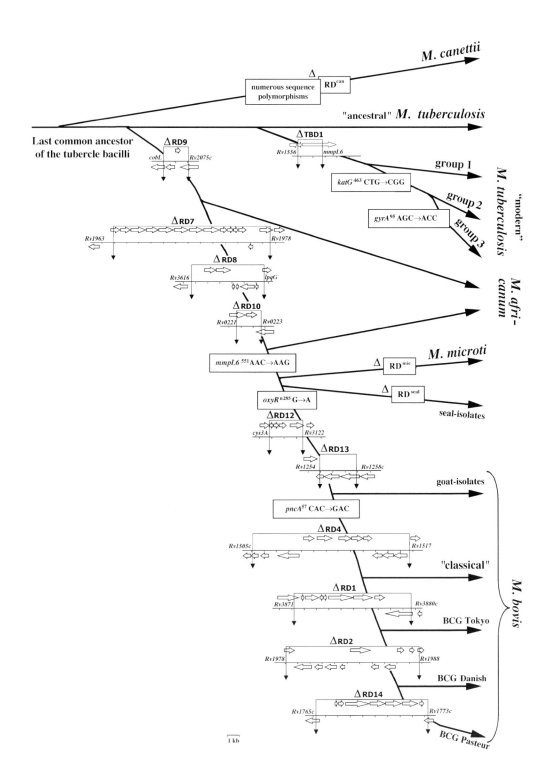

from most slow-growing mycobacteria (~4.4 Mb), approaching that of the distantly related actinomycete *S. coelicolor (51)*, which is described in detail in a separate section.

The larger genome of *M. smegmatis*, together with 16S rRNA-based phylogeny, suggests that the branch of slow-growing mycobacteria represents the most recently evolved part of the genus *(66)*. With the exception of a few genes, like *mgtC (67)*, found in slow-growing mycobacteria but not in fast growers, it seems that the loss of genetic material, rather than its acquisition by horizontal transfer, has been an important factor in the slow growers becoming pathogens. As described above, this trend continues, in extreme form, in the genome of *M. leprae (3)*.

Functional Aspects Uncovered by Comparative Genomics

Apart from completely revising the evolutionary history of the tubercle bacilli *(42)*, comparative genomics has the potential to provide new insight into the original attenuation process of the BCG vaccine strain and other virulence-related aspects. One of the most interesting regions in this respect is the RD1 locus (Fig. 4), which is absent from all BCG strains *(31,61)*, but present in all virulent *M. tuberculosis* and *M. bovis* strains tested so far *(42)*. Among other functions, RD1 encodes the potent T-cell antigens CFP-10 and ESAT-6, and the emerging findings of functional genomics suggest that loss of this locus contributed to the attenuation of BCG.

Comparative genomics has not only uncovered deletions, but also discovered two large tandem duplications of 29 and 36 kb (DU1, DU2) in the genome of BCG Pasteur *(68)*. These seem to have arisen independently as their presence or their size varies among the different BCG substrains. While DU1 appears to be restricted to BCG Pasteur, DU2 has been detected in all BCG substrains tested so far (our unpublished observations). Interestingly, DU1 comprises the *oriC* locus, indicating that BCG Pasteur is diploid for *oriC* and several genes involved in replication *(68)*. For DU2, it is known that the tandem duplication resulted in diploidy for 30 genes, including *sigH*, a sigma factor implicated in the heat shock response *(69)*. Gene duplications are a common evolutionary response in bacteria exposed to different selection pressures in the laboratory and presumably in nature *(70)* because they provide a means for increasing gene dosage, for generating novel functions from potential gene fusion events at duplication end points, and represent a source of redundant DNA for divergence. Much gene duplication and subsequent divergence occurred in the leprosy and tubercle bacilli, and about a 250-kb segment of the *M. smegmatis* genome also appears to have been duplicated *(71)*. It will be interesting to learn how common duplications are in a broader spectrum of actinobacteria.

GENOME SEQUENCE OF *STREPTOMYCES COELICOLOR*

Introduction to Streptomycetes

The successful adaptation to their niche is evidenced by the fact that streptomycetes are almost ubiquitous in soil. Nutritionally, physically, and biologically, soil is a complex

---

Fig. 4. (*Opposite page*) Evolutionary scheme for the tubercle bacilli based on conserved deletions and selected single-nucleotide polymorphisms identified in certain members of the *M. tuberculosis* complex. The deleted regions are delimited by arrows. Open reading frames are represented as pointed boxes showing the direction of transcription and their position in the *M. tuberculosis* H37Rv genome.

and variable environment. The streptomycetes colonize this environment by growing as branching hyphae to form mycelium, a feature that led to their early grouping with fungi. In response to certain changes, probably slowing of the growth rate because of exhaustion of nutrients, the mycelium will differentiate to form aerial hyphae. These protrude from the soil particles on which the organisms grow and are topped with curled chains of spores that serve as dispersal units. *Streptomyces coelicolor* is genetically the best-studied member of the genus and as such was an ideal candidate for genome sequencing *(72)*.

A Large Chromosome with Many Genes

The major distinguishing features of the *S. coelicolor* genome are its size and structure. Like most streptomycetes, it has a large linear chromosome, and at 8,667,507 bp, it is the largest complete bacterial genome sequence currently available *(51)*. With a typical coding density for bacteria of 88.9%, the genome contains 7825 genes (including only 55 pseudogenes; Table 1). To put this figure in perspective, it is about twice as many as *M. tuberculosis*, 3000 more than the lower eukaryote *Schizosaccharomyces pombe (73)* and about one-quarter of the number predicted in humans *(74,75)*. This plethora of genes provokes two immediate questions: Why does *S. coelicolor* have so many genes, and what do those genes encode? The answer to both questions can be explained by considering the complex environment and lifestyle of the organism.

With the essential and housekeeping functions easily accommodated, *S. coelicolor* can dedicate much of the remainder of its proteome to interactions with the world beyond its cell wall. Soil presents a multitude of possible nutrient sources in the form of biopolymers derived from the decay of a wide variety of organisms (plants, animals, insects, fungi, bacteria). To exploit these nutrient sources, *S. coelicolor* secretes a predicted 819 proteins (10.5% of the proteome). Many of the secreted proteins are hydrolases and include proteases, chitinases, cellulases, amylases, and pectinases. The breakdown products of such hydrolytic reactions are imported into the cell cytoplasm along with metal and other ions, amino acids, and peptides. Accordingly, *S. coelicolor* also dedicates a major proportion of its genome to transport, both in and out of the cell, with 614 proteins (7.8%) predicted to have transport function.

The most famous exported products of *S. coelicolor* are its antibiotics. These compounds are synthesized via so-called secondary metabolic pathways. The genome contains 22 clusters of genes likely to be involved in the production of secondary metabolites that belong to diverse chemical classes and have a range of predicted roles *(51)*. Only 4 of these clusters were known prior to the genome project, 3 encoding enzymes for antibiotic production *(76)* and 1 for production of a spore pigment *(77)*. The discovery of an additional 18 clusters brings the total number of genes to approximately 220 (the precise boundaries of some of the clusters will need to be defined experimentally), demonstrating a clear dedication of coding capacity to secondary metabolism. Such an investment would be expected to confer a considerable competitive advantage. The cluster for production of the calcium-dependent antibiotic includes 41 genes *(78)*. The return for this investment is possession of a compound that presumably plays a role in suppressing competing bacteria. Predicted functions for the novel clusters include resistance to desiccation, adaptation to low temperature, and sequestration of iron from the environment.

Like *M. tuberculosis*, *S. coelicolor* also has a large complement of cytochrome P450s. There were 18 such enzymes predicted from the genome sequence and strikingly con-

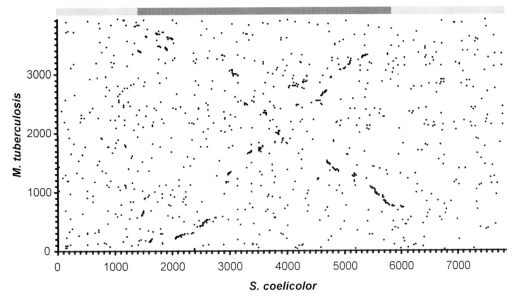

Fig. 5. Comparison of chromosome structure for *S. coelicolor* vs *M. tuberculosis*. Axes represent the proteins encoded in the order in which they occur on the chromosomes. For each genome, DnaA is centrally located. Dots represent a reciprocal best match (by FASTA [fast-all] comparison) between protein sets. The bar above the plot indicates the core (solid, SCO1440-5869) and arm (hatched) regions of the *S. coelicolor* chromosome.

firmed by cloning and expression in *E. coli (79)*. This provides another potential adaptation to the utilization or elimination of diverse organic molecules in the soil.

To control this huge arsenal of genes, the *S. coelicolor* genome has an unprecedented proportion of regulatory proteins (12.3%), extending the observation that, for bacteria in general, the proportion of regulatory genes increases with genome size *(80)*. The large numbers of extra cytoplasmic function sigma factors *(81)* and two-component sensor/regulator pairs–53- indicate a particular emphasis on the detection of, and response to, extracellular stimuli. The genome also encodes members of many other known regulator groups, as well as a family of 25 *S. coelicolor*–specific DNA-binding proteins that may constitute a novel type of regulator.

Biphasic Chromosome Structure

The linear chromosome of *S. coelicolor* can be divided into three main zones with a central core region flanked by two arms. The core region appears to be inherited from a common ancestor of actinomycetes; the arm regions are derived from another source. Figure 5 shows a comparison with the *M. tuberculosis* genome for which each dot represents a reciprocal best match between proteins and indicates the relative position of the corresponding genes. The broken X pattern *(82)* reflects the synteny, conservation of gene sequence and gene order, between the two chromosomes.

The syntenic regions are spread throughout the *M. tuberculosis* chromosome, but are restricted to the core of the *S. coelicolor* chromosome. The major regions of synteny contain genes concerned with primary cellular functions such as cell division, nucleotide and amino acid biosynthesis, central metabolism, and ribosome production. This observation can also be related to the biphasic structure of the *S. coelicolor* chromosome. Nearly all the genes likely to have an essential function are located in the core; those with nonessential contingency functions, such as production of secondary metabolites or the "specialist" hydrolytic exoenzymes, are usually found in the arms. It is known that, under laboratory conditions, streptomycetes can tolerate deletion of 1 Mb or more of DNA from the ends of their chromosomes *(83)*. Perhaps the core region of the chromosome alone is sufficient for generation of streptomycetes under favourable conditions, and the arms are simply a reservoir of genes necessary for success under specialist conditions.

The biphasic structure of the *S. coelicolor* chromosome may provide clues how the large linear chromosomes of streptomycetes have evolved. Assuming that the actinomycete common ancestral chromosome was circular *(84)*, at some point the ability to exist in the linear form was acquired. This may have been via the integration of, or recombination with, a linear plasmid encoding the machinery necessary to produce the stable linear DNA ends or bacterial "telomeres" that are seen in streptomycetes *(85)*. The subtelomeric regions of the *S. coelicolor* chromosome have a higher proportion of transposases and pseudogenes, suggesting a local increased tolerance to insertion events. Perhaps the newly linear ancestor had a similar tolerance, providing a route for end-specific accumulation of DNA.

Clearly, acquired genes that gave competitive advantage would be selectively maintained, making them intolerant to disruption and causing an increase in genome size. The arm regions in *S. coelicolor* appear to have expanded to the point at which they represent approximately half the genome. Having this large reservoir of "contingency" genes has allowed some seemingly extravagant inclusions in the *S. coelicolor* chromosome arms. One example is the presence of two separate (actually on opposite arms) operons that seem to have the potential to code for the synthesis of gas vesicles *(51)*. Another is the scattering of 13 conservons, divergent copies of a 4-gene operon of unknown function for which the only other known occurrence is a single copy in *M. tuberculosis* (Rv3362c–Rv3365c). The existence of larger circular genomes in *Myxobacteria (84)* implies that linearity is not a prerequisite for large chromosome size, but it does seem that linearity in *Streptomyces* chromosomes provides an effective route for genome expansion.

COMPARATIVE ACTINOBACTERIAL GENOMICS

As indicated in the previous section, comparative genomics is uncovering conserved genes and operons that so far appear to be confined to actinobacteria. Some of these may confer specific properties on actinomycetes and are thus likely to be the subject of intense research in the coming years. Although still at a preliminary stage, several themes are emerging, and some of these are discussed here. In mycobacteria, proteins belonging to the ESAT-6 family have attracted great interest owing to their antigenicity for T cells and potential as subunit vaccines or diagnostic reagents. There are 11 pairs of ESAT-6 genes in *M. tuberculosis* H37Rv *(10,86)*, with the genetic arrangement reflecting the probable interaction between their protein products *(39)*.

The gene for the prototype family member, ESAT-6 *(38)*, is located in the RD1 locus near the origin of replication of the tubercle bacilli *(61)*. The biological importance of the RD1 genes for mycobacteria is underlined by the fact that they are also conserved in *M. leprae (3*; Fig. 6) despite the radical gene decay observed in this organism. How-

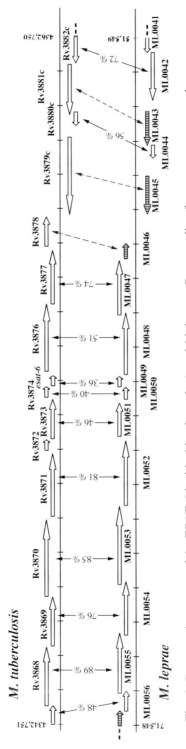

Fig. 6. Conserved structure of the ESAT-6 loci in *M. tuberculosis* and *M. leprae*. Open reading frames are represented as open arrows showing the direction of transcription and percentage amino acid similarities between orthologs. Numbers refer to the genomic position in *M. tuberculosis* H37Rv and *M. leprae*. Note that the ESAT-6 operon in *M. leprae* is on the complementary strand relative to *M. tuberculosis* H37Rv, but situated on the other side of *oriC*. Pseudogenes are shown as shaded arrows.

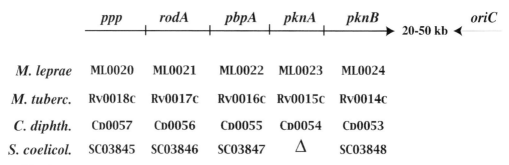

Fig. 7. Conserved structure of the *ppp–pknB* gene cluster in actinobacteria. This cluster comprises five genes, except in *S. coelicolor*, for which *pknA* appears to be missing: *ppp*, phosphoprotein phosphatase; *rodA*, cell division protein; *pbpA*, peptidoglycan biosynthesis protein; *pknA*, *pknB*, serine–threonine–protein kinases. In all four bacteria, the cluster is situated near *oriC*, but transcribed convergently with respect to the direction of replication.

ever, ESAT-6 genes are not restricted to pathogenic mycobacteria. *In silico* searches revealed orthologs in the unfinished genome sequences of the environmental species *M. smegmatis* and *M. marinum*. Interestingly, ESAT-6 family members were also identified in the genomes of the more distantly related *C. diphtheriae* and *S. coelicolor*, although the level of similarity was low *(86)*.

There is another block of five conserved genes found near *oriC* (Fig. 7) that encode a set of proteins that may be involved in signal transduction pathways and cell division in mycobacteria, corynebacteria, and streptomycetes *(3,4,87,88)*. In all cases, the operon is transcribed counter to the direction of replication, and in *S. coelicolor*, the putative operon contains one fewer serine–threonine protein kinase gene *(51)*.

Another unusual feature of the actinomycetes concerns the mismatch DNA repair system. Unlike Gram-negative bacteria, the actinomycetes apparently lack the *mutL–mutS* system that normally fulfills this function *(89)*, but contain multiple *mutT* orthologs instead. Further comparative analyses will certainly identify many more interesting genomic regions that can then be analyzed by detailed functional studies.

CONCLUDING REMARKS

Here, we have described the spectacular advances that have been made in the genomics of the actinobacteria and emphasized how this new information and knowledge are helping unravel the biology of this broad group of organisms and promote their functional exploration. As more genome sequences become available, we can expect deeper understanding of phylogeny, rich insight into behavior and developmental processes, and clearer demarcation of the biological traits that distinguish the actinobacteria from other prokaryotes.

ACKNOWLEDGMENTS

We gratefully acknowledge the contributions made by the actinobacterial genomics community and the financial support of the Institut Pasteur, the Association Française Raoul Follereau, ILEP, the New York Community Trust, the Wellcome Trust, the European Union (QLRT-2000-02018), BBSRC, and DEFRA.

REFERENCES

1. Woese CR, Kandler O, Wheelis ML. Towards a natural system of organisms: proposal for the domains Archaea, Bacteria, and Eucarya. Proc Natl Acad Sci USA 1990; 87:4576–4579.
2. Stackebrandt E, Raincy FA, Ward-Rainey NL. Proposal for a new hierarchic classification system, *Actinobacteria* classis nov. Int J Syst Bacteriol 1997; 47:479–491.
3. Cole ST, Eiglmeier K, Parkhill J, et al. Massive gene decay in the leprosy bacillus. Nature 2001; 409:1007–1011.
4. Cole ST, Brosch R, Parkhill J, et al. Deciphering the biology of *Mycobacterium tuberculosis* from the complete genome sequence. Nature 1998; 393:537–544.
5. Sreevatsan S, Pan X, Stockbauer KE, et al. Restricted structural gene polymorphism in the *Mycobacterium tuberculosis* complex indicates evolutionarily recent global dissemination. Proc Natl Acad Sci USA 1997; 94:9869–9874.
6. Kapur V, Whittam TS, Musser J. Is *Mycobacterium tuberculosis* 15,000 years old? J Infect Dis 1994; 170:1348–1349.
7. Camus J-C, Pryor MJ, Médigue C, Cole ST. Re-annotation of the genome sequence of *Mycobacterium tuberculosis* H37Rv. Microbiology 2002; 148(Pt 10):2967–2973.
8. Fleischmann RD, Alland D, Eisen JA, et al. Whole-genome comparison of *Mycobacterium tuberculosis* clinical and laboratory strains. J Bacteriol 2002; 184:5479–5490.
9. Valway SE, Sanchez MP, Shinnick TF, et al. An outbreak involving extensive transmission of a virulent strain of *Mycobacterium tuberculosis*. N Engl J Med 1998; 338:633–639.
10. Tekaia F, Gordon SV, Garnier T, Brosch R, Barrell BG, Cole ST. Analysis of the proteome of *Mycobacterium tuberculosis* in silico. Tuber Lung Dis 1999; 79:329–342.
11. Gordon SV, Heym B, Parkhill J, Barrell B, Cole ST. New insertion sequences and a novel repeated sequence in the genome of *Mycobacterium tuberculosis* H37Rv. Microbiology 1999; 145:881–892.
12. Lee TY, Lee TJ, Belisle JT, Brennan PJ, Kim SK. A novel repeat sequence specific to *Mycobacterium tuberculosis* complex and its implications. Tuber Lung Dis 1997; 78:13–19.
13. van Embden JDA, Cave WM, Crawford JT, et al. Strain identification of *Mycobacterium tuberculosis* by DNA fingerprinting: recommendations for a standardized methodology. J Clin Microbiol 1993; 31:406–409.
14. Fang Z, Doig C, Kenna DT, et al. IS*6110*-mediated deletions of wild-type chromosomes of *Mycobacterium tuberculosis*. J Bacteriol 1999; 181:1014–1020.
15. Brosch R, Philipp W, Stavropolous E, Colston MJ, Cole ST, Gordon SV. Genomic analysis reveals variation between *Mycobacterium tuberculosis* H37Rv and the attenuated *M. tuberculosis* H37Ra. Infect Immun 1999; 67:5768–5774.
16. Ho TBL, Robertson BD, Taylor GM, Shaw RJ, Young DB. Comparison of *Mycobacterium tuberculosis* genomes reveals frequent deletions in a 20 kb variable region in clinical isolates. Comparative Functional Genomics (Yeast) 2001; 17:272–282.
17. Daffe M, Draper P. The envelope layers of mycobacteria with reference to their pathogenicity. Adv Microb Physiol 1998; 39:131–203.
18. Peterson JA, Graham SE. A close family resemblance: the importance of structure in understanding cytochromes P450. Structure 1998; 6:1079–1085.
19. Aoyama Y, Horiuchi T, Gotoh O, Noshiro M, Yoshida Y. CYP51-like gene of *Mycobacterium tuberculosis* actually encodes a P450 similar to eukaryotic CYP51. J Biochem 1998; 124:694–696.
20. Podust LM, Poulos TL, Waterman MR. Crystal structure of cytochrome P450 14alpha-sterol demethylase (CYP51) from *Mycobacterium tuberculosis* in complex with azole inhibitors. Proc Natl Acad Sci USA 2001; 98:3068–3073.

21. Bellamine A, Mangla AT, Nes WD, Waterman MR. Characterization and catalytic properties of the sterol 14alpha-demethylase from *Mycobacterium tuberculosis*. Proc Natl Acad Sci USA 1999; 96:8937–8942.

22. Weber I, Fritz C, Ruttkowski S, Kreft A, Bange FC. Anaerobic nitrate reductase (narGHJI) activity of *Mycobacterium bovis* BCG in vitro and its contribution to virulence in immunodeficient mice. Mol Microbiol 2000; 35:1017–1025.

23. Cole ST. Learning from the genome sequence of *Mycobacterium tuberculosis* H37Rv. FEBS Lett 1999; 452:7–10.

24. Banu S, Honoré N, Saint-Joanis B, Philpott D, Prévost M-C, Cole ST. Are the PE-PGRS proteins of *Mycobacterium tuberculosis* variable surface antigens? Mol Microbiol 2002; 44: 9–19.

25. Brennan MJ, Delogu G. The PE multigene family: a molecular mantra for mycobacteria. Trends Microbiol 2002; 10:246–249.

26. Brennan MJ, Delogu G, Chen Y, et al. Evidence that mycobacterial PE_PGRS proteins are cell surface constituents that influence interactions with other cells. Infect Immun 2001; 69: 7326–7333.

27. Delogu G, Brennan MJ. Comparative immune response to PE and PE_PGRS antigens of *Mycobacterium tuberculosis*. Infect Immun 2001; 69:5606–5611.

28. Espitia C, Laclette JP, Mondragon-Palomino M, et al. The PE-PGRS glycine-rich proteins of *Mycobacterium tuberculosis*: a new family of fibronectin-binding proteins? Microbiology 1999; 145:3487–3495.

29. Camacho LR, Ensergueix D, Perez E, Gicquel B, Guilhot C. Identification of a virulence gene cluster of *Mycobacterium tuberculosis* by signature-tagged transposon mutagenesis. Mol Microbiol 1999; 34:257–267.

30. Ramakrishnan L, Federspiel NA, Falkow S. Granuloma-specific expression of mycobacterium virulence proteins from the glycine-rich PE-PGRS family. Science 2000; 288:1436–1439.

31. Behr MA, Wilson MA, Gill WP, et al. Comparative genomics of BCG vaccines by whole-genome DNA microarrays. Science 1999; 284:1520–1523.

32. Gordon SV, Brosch R, Billault A, Garnier T, Eiglmeier K, Cole ST. Identification of variable regions in the genomes of tubercle bacilli using bacterial artificial chromosome arrays. Mol Microbiol 1999; 32:643–656.

33. Titball RW. Bacterial phospholipases. Symp Ser Soc Appl Microbiol 1998; 27:127S–137S.

34. Arruda S, Bomfim G, Knights R, Huima-Byron T, Riley LW. Cloning of an *M. tuberculosis* DNA fragment associated with entry and survival inside cells. Science 1993; 261:1454–1457.

35. Scorpio A, Zhang Y. Mutations in *pncA*, a gene encoding pyrazinamidase/nicotinamidase, cause resistance to the antituberculous drug pyrazinamide in tubercle bacillus. Nat Med 1996; 2:662–667.

36. Collins DM, Kwakami RP, de Lisle GW, Pascopella L, Bloom BR, Jacobs JWR. Mutation of the principal s factor causes loss of virulence in a strain of the *Mycobacterium tuberculosis* complex. Proc Natl Acad Sci USA 1995; 92:8036–8040.

37. Wiker HG, Nagai S, Hewinson RG, Russell WP, Harboe M. Heterogenous expression of the related MPB70 and MPB83 proteins distinguish various substrains of *Mycobacterium bovis* BCG and *Mycobacterium tuberculosis* H37Rv. Scand J Immunol 1996; 43:374–380.

38. Sorensen AL, Nagai S, Houen G, Andersen P, Andersen Å. Purification and characterization of a low molecular mass T-cell antigen secreted by *Mycobacterium tuberculosis*. Infect Immun 1995; 63:1710–1717.

39. Renshaw PS, Panagiotidou P, Whelan A, et al. Conclusive evidence that the major T-cell antigens of the *M. tuberculosis* complex ESAT-6 and CFP-10 form a tight, 1:1 complex and characterisation of the structural properties of ESAT-6, CFP-10 and the ESAT-6-CFP-10 complex: implications for pathogenesis and virulence. J Biol Chem 2002; 8:8.

40. Li J, Ochman H, Groisman EA, et al. Relationship between evolutionary rate and cellular location among the Inv/Spa invasion proteins of *Salmonella enterica*. Proc Natl Acad Sci USA 1995; 92:7252–7256.

41. Sokurenko EV, Chesnokova V, Dykhuizen DE, et al. Pathogenic adaptation of *Escherichia coli* by natural variation of the FimH adhesin. Proc Natl Acad Sci USA 1998; 95:8922–8926.

42. Brosch R, Gordon SV, Marmiesse M, et al. A new evolutionary scenario for the *Mycobacterium tuberculosis* complex. Proc Natl Acad Sci USA 2002; 99:3684–3689.

43. Tseng T-T, Gratwick KS, Kollmann J, et al. The RND permease superfamily: an ancient, ubiquitous and diverse family that includes human disease and develoment proteins. J Mol Microbiol Biotechnol 1999; 1:107–125.

44. Wooff E, Michell SL, Gordon SV, et al. Functional genomics reveals the sole sulphate transporter of the *Mycobacterium tuberculosis* complex and its relevance to the acquisition of sulphur in vivo. Mol Microbiol 2002; 43:653–663.

45. Tamas I, Klasson LM, Sandstrom JP, Andersson SG. Mutualists and parasites: how to paint yourself into a (metabolic) corner. FEBS Lett 2001; 498:135–139.

46. Andersson JO, Andersson SGE. Insights into the evolutionary process of genome degradation. Curr Opin Genet Dev 1999; 9:664–671.

47. Shigenobu S, Watanabe H, Hattori M, Sakaki Y, Ishikawa H. Genome sequence of the endocellular bacterial symbiont of aphids *Buchnera* sp APS. Nature 2000; 407:81–86.

48. Anderssen SGE, Zomorodipour A, Andersson JO, et al. The complete genome sequence of the obligate intracellular parasite *Rickettsia prowazekii*. Nature 1998; 396:133–140.

49. Wolf YI, Aravind L, Grishin NV, Koonin EV. Evolution of amino-acyl–tRNA synthetases—analysis of unique domain architectures and phylogenetic trees reveals a complex history of horizontal gene transfer events. Genome Res 1999; 9:689–710.

50. Cole ST, Supply P, Honoré N. Repetitive sequences in *Mycobacterium leprae* and their impact on genome plasticity. Lepr Rev 2001; 72:449–461.

51. Bentley SD, Chater KF, Cerdeno-Tarraga AM, et al. Complete genome sequence of the model actinomycete *Streptomyces coelicolor* A3(2). Nature 2002; 417:141–147.

52. Eiglmeier K, Parkhill J, Honoré N, et al. The decaying genome of *Mycobacterium leprae*. Lepr Rev 2001; 72:387–398.

53. Jormakka M, Tornroth S, Byrne B, Iwata S. Molecular basis of proton motive force generation: structure of formate dehydrogenase-N. Science 2002; 295:1863–1868.

54. Jacobs WR Jr, Kalpana GV, Cirillo JD, et al. Genetic systems for mycobacteria. Methods Enzymol 1991; 204:537–555.

55. Snapper SB, Lugosi L, Jekkel A, et al. Lysogeny and transformation in mycobacteria: stable expression of foreign genes. Proc Natl Acad Sci USA 1988; 85:6987–6991.

56. Snapper SB, Melton RE, Mustafa S, Kieser T, Jacobs WR. Isolation and characterization of efficient plasmid transformation mutants of *Mycobacterium smegmatis*. Mol Microbiol 1990; 4: 1911–1919.

57. Bardarov S, Kriakov J, Carriere C, et al. Conditionally replicating mycobacteriophages: a system for transposon delivery to *Mycobacterium tuberculosis*. Proc Natl Acad Sci USA 1997; 94: 10,961 10,966.

58. Pelicic V, Jackson M, Reyrat JM, Jacobs WR Jr, Gicquel B, Guilhot C. Efficient allelic exchange and transposon mutagenesis in *Mycobacterium tuberculosis*. Proc Natl Acad Sci USA 1997; 94: 10,955–10,960.

59. Parish T, Stoker NG. Use of a flexible cassette method to generate a double unmarked *Mycobacterium tuberculosis tlyA plcABC* mutant by gene replacement. Microbiology 2000; 146: 1969–1975.

60. Hinds J, Mahenthiralingam E, Kempsell KE, et al. Enhanced gene replacement in mycobacteria. Microbiology 1999; 145:519–527.

61. Mahairas GG, Sabo PJ, Hickey MJ, Singh DC, Stover CK. Molecular analysis of genetic differences between *Mycobacterium bovis* BCG and virulent *M. bovis*. J Bacteriol 1996; 178: 1274–1282.

62. Salamon H, Kato-Maeda M, Small PM, Drenkow J, Gingeras TR. Detection of deleted genomic DNA using a semiautomated computational analysis of GeneChip data. Genome Res 2000; 10: 2044–2054.

63. Brosch R, Pym AS, Gordon SV, Cole ST. The evolution of mycobacterial pathogenicity: clues from comparative genomics. Trends Microbiol 2001; 9:452–458.

64. Bates JH, Stead WW. The history of tuberculosis as a global epidemic. Med Clin North Am 1993; 77:1205–1217.

65. Stead WW, Eisenach KD, Cave MD, et al. When did *Mycobacterium tuberculosis* infection first occur in the New World? An important question with public health implications. Am J Respir Crit Care Med 1995; 151:1267–1268.

66. Pitulle C, Dorsch M, Kazda J, Wolters J, Stackebrandt E. Phylogeny of rapidly growing members of the genus *Mycobacterium*. Int J Syst Bacteriol 1992; 42:337–343.

67. Buchmeier N, Blanc-Potard A, Ehrt S, Piddington D, Riley L, Groisman EA. A parallel intraphagosomal survival strategy shared by *Mycobacterium tuberculosis* and *Salmonella enterica*. Mol Microbiol 2000; 35:1375–1382.

68. Brosch R, Gordon SV, Buchrieser C, Pym A, Garnier T, Cole ST. Comparative genomics uncovers tandem chromosomal duplications in some strains of *Mycobacterium bovis* BCG: implications for vaccination. Comparative Functional Genomics (Yeast) 2000; 17:111–123.

69. Fernandes ND, Wu QL, Kong D, Puyang X, Garg S, Husson RN. A mycobacterial extracytoplasmic sigma factor involved in survival following heat shock and oxidative stress. J Bacteriol 1999; 181:4266–4274.

70. Lupski JR, Roth JR, Weinstock GM. Chromosomal duplications in bacteria, fruit flies, and humans. Am J Hum Genet 1996; 58:21–27.

71. Galamba A, Soetaert K, Wang XM, De Bruyn J, Jacobs P, Content J. Disruption of *adhC* reveals a large duplication in the *Mycobacterium smegmatis* mc(2)155 genome. Microbiology 2001; 147: 3281–3294.

72. Redenbach M, Kieser HM, Denapaite D, et al. A set of ordered cosmids and a detailed genetic and physical map for the 8 Mb *Streptomyces coelicolor* A3(2) chromosome. Mol Microbiol 1996; 21:77–96.

73. Wood V, Gwilliam R, Rajandream MA, et al. The genome sequence of *Schizosaccharomyces pombe*. Nature 2002; 415:871–880.

74. Venter JC, Adams MD, Myers EW, et al. The sequence of the human genome. Science 2001; 291: 1304–1351.

75. Consortium IHGS. Initial sequencing and analysis of the human genome. Nature 2001; 409: 860–921.

76. Hopwood DA, Chater KE, Bibb MJ. Genetics of antibiotic production in *Streptomyces coelicolor* A3(2), a model streptomycete. In: Vining LC, Stuttard C. (eds). Genetics and Biochemistry of Antibiotic Production. Newton, MA: Butterworth-Heinemann, 1995, pp. 65–102.

77. Davis NK, Chater KF. Spore colour in *Streptomyces coelicolor* A3(2) involves the developmentally regulated synthesis of a compound biosynthetically regulated to polyketide antibiotics. Mol Microbiol 1990; 4:1679–1691.

78. Chong PP, Podmore SM, Kieser HM, et al. Physical identification of a chromosomal locus encoding biosynthetic genes for the lipopeptide calcium-dependent antibiotic (CDA) of *Streptomyces coelicolor* A3(2). Microbiology 1998; 144:193–199.

79. Lamb DC, Skaug T, Song HL, et al. The cytochrome P450 complement (CYPome) of *Streptomyces coelicolor* A3(2). J Biol Chem 2002; 9:9.

80. Stover CK, Pham XQ, Erwin AL, et al. Complete genome sequence of *Pseudomonas aeruginosa* PA01, an opportunistic pathogen. Nature 2000; 406:959–964.

81. Lonetto MA, Brown KL, Rudd KE, Buttner MJ. Analysis of the *Streptomyces coelicolor sigE* gene reveals the existence of a subfamily of eubacterial RNA polymerase sigma factors involved in the regulation of extracytoplasmic functions. Proc Natl Acad Sci USA 1994; 91:7573–7577.

82. Eisen JA, Heidelberg JF, White O, Salzberg SL. Evidence for symmetric chromosomal inversions around the replication origin in bacteria. Genome Biol 2000; 1, Genome Biology (RESEARCH) 0011.

83. Volff JN, Altenbuchner, J. Genetic instability of the *Streptomyces* chromosome. Mol Microbiol 1998; 27:239–246.

84. Casjens S. The diverse and dynamic structure of bacterial genomes. Annu Rev Genet 1998; 32: 339–377.

85. Chen CW, Huang CH, Lee HH, Tsai HH, Kirby R. Once the circle has been broken: dynamics and evolution of Streptomyces chromosomes. Trends Genet 2002; 18:522–529.

86. Gey Van Pittius NC, Gamieldien J, Hide W, Brown GD, Siezen RJ, Beyers AD. The ESAT-6 gene cluster of *Mycobacterium tuberculosis* and other high G+C Gram-positive bacteria. Genome Biol 2001; 2, RESEARCH0044.

87. Fsihi H, De Rossi E, Salazar L, et al. Gene arrangement and organisation in a ~76 kilobase fragment encompassing the *oriC* region of the chromosome of *Mycobacterium leprae*. Microbiology 1996; 142:3147–3161.

88. Av-Gay Y, Davies J. Components of eukaryotic-like protein signaling pathways in *Mycobacterium tuberculosis*. Microb Comp Genomics 1997; 2:63–73.

89. Mizrahi V, Andersen SJ. DNA repair in *Mycobacterium tuberculosis*. What have we learnt from the genome sequence? Mol Microbiol 1998; 29:1331–1339.

20

Parasite Genomics

Malcolm J. Gardner

INTRODUCTION

Parasites are a very diverse group of organisms, ranging from protists such as the malaria parasite *Plasmodium*, to multicellular worms such as *Schistosoma*, the cause of bilharzia. They cause a variety of diseases in humans and are most prevalent in the tropical and subtropical regions of the world. The impact of these diseases on the economies of developing countries is severe; it has been estimated that, from 1965 to 1990, malaria alone may have limited the gross domestic product of malaria-endemic developing countries to one-half that of comparable nonmalarious countries (*1*). In addition, these diseases exact social costs that are more difficult to measure, such as reductions in school attendance and academic performance. These far-ranging negative impacts of disease on health and economic development have been more widely recognized in recent years and have spurred efforts to develop effective countermeasures, such as new drugs, vaccines, and vector control methods.

Parasites also infect many species of domesticated animals in both developed and developing countries. The coccidian parasite *Eimeria*, for example, infects chickens and other poultry. *Theileria parva*, a relative of the malaria parasite, infects cattle in large parts of sub-Saharan Africa and causes a lethal disease known as East Coast fever (ECF). It is a major impediment to the development of agriculture in Africa and is especially difficult for smallholder farmers. These and many other parasites reduce agricultural efficiency and productivity essential for feeding growing populations.

Most parasites are difficult to study in the laboratory; they are unlike the so-called model organisms such as yeast, the fruit fly, or the nematode *Caenorhabditis elegans*, which were selected for study because they could be grown in the laboratory, had fast generation times, and were amenable to genetic approaches. Many parasites, such as the malaria parasites, have complex life cycles that involve both vertebrate and invertebrate hosts and are difficult, expensive, or impossible to maintain in the laboratory. Of the four species of malaria parasite that infect humans, *Plasmodium falciparum* is the only one that can be grown continuously in vitro, and this is only possible for the erythrocytic stages. All other rodent and primate malaria parasites must be maintained in animal models, although recent reports suggest that some of the sexual stages of *Plasmodium berghei* can be cultured in vitro (*2*) and erythrocytic stages of the primate parasite *Plas-

From: *Microbial Genomes*
Edited by: C. M. Fraser, T. D. Read, and K. E. Nelson © Humana Press Inc., Totowa, NJ

modium knowlesi can be adapted to growth in culture *(3)*. The erythrocytic stages of all other human and rodent malaria parasites must be maintained in animal models.

These difficulties inherent in the study of most parasites have retarded progress in research in comparison to what has been achieved in the study of other model eukaryotes. Genomics, however, has begun to level the playing field for parasitologists. Early genomics efforts accelerated gene discovery by the sequencing of randomly selected complementary deoxyribonucleic acids (cDNAs) to generate expressed sequence tags (ESTs), which provided partial or complete gene sequences and insights into the patterns of stage-specific gene expression for many parasites; many of these early projects were initially coordinated by the World Health Organization Tropical Disease Research (WHO/TDR; http://www.who.int/tdr/) program *(4)*. Because the genomes of many parasites have high gene density and less intervening sequence than mammalian genomes, random sequencing of genomic fragments (genome survey sequences, GSSs) was also helpful for gene discovery.

As the technology for sequencing and analysis of genome data improved and the costs of sequencing declined, efforts to sequence single chromosomes and the entire genomes of many parasites were launched. These projects are now beginning to bear fruit. Provision of parasite genome sequence information has invigorated the field by providing many new starting points for research, and the abundance of information has attracted investigators to the field. Functional genomics, primarily based on microarrays *(5–10)* and proteomics approaches *(11–13)*, is being used to overlay gene and protein expression information on genome sequences. Fresh insights into the workings of these organisms and their interactions with their hosts are beginning to be gained.

In this chapter, I review the status of the genome-sequencing efforts for many of the most important human and animal parasites. The focus here is on genome-sequencing efforts rather than EST projects and on those parasite genome-sequencing projects that are beginning to produce results. A convenient listing of most of the EST and genome-sequencing projects for parasites and other eukaryotes, with links to the appropriate Web sites, can be found at the National Center for Biotechnology Information Web site (http://www.ncbi.nlm.nih.gov/PMGifs/Genomes/EG_T.html).

PARASITE GENOME PROJECTS

The genomes of parasites vary tremendously in terms of size, conformation, gene density, number of chromosomes, ploidy, base composition, repetitiveness, and other factors. For example, protists such as *Plasmodium* spp are haploid organisms with nuclear genomes less than 25 Mb and 14 linear chromosomes; a metazoan parasite such as the worm *Schistosoma mansoni* is diploid and has a genome size of about 270 Mb.

The sequencing strategy used to determine a genome sequence must be appropriate for the genome and organisms to be sequenced. The selection of the sequencing strategy may be influenced by technical factors, such as the stability of the DNA in *Escherichia coli*, the availability of large insert libraries, the sequencing or genome assembly technologies that exist on the initiation of the project, the amount of funds available, and strategic factors such as the importance of the organism in terms of its impact on human or animal health. Fortunately, cost is becoming less of an issue because of the impressive cost reductions achieved in recent years through improvements in instrumentation, primarily the switch from gel to capillary sequencers, increased automation, and improved laboratory protocols. For example, at The Institute for Genomic Research (TIGR) the direct cost per sequence has declined from $7.80 in 1996 to less than $1.00, and the read length and success rates have improved markedly. This makes the sequencing of any one genome less expensive, freeing resources to sequence other genomes (e.g., *Plasmodium yoelii yoelii* and *Plasmodium vivax*, discussed in separate sections) or to sequence larger genomes that would have been prohibitively expensive a few years ago. Costs are likely to decline even further as new sequencing technologies are developed. In the future, the genome of virtually any parasite of interest will be sequenced to completion, as will multiple clones or isolates of the most important pathogens.

APICOMPLEXA

Plasmodium falciparum

The complete genome sequence of the human malaria parasite *P. falciparum*, which kills 1–3 million people every year, was recently determined in a collaborative effort by the Wellcome Trust Sanger Institute, TIGR, and Stanford University *(14–19)*. This effort began in 1996 *(20)* with the formation of an international consortium of funding agencies and sequencing centers. The genome of clone 3D7, a clone derived from an isolate adapted to culture in 1976 and that is able to complete the entire life cycle *(21)*, was sequenced using a chromosome-by-chromosome shotgun sequencing strategy that was appropriate for the sequencing technology, assembly and genome closure protocols, and the funds available for the project.

Of all the microbial genomes that have been sequenced to date, *P. falciparum* was probably the most difficult genome to complete, primarily because of the extremely high adenosine and thymidine (A+T) content (80%) of the nuclear DNA. Determination of the genome sequence required modifications to almost every aspect of the standard shotgun sequencing procedure *(17–19)*. For example, the protocols used for construction of shotgun libraries were altered to prevent the denaturation of the A+T-rich DNA during gel purification, transposon insertion techniques were developed to allow accurate determination of the sequence in areas containing long homopolymer runs of A and T residues, and HAPPY mapping *(16)* and optical restriction mapping *(22,23)* were used to order contigs and validate the final chromosome sequences. In addition, two new gene finders were developed because existing gene finders were not optimal for the prediction of genes in eukaryotic genomes of high gene density *(24,25)*. Completion of the genome sequence took over 6 years, primarily because of the difficulty of closing hundreds of A+T-rich gaps in the sequence. The published genome sequence contained approximately 93 gaps (most < 2500 nt), but efforts to close the remaining gaps are under way.

The *P. falciparum* genome is approximately 23 Mb in size and encodes about 5300 genes *(14)*. Of the genes, 54% contained introns. On average, the coding regions of *P. falciparum* genes were longer than in other sequenced eukaryotes, such as *Schizosaccaromyces pombe* (2.3 kb vs 1.4 kb). The reason for this increased gene length is unknown. Fully 60% of the encoded proteins have little or no similarity to proteins in other organisms and are of unknown function. The proportion of genes encoding these so-called hypothetical proteins is higher in *P. falciparum* than in other sequenced organisms, which is probably a reflection of the greater evolutionary distance between *Plasmodium* and other well-studied eukaryotes.

Analysis of the predicted proteome provided an overview of metabolism and transport in malaria parasites. However, some features of parasite metabolism remained unclear because of the absence of some enzymes or enzyme subunits, and the predicted subcellular localization of some enzymes differed from the known localization of the enzymes in other organisms, making it difficult to reconstruct the metabolism of the parasite with certainty. Nevertheless, analysis of the genome sequence provided valuable insights into the biochemistry of the parasite and pinpointed areas that require further investigation in the laboratory (see Chapter 6).

Plasmodium falciparum appears to have very reduced capacities for metabolism and for the transport of organic nutrients and ions than free-living organisms. Sequence similarity searches with sequences of known enzymes revealed that only 14% of the proteins encoded enzymes; this is a much lower proportion than observed in other sequenced eukaryotes. Similarly, the *P. falciparum* genome encoded a smaller repertoire of membrane transporters in comparison to other free-living eukaryotic microbes such as *Saccharomyces cerevisiae* and *S. pombe*. Earlier biochemical studies suggested that erythrocytic stage parasites rely primarily on glycolysis for adenosine triphosphate (ATP) production. The *P. falciparum* genome encoded enzymes for the complete glycolytic pathway from glucose-6-phosphate to pyruvate and for the conversion of pyruvate to lactate.

All of the enzymes of the tricarboxylic acid (TCA) cycle were identified, but the function of the TCA cycle was unclear. Pyruvate dehyrogenase, which is usually localized in the mitochondrion and converts pyruvate to acetyl–coenzyme A required for the first step of the TCA cycle, was predicted to be located within the apicoplast, a relict plastid, in *P. falciparum*. Moreover, malate dehydrogenase appeared to be localized in the cytoplasm, rather than the mitochondrion as in other eukaryotes, and was thought to be replaced in the TCA cycle by mitochondrial malate–quinone oxidoreductase. These unusual features suggested that, at least in erythrocytic stages, the TCA cycle may be used to supply intermediates for biosynthetic pathways such as heme biosynthesis rather than for the complete oxidation of the products of glycolysis.

The role of the TCA cycle in other stages of the life cycle is not known with certainty, but proteomic studies have revealed that some TCA cycle enzymes seem to be more abundant in gametocytes than in erythrocytic parasites, suggesting that the TCA cycle may be more important in the sexual stages in the mosquito *(11,12)*. Also unexpected was the apparent absence of the F_o a and b subunits of ATP synthase. This implied that the ATP synthase might be nonfunctional in *Plasmodium*, although the genes encoding these proteins are short and may be located within the unsequenced regions of the genome. Alternatively, these subunits may be quite divergent and not easily recognized by sequence similarity approaches.

Other unusual features of *P. falciparum* metabolism that could be inferred from the genome sequence were the absence of gluconeogenesis and the lack of any enzymes for the biosynthesis of amino acids, apart from enzymes required for amino acid interconversions. The lack of amino acid biosynthetic pathways, and the apparent absence of clear homologs of known amino acid transporters, emphasized the parasite's dependence on the host for amino acids, at least in the erythrocytic stages in which amino acids are obtained by the digestion of hemoglobin in the food vacuole. How the parasite obtains amino acids during the mosquito stages of the life cycle is not known. Biosynthesis of fatty acids and isoprenoids occurs in the apicoplast. Both of these pathways resemble the pathways used by plants and bacteria rather than animals and offer multiple potential targets for novel antimalarials *(26–28)*. Overall, the metabolic and transport capabilities of *Plasmodium* are less than that of free-living organisms, which may be a reflection of its parasitic lifestyle (see Fig. 3 in Chapter 6).

Another clear difference between the genomes of *P. falciparum* and other sequenced eukaryotes was the abundance of genes in the malaria parasite that are involved in immune evasion and other host–parasite interactions (3.9 and 1.3% of all genes, respectively). The 3D7 genome contains 59 *var* genes that encode highly polymorphic proteins known as *P. falciparum* erythrocyte membrane protein 1 (PfEMP1). These proteins are expressed on the surface of infected red blood cells, mediate cytoadherence of the infected cells to host capillary endothelium, and cause the sequestration of the infected cells in many organs, including the brain. The PfEMP1 proteins are the targets of protective antibody responses, but transcriptional switching between different *var* genes results in antigenic variation and the evasion of the host immune response. The *rif* genes, of which there were 149 in the 3D7 genome, encode another group of proteins called rifins. These are also expressed on the surface of the infected erythrocyte and undergo antigenic variation, but their exact function is unclear. A third group of proteins called STEVORs (28 in the 3D7 genome) are similar in sequence to the rifins.

Members of the PfEMP1, rifin, and STEVOR families exhibit extensive sequence diversity, and the genes encoding these proteins occur in clusters, most of which are located in subtelomeric regions in association with several kinds of repetitive sequences. The repetitive sequences are thought to facilitate recombination between different alleles of these highly polymorphic proteins and contribute to the generation of antigen diversity.

The sequenced 3D7 clone of *P. falciparum* has been cultured for years and has not faced immune pressure. Cultured parasites have been known to suffer chromosome breakage and healing events that result in subtelomeric deletions and gene loss. Several chromosomes in 3D7 seem to have been truncated in this manner. It would be interesting to compare the genome of the 3D7 parasite to the genome of one or more recent clinical isolates to see whether the repertoire of genes involved in immune evasion and host-parasite interactions in this clone is representative of parasites found in the wild. This may be done in the near future (N. Hall, Sanger Institute, personal communication, 2003).

The *P. falciparum* genome project was the first project devoted to the completion of the entire genome of an important human parasite. An important aspect of this project was the involvement of the broader malaria research community in the project via semiannual meetings organized by representatives of the four private and public agencies that supported the sequencing effort. These meetings were used to discuss the progress as well as the technical difficulties encountered by the sequencers, to look forward to the eventual completion of the genome, and to plan future research activities, such as functional genomics and proteomics. A data release policy was also developed.

Although the data release policy that was adopted in 1997 did not please everyone *(29, 30)*, it did establish a process by which preliminary sequence data and annotation were released to the community to "jump start biological experimentation" and recognize the desire of the sequencing centers to assemble, annotate, and publish the genome sequence. After the establishment of this policy, dozens of papers were published in which the use of preliminary sequence data were acknowledged.

Another issue discussed at length in these meetings was the need for a centralized, user-friendly database from which malaria investigators could obtain *P. falciparum* genome sequence information and access sequence analysis tools without having to visit several different Web sites. This desire resulted in the establishment of PlasmoDB *(31,32)* at the University of Pennsylania in June 2000. PlasmoDB was subsequently expanded to include sequence data from other species of *Plasmodium*, including rodent, primate, and human malaria parasites, and contains gene expression information from ESTs *(33)*, SAGE *(34,35)*, and microarray *(7,8)* data sets.

Plasmodium vivax

Plasmodium vivax is the second most important human malaria parasite, causing an estimated 70–80 million cases of malaria annually *(36)*. It is prevalent in Asia, Oceania, the Americas, and several regions of Africa and accounts for half of the malaria cases outside the African continent. *Plasmodium vivax* is also a serious threat to travelers and military personnel from countries not endemic for malaria. Although infrequently fatal, *P. vivax* malaria is a very debilitating disease that impairs quality of life and economic productivity. Chloroquine resistance of *P. vivax* has been documented in several countries *(37–39)*, which may lead to an increase in the worldwide incidence and prevalence of *P. vivax* malaria.

Multiple features of this parasite differentiate it from *P. falciparum*. Unlike *P. falciparum*, *P. vivax* produces dormant forms in the liver called hypnozoites; the activation and subsequent development of the hypnozoites long after the initial infection cause relapses of the disease. During the erythrocytic phase, *P. vivax* can invade reticulocytes, but not mature erythrocytes, which results in lower parasitemias than those observed in *P. falciparum* infections. In addition, *P. vivax* invasion of reticulocytes requires interaction between the Duffy antigen/receptor for chemokines *(40)*, which is expressed on the surface of the reticulocyte, and the Duffy binding protein expressed on the merozoite surface *(41)*. *Plasmodium vivax* also differs from *P. falciparum* in that it does not sequester in the capillaries; sequestration is thought to be involved in cerebral malaria and other syndromes characteristic of severe *P. falciparum* malaria.

Despite its importance as a major human pathogen, *P. vivax* is relatively little studied compared to *P. falciparum*. Unlike *P. falciparum (42,43)*, large scale in vitro culture of erythrocytic *P. vivax* parasites has not been achieved. Consequently, maintenance of *P. vivax* in the laboratory requires the use of nonhuman primates for the production of parasite material, a prohibitively expensive proposition for most laboratories. Using funds remaining from the *P. falciparum* genome project, TIGR and the Naval Medical Research Center are sequencing the *P. vivax* genome using a whole genome shotgun strategy.

The *P. vivax* nuclear genome consists of 12–14 linear chromosomes that range from 1.2 to 3.5 Mb *(44)*. Initial estimates put the genome size at 35–40 Mb, but this is likely to be an overestimate because the genomes of other *Plasmodium* species *(45)* and preliminary sequencing results suggest that the genome is about 23–25 Mb. A large-scale sampling of the *P. vivax* genome (11,000 genome survey sequences from mung bean nuclease-digested genomic DNA libraries) has given some insight into the coding potential of the parasite *(46)*. Homologs of previously identified *Plasmodium* genes were identified, and similar numbers of proteins were common between the *P. vivax* and the *P. falciparum* proteomes. A large multigene family of 600–1000 variant proteins termed the *vir* family has also recently been identified in *P. vivax (47)*, orthologs of which have been found in several rodent malaria species *(45,48)*, but not *P. falciparum*.

Gene-mapping studies *(45,49,50)*, small-scale sequencing projects *(47,51)*, and genome-wide synteny maps *(45)* show that there is extensive conservation of gene synteny within the *Plasmodium* genus. Gene content and order appear to be highly conserved across hundreds of kilobases of DNA. For example, a 200-kb segment of *P. vivax* DNA contains 36 contiguous genes in the same order and orientation and with the identical structures as the genes on the orthologous segment of *P. falciparum* chromosome 3 *(51)*. The degree of synteny between *Plasmodium* species decreases as the phylogenetic distance between them increases, as has been shown in other systems.

The Salvador I strain of *P. vivax*, isolated from a naturally acquired infection in a patient from El Salvador *(52)*, is currently in the random phase of whole genome shotgun sequencing at TIGR. Small- and medium-size insert genomic DNA libraries were constructed, and sequencing to ninefold coverage has been completed *(53)*. After assembly of the genome, a comparison of the predicted *P. vivax* open reading frames (ORFs) in the preliminary contigs against protein databases revealed that 37% of ORFs have sequence similarity to known proteins, and 78% of these hits are to *P. falciparum* proteins. The remaining 63% of ORFs have no apparent similarity to any known protein, a proportion similar to that found in the analysis of the *P. falciparum* genome *(14)*. Gap closure is now under way, and completion of the genome sequence is expected in 2004. Fortunately, the initial gap closure procedures suggested that completion of the *P. vivax* genome will not be as difficult as it was to finish *P. falciparum* because of the higher guanine and cytosine content of *P. vivax* genomic DNA (Jane Carlton, personal communication, 2003).

Plasmodium yoelii yoelii *and Other Rodent Malaria Parasites*

Plasmodium y. yoelii is a rodent malaria parasite that is used as an animal model in many laboratories. It has been used extensively in studies of antigens expressed by sporozoites and exoerythrocytic stages and for the testing of vaccines against these antigens. Using funds remaining from the *P. falciparum* genome project, TIGR and the Naval Medical Research Center sequenced the *P. y. yoelii* genome to 5× coverage and performed a comparative analysis of the *P. y. yoelii* and *P. falciparum* genomes *(45)*. The 23-Mb *P. y. yoelii* genome encodes approximately 5900 genes, more than *P. falciparum*, probably because of 838 small genes *(yir)* in *P. y. yoelii* that encode variant antigens. Just over half of the predicted *P. y. yoelii* proteins had orthologs in *P. falciparum*. Almost all of these were located in the central portions of the chromosomes, whereas species-specific genes like the *yir* genes in *P. y. yoelii* and the *rif*, *stevor*, and *var* genes in *P. falciparum* were located in the subtelomeric regions of chromosomes.

Previous studies using pulsed field gels of the conservation of gene synteny in rodent and human malaria parasites revealed extensive regions of gene synteny. The *P. falciparum* and *P. y. yoelii* genome sequences enabled studies of gene synteny to be performed using bioinformatic techniques at a much higher resolution across both genomes. As a first step, unordered *P. y. yoelii* contigs and the *P. falciparum* genome were translated in all six reading frames, and all unique exact matches longer than five amino acids between the species were computed using the minimal unique matches (MUMmer) program. In this way, contigs representing over 70% of the *P. y. yoelii* genome could be tiled against the *P. falciparum* genome, suggesting that there is extensive conservation of gene order between the two species. Most of the conserved *P. y. yoelii* contigs mapped to the central regions of the *P. falciparum* chromosomes.

To identify long stretches of conservation of gene synteny and syntenic break points, the linkage of tiled *P. y. yoelii* contigs was determined by mate–pair information, identification of ESTs that spanned gaps between contigs, and by polymerase chain reaction amplification of the sequence between linked contigs from genomic DNA. These studies identified 457 groups of linked *P. y. yoelii* contigs, up to 800 kb long, which were then mapped to *P. y. yoelii* chromosomes via chromosome-specific markers. Mapping of the linked *P. y. yoelii* contigs to the *P. falciparum* genome provided a map of conserved gene synteny and identified break points in synteny between the two parasites *(45)*. About 70% of *P. y. yoelii* genes occurred in the same order and orientation as the putative orthologous genes in *P. falciparum*. Syntenic break points in the *P. y. yoelii* genome were frequently associated with loci encoding ribosomal ribonucleic acid (rRNA) genes, suggesting that chromosome breakage and recombination at the rRNA loci may be responsible for the disruption of synteny in malaria parasites. As well as highlighting syntenic regions of the *P. falciparum* and *P. y. yoelii* genomes, analysis of the structures of orthologous genes in syntenic regions shared by the two species revealed remarkable conservation of exon structures in genes. Thus, as noted here, alignments of orthologous genes in *Plasmodium* species may be very helpful in the elucidation of gene structures *(54,55)*.

Although this medium-coverage sequencing effort provided a great deal of useful information regarding the *P. y. yoelii* genome and its differences and similarites with the genome of *P. falciparum*, the fragmented nature of the sequence data prevented a more thorough analysis. First, the draft sequence was unedited, and because genome closure efforts frequently uncover and correct misassemblies caused by repetitive sequences, the *P. y. yoelii* sequence may contain more errors and uncertainties than is generally the case with "complete" genome sequences. Second, many of the *P. y. yoelii* gene models are incomplete. Third, the lack of complete chromosome sequences made the analysis of the conservation of gene synteny between *P. y. yoelii* and *P. falciparum* very difficult. On the other hand, low- or medium-coverage data are very useful for gene discovery in organisms of high gene density like *Plasmodium*; this facilitates many laboratory investigations.

The genomes of two other widely studied rodent malaria parasites, *P. berghei* and *Plasmodium chabaudi*, are currently being sequenced to 5× coverage, which should allow the identification of at least 90% of the genes. In addition, the Sanger Institute is also sequencing the genome of the primate malaria parasite *P. knowlesi*. *Plasmodium knowlesi* is closely related to the human parasite *P. vivax* and is frequently used to study the parasite molecules involved in the invasion of red cells and for the testing of vaccine candidate antigens and vaccine delivery systems *(56)*. Techniques allowing long-term (up to 18 months) in vitro cultivation of *P. knowlesi* have been developed for this parasite, which is normally grown in rhesus monkeys. Parasites grown in vitro for extended periods can be readapted to in vivo growth. In addition, procedures for transfection of *P. knowlesi* in vitro and the rapid generation of gene knockouts are available. These features of the *P. knowlesi* model, combined with information to be gleaned from the

P. knowlesi genome sequence and its comparison to other malaria parasites, will permit studies of parasite gene function and host–parasite interactions in a system closely related to humans.

Piroplasms

Piroplasms are tick-borne parasites that invade the erythrocytes, and sometimes other cells, of their mammalian hosts. Unlike their close relatives the malaria parasites, piroplasms do not form pigment from hemoglobin. Two of the most important piroplasms are *Babesia* and *Theileria*. They are parasites of animals, although *Babesia* does occasionally infect elderly, immunocompromised, or splenectomized humans.

Babesia bovis and *Babesia bigemina* are the causative agents of bovine babesiosis (tick fever) in the tropical and subtropical regions of the world; *Babesia divergens* is found in temperate climates. *Babesia* causes anemia and fever in infected cattle, resulting in a loss of productivity, and it can induce abortions in pregnant animals. The prevention of *Babesia* infections is often accomplished by vaccination with live attenuated parasites or by treatment of cattle with acaracides that prevent tick infestation and the transmission of the parasite to cattle. These measures are expensive and difficult to maintain over the long term, even in highly developed countries.

The *B. bovis* genome was studied using pulsed field gel electrophoresis and consists of four chromosomes ranging from 1.4 to 3.2 Mb, with a cumulative length of 9.4 Mb *(57)*. Efforts to sequence the genome using a whole genome shotgun strategy are expected to begin soon. ESTs are also being generated from cyclic DNA libraries made from infected red cells (http://www.sanger.ac.uk/Projects/B_bovis/).

Parasites of the genus *Theileria* infect a wide variety of domesticated and wild ruminants, including cows, sheep, buffalo, and deer. Perhaps the two most economically important species are *T. parva* and *Theileria annulata*. *Theileria parva* is the causative agent of East Coast Fever (ECF) in cattle, an acute disease that results in high rates of livestock morbidity and mortality in large areas of sub-Saharan Africa *(58)*. One million cattle die each year from ECF, with annual economic costs estimated to be greater than $168 million *(59)*. Infection is intitiated when sporozoites are introduced into the host by the feeding of infected brown ear ticks (*Rhipicephalus appendiculatus*). The sporozoites invade host lymphocytes and develop into multinucleated schizonts. The presence of the parasite within the lymphocyte cytoplasm induces the malignant transformation of the infected cell. Schizont-infected lymphocytes proliferate so extensively that infected animals die within 1 month of a leukemia-like disease. The transformed cells infected with *T. parva* can be maintained indefinitely in vitro and are tumorigenic in nude mice, but the transformed phenotype can be reversed by treatment with antiparasitic agents. *Theileria* spp therefore provide a unique model system in which to study the induction, maintenance, and reversal of cell transformation *(60)*. Study of the mechanisms used by *T. parva* to transform mammalian cells may reveal phenomena with relevance to cancer in humans.

Like *Babesia*, the main control measures used against *T. parva* are live attenuated vaccines and treatment of cattle with acaracides, measures that are difficult to maintain and afford in underdeveloped regions of Africa. Development of subunit vaccines, potentially the most effective method of ECF control, is a major research objective.

Studies of the immune responses in animals protected by the live attenuated vaccines have shown that the protection is caused by the killing of schizont-infected cells by CD8[+], major histocompatibility–restricted T cells.

Over the past several years, the International Livestock Research Institute (ILRI, Nairobi, Kenya) has been trying without success to identify parasite antigens expressed in schizont-infected cells that are the targets of protective CD8[+] T cells; the lack of success is primarily because of the intracellular lifestyle of the parasite. ILRI and TIGR are collaborating to sequence the genome of *T. parva* to assist in the identification of antigens expressed in schizont-infected cells. A whole genome shotgun approach is being used. All four of the parasite chromosomes have been completed, although an approximately 120-kb region of chromosome 3 containing the highly repetitive Tpr sequences could not be solved. The genome is approximately 8.4 Mb long, and annotation of the sequence is under way. Preliminary sequences of novel *T. parva* genes are being used at ILRI to identify antigens expressed in schizont-infected cells, and microarray studies of parasite gene expression throughout the life cycle are planned.

Theileria annulata is prevalent in countries surrounding the Mediterranean Sea and in Asia. Like *T. parva*, it also infects cattle, causing a fatal proliferative disorder, but transforms macrophages rather than lymphocytes *(60)*. Thus, *T. parva* and *T. annulata* share many characteristics, but transform different host cells. The *T. annulata* genome is being sequenced by the Sanger Institute. Comparison of the *T. parva* and *T. annulata* genomes to identify species-specific genes and conserved genes may shed light on the mechanisms by which these parasites invade and tranform different host cell types.

Toxoplasma gondii

Toxoplasma gondii is an ubiquitous parasite that is able to infect a wide variety of animals. In humans, *T. gondii* is a common opportunistic pathogen in patients with HIV/AIDS (human immunodeficiency virus/acquired immodeficiency syndrome) or other immunocompromised individuals. It can also cause severe congenital birth defects in children infected *in utero*. *Toxoplasma gondii* is also an important veterinary pathogen. It is a widely studied parasite because of its pathogenicity to humans and animals, as well the fact that it is quite easy to maintain in the laboratory, in vitro or in animals, and thus can serve as an easily manipulated model system for its relatives, the malaria parasites and other apicomplexans. It is possible to generate and isolate mutants of *Toxoplasma*, perform genetic crosses, transfect both stably and transiently with foreign DNA, generate knockouts and allelic replacements, and perform insertional mutagenesis (for review, see ref. *61*).

The haploid nuclear genome of *T. gondii* consists of 11 chromosomes, ranging from 2 to more than 10 Mb, that total approx 80 Mb *(62)*. Like *Plasmodium*, *T. gondii* also possesses a 6-kb mitochondrial genome and a 35-kb apicoplast genome *(63)*. Two parallel genome sequencing efforts are under way (Table 1). One project under way at TIGR in collaboration with the University of Pennsylvania is utilizing a shotgun strategy to determine the genome sequence of a strain (type II) most commonly associated with patients with HIV/AIDS. Initially, the goal of this project was to sequence the genome to 3× coverage, but subsequently the scope was expanded to include 8× coverage and annotation because of declining costs and the availability of additional funds. This project recently reached 7× coverage, and preliminary contigs have been released. The second

Table 1
Selected Parasite Genome Projects[a]

Organism	Disease	Genome characteristics
Apicomplexa		
Plasmodium falciparum	Malaria (humans)	23 Mb, 20% G+C
Plasmodium vivax	Malaria (humans)	25 Mb, 38% G+C
Plasmodium yoelii	Malaria (rodents)	23 Mb, 23% G+C
Plasmodium berghei	Malaria (rodents)	25 Mb, 20% G+C
Plasmodium chabaudi	Malaria (rodents)	25 Mb, 20% G+C
Plasmodium vinckei	Malaria (rodents)	25 Mb, 20% G+C
Plasmodium knowlesi	Malaria (primates)	25 Mb, 38% G+C
Plasmodium gallinaceum	Malaria (chickens)	25 Mb, 20% G+C
Theileria parva	East Coast fever (cattle)	8.5 Mb, 34% G+C
Theileria annulata	Theileriosis (cattle)	8.5 Mb, 32.5% G+C
Toxoplasma gondii	Toxoplasmosis	80 Mb, 52% G+C
Cryptosporidium parvum	Diarrhea	9.4 Mb, 68% G+C
Kinetoplastida		
Leishmania major	Leishmaniasis	34 Mb, 63% G+C
Trypanosoma brucei	African sleeping sickness	35 Mb, 50% G+C
Trypanosoma cruzi	Chagas disease	40 Mb, 50% G+C
Microsporidia		
Encephalitozoon cuniculi	Gastrointestinal infections	2.9 Mb, 47% G+C
Hexamitidae		
Giardia lamblia	Giardiasis	12 Mb, 46% G+C
Entamoebidae		
Entamoeba histolytica	Amoebic dysentery	20 Mb, 24% G+C
Trematoda		
Schistosoma mansoni	Bilharzia	270 Mb, 37% G+C
Nematoda		
Brugia malayi	Lymphatic filariasis	110 Mb, 71% G+C

[a]Genome sequencing projects only; EST projects are not included. A convenient list of eukarytic genome, EST, and GSS projects is available at the National Center for Biotechnology Information Web site: http://www.ncbi.nlm.nih.gov/PMGifs/Genomes/EG_T.html. See the text for references.

project is being conducted at the Sanger Institute (Table 1) and is to determine the sequence of chromosome 1 and produce end sequences from a genomic bacterial artificial chromosome (BAC) library.

A resource that will prove very useful during the annotation of the *T. gondii* genome is the large collection of ESTs available from the different life cycle stages of the organism (currently ~64,000 ESTs in GenBank) *(64–66)*. The ESTs have been clustered *(65, 67)* and used to produce gene indices (http://www.tigr.org/tdb/tgi/tggi/). These assembled ESTs can be used to produce large training sets for gene-finding software, which should provide for more accurate and complete gene predictions than has been possible to date for *Plasmodium*, which had a much smaller training set of experimentally verified gene sequences for training of gene finders. A *Toxoplasma* database, ToxoDB (http://toxodb.org/ToxoDB.shtml), derived from PlasmoDB, has also been established to provide access to genome sequence information *(68)*.

Cryptosporidium parvum

Cryptosporidium parvum causes acute gastrointestinal illness in humans and animals. It is one of the most widespread enteric pathogens in many developing countries, and in developed countries it has been thought of as an "emerging" pathogen because of its prevalence in immunocompromised individuals *(69)*. Despite its medical importance, it is little studied, primarily because it cannot be grown efficiently in vitro. The karyotype of *C. parvum* was determined by pulsed field gel electrophoresis, and it is comprised of eight chromosomes, ranging from 0.94 to 1.44 Mb, totaling approximately 9.4 Mb *(70,71)*. Genome-sequencing efforts began with EST- and GSS-based projects *(72–74)*, and more recently, two efforts to sequence the entire *C. parvum* genome began (Table 1). One group is sequencing a genotype I isolate, which causes most human infections; the second group is sequencing a genotype 2 isolate, which is responsible for all animal infections and some human infections.

Unlike the other apicomplexans studied to date, *C. parvum* does not appear to contain an apicoplast genome *(75)*, and it has been proposed that *C. parvum* may have "lost" the apicoplast at some point during evolution. If this is the case, it would be very interesting to compare the *C. parvum* genome to the genomes of other apicomplexans, particularly to that of *P. falciparum*. The apicoplast plays essential biochemical roles in other apicomplexans, and in *P. falciparum*, up to 10% of the predicted nuclear genes encode proteins targeted to the apicoplast *(14,76)*.

KINETOPLASTIDA

The kinetoplastidae are a large group of flagellated protists that cause diseases in many different organisms, including both plants and animals. In humans, kinetoplastids cause diseases that include African sleeping sickness, Chagas disease, and cutaneous and visceral leishmaniasis. Hundreds of millions of people are afflicted with these diseases every year. As well as their importance as pathogens, these parasites exhibit very interesting biological features, such as antigenic variation, the kinetoplast, trans-splicing, RNA-editing, and glycophosphatidylinositol-anchored proteins. Unlike many other parasites, the kinetoplastidae are relatively easy to study in the laboratory. This and their importance as pathogens and their novel biological characteristics have made them popular model systems. Several species of kinetoplastidids are being sequenced, which will provide abundant opportunities for comparative genomics.

Leishmania major

Leishmania major is transmitted by sandflies; it invades and resides within host macrophages and causes cutaneous and visceral leishmaniasis. The parasite is diploid, and the nuclear genome is approximately 34 Mb, consists of 36 pairs of chromosomes, and is quite G+C rich (~63%). The genome also contains repetitive sequences, including hexameric telomeric repeats, simple sequence repeats, and transposable elements. The Leishmania Genome Network (http://www.sanger.ac.uk/Projects/L_major/), sponsored by WHO/TDR, is coordinating the sequencing of the *L. major* genome. Because of the complex genome structure, a multiprong approach has been adopted, including the sequencing of overlapping cosmid clones *(77)*, shotgun sequencing of chromosomes isolated from pulsed field gels, and BAC end sequencing.

The sequence of the smallest chromosome, chromosome 1 (269 kb), has been determined *(78)*. This chromosome encoded 79 protein-encoding genes located in a central 257-kb region; this region was flanked by 4- and 8-kb regions devoid of coding sequences and by subtelomeric and telomeric repetitive sequences. An interesting feature of the chromosome was the organization of the protein-encoding genes. The first 29 genes, covering 73 kb, were located on one strand of DNA, and the other 50 genes (182 kb) were on the opposite strand, resulting in a head-to-head arrangement of two polycistronic gene clusters with a 1.6-kb region (the switch region) in between. The switch region was proposed to function in transcription initiation, as an origin of replication or as a centromere. However, deletion of the chromosome 1 switch region did not appear to affect the transcription of a transgene inserted nearby or alter mitotic stability of the chromosome. However, deletion of the switch region in all copies of chromosome 1 resulted in the appearance of another copy of the chromosome in all of the clones analyzed *(79)*. Thus, the switch region does not appear essential for gene expression or chromosome stability, but it is essential for parasite survival. Many of the chromosomes have been completed, and the sequencing of the remaining chromosomes is under way (http://www.sanger.ac.uk/Projects/L_major/progress.shtml). Large polycistronic gene clusters as found on chromosome 1 appear to be a common feature of the *L. major* genome.

Trypanosoma brucei

African sleeping sickness in humans is caused by infection with the parasites *Trypanosoma brucei gambiense* and *Trypanosoma brucei rhodesiense*, which are transmitted by the tsetse fly in large parts of sub-Saharan Africa. Upward of 500 million people live in areas where the tsetse fly is found, and there are an estimated several hundred thousand cases of sleeping sickness every year. Approximately 15,000 deaths are attributed to sleeping sickness annually; there is no vaccine available to prevent infection, and most of the existing drug treatments are toxic or expensive. A third subspecies, *Trypanosoma brucei brucei* parasite, causes the fatal disease ngama in domesticated livestock. Large regions of Africa cannot be used to raise animals susceptible to this infection, which has prevented the development of systems for the production of meat and other products in these areas.

The *T. brucei* genome is being sequenced by TIGR and the Sanger Institute (Table 1). The genome organization is quite complex because there are at least 11 megabase-size chromosome pairs (~1 to 6 Mb), many chromosomes of intermediate size (0.2–0.9 Mb), and 50–100 linear minichromosomes (0.050–0.150 Mb). A BAC-by-BAC, map-as-you-go approach is being used. The repetitive nature of the genome has proved troublesome, but as of this writing, all of the chromosomes have been completely sequenced and annotated *(80,81)*. Similar to what was found in *L. major*, genes in the first two annotated *T. brucei* chromosomes are arranged in long polycistronic unidirectional clusters, some of which are larger than 250 kb and contain up to 100 genes. Strand switching between the clusters is associated with a change in the sign of the G+C skew statistic (a measure of base composition asymmetry between strands), suggesting that the observed base composition asymmetry may be linked to transcription or transcription-coupled repair *(80)*. Other features include novel gene families, including some in which members of the gene family are in tandem arrays; numerous retroelements; and duplications of chromosome segments. These studies also provided an examination of the subtelomeric expression sites involved in the expression of variant surface glycoprotein genes, which play a role in antigenic variation.

Trypanosoma cruzi

Trypanosoma cruzi is the causative agent of Chagas disease, a disease found in large parts of South and Central America. Up to 18 million people may become infected, and up to 50,000 deaths occur each year. The parasite is transmitted by reduviid (kissing) bugs, which become infected by feeding on infected humans and then pass on the infection to other hosts via feces. The parasite-containing feces can be rubbed into an open wound or the eyes to establish an infection. Once inside the human host, the parasite invades numerous cell types. Acute infections are usually mildly symptomatic, but can develop into chronic infections that lead to organ damage, particularly of the heart and digestive tract. Up to one-third of chronically infected individuals will die from the disease.

The *T. cruzi* genome is composed of up to 40 chromosome pairs and is 44 Mb long *(82)*. An international genome project, the *T. cruzi* Genome Initiative (http://www.dbbm. fiocruz.br/TcruziDB/, is focused on the generation of yeast artificial chromosome (YAC), BAC, cosmid libraries, and ESTs has been under way for several years *(83–85)*. A full genome project involving the University of Uppsala, the Seattle Biomedical Research Institute, and TIGR was intiated (Table 1). Initially, the strategy for this project was to use a BAC-based sequencing approach, but the highly repetitive nature of the genome interfered with fingerprinting and the selection of BACs for sequencing, so the approach was changed to a whole genome shotgun strategy. The aim is to sequence to 20× coverage because the isolate being sequenced is composed of two haplotypes. The random sequencing phase has reached 14× coverage. Accurate assembly of the genome is also challenging because of the abundance of repeated sequences and the presence of two haplotypes.

MICROSPORIDIA

Microsporidia are obligate intracellular eukaryotic parasites that cause digestive tract and neurological infections in humans, particularly in immunocompromised individuals. They have a simple, but remarkable, life cycle involving multiplication of the parasite within host cells (merogony) and the formation of spores (sporogony) for transmission of the parasite to other host cells. They do not contain mitochondria or peroxysomes, and their phylogenetic status has been controversial. Originally considered amitochondriate protists, they were subsequently proposed to be fungi that had lost their mitochondria.

The genome sequence of one microsporidian, *Encephalitozoon cuniculi*, was recently completed, revealing a genome that is highly compacted in comparison to those of free-living eukaryotes *(86)*. The genome is only 2.9 Mb, consists of 11 chromosomes, and encodes approximately 2000 genes. The gene density is thus only 1 gene per 1.25 kb, four times that of the yeasts *S. pombe* and *S. cerevisiae*. This high gene density is reflected in the very short intergenic regions (mean 129 nt), very few introns, coding sequences that, at an average size of 1077 nt, are smaller than in other sequenced organisms. For example, 350 *E. cuniculi* proteins were an average of 15% shorter than their orthologs in *S. cerevisiae*. This is quite different from what was found in *P. falciparum*, for which the mean length of coding sequences (2283 nt) was much greater than in other sequenced eukaryotes (from 1300 to 1600 nt).

Analysis of the genome sequence provided many insights into the biochemistry and cell biology of the organism. It appears to lack enzymes for the TCA cycle, respiratory electron transport, amino acid biosynthesis, *de novo* purine and pyrimidine biosynthesis, and fatty acid synthase and other pathways and contains just a handful of membrane transporters. These features confirm that the parasite is dependent on the host for the provision of important biochemical substrates. Although *E. cuniculi* parasites do not contain recognizable mitochondria or mitochondrial DNA, the identification of *E. cuniculi* orthologs of *S. cerevisiae* mitochondrial proteins involved in iron–sulfur cluster assembly led to the proposal that these organisms may contain cryptic organelles, called *mitosomes*, that may be involved in protection against oxidative stress.

HEXAMITIDAE

Giardia lamblia is a flagellated eukaryotic parasite that infects the small intestine of humans and other mammals and can cause a variety of gastrointestinal symptoms, particulary diarrhea. An infection is initiated by the ingestion of cysts, which develop into the trophozoite form of the organism in the small intestine. *Giardia* and related parasites such as *Trichomomas* are thought to be among the most primitive eukaryotes. *Giardia* lacks mitochondria, nucleoli, and peroxisomes, but retains other features typical of eukaryotic cells. The organism is polyploid, and each trophozoite contains two apparently identical or similar nuclei, each with complete copies of the genome *(87)*. The genome is estimated to be 12 Mb and composed of five chromosomes ranging from 1.6 to 3.8 Mb.

The Marine Biological Laboratory is sequencing the *G. lamblia* genome using a whole genome shotgun strategy *(88)*. The random sequencing phase has attained almost 7× coverage, and contigs representing at least 90% of the genome are available (http://jbpc.mbl.edu/Giardia-HTML/). The genome of *Trichomonas vaginalis*, which causes a sexually transmitted disease that is one of the most common forms of infertility, is also being sequenced at TIGR (Jane Carlton, personal communication, 2003). Besides their utility in facilitating laboratory investigations of these parasites, the genome sequences of these organisms should shed light on the phylogenetic relationships between these primitive parasites and higher eukaryotes.

ENTAMOEBIDAE

Entamoeba histolytica is a protozoan parasite that causes dysentary and liver abscesses in infected individuals. It is transmitted by the ingestion of parasites in contaminated water supplies and affects approximately 40–50 million people each year. The *E. histolytica* genome is approximately 20 Mb and consists of approximately 14 linear chromosomes as well as circular DNA molecules, and it contains abundant repetitive sequences *(89)*. The genome is being sequenced by TIGR and the Sanger Institute in a collaborative effort, with shotgun sequence reads from both centers combined for the assembly of the genome sequence.

An unusual problem encountered during the sequencing of this genome was the presence of episomes encoding rRNA genes. Sequences from the episome comprised 15% of the total sequences from the initial shotgun libraries. Cleavage of the rRNA-encoding episomes with a restriction enzyme and removal of the linearized episomes by electrophoresis enabled the construction of genomic libraries with fewer rRNA sequences *(90)*. Random sequencing has reached approximately ninefold coverage (http://www.tigr.org/tdb/e2k1/eha1/; http://www.sanger.ac.uk/Projects/E_histolytica/).

To facilitate comparative studies, partial genome sequencing of several other species of *Entamoeba* are under way, including *Entamoeba dispar*, a noninvasive species that infects humans but is morphologically indistinguishable from *E. histolytica*. These studies may help identify virulence factors that contribute to pathogenicity.

TREMATODA

Schistosomes, commonly known as the blood flukes, are found in Africa, Asia, and the Americas. These parasites are transmitted by snails, which release motile, invasive forms of the parasite called cercaria into the water. The cercaria burrow through the skin of the mammalian host, transform into schistosomula, and enter the circulatory system. The schistosomula develop into male and female adult worms that form pairs, reside in blood vessels of the host, and produce eggs that are released into the environment in the feces or urine, depending on the species of schistosome. Miracidia develop from the eggs and infect snails to complete the life cycle. Several species of schistosomes infect humans, including *S. mansoni*, *Schistosoma haematobium*, and *Schistosoma japonicum*. They can cause chronic infections and debilitating disease, including damage to the liver and intestine (*S. mansoni*) or urinary tract (*S. haematobium, S. japonicum*).

Schistosoma mansoni was the target of one of the first WHO/TDR-sponsored parasite sequencing projects, the *Schistosoma* Genome Network (http://www.nhm.ac.uk/hosted_sites/schisto/). Because of the large size of the *Schistosoma* genome (270 Mb), this effort was primarily devoted to the production of ESTs for gene discovery, but it was notable for the early involvement of scientists from countries where schistosomiasis is endemic *(91)*. The *S. mansoni* genome (270 Mb) is currently being sequenced at TIGR using a whole genome shotgun strategy employing large (BAC), medium (12- to 15-kb), and small (2- to 3-kb) insert libraries. To date, approximately 31,000 BAC end sequences have been generated (http://www.tigr.org/tdb/e2k1/sma1/), and sequencing of the smaller clones has reached about 2× coverage (Najib El-Sayed, personal communication). Funds are available for 3× sequence coverage of the genome, which should provide partial or full-length sequences of more than 90% of parasite genes. The large EST data set and the clustered and assembled ESTs generated by the *Schistosoma* Genome Network will be invaluable for the annotation of the genome sequence *(92)*.

NEMATODA

The nematodes *Brugia malayi* and *Wucheria bancrofti* cause lymphatic filariasis, a disfiguring and debilitating disease that infects over 100 million people. Another nematode, *Onchocerca volvulus*, causes river blindness in large swaths of sub-Saharan Africa. These parasitic worms were targeted for genomic analysis, primarily ESTs, by the WHO/TDR-sponsored Filarial Genome Project (http://nema.cap.ed.ac.uk/fgn/filgen1.html;

93). Most of the work has been conducted on *B. malayi* because all stages of the life cycle, including the stages in the mosquito vector, can be maintained in the laboratory, but some data have also been collected for *O. volvulus (94)*.

An effort to sequence the 110-Mb genome of *B. malayi* was initiated (http://www.tigr.org/tdb/e2k1/bma1/). A whole genome strategy using small, medium, and large (BAC) insert libraries is being used, and 1× coverage has been attained. Another effort is under way at the Sanger Institute to sequence a 150-kb region of the genome (http://www.sanger.ac.uk/Projects/B_malayi/).

CONCLUDING REMARKS

Genome-sequencing efforts have provided vast amounts of information from organisms ranging from the simplest prokaryotes to humans. Genome sequences have already had a major impact on parasitology by providing insights into parasite biology and host–parasite interactions; these insights could not have been obtained easily using other methods. They have provided literally thousands of new starting points for laboratory investigations.

However, genome sequence data can be difficult for bench scientists to view, query, and apply to experimental work under way in their laboratories. To address these problems, a wide variety of databases and software have been developed to facilitate the storage and manipulation of genome sequence information; most of the databases and software are publicly accessible over the Internet. Organism-specific databases such as FlyBase (http://flybase.bio.indiana.edu/; *(95)* and *Saccharomyces* Genome Database; (http://www.yeastgenome.org/) provide one-stop shopping for genome sequence data and information on strains, mutants, reagents, and the like for model organisms. Several such databases exist for parasitic organisms and have proven extremely popular to and useful for the scientific community. PlasmoDB (www.plasmodb.org), for example, contains the genome sequence data and annotation, finished and preliminary, for all of the *Plasmodium* genome-sequencing projects.

A large investment has been made to generate parasite genome sequences. To capitalize on this investment and ensure that the information can be effectively applied to basic research and the development of new control measures for parasitic diseases, it is imperative that these valuable scientific resources continue to be developed for all parasites. Procedures must be established to ensure that the information in these databases is continually updated to reflect new experimental findings and improvements in bioinformatic methods for gene finding and annotation.

Finally, training in bioinformatics, such as that offered by the WHO/TDR and the Malaria Research and Reference Reagent Resource Center (http://www.malaria.mr4.org/index.html), is essential to ensure that scientists working in both developed and developing countries are able to obtain and apply genome information to fundamental and applied research in parasitology.

ACKNOWLEDGMENTS

I thank my colleagues at TIGR, particularly the members of the Parasite Genomics Group (Vish Nene, Brendan Loftus, Jane Carlton, Najib El-Sayed, Elodie Ghedin, and

Ruobing Wang) for their support and encouragement. Funding of parasite genome-sequencing efforts at TIGR have been supported by the National Institute for Allergy and Infectious Diseases, The Burroughs Wellcome Fund, the U.S. Department of Defense, the International Livestock Research Institute, and J. Craig Venter.

REFERENCES

1. Sachs J, Malaney P. The economic and social burden of malaria. Nature 2002; 415:680–685.
2. Al-Olayan EM, Beetsma AL, Butcher GA, Sinden RE, Hurd H. Complete development of mosquito phases of the malaria parasite in vitro. Science 2002; 295:677–679.
3. Kocken CH, Ozwara H, van der Wel A, Beetsma AL, Mwenda JM, Thomas AW. *Plasmodium knowlesi* provides a rapid in vitro and in vivo transfection system that enables double-crossover gene knockout studies. Infect Immun 2002; 70:655–660.
4. Johnston DA, Blaxter ML, Degrave WM, Foster J, Ivens AC, Melville SE. Genomics and the biology of parasites. Bioessays 1999; 21:131–147.
5. Singh U, Brewer JL, Boothroyd JC. Genetic analysis of tachyzoite to bradyzoite differentiation mutants in *Toxoplasma gondii* reveals a hierarchy of gene induction. Mol Microbiol 2002; 44: 721–733.
6. Ben Mamoun C, Gluzman IY, Hott C, et al. Co-ordinated programme of gene expression during asexual intraerythrocytic development of the human malaria parasite *Plasmodium falciparum* revealed by microarray analysis. Mol Microbiol 2001; 39:26–36.
7. Hayward RE, Derisi JL, Alfadhli S, Kaslow DC, Brown PO, Rathod PK. Shotgun DNA microarrays and stage-specific gene expression in *Plasmodium falciparum* malaria. Mol Microbiol 2000; 35:6–14.
8. Bozdech Z, Zhu J, Joachimiak MP, Cohen FE, Pulliam B, DeRisi JL. Expression profiling of the schizont and trophozoite stages of *Plasmodium falciparum* with a long-oligonucleotide microarray. Genome Biol 2003; 4:R9.
9. Blader IJ, Manger ID, Boothroyd JC. Microarray analysis reveals previously unknown changes in *Toxoplasma gondii*-infected human cells. J Biol Chem 2001; 276:24,223–24,231.
10. Cleary MD, Singh U, Blader IJ, Brewer JL, Boothroyd JC. *Toxoplasma gondii* asexual development: identification of developmentally regulated genes and distinct patterns of gene expression. Eukaryot Cell 2002; 1:329–340.
11. Florens L, Washburn MP, Raine JD, et al. A proteomic view of the *Plasmodium falciparum* life cycle. Nature 2002; 419:520–526.
12. Lasonder E, Ishihama Y, Andersen JS, et al. Analysis of the *Plasmodium falciparum* proteome by high-accuracy mass spectrometry. Nature 2002; 419:537–542.
13. Cohen AM, Rumpel K, Coombs GH, Wastling JM. Characterisation of global protein expression by two-dimensional electrophoresis and mass spectrometry: proteomics of *Toxoplasma gondii*. Int J Parasitol 2002; 32:39–51.
14. Gardner MJ, Hall N, Fung E, et al. Genome sequence of the human malaria parasite *Plasmodium falciparum*. Nature 2002; 419:498–511.
15. Gardner MJ, Shallom S, Carlton JM, et al. Sequence of *Plasmodium falciparum* chromosomes 2, 10, 11, and 14. Nature 2002; 419:531–534.
16. Hall N, Pain A, Berriman M, et al. Sequence of *Plasmodium falciparum* chromosomes 1, 3–9 and 13. Nature 2002; 419:527–531.
17. Hyman RW, Fung E, Conway et al. Sequence of *Plasmodium falciparum* chromosome 12. Nature 2002; 419:534–537.
18. Gardner MJ, Tettelin H, Carucci DJ, et al. Chromosome 2 sequence of the human malaria parasite *Plasmodium falciparum*. Science 1998; 282:1126–1132.

19. Bowman S, Lawson D, Basham D, et al. The complete nucleotide sequence of chromosome 3 of *Plasmodium falciparum*. Nature 1999; 400:532–538.

20. Hoffman SL, Bancroft WH, Gottlieb M, et al. Funding for malaria genome sequencing. Nature 1997; 387:647.

21. Walliker D, Quayki I, Wellems TE, McCutchan TF. Genetic analysis of the human malaria parasite *Plasmodium falciparum*. Science 1987; 236:1661–1666.

22. Jing J, Aston C, Zhongwu L, et al. Optical mapping of *Plasmodium falciparum* chromosome 2. Genome Res 1999; 9:175–181.

23. Lai Z, Jing J, Aston C, et al. A shotgun optical map of the entire *Plasmodium falciparum* genome. Nature Genetics 1999; 23:309–313.

24. Salzberg SL, Pertea M, Delcher A, Gardner MJ, Tettelin H. Interpolated Markov models for eukaryotic gene finding. Genomics 1999; 59:24–31.

25. Cawley SE, Wirth AI, Speed TP. Phat—a gene finding program for *Plasmodium falciparum*. Mol Biochem Parasitol 2001; 118:167–174.

26. Jomaa H, Wiesner J, Sanderbrand S, et al. Inhibitors of the nonmevalonate pathway of isoprenoid biosynthesis as antimalarial drugs. Science 1999; 285:1573–1576.

27. Surolia N, Surolia A. Triclosan offers protection against blood stages of malaria by inhibiting enoyl-ACP reductase of *Plasmodium falciparum*. Nat Med 2001; 7:167–173.

28. Surolia N, RamachandraRao SP, Surolia A. Paradigm shifts in malaria parasite biochemistry and anti-malarial chemotherapy. Bioessays 2002; 24:192–196.

29. Macilwain C. Biologists challenge sequencers on parasite genome publication. Nature 2000; 405: 601–602.

30. Gottlieb M, McGovern V, Goodwin P, Hoffman S, Oduola A. Please don't downgrade the sequencers' role. Nature 2000; 406:121–122.

31. Kissinger JC, Brunk BP, Crabtree J, et al. The *Plasmodium* genome database. Nature 2002; 419: 490–492.

32. Bahl A, Brunk B, Crabtree J, et al. PlasmoDB: the *Plasmodium* genome resource. A database integrating experimental and computational data. Nucleic Acids Res 2003; 31:212–215.

33. Chakrabarti D, Reddy GR, Dame JB, et al. Analysis of expressed sequence tags from *Plasmodium falciparum*. Mol Biochem Parasitol 1994; 66:97–104.

34. Patankar S, Munasinghe A, Shoaibi A, Cummings LM, Wirth DF. Serial analysis of gene expression in *Plasmodium falciparum* reveals the global expression profile of erythrocytic stages and the presence of anti-sense transcripts in the malarial parasite. Mol Biol Cell 2001; 12:3114–3125.

35. Munasinghe A, Patankar S, Cook BP, et al. Serial analysis of gene expression (SAGE) in *Plasmodium falciparum*: application of the technique to A-T rich genomes. Mol Biochem Parasitol 2001; 113:23–34.

36. Mendis K, Sina BJ, Marchesini P, Carter R. The neglected burden of *Plasmodium vivax* malaria. Am J Trop Med Hyg 2001; 64:97–106.

37. Baird JK, Basri H, Purnomo, et al. Resistance to chloroquine by *Plasmodium vivax* in Irian Jaya, Indonesia. Am J Trop Med Hyg 1991; 44:547–552.

38. Schuurkamp GJ, Spicer PE, Kereu RK, Bulungol PK, Rieckmann KH. Chloroquine-resistant *Plasmodium vivax* in Papua New Guinea. Trans R Soc Trop Med Hyg 1992; 86:121–122.

39. Phillips EJ, Keystone JS, Kain KC. Failure of combined chloroquine and high-dose primaquine therapy for *Plasmodium vivax* malaria acquired in Guyana, South America. Clin Infect Dis 1996; 23:1171–1173.

40. Horuk R, Chitnis CE, Darbonne WC, et al. A receptor for the malarial parasite *Plasmodium vivax*: the erythrocyte chemokine receptor. Science 1993; 261:1182–1184.

41. Fraser T, Michon P, Barnwell JW, et al. Expression and serologic activity of a soluble recombinant *Plasmodium vivax* Duffy binding protein. Infect Immun 1997; 65:2772–2777.

42. Trager W, Jensen W. Cultivation of malaria parasites. Nature 1978; 273:621–622.
43. Haynes JD, Diggs CL, Hines FA, Desjardins RE. Culture of human malaria parasites *Plasmodium falciparum*. Nature 1976; 263:767–769.
44. Carlton JM-R, Galinski MR, Barnwell JW, Dame JB. Karyotype and synteny among the chromosomes of all four species of human malaria parasite. Mol Biochem Parasitol 1999; 101:23–32.
45. Carlton JM, Angiuoli SV, Suh BB, et al. Genome sequence and comparative analysis of the model rodent malaria parasite *Plasmodium yoelii yoelii*. Nature 2002; 419:512–519.
46. Carlton JM, Muller R, Yowell CA, et al. Profiling the malaria genome: a gene survey of three species of malaria parasite with comparison to other apicomplexan species. Mol Biochem Parasitol 2001; 118:201–210.
47. del Portillo HA, Fernandez-Becerra C, Bowman S, et al. A superfamily of variant genes encoded in the subtelomeric region of *Plasmodium vivax*. Nature 2001; 410:839–842.
48. Janssen CS, Barrett MP, Lawson D, et al. Gene discovery in *Plasmodium chabaudi* by genome survey sequencing. Mol Biochem Parasitol 2001; 113:251–260.
49. Janse CJ, Carlton JM-R, Walliker D, Waters AP. Conserved location of genes on polymorphic chromosomes of four species of malaria parasites. Mol Biochem Parasitol 1994; 68:285–296.
50. Carlton JMR, Vinkenoog R, Waters AP, Walliker D. Gene synteny in species of *Plasmodium*. Mol Biochem Parasitol 1998; 93:285–294.
51. Tchavtchitch M, Fischer K, Huestis R, Saul A. The sequence of a 200 kb portion of a *Plasmodium vivax* chromosome reveals a high degree of conservation with *Plasmodium falciparum* chromosome 3. Mol Biochem Parasitol 2001; 118:211–222.
52. Collins WE, Contacos PG, Krotoski WA, Howard WA. Transmission of four Central American strains of *Plasmodium vivax* from monkey to man. J Parasitol 1972; 58:332–335.
53. Carlton J. The *Plasmodium vivax* genome sequencing project. Trends Parasitol 2003; 19:227–231.
54. van Lin LH, Janse CJ, Waters AP. The conserved genome organisation of non-falciparum malaria species: the need to know more. Int J Parasitol 2000; 30:357–370.
55. van Lin LH, Pace T, Janse CJ, et al. Interspecies conservation of gene order and intron-exon structure in a genomic locus of high gene density and complexity in *Plasmodium*. Nucleic Acids Res 2001; 29:2059–2068.
56. Rogers WO, Weiss WR, Kumar A, et al. Protection of rhesus macaques against lethal *Plasmodium knowlesi* malaria by a heterologous DNA priming and poxvirus boosting immunization regimen. Infect Immun 2002; 70:4329–4335.
57. Jones SH, Lew AE, Jorgensen WK, Barker SC. *Babesia bovis*: genome size, number of chromosomes and telomeric probe hybridisation. Int J Parasitol 1997; 27:1569–1573.
58. Norval RAI, Perry BD, Young AS. The Epidemiology of Theileriosis in Africa. New York: Academic, 1992.
59. Mukhebi A, Perry BD, Kruska R. Estimated economics of theileriosis in Africa. Prevent Vet Med 1992; 12:73–85.
60. Dobbelaere D, Heussler V. Transformation of leukocytes by *Theileria parva* and *T. annulata*. Annu Rev Microbiol 1999; 53:1–42.
61. Roos DS, Darling J, Reynolds MG, Hager KM, Striepen B, Kissinger JC. *Toxoplasma* as a model parasite: apicomplexan biochemistry, cell biology, molecular genetics, genomics and beyond. In: Tschudi C, Pearce EJ (eds). Biology of Parasitism. Boston: Kluwer, 2000, pp. 143–167.
62. Sibley LD, Boothroyd JC. Construction of a molecular karyotype for *Toxoplasma gondii*. Mol Biochem Parasitol 1992; 51:291–300.
63. Kohler S, Delwiche CF, Denny PW, et al. A plastid of probable green algal origin in apicomplexan parasites. Science 1997; 275:1485–1489.
64. Manger ID, Hehl A, Parmley S, et al. Expressed sequence tag analysis of the bradyzoite stage of *Toxoplasma gondii*: identification of developmentally regulated genes. Infect Immun 1998; 66:1632–1637.

65. Ajioka JW. *Toxoplasma gondii*: ESTs and gene discovery. Int J Parasitol 1998; 28:1025–1031.

66. Wan KL, Blackwell JM, Ajioka JW. *Toxoplasma gondii* expressed sequence tags: insight into tachyzoite gene expression. Mol Biochem Parasitol 1996; 75:179–186.

67. Ajioka JW, Boothroyd JC, Brunk BP, et al. Gene discovery by EST sequencing in *Toxoplasma gondii* reveals sequences restricted to the Apicomplexa. Genome Res 1998; 8:18–28.

68. Kissinger JC, Gajria B, Li L, Paulsen IT, Roos DS. ToxoDB: accessing the *Toxoplasma gondii* genome. Nucleic Acids Res 2003; 31:234–236.

69. Spano F, Crisanti A. *Cryptosporidium parvum*: the many secrets of a small genome. Int J Parasitol 2000; 30:553–565.

70. Caccio S, Camilli R, La Rosa G, Pozio E. Establishing the *Cryptosporidium parvum* karyotype by *Not*I and *Sfi*I restriction analysis and Southern hybridization. Gene 1998; 219:73–79.

71. Blunt DS, Khramtsov NV, Upton SJ, Montelone BA. Molecular karyotype analysis of *Cryptosporidium parvum*: evidence for eight chromosomes and a low-molecular-size molecule. Clin Diagn Lab Immunol 1997; 4:11–13.

72. Strong WB, Nelson RG. Preliminary profile of the *Cryptosporidium parvum* genome: an expressed sequence tag and genome survey sequence analysis. Mol Biochem Parasitol 2000; 107:1–32.

73. Liu C, Vigdorovich V, Kapur V, Abrahamsen MS. A random survey of the *Cryptosporidium parvum* genome. Infect Immun 1999; 67:3960–3969.

74. Abrahamsen MS. *Cryptosporidium parvum* gene discovery. Adv Exp Med Biol 1999; 473:241–247.

75. Zhu G, Marchewka MJ, Keithly JS. *Cryptosporidium parvum* appears to lack a plastid genome. Microbiology 2000; 146(Pt 2):315–321.

76. Foth BJ, Ralph SA, Tonkin CJ, et al. Dissecting apicoplast targeting in the malaria parasite *Plasmodium falciparum*. Science 2003; 299:705–708.

77. Ivens AC, Lewis SM, Bagherzadeh A, Zhang L, Chan HM, Smith DF. A physical map of the *Leishmania major* Friedlin genome. Genome Res 1998; 8:135–145.

78. Myler PJ, Audleman L, deVos T, et al. *Leishmania major* Friedlin chromosome 1 has an unusual distribution of protein-coding genes. Proc Natl Acad Sci USA 1999; 96:2902–2906.

79. Dubessay P, Ravel C, Bastien P, et al. The switch region on *Leishmania major* chromosome 1 is not required for mitotic stability or gene expression, but appears to be essential. Nucleic Acids Res 2002; 30:3692–3697.

80. El-Sayed NM, Ghedin E, Song J, et al. The sequence and analysis of *Trypanosoma brucei* chromosome II. Nucleic Acids Res 2003; 31:4856–4863.

81. Hall N, Berriman M, Lennard NJ, et al. The DNA sequence of chromosome 1 of an African trypanosome: gene content, chromosome organisation, recombination and polymophism. Nucleic Acids Res 2003; 31:4864–4873.

82. Gull K. The biology of kinetoplastid parasites: insights and challenges from genomics and post-genomics. Int J Parasitol 2001; 31:443–452.

83. Zingales B, Rondinelli E, Degrave W, et al. The *Trypanosoma cruzi* genome initiative. Parasitol Today 1997; 13:16–22.

84. Brandao A, Urmenyi T, Rondinelli E, Gonzalez A, de Miranda AB, Degrave W. Identification of transcribed sequences (ESTs) in the *Trypanosoma cruzi* genome project. Mem Inst Oswaldo Cruz 1997; 92:863–866.

85. Aguero F, Verdun RE, Frasch AC, Sanchez DO. A random sequencing approach for the analysis of the *Trypanosoma cruzi* genome: general structure, large gene and repetitive DNA families, and gene discovery. Genome Res 2000; 10:1996–2005.

86. Katinka MD, Duprat S, Cornillot E, et al. Genome sequence and gene compaction of the eukaryote parasite *Encephalitozoon cuniculi*. Nature 2001; 414:450–453.

87. Yu LZ, Birky CW Jr, Adam RD. The two nuclei of *Giardia* each have complete copies of the genome and are partitioned equationally at cytokinesis. Eukaryot Cell 2002; 1:191–199.

88. McArthur AG, Morrison HG, Nixon JE, et al. The *Giardia* genome project database. FEMS Microbiol Lett 2000; 189:271–273.

89. Bhattacharya A, Satish S, Bagchi A, Bhattacharya S. The genome of *Entamoeba histolytica*. Int J Parasitol 2000; 30:401–410.

90. Mann BJ. *Entamoeba histolytica* Genome Project: an update. Trends Parasitol 2002; 18:147–148.

91. Franco GR, Valadao AF, Azevedo V, Rabelo EM. The *Schistosoma* gene discovery program: state of the art. Int J Parasitol 2000; 30:453–463.

92. Oliveira G, Johnston DA. Mining the schistosome DNA sequence database. Trends Parasitol 2001; 17:501–503.

93. Williams SA, Lizotte-Waniewski MR, Foster J, et al. The filarial genome project: analysis of the nuclear, mitochondrial and endosymbiont genomes of *Brugia malayi*. Int J Parasitol 2000; 30:411–419.

94. Williams SA, Laney SJ, Lizotte-Waniewski M, Bierwert LA, Unnasch TR. The river blindness genome project. Trends Parasitol 2002; 18:86–90.

95. FlyBase Consortium. The FlyBase database of the *Drosophila* genome projects and community literature. Nucleic Acids Res 2003; 31:172–175.

21

Genome Sequence
of an Extremely Halophilic Archaeon

Shiladitya DasSarma

INTRODUCTION

Extreme halophiles are novel microorganisms that require 5–10 times the salinity of seawater (ca. 3–5M NaCl) for optimal growth *(1,2)*. They include diverse prokaryotic species, both archaeal and bacterial, and some eukaryotic organisms. Extreme halophiles are found in hypersaline environments near the sea or salt deposits of marine or nonmarine origin. Two of the largest hypersaline lakes supporting a variety of halophilic species are the Great Salt Lake in the western United States and the Dead Sea in the Middle East. Some of the most interesting hypersaline environments are small artificial solar salterns used for producing salt from the sea, which are distributed throughout the world. Many hypersaline environments exhibit gradients of increasing salinity temporally and produce sequential growth of progressively more halophilic species, including complex microbial mats and spectacular blooms of bright red and red-orange colored species. These environments are important ecologically, frequently supporting entire populations of such exotic birds as pink flamingoes, which obtain their color from the pigmented halophilic microorganisms. A critical feature of halophilic microbes that prevents cell lysis in hypersaline environments is their high internal concentration of compatible solutes (e.g., amino acids, polyols, and salts), which act as osmoprotectants.

Although a wide variety of halophiles has been cultured, the genome of only a single extreme halophile, *Halobacterium* sp NRC-1, has been completely sequenced thus far *(3,4)*. This species is a typical halophile commonly found in many hypersaline environments, including the Great Salt Lake and solar salterns. Phylogenetically, it is classified as an archaeon, a member of the third branch of life (Fig. 1). It has a growth optimum of 4.5M NaCl, close to the saturation point, and a high concentration of K$^+$ salts internally. *Halobacterium* NRC-1 is a mesophilic archaeon, with a temperature optimum of 42°C for growth. Alhough *Halobacterium* species are thought to have limited physiological capabilities, strain NRC-1 is metabolically quite versatile, growing aerobically, anaerobically, and phototrophically. Phototrophic growth is mediated by the light-driven proton pumping of bacteriorhodopsin, which forms a two-dimensional crystalline lattice in the purple membrane. *Halobacterium* NRC-1 is also highly resistant to ultraviolet and γ-radiation and displays sophisticated motility responses, including phototaxis, chemotaxis,

From: *Microbial Genomes*
Edited by: C. M. Fraser, T. D. Read, and K. E. Nelson © Humana Press Inc., Totowa, NJ

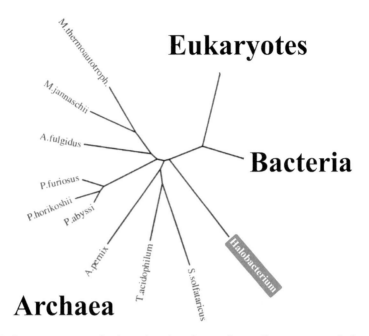

Fig. 1. Whole genome tree of selected archaeal organisms. Gene content phylogeny done by neighbor-joining using the SHOT web server *(19)* indicates that *Halobacterium* is located at the base of the archaeal branch of the phylogenetic tree.

and gas vesicle-mediated flotation. One of the most notable features of *Halobacterium* NRC-1, revealed by genome sequencing, is a highly acidic proteome, which is likely essential to maintain protein solubility and function in high salinity. Significantly, this organism is amenable to analysis using well-developed genetic methodology, including gene knockouts, expression vectors, and complementation systems, which make *Halobacterium* NRC-1 a good model for functional genomic studies among extremophiles and archaea *(2)*.

In addition to *Halobacterium* NRC-1, several other halophiles are the subject of ongoing genome projects. The most notable among these are two Dead Sea archaea, *Haloarcula marismortui* and *Haloferax volcanii (1)*, which are slightly less halophilic than *Halobacterium* NRC-1, with an optimum salinity of 2–3*M* NaCl and a high magnesium ion tolerance, reflecting the salt composition of their environment. They also display metabolic capability for growth in media containing simple sugars and carbohydrates as carbon and energy sources. Several other interesting categories of halophiles worthy of genomic studies include alkaliphilic halophiles, which grow in soda lakes with pH of 9.0–11.0; psychrotrophic halophiles, which grow at freezing temperatures in Antarctic lakes; bacterial halophiles, which tolerate a wide range of salinity; and eukaryotic halophiles, such as the green algae, *Dunaliella salina*. Finally, sequencing of a haloarchaeal strain with a nearly identical chromosome to strain NRC-1 is also in progress. A listing of current genome projects on halophiles is maintained on the Halophile Genomes Web site at the University of Maryland Biotechnology Institute, Center of Marine Biotechnology (http://zdna2.umbi.umd.edu).

THE *HALOBACTERIUM* GENOME

The genomes of *Halobacterium* species were originally studied a half-century ago; they are composed of two components, a major fraction that is G+C-rich and a relatively A+T-rich (58% G+C) satellite *(5)*. Subsequent studies showed that the satellite deoxyribonucleic acid (DNA) corresponded mainly to large heterogeneous extrachromosomal replicons containing many transposable insertion sequence (IS) elements *(6)*. For *Halobacterium* NRC-1, extensive mapping revealed the presence of three replicons: pNRC100, about 200 kbp; pNRC200, nearly twice the size of pNRC100; and a 2-Mbp chromosome (Fig. 2) *(7,8)*. The pNRC100 replicon was found to be partly identical to pNRC200 and to exist as inversion isomers *(7)*. The chromosomes of strain NRC-1 and another wild-type strain, GRB, were compared by restriction mapping, which showed extensive regions of similarity and a few regions with differences, including a large inversion and an insertion. Ordered cosmid libraries representing the genomes of *Halobacterium* species GRB and *H. volcanii* were also constructed and compared by hybridization, which indicated the lack of any detectable conserved gene organization *(9)*. These and other mapping projects suggest that significant diversity exists within the genomes of halophilic archaea.

Genome Sequencing and Analysis

Because of the high G+C composition and the large number of IS elements, the *Halobacterium* NRC-1 genome was sequenced in two stages. Initially, the pNRC100 replicon was sequenced by a combination of random shotgun sequencing of libraries made from purified covalently closed circular DNA and directed sequencing of cloned and mapped *Hin*dIII fragments *(3,7)*. This approach permitted the assembly of an unstable replicon that undergoes frequent DNA rearrangements, including inversion isomerization, and contains many IS elements. Subsequently, whole genome random shotgun sequencing was performed, providing 7.5× coverage of the relatively stable large chromosome *(4)*. Remaining lower-quality regions were sequenced using polymerase chain reaction fragments and by primer walking. The NRC-1 genome was assembled using the *Phred, Phrap,* and *Consed* programs, initially masking all the known and putative new IS elements, to avoid the formation of chimeric contigs *(4,10)*.

The complete genome sequence of *Halobacterium* NRC-1 revealed a 2,571,010-bp genome, including the 2,014,239-bp G+C-rich chromosome, and two smaller circles, 191,346-bp pNRC100, and 365,425-bp pNRC200 (Table 1; Fig. 2) *(3,4)*. Interestingly, pNRC100 and pNRC200 contained a 145,428-bp region of 100% identity, including 33- to 39-kb inverted repeats, which mediate inversion isomerization; the small single copy region; and a part of the large single copy regions (Fig. 2) *(7)*. The unique regions of the large single copy region contained 45,918 bp for pNRC100 and 219,997 bp for pNRC200. Glimmer (Gene Locator and Interpolated Markov Modeler) was used to identify 2,630 likely genes in the genome, of which 64% coded for proteins with significant matches to the databases *(4)*. In addition, 52 ribonucleic acid (RNA) genes were identified. About 40 genes in pNRC100 and pNRC200 coded for proteins likely to be essential or important for cell viability, such as a DNA polymerase, TBP and TFB transcription factors, and the arginyl–tRNA (transfer RNA) synthetase, suggesting that these replicons should be classified as minichromosomes rather than megaplasmids *(3,4)*.

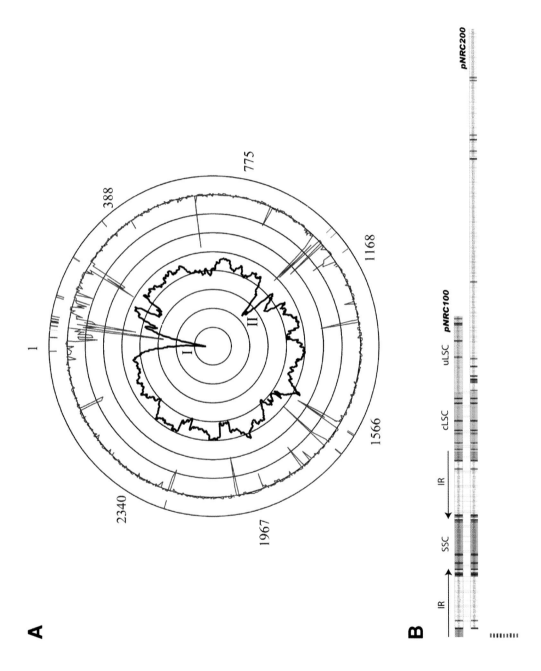

Proteome Analysis

One of the most dramatic results of genome sequencing of *Halobacterium* NRC-1 was the finding of an extremely acidic complement of encoded proteins, which is likely directly related to protein function in its hypersaline (>4*M* KCl) cytoplasm *(11)*. Calculated isoelectric points (p*I*s) for predicted proteins showed an average p*I* of approx 5, a prediction confirmed by proteomic analysis (Fig. 3). Similarly, acidic proteomes were predicted from partial genome sequences of two other halophiles, *H. marismortui* and *H. volcanii*. In contrast, the average p*I*s of nearly all other proteomes are close to neutral. Notable exceptions are *Methanobacterium thermoautotrophicum*, which also contains both an acidic proteome and a relatively high (~1*M*) internal concentration of K$^+$ ions, and three hyperthermophiles (*Pyrobaculum aerophilum*, *Pyrococcus furiosus*, and *Sulfolobus solfataricus*), which have relatively basic proteomes. Homology modeling has shown that the acidic p*I* of *Halobacterium* NRC-1 proteins is correlated with a high concentration of surface negative charge *(11)*. For example, a transcription factor (TbpE) and a topoisomerase subunit (GyrA) showed a marked increase in surface negative charge when compared to their homologs in nonhalophilic organisms *(11)*.

G+C Composition and IS Elements

Common characteristics of halophile genomes are their high G+C composition major fraction, low G+C satellite fraction, and a preponderance of IS elements *(6)*. For *Halobacterium* NRC-1, the two pNRC replicons, which represent only 22% of the genome, are substantially less G+C rich (58–59% G+C) than the large chromosome (68% G+C) and contain a majority (69/91 or 76%) of the IS elements in the genome (Fig. 2). In addition, two regions of the chromosome are less G+C rich than average, with one 270-kbp region (region I) containing 65% G+C and 13 IS elements and a second 150-kbp region (region II) with 66% G+C and 4 IS elements (Fig. 2) *(11)*. Interestingly, a 15-kbp region

---

Fig. 2. (*Opposite page*) (**A**) Circular map of the *Halobacterium* NRC-1 large chromosome and (**B**) aligned linear genetic maps of pNRC100 and pNRC200 replicons. (A) The circular map of the large chromsome plots contains locations of IS elements (outer scale), χ-squared analysis (red line), and G+C composition of open reading frames (black line). Colored bars associated with the outermost circle indicate the position of the chromosomal IS elements (ISH1, beige; ISH2, purple; ISH3, green; ISH4, yellow; ISH6, pink; ISH8, blue; ISH10, red). Roman numerals I and II indicate AT-rich islands. (B) The circular replicons are depicted in linear forms, with the genes and IS elements represented as blocks. The two replicons contain 145,428 bp of identity and either 45,918 bp or 219,997 bp of unique DNA for pNRC100 and pNRC200, respectively *(3,4)*. The 33- to 39-kb inverted repeats are shown in yellow (conserved in all copies) and orange (conserved in some, but not all, copies); the small single copy regions are in purple; the common large single copy regions are in bright green; and the unique large single copy regions are in tan (pNRC100) and light green (pNRC100). The IS elements are shown in dark orange (ISH2), brown (ISH3), indigo (ISH5), blue (ISH7), dark green (ISH8), teal (ISH9), red (ISH10), and blue-gray (ISH11). The pNRC replicons contain 69 IS elements (44 unique), 29 on pNRC100 and 40 on pNRC200; with 6 elements in the inverted repeats (repeated twice in both pNRC100 and pNR200 each), 4 elements in the SSC region in both pNRC100 and pNRC200, 7 elements in the common large single copy region in both pNRC100 and pNRC200; and 23 elements in the unique large single copy regions, 6 in pNRC100 and 17 in pNRC200. (Figure 2A reproduced with permission from Cold Spring Harbor Laboratory Press, ref. *11*.)

Table 1
Halobacterium **NRC-1 Genome Statistics**

	Total	Chromosome	pNRC200	pNRC100
Size (bp)	2,571,010	2,014,239	365,425	191,346
G+C composition (%)	65.9	67.9	59.2	57.9
Number of predicted genes	2,682	2,111	374	197
Coding (%)	84	87	76	71
Number of IS elements	91	22	40	29
ISH1	1	1	0	0
ISH2	13	4	5	4
ISH3	23	5	10	8
ISH4	2	1	0	1
ISH5	6	0	4	2
ISH6	2	1	1	0
ISH7	4	0	2	2
ISH8	21	5	10	6
ISH9	4	0	2	2
ISH10	6	2	2	2
ISH11	7	2	3	2
ISH12	2	1	1	0

on the pNRC inverted repeats is higher in G+C content (64%) than pNRC100 as a whole (58%) and lacks any IS elements *(3)*, indicating the occurrence of genomic regions with diverse character in all three replicons. All together, there are 91 IS elements, which represent 12 families in the NRC-1 genome (Table 1) *(4)*. These findings suggest the involvement of IS elements in DNA exchange between the replicons of *Halobacterium* NRC-1.

The high G+C composition of *Halobacterium* NRC-1 is likely an adaptation to survival under intense solar radiation (e.g., to minimize targets for thymine dimer formation). Statistically, the number of thymine dimer sites is expected to be nearly 60% lower for the NRC-1 large chromosome compared to a comparable size replicon of 50% G+C. However, dinucleotide analysis indicated even fewer sites, by an additional 20%, than predicted from the G+C content *(11)*. The high G+C composition also results in an extreme third-position G+C bias in the codon usage (86% G+C vs 70% and 46% in the first two positions) *(11)*.

ANNOTATION OF THE *HALOBACTERIUM* GENOME

The *Halobacterium* Genome Consortium, an international group representing 12 institutions, conducted annotation of the NRC-1 genome from summer 1999 to summer 2000. Data were released starting at 3× coverage periodically until completion, with a workshop held in Amherst, Massachusetts, in January 2000. This effort led to a thorough analysis of this first halophile sequence and made it maximally useful to the community. In the subsequent 2-year period, numerous additional genes have been identified. The high points of the current annotation are summarized here, and a comprehensive database is available at the Halophile Genomes web site (http://zdna2.umbi.umd.edu).

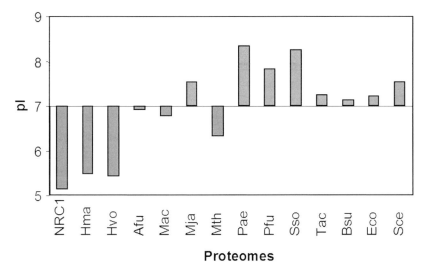

Fig. 3. Average p*I* profiles of proteomes predicted from genome sequences: *Halobacterium* sp NRC-1 (NRC1), *Haloarcula marismortui* (Hma), *Haloferax volcanii* (Hvo), *Archaeoglobus fulgidus* (Afu), *Methanosarcina acetivorans* (Mac), *Methanococcus jannaschii* (Mja), *Methanobacter thermoautotrophicum* (Mth), *Pyrobaculum aerophilum* (Pae), *Pyrococcus furiosus* (Pfu), *Sulfolobus solfataricus* (Sso), *Thermoplasma acidophilum* (Tac), *Bacillus subtilis* (Bsu), *Escherichia coli* K12 (Eco), *Saccharomyces cerevisiae* (Sce).

DNA Replication

The *Halobacterium* NRC-1 genome codes for a heterodimeric family D DNA polymerase found in Archaea; many eukaryoticlike replication proteins; 2 family B DNA polymerases, one coded by pNRC200; origin recognition and helicase recruiters (10 Orc1/Cdc6); replicative helicase (MCM); ssDNA binding proteins (6 Rfa); primases (2 Pri); clamp loaders (RfcABC); processivity clamp (2 proliferating cell nuclear antigen homologs); type I topoisomerase (TopA); type II topoisomerases (Top6A and Top6B); RNA primer removal (Rad2 and RNaseH); and a few bacterial genes involved in replication, a primase (DnaG) and topoisomerase (GyrA and GyrB). Interestingly, multiple copies of genes coding for eukaryotic origin recognition complex proteins Orc1/Cdc6 were found, including 3 scattered on the large chromosome, suggesting the possibility of multiple replication origins *(11)*. When analyzed for strand-specific G+C nucleotide variation or G+C skew, the large chromosome of *Halobacterium* NRC-1 was found to contain 4 inflection points. Two of the three *orc1/cdc6* genes were located near the inflection points, suggesting that *Halobacterium* NRC-1 has a novel replication system with two separate origins of replication on the large chromosome *(11)*.

DNA Repair

The *Halobacterium* NRC-1 genome contains many DNA repair genes (Fig. 4), likely necessary to repair DNA damage resulting from intense solar radiation in its environment *(12)*. Consistent with expectations, NRC-1 displays high levels of resistance to both ultraviolet and γ-radiation. Photoreactivation is a very efficient process in *Halobacterium*, and two photolyase/cryptochrome homologs are encoded in the genome,

the genome sequence and saturation mutagenesis of the *bop* promoter provided evidence for alternate TATA box sequences *(14)*. Nearly 100 transcriptional regulators, mostly bacterial type, have also been identified.

Protein Synthesis

The translation system of *Halobacterium* NRC-1 has hybrid eukaryotic and bacterial character, but like other Archaea, all of its ribosomal proteins have eukaryotic homologs. Interestingly, the ribosomal protein genes of *Halobacterium* NRC-1 are organized into multigene clusters that resemble operons of bacteria. In addition to the 52 RNAs (16S, 23S, and 5S rRNAs, 47 tRNAs [transfer RNAs], 7S RNA, and RNaseP), NRC-1 has 18 different aminoacyl–tRNA synthetases coded in the genome plus the GatABC amidotransferases for charging with glutamine and asparagine *(4)*. Interestingly, one aminoacyl–tRNA synthetase, ArgRS, closely related to the bacterial and yeast mitochondrial enzymes, is coded by pNRC200.

For protein secretion, the *Halobacterium* NRC-1 general secretory (Sec) machinery is a hybrid of eukaryotic and bacterial systems. Sec61α, Sec61γ, SRP54, SRP19, and the 7S RNA are related to the corresponding eukaryotic factors, while FtsY, SecD, and SecF (but not SecA) are related to the bacterial factors *(4)*. In addition to the Sec system, recent bioinformatic analysis has suggested that the twin-arginine (Tat) protein export pathway used for secretion of mainly redox proteins in bacteria is also present in NRC-1 and may be commonly used in this archaeon *(15,16)*.

Cell Envelope

Halobacterium NRC-1 cells are surrounded by a single lipid bilayer membrane and an S layer assembled from the cell surface glycoprotein. The cytoplasm is in osmotic equilibrium with the hypersaline environment, with a correspondingly high intracellular K^+ concentration that may be equivalent to the external Na^+ concentration. Like other Archaea, the polar lipids are based on archaeol, a glycerol diether lipid containing phytanyl chains derived from C_{20} isoprenoids. The *Halobacterium* NRC-1 genome contained all of the key enzyme genes of isoprenoid synthesis, including HMG–coenzyme A reductase (MvaA), the target of the growth inhibitor mevinolin *(4)*. To maintain the ionic balance, NRC-1 encodes multiple K^+ transporters, including KdpABC, an ATP-driven K^+ transport system, and TrkAH, a low-affinity K^+ transporter driven by the membrane potential (Fig. 4). Active Na^+ efflux is likely mediated by NhaC proteins coding for unidirectional Na^+/H^+ antiporters. Interestingly, genes coding KdpABC and copies of TrkA (three of five) and NhaC (one of three) are found on pNRC200. In addition, active transporters for nutrient uptake were identified for cationic amino acids (Cat) and proline (PutP), dipeptides (DppABCDF), oligopeptides (AppACF), a sugar transporter (Rbs), removal of heavy metals (arsenite and cadmium) and other toxic compounds (multidrug resistance homologs), and multiple copies of phosphate transporter systems, PstABC, and phosphate permease.

Purple Membrane

Halobacterium NRC-1 contains purple membrane, a two-dimensional crystalline lattice of the light-driven proton pump, bacteriorhodopsin, a complex of a protein, bacterio-opsin, and a chromophore, retinal (Fig. 4). Under high-illumination conditions, cells can grow phototrophically, a capability recently recognized in planktonic bacteria *(12)*. Five purple membrane regulon genes, which are clustered on the chromosome and coordinately regulated, were identified, including *bop*, specifying bacteriorhodopsin; *crt*B1 and *brp*, coding the first and last committed steps of retinal synthesis, respectively, *blp*, a gene of unknown function; and *bat*, the sensor–regulator *(14)*. The *bat* gene product (Bat) is a member of a small gene family, containing a GAF (cGMP-binding) domain, PAS/PAC (redox-sensing) domain, and DNA-binding helix-turn-helix motif, which likely binds an UAS (upstream activator protein) sequence for gene activation. The *bop* gene TATA box sequence deviates from the consensus archaeal promoter sequence, suggesting the involvement of novel factors, such as alternate TBP and TFB proteins, in its transcription *(14)*.

Taxis and Signal Transduction

Halobacterium species are highly chemotactic and phototactic, with both chemical gradients and gradients of light intensity or color modulating their swimming behavior. A large number of taxis genes have been identified, including *sop*I and *sop*II, coding for the phototaxis receptors; SRI and SRII, which are in the bacteriorhodopsin family (and also including halorhodopsin, a chloride pump) (Fig. 4) *(12)*. SRI mediates attractant responses to orange light and repellent responses to near-ultraviolet light, while SRII is a blue light repellent photoreceptor. Interestingly, homologs of haloarchaeal rhodopsins have recently been found in the genomes of fungi, algae, marine bacteria, and cyanobacteria *(12)*. A total of 17 *htr* genes coding for integral membrane proteins homologous to bacterial chemotaxis receptors were found, as were a complete set of *che* genes encoding chemotaxis determinants. There are 6 flagellin genes and an archaeal-type flagellar apparatus *(16)*. A large gene cluster, *fla*D-K, codes the archaeal flagellar apparatus, with *fla*D, *fla*E, *fla*G, *fla*H, *fla*I, and *fla*J similar to other archaea and only *fla*K resembling a bacterial flagellar regulator. Two-component regulatory systems are evident in the *Halobacterium* NRC-1 genome, including 6 response regulator genes and 14 histidine kinases. The *Halobacterium* NRC-1 genome revealed the presence of several possible circadian photoregulators, including a eukaryotic cryptochrome and a cyanobacterial KaiC-like protein, consistent with a circadian rhythm in this phototrophic microbe *(12)*.

Gas Vesicles

Halobacterium species, like many photosynthetic aquatic prokaryotes, possess the ability to regulate buoyancy by the synthesis of gas-filled vesicles (Fig. 4). The requirements for gas vesicle formation have been extensively studied in NRC-1 by genetic analysis *(17)*. A cluster of genes, *gvp*MLKJIHGFEDACN(O), present on both pNRC100 and pNRC200 in NRC-1 was shown to be necessary and sufficient for wild-type gas vesicle synthesis. Interestingly, the genome sequence of *Halobacterium* NRC-1 also revealed a silent, but nearly complete, *gvp* gene cluster, lacking only *gvp*M, on pNRC200 *(4,12)*.

Carotenoids and Retinal

Halobacterium produces red-orange carotenoids that are essential for phototransduction and protection against photodamage, the most abundant being bacterioruberins

(Fig. 4). Genes encoding bacterial phytoene synthases have been identified in *Halobacterium* NRC-1, *crt*B1, and *crt*B2, and several genes coding for subsequent desaturation steps are likely coded by *crt*I1, *crt*I2, and *crt*I3 *(4)*. Genes that catalyze subsequent conversion to bacterioruberin have not yet been identified. In a branch of the carotenoid pathway, lycopene is cyclized by the *crt*Y gene product to form β-carotene, which is oxidatively cleaved to form retinal by the *brp* and *blh* gene products (Fig. 4) *(18)*. For certain steps of the carotenoid biosynthetic pathway, multiple genes may exist in *Halobacterium* NRC-1, and these may be differentially regulated by light or oxygen.

Energy Metabolism

Halobacterium NRC-1 can grow chemoorganotrophically, either aerobically or anerobically, and has phototrophic capability using bacteriorhodopsin. *Halobacterium* requires all but 5 of the 20 amino acids for growth, and several amino acids may be used as a source of energy. Aerobically, arginine and aspartate can be used via the citric acid cycle; anaerobically, arginine can be used via the arginine deiminase pathway, coded by the *arc*RACB genes on pNRC200 (Fig. 4) *(3)*. Genes for a gluconeogenic pathway for carbohydrate synthesis during growth on amino acids and nearly all genes for a reverse Embden–Meyerhof glycolytic pathway are present. Although *Halobacterium* is reported to be unable to metabolize sugars, a sugar uptake transporter and genes coding for glucose dehydrogenase and 2-keto-3-deoxygluconate kinase, a semi-phosphorylated Entner–Doudoroff pathway, are present in *Halobacterium* NRC-1. The genes for gluconeogenesis and catabolism of glyceraldeyde 3-phosphate (produced by glucose catabolism) to pyruvate are also present. *Halobacterium* NRC-1 also possesses genes encoding enzymes of the bacterial-like fatty acid β-oxidation pathway and a 2-oxoacid dehydrogenase complex.

EVOLUTION AND LATERAL GENE TRANSFERS

Halobacterium NRC-1 is an organism of evolutionary interest that is distantly related to some methanogens and is classified as a euryarchaeote based on the 16S rRNA sequence. After complete sequencing, the *Halobacterium* NRC-1 genome was compared to 11 other complete genomes by gene content analysis using the DARWIN suite of programs *(4)*. The results confirmed the archaeal status of NRC-1, with the closest relatives being *Archeoglobus fulgidus* and *Methanococcus jannaschii*. Interestingly, however, similarities were also noted to the Gram-positive bacterium, *Bacillus subtilis*, and the radiation-resistant bacterium, *Deinococcus radiodurans*. More recently, whole genome analysis using a larger number of completed genomes showed *Halobacterium* NRC-1 to branch at the root of the archaeal tree (Fig. 1) *(19)*. The discrepancy between the 16S rRNA and whole genome trees requires a more detailed investigation because it suggests the possibility for the appearance of halophiles at a very early point in evolution. However, an additional possibility is that the position of NRC-1 in whole genome trees is distorted, with *Halobacterium* pulled away from the other archaea and toward the bacteria as a consequence of many lateral gene transfers from bacteria.

A comprehensive analysis of gene histories of *Halobacterium* NRC-1 has recently been conducted (S. P. Kennedy and S. DasSarma, unpublished). Detailed phylogenetic analysis of proteins catalogued as having bacterial phylogenies in the National Center for Biotechnology Information Clusters of Orthologous Groups database was carried out. In addition bacterial-like genes clustered together in the genome and coding specific metabolic pathways were also subjected to phylogenetic analysis. Based on this analysis, several hundred proteins, including biosynthetic, transport, and energy systems (e.g., histidine utilization, purine metabolism, glycerol utilization) and components of the electron transport chain were found to display clear bacterial histories. These genes are likely to have been acquired in this halophile by lateral gene transfers. Surprisingly, no physical link was observed with IS elements for these bacterial genes, suggesting that the genes were acquired at an early point in evolution, and any vestige of the underlying acquisition recombinational activity has been ameliorated. Although the mechanisms responsible for interdomain genetic exchanges are unknown, the finding of hundreds of bacterial genes in NRC-1 likely reflects the long-term opportunity for exchanges between halophilic bacteria and archaea cohabiting hypersaline environments over evolutionary time. In this respect, NRC-1 is similar to some other mesophilic archaea *(20)* and hyperthermophilic bacteria *(21)* in having large numbers of horizontally acquired genes in its genome.

Acquisition of Respiratory Chain Components

Two of the most interesting cases of possible lateral gene transfers into *Halobacterium* NRC-1 are the genes encoding electron transport chain factors and biosynthetic proteins *(11)*. Ten *nuo* genes, encoding subunits of NADH dehydrogenase, along with 3 *cox* genes, encoding subunits of cytochrome-*c* oxidase, are clustered together into probable operons, as are 6 *men* genes, for menaquinone biosynthesis. Interestingly, the *nuo* gene order is conserved with respect to *Escherichia coli,* with closest branching to *Synechocystis* sp PCC6803; the *men* gene order is conserved with respect to both *E. coli* and *D. radiodurans,* with closest branching to *B. subtilis.* Moreover, the G+C analysis of these two groups of genes showed they were distinguishable from the average chromosomal genes (64 or 73% compared with 68%). These results point to the interesting possibility that adaptation of halophiles to an oxidizing atmosphere occurred via the acquisition of electron transport chain components from aerobic bacteria through lateral transfer events. Further analysis is necessary to determine whether such transfers of respiratory genes have occurred once or repeatedly in the evolution of the diversity of modern halophiles.

Evolution of Purple Membrane

Retinal-containing chromoproteins like bacteriorhodopsin in purple membrane and sensory rhodopsins have recently been discovered in diverse bacteria and eukaryotes and are therefore present in all three branches of life, Archaea, Bacteria, and Eukarya *(12,22).* Although the evolutionary origin of retinal chromoproteins is unclear at present, their wide distribution in nature is consistent with horizontal transmission. An interesting further speculation is that primordial rhodopsins were an early evolutionary invention and may have been responsible for the original dominant form of phototrophy in the sea, pre-dating chlorophyll-based photosynthesis. Such early phototrophs, with the relatively simple capacity for coupling transmembrane light-driven proton pumping to adenosine triphosphate synthesis *(22,23),* could have arisen in a reducing atmosphere (although a small quantity of oxygen would have been necessary for the synthesis of retinal). Evolution of organisms with more complex chlorophyll-based photosynthetic systems operating

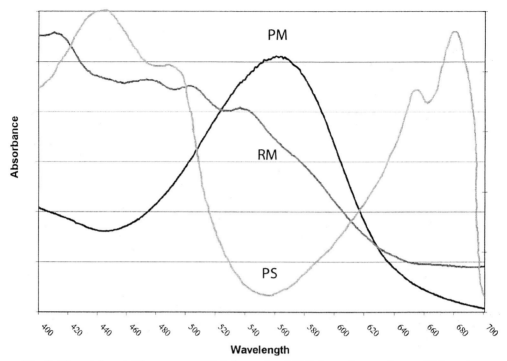

Fig. 5. Ultraviolet-visible spectra of *Halobacterium* NRC-1 purple membrane (PM), red membrane (RM), and photosynthetic membrane (PM). Purple membrane and red membrane were separated on a sucrose gradient, and spectra were plotted with photosynthetic membrane. The complementarity of purple and photosynthetic membrane spectra is apparent, consistent with coevolution of the two membranes.

with great efficiency could subsequently have displaced purple membrane–containing organisms from most environments. Interestingly, the complementarity of the spectra for purple membrane, with a peak at 568 nm, and photosynthetic membranes, with a trough in this same wavelength, is striking (Fig. 5) and is consistent with coevolution of the two types of membranes. Moreover, both chlorophyll-based cyanobacteria and purple membrane–based haloarchaea still coexist in modern hypersaline environments, with the former dominating at relatively lower salinity and the latter dominating at saturating salinity.

Evolution of pNRC Replicons

The genome organization of *Halobacterium* NRC-1, with a large chromosome and two related extrachromosomal replicons, is both complex and intriguing. One possible reason for the maintenance of multiple replicons, including pNRC100 and pNRC200, is that they have captured some essential genes and are therefore required for viability. The compatibility between these related replicons may be explained by the presence of multiple origins of replication of different compatibility groups *(3,24)*. Because dozens of copies of IS elements are present on these replicons, the transposable elements are likely responsible for frequently promoting exchanges of DNA between them. Moreover, once an extrachromosomal replicon is established in two or more copies, continued DNA exchanges between individual copies of the smaller replicons and the large chromosome could result in generation of additional genomic diversity.

Such a possible scheme has been proposed for the evolution of pNRC100, including multiple replicon fusions of precursor plasmids, followed by the acquisition of chromosomal genes, with both processes mediated by IS elements *(3)*. The duplication of a portion of a pNRC100 precursor replicon through unequal crossing over of two IS element pairs would have resulted in the formation of inverted repeats, which subsequently would serve to stabilize the region within the repeats and create inversion isomers. Through such processes, essential genes may have been captured from the chromosome, stabilized on the pNRC100 and pNRC200 replicons, and resulted in their achievement of minichromosome status.

The existence of multiple minichromosome replicons with the capability to acquire new genes and harboring multiple essential genes is a highly novel character of the *Halobacterium* NRC-1 genome. As a result, the NRC-1 genomic condition may be one of a competitive dynamic equilibrium between several essential replicons in the genome. Such a condition may arise from time to time in evolution and subside for intervening periods through reduction in numbers by replicon fusions. The heterogeneity of minichromosomes among *Halobacterium* strains is testament to such underlying dynamic processes *(25)*. Given these findings in *Halobacterium* NRC-1, it is not inconceivable that competition between replicons is a general phenomenon in evolution and may play an important role in shaping the long-term evolution of prokaryotic genomes, including the evolution of new chromosomes from plasmids.

FUTURE PROSPECTS

The complete sequence of *Halobacterium* NRC-1 has provided an excellent platform for evolutionary and comparative genomic analysis of an extremely halophilic archaeon *(4,11)*. As one of the few sequenced mesophilic archaea, which coinhabits a dynamic environment populated by a multitude of bacteria, hundred of genes with bacterial or uncertain histories have been uncovered. Additional genomic studies of diverse halophiles *(1)* (e.g., marine haloarchaea) are necessary to provide a significantly better understanding of the evolutionary position of these novel microorganisms. The finding of large dynamic extrachromosomal replicons, containing both essential genes and a large number of IS elements, has suggested the occurrence of multiple chromosomes that may compete for genes *(3)*.

In addition to evolutionary insights, the ease of culture and the wide range of biological responses of halophiles promise significant opportunities in functional genomics and biotechnology. DNA arrays, proteomics, and gene knockouts are all approaches available for further studies of *Halobacterium* biology *(2,26)*. The recent use of a whole genome microarray to study purple membrane expression illustrates the power of functional genomic approaches and remind us of the need to adhere to established rigorous genetic practices in the postgenomic era *(27,28)*. Significantly, halophilic archaea serve as excellent models for fundamental aspects of eukaryotic biology (e.g., DNA replication, transcription, and translation). Finally, halophilic proteins and complexes, many of which are extremely novel, provide genuine future opportunities for biotechnology, including the development of new vaccines and antibiotics *(29,30)*.

ACKNOWLEDGMENTS

Studies of haloarchaeal genomics in my laboratory have been generously supported by the National Science Foundation. I wish to thank many current and former students and associates and collaborators in the *Halobacterium* Genome Consortium who provided much of the information collected in this chapter. Special thanks are given to Dr. Philip Harriman for support and encouragement.

REFERENCES

1. DasSarma S, Arora P. Halophiles. In: Encyclopedia of Life Sciences. London: Macmillan, 2000, pp. 458–466.
2. DasSarma S, Robb FT, Place AR, et al. (eds). Archaea: A Laboratory Manual—Halophiles. Cold Spring Harbor, NY: Cold Spring Harbor, Laboratory Press, 1995.
3. Ng W-L, Ciufo SA, Smith TM, et al. Snapshot of a large dynamic replicon from a halophilic archaeon: megaplasmid or minichromosome? Genome Res 1998; 8:1131–1141.
4. Ng WV, Kennedy SP, Mahairas GG, et al. Genome sequence of *Halobacterium* species NRC-1. Proc Natl Acad Sci USA 2000; 97:12,176–12,181.
5. Joshi JG, Guild WR, Handler P. The presence of two species of DNA in some halobacteria. J Mol Biol 1963; 6:34–38.
6. Charlebois RL, Doolittle WF. Transposable elements and genome structure in halobacteria. In: Berg DE, Howe MM (eds). Mobile DNA. Washington, DC: American Society for Microbiology, 1989, pp. 297–307.
7. Ng W-L, Kothakota S, DasSarma S. Structure of the large gas vesicle plasmid in *Halobacterium halobium*: inversion isomers, inverted repeats, and insertion sequences. J Bacteriol 1991; 173: 1958–1964.
8. Hackett NR, Bobovnikova Y, Heyrovska N. Conservation of chromosomal arrangement among three strains of the genetically unstable archaeon *Halobacterium* species. J Bacteriol 1994; 176: 7711–7718.
9. St Jean A, Charlebois RL. Comparative genomic analysis of the *Haloferax volcanii* DS2 and *Halobacterium* sp GRB contig maps reveals extensive rearrangement. J Bacteriol 1996; 178: 3860–3868.
10. Gordon D, Abajian C, Green P. Consed: a graphical tool for sequence finishing. Genome Res 1998; 8:195–202.
11. Kennedy SP, Ng WV, Salzberg SL, Hood L, DasSarma S. Understanding the adaptation of *Halobacterium* species NRC-1 to its extreme environment through computational analysis of its genome sequence. Genome Res 2001; 11:1641–1650.
12. DasSarma S, Kennedy SP, Berquist B, et al. Genomic perspective on the photobiology of *Halobacterium* species NRC-1, a phototrophic, phototactic, and UV-tolerant haloarchaeon. Photosyn Res 2001; 70:3–17.
13. Baliga NS, Goo YA, Ng WV, Hood L, Daniels CJ, DasSarma S. Is gene expression in *Halobacterium* NRC-1 regulated by multiple TBP and TFB transcription factors? Mol Microbiol 2000; 36:1184–1185.
14. Baliga NS, Kennedy SP, Ng WV, Hood L, DasSarma S. Genomic and genetic dissection of an archaeal regulon. Proc Natl Acad Sci USA 2001; 98:2521–2525.
15. Bolhuis A. Protein transport in the halophilic archaeon *Halobacterium* sp NRC-1: a major role for the twin-arginine translocation pathway? Microbiology 2002; 148:3335–3346.
16. Patenge N, Berendes A, Engelhardt H, Schuster SC, Oesterhelt D. The *fla* gene cluster is involved in the biogenesis of flagella in *Halobacterium*. Mol Microbiol 2001; 41:653-663.

17. DasSarma S, Arora P. Genetic analysis of the gas vesicle gene cluster in haloarchaea. FEMS Microbiol Lett 1997; 153:1–10.
18. Peck RF, Echavarri-Erasun C, Johnson EA, et al. *brp* and *blh* are required for synthesis of the retinal cofactor of bacteriorhodopsin in *Halobacterium*. *J Biol Chem* 2001; 276:5739–5744.
19. Korbel JO, Snel B, Huynen MA, Bork P. SHOT: a web server for the construction of genome phylogenies. Trends Genet 2002; 18:158–162.
20. Deppenmeier U, Johann A, Hartsch T, et al. The genome of *Methanosarcina mazei*: evidence for lateral gene transfer between bacteria and archaea. J Mol Microbiol Biotechnol 2002; 4:453–461.
21. Nelson KE, Clayton RA, Gill SR, et al. Evidence for lateral gene transfer between archaea and bacteria from genome sequence of *Thermotoga maritima*. Nature 1999; 399:323–329.
22. Beja O, Aravind L, Koonin EV, et al. Bacterial rhodopsin: evidence for a new type of photo-trophy in the sea. Science 2000; 289:1902–1906.
23. Racker E, Stoeckenius W. Reconstitution of purple membrane vesicles catalyzing light-driven proton uptake and adenosine triphosphate formation. J Biol Chem 1974; 249:662–663.
24. Ng WL, DasSarma S. Minimal replication origin of the 200-kilobase *Halobacterium* plasmid pNRC100. J Bacteriol 1993; 175:4584–4596.
25. Ng W-L, Arora P, DasSarma S. Large deletions in class III gas-vesicles deficient mutants of *Halobacterium*. Sys Appl Microbiol 1994; 16:560-568.
26. Peck RF, DasSarma S, Krebs MP. Homologous gene knockout in the archaeon *Halobacterium* with *ura3* as a counterselectable marker. Mol Microbiol 2000; 35:667–676.
27. Baliga NS, Pan M, Goo YA, et al. Coordinate regulation of energy transduction modules in *Halobacterium* sp analyzed by a global systems approach. Proc Natl Acad Sci USA 2003; 99: 14,913–14,918.
28. DasSarma S. Biology reports Ltd. faculty of 1000 commentary. Available at: http://www.faculty of1000.com/article/12403819. Accessed January 8, 2003.
29. Stuart ES, Morshed F, Sremac M, DasSarma S. Antigen presentation using novel particulate organelles from halophilic archaea. J Biotechnol 2001; 88:119–128.
30. Hansen JL, Ippolito JA, Ban N, Nissen P, Moore PB, Steitz TA. The structures of four macro-lide antibiotics bound to the large ribosomal subunit. Mol Cell 2002; 10:117–128.

VI   APPLICATIONS OF GENOMIC DATA

22

Microarrays, Expression Analysis, and Bacterial Genomes

Carsten Rosenow and Brian Tjaden

INTRODUCTION

The fast accumulation of genomic information and the concurrent development of microarray technology provide the researcher with the unique opportunity to investigate the complexities of biology at an unprecedented rate. The large number of sequenced genomes can be utilized to develop microarrays for a variety of applications, ranging from ribonucleic acid (RNA) expression analysis, comparative genome hybridization, and large-scale genotyping to transcriptome analysis and RNA decay studies. In addition, microarrays have the potential to accelerate the drug development process and will play an important role in the diagnostic market (Fig. 1). The development of new technologies together with their combination with high-throughput techniques and powerful computing capabilities are some of the great advances in the genomics era.

TECHNOLOGY

Microarrays consist of deoxyribonucleic acid (DNA) attached to a solid support, usually glass. This DNA can be a polymerase chain reaction product, cDNA (complementary DNA) transcript, or oligonucleotides, which are complementary to a target sequence that needs to be analyzed. Because of the high density with which these microarrays can be produced, they allow massively parallel data acquisition. In addition, advances in bioinformatics and data analysis applications allow the subsequent translation of these large data sets into meaningful results (described in the section on data analysis).

Currently, there are two prominent types of array technologies: spotted microarrays and oligonucleotide probe arrays. The spotted microarrays utilize presynthesized cDNA or polymerase chain reaction products, which are "spotted" onto a glass surface. Oligonucleotide probe arrays use *in situ* DNA synthesis to build sets of oligomers with known sequences on a glass surface. The synthesis of the oligonucleotides is accomplished using either ink jet printing or photolithographic methods similar to those used in the semiconductor industry. However, because of the fundamental differences between the spotting and the *in situ* synthesis technologies, they show very different levels of reliability, density, reproducibility, and expression results are usually not comparable with each other. For a comprehensive review of both technologies, refer to recent reviews (1–3).

From: *Microbial Genomes*
Edited by: C. M. Fraser, T. D. Read, and K. E. Nelson © Humana Press Inc., Totowa, NJ

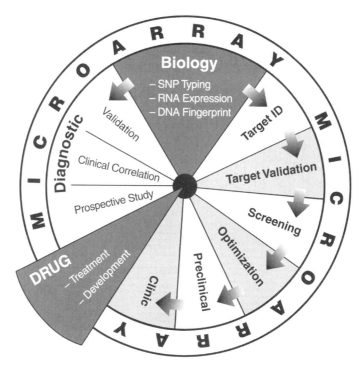

Fig. 1. Microarray applications have an impact on all areas of the drug development process and have potential for the diagnostic market.

EXPRESSION ANALYSIS

Genomics is changing the understanding of biology. DNA sequencing projects provide a global view of the organizational content of an organism. However, understanding the cellular processes requires more than just the knowledge of genes and genomes; their interaction and function must be understood. Microarrays are being used to detect simultaneously the expression levels of large numbers of messenger RNAs (mRNAs) in an organism. For this application, cRNA *(4)*, RNA, or cDNA *(5,6)* is fluorescently labeled and hybridized to microarrays. Based on the labeling technology used, RNA levels can be determined relative to a control sample (usually with spotted arrays) or as an absolute measure of RNA quantity in a given cell (usually with high-density oligonucleotide probe arrays). Initial obstacles for the study of prokaryotic RNA expression have been overcome through the development of a direct-RNA labeling protocol and a cDNA synthesis approach *(5,7)*.

Early studies looked at expression changes in microorganisms grown under different growth conditions *(8,9)*, the response of bacteria to drugs *(10)*, or the influence that a variety of environmental changes have on gene expression *(6)*. The availability of microarrays for bacterial pathogens together with the ability to study the expression changes in the affected human cell simultaneously can help us understand host–pathogen interactions and subsequently can result in the identification of novel drug targets *(11,12)*.

The most obvious choice for a drug target in microorganisms are proteins essential for in vivo growth or proteins related to virulence of the microorganisms. The identification of genes expressed only in vivo or during an infection process can be greatly enhanced using microarrays. In addition, understanding the host response can provide additional drug targets on the cell surface of the host (adhesion molecules) or inside the cells necessary for invasion or bacterial survival.

A study by Cohen et al. *(13)* identified 97 human genes differentially expressed after *Listeria monocytogenes* infection of a human promyelocytic cell line. The authors stressed the fact that these studies not only enhance the understanding of the pathogenesis mechanism and molecular physiology, but also help identify new therapeutic targets. This becomes especially important in light of the recent emergence of bacteria with multiple drug resistance mechanisms and the increasing number of drug-resistant pathogens *(14,15)*.

One of the important short-term tasks for the microarray community is the creation of a public database to help identify metabolic and regulatory pathways, define transcription regulators, and simulate cellular processes and relationships based on microarray experiments. To achieve these goals, the collection and availability of primary microarray data are crucial. Some of the large institutions have already established public databases *(16–19)* (www.ncbi.nlm.nih.gov/geo), but data otherwise are only published on either Web pages from peer-reviewed journals or on the author's own Web page.

A push toward standardization of microarray data and the disclosure of detailed experimental information came through the MIAME (Minimum Information About a Microarray Experiment) effort *(17)*. This proposal describes the minimum information required to ensure that microarray data can be easily interpreted, and that results derived from its analysis can be independently verified.

The advances in the field of functional genomics will partly depend on the participation of the scientists in their data disclosure. Additional value can be added to these databases by combining the expression data with other means of analysis, such as cluster analysis, comparative genome hybridization, primary DNA sequence annotation, proteomics data, or the identification of homologous regions in other organisms (Fig. 2).

COMPARATIVE GENOME HYBRIDIZATION, RESEQUENCING, AND GENOTYPING USING MICROARRAYS

Comparative genome hybridization and the study of genetic variability become increasingly important in biology for understanding the mechanisms of microbial pathogenesis and for tracking infectious disease outbreaks. High-density oligonucleotide probe arrays enable rapid comparative genome hybridizations, resequencing, and subsequent genotyping of any organism. Resequencing applications become increasingly important to determine the genotypic differences between strains with different phenotypic characteristics. To determine the sequence at a given site compared to a known sequence using microarrays, four oligonucleotide probes are tiled on the microarray. These four probes differ only in the central position or the 13th base and contain each of the four possible nucleotides (Fig. 3). For increased redundancy, each target sequence can be queried for both the forward and the reverse strand. Depending on the density and feature size of the microarray used, contiguous or noncontiguous sequence information up to 120 kb can be collected in one hybridization step. The average call rate for these arrays has been shown to be greater than 80%, with an accuracy of 99.999% *(20)*.

For applications with no need for the specific nucleotide sequence, a more global view of genome content can be achieved through comparative genome hybridizations.

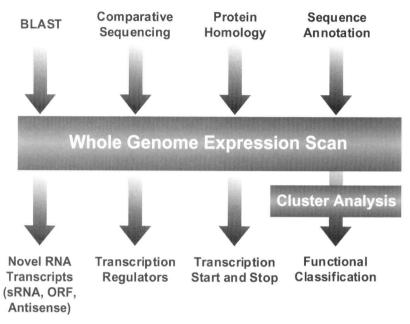

Fig. 2. Whole genome expression scans combined with additional database information and sequence analysis tools can accelerate gene characterization and the identification of new pathways and regulatory mechanisms inside the cell.

Comparison of genomes can help us better understand the evolution of bacterial strains, differentiate pathogens from nonpathogens, and identify differences in gene content between bacterial strains or their subtypes. All this information, coupled with whole genome expression analysis provides genotypic and phenotypic information that can be used to understand better the underlying differences in pathogenesis between two related strains and may facilitate the characterization of genes and their function (functional genomics).

One example for this approach was demonstrated by Kato-Maeda et al. using a *Mycobacterium tuberculosis* high-density oligonucleotide array *(21)*. They were able to detect a pattern of deletions in 19 clinically and epidemiologically well-characterized isolates of *M. tuberculosis* that was identical within mycobacterial clones, but different between different clones. These data suggest that, similar to changes in human mitochondrial DNA, genomic deletions can be used to reconstruct phylogenetic trees.

Another application for comparative genome hybridization includes the identification of chromosome rearrangements, which are common in different isolates of bacterial species *(7,22–24)*. These rearrangements are created by homologous recombination and include deletions, duplications, and inversions. In addition, repetitive elements in the genome of *Neisseria meningitidis* have been shown to be involved in antigenic variation and genome fluidity by facilitating recombination *(25,26)*.

Another source of genetic variation in bacterial species is horizontal gene transfer, which is an important process for adaptation to new environmental niches *(27)* and can be studied using microarrays. In *Staphylococcus aureus*, many virulence-associated as

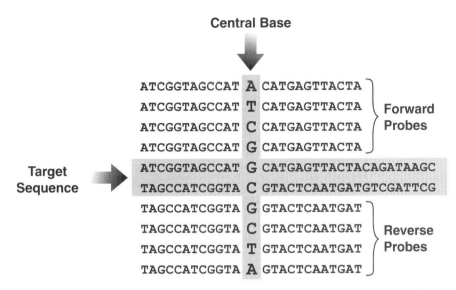

Fig. 3. Comparative sequence analysis Arrays: design strategy. Sequence analysis is accomplished using a four-probe interrogation strategy. Four 25-mer oligonucleotide probes are used to determine the identity of the central base in the sequence. The most stable hybrid will provide the highest fluorescent signal among the four probes. An objective statistical framework is applied to assign each base call a quality score using both the forward and the reverse strand.

well as antibiotic resistance genes are horizontally acquired *(28,29)*. The comparison of five different nonpathogenic *Escherichia coli* strains with the sequence of the *E. coli* MG1655 genome using DNA arrays confirmed a high degree of chromosomal variation over the whole chromosome *(7)*. Surprisingly, the identified regions consist of an unusually high number of putative open reading frames and mobile genetic elements *(30)*. Another group has used comparative genome hybridization with a *Streptococcus pneumoniae* DNA array and produced similar results. They concluded that variable genomic regions are usually associated with transposases or a prophage *(31)*.

Combining *in silico* analyses with comparative genome hybridization can determine the genetic heterogeneity of an organism and provides a powerful tool for the microbiology laboratory *(32)*.

DATA ANALYSIS

One of the great advantages of microarray experiments is their ability to assay the expression levels of a large number of genes simultaneously. This provides scientists with a powerful tool for investigating trends across groups of genes; however, when considering expression of single genes as evidenced from an array experiment, a scientist must proceed with considerable caution. Biology is complex, and microarray experiments are fraught with uncertainty, error, and noise. Cell growth, mRNA extraction, and fluorescent labeling are imprecise procedures that potentially bias results. Hybridization is a stochastic process, and cross-hybridization as well as nonspecific hybridization events must be managed and minimized in array experiments. Also, laser scanning

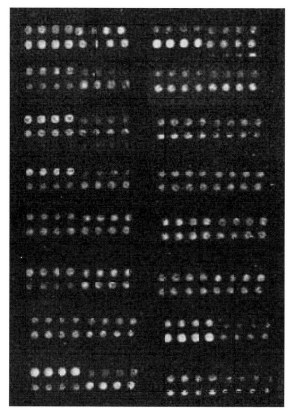

Fig. 4. A section of a cDNA array assaying genes in *Haemophilus influenzae*. There are four identical spots to assay each gene.

of an array to quantify spot intensities and their backgrounds can introduce noise into expression data.

Although improving technology is continuously mitigating these sources of error, most researchers are strong advocates of replicate experiments to ensure significance and reliability of results. Of course, it is also desirable to gather expression data for cells grown under a battery of different conditions to explore the best range of expression profiles. So, for a fixed budget of microarray experiments, a researcher must decide how to balance the trade-off between replicate experiments for the same condition providing more statistically significant results vs varying experiments to explore better the expression space.

Repeat measurements can be performed at any stage in the array experimental process, depending on which cause of error is the target, but most often, arrays contain multiple spots per gene, allowing for repeat measurements, and repeat experiments are performed on identical microarray chips. For example, Fig. 4 shows a section of a cDNA array assaying the transcript expression of genes in *Haemophilus influenzae*. The array contains four identical spots for each gene.

Other arrays, such as oligonucleotide arrays, may contain multiple probes per gene, and the probes may assay different regions of the gene's transcript. Figure 5 shows the

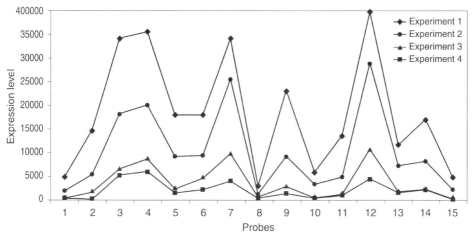

Fig. 5. The expression levels, from multiple experiments, of the *rpsB* gene from *Escherichia coli* as measured by oligonucleotide microarrays, each with 15 probes assaying the gene.

expression level of the *rpsB* gene from *E. coli*. The array contains 15 different oligonucleotide probes interrogating the gene. To estimate the expression of the gene using data from a single experiment, a simple solution is to take the average of the 15 expression values. With this approach, each probe contributes equally to the average expression level.

However, one of the challenges in using oligonucleotide probes to measure gene expression is that different oligonucleotides have different hybridization affinities, so that different probes assaying the same transcript often yield significantly different expression levels. If data are available from multiple experiments, this challenge can be overcome by weighting each probe differently, based on an estimation of the probe's hybridization affinity, so that the expression level of the gene is a weighted average of the 15 probes (Fig. 5).

This can be achieved using an expectation maximization (EM) approach, which captures the unique responsiveness of each probe and provides a more precise estimate for the expression of the gene than the simple averaging approach *(33)*. The EM algorithm iteratively estimates the transcript expression level of the *rpsB* gene in each of the four experiments (expectation step), and then, based on these estimated expression levels, it calculates optimal values for a set of parameters that capture the unique responsiveness of each probe (maximization step). Rather than using an array with multiple varying oligonucleotide probes per gene, multiple expression measurements are obtained for a gene either by using replicate arrays or by using arrays that contain exact replicate oligonucleotides or cDNA spots for a gene, and the EM approach is modified appropriately. In the maximization step, instead of calculating optimal values for parameters, which capture the responsiveness of probes, optimal values are calculated for parameters, which capture the background and error of individual spots or oligonucleotides. By using replicate spots per gene or replicate arrays per condition, the statistical significance of a gene's expression level can be quantified and ultimately more reliable results can be obtained *(34,35)*.

	gene 1	gene 2	gene 3	...	gene n
experiment 1	150	211	478	. . .	982
experiment 2	817	300	525	. . .	700
. . .	. . .	. . .	. . .	. . .	. . .
experiment m	073	009	112	. . .	213

Fig. 6. An example expression matrix for *n* genes and *m* experiments. Each column in the matrix corresponds to the expression of a particular gene in each of the *m* experiments, and each row corresponds to the expression of all *n* genes in a particular experiment.

Once expression values have been calculated for each gene in each experiment, the data are generally viewed as a matrix of expression values. As shown in Fig. 6, each column of the matrix corresponds to a gene and each row to an experiment. If *n* genes are assayed across *m* experiments, there is an $n \times m$ expression matrix. At this point, the similarity of genes may be investigated based on their expression profiles. There are many different similarities or distance metrics, including Pearson correlation, Euclidean distance, chi square, Spearman rank correlation, and Kendall's tau. One of the more common measures, the Pearson correlation coefficient, is illustrated in Fig. 7 for two genes *a* and *b*, where *a* is the column vector $(a_1, a_2, ..., a_m)$, and *b* is the column vector $(b_1, b_2, ..., b_m)$. Thus, the correlation *r* of the expression vectors of two genes (or two experiments) can be calculated by this method. If the correlation is calculated for all possible pairs of genes, an $n \times n$ correlation matrix can be determined for which each entry represents the similarity of expression of two genes. In this way, for any given gene, the other genes that are similarly expressed in experiments can be found.

Once a correlation matrix is calculated, the analysis can be extended via clustering to identify groups of genes, which are similarly expressed. Clustering expression profiles is a particularly interesting problem from a computational point of view because the data points (the expression vectors of genes) reside in *m*-dimensional space, which, depending on the number of experiments performed, is generally quite large. There are many different algorithms employed for clustering gene expression data, including hierarchical clustering *(36)*, *k* means *(37)*, CAST (cluster affinity research technique) *(38)*, self-organizing maps *(39)*, and model-based clustering *(40)*. A few of the more common approaches are summarized next.

Hierarchical clustering is a greedy clustering approach that has the advantage of being simple and producing a final clustering organized into a hierarchical tree that is easy to view graphically. The results of hierarchical clustering can vary depending on how the distance between two clusters of points is determined. In *single-link clustering*, the distance between two clusters is the shortest distance between any point in one cluster and any point in the other cluster. In *average-link clustering*, the distance between clusters is the distance between cluster centroids. In *complete-link clustering*, the dis-

$$r = \frac{m\sum_i a_i b_i - \sum_i a_i \sum_i b_i}{\sqrt{\left[m\sum_i a_i^2 - \left(\sum_i a_i\right)^2 \right] \times \left[m\sum_i b_i^2 - \left(\sum_i b_i\right)^2 \right]}}$$

Fig. 7. For two genes with expression vectors $(a_1, a_2, ..., a_m)$ and $(b_1, b_2, ..., b_m)$, the formula represents the Pearson correlation coefficient across the m experiments. The correlation r ranges $-1 \leq 0 \leq 1$, where values of r near 1 represent genes with an expression that is highly correlated, values near 0 represent genes with an expression that is uncorrelated, and values near -1 represent genes with an expression that is negatively correlated.

tance between clusters is the farthest distance between any point in one cluster and any point in the other cluster. Different distance metrics require different amounts of computation and affect the final topography of the clustering hierarchy.

1. Let each expression vector be a cluster containing one data point.
2. Find the two clusters with the smallest distance (single link, average link, or complete link) between them and merge them into one cluster.
3. Repeat step 2 until all clusters are merged into a single cluster (the root of the hierarchical tree).

The k-means method is a common and relatively simple heuristic method for clustering data points into k clusters. This is actually a special case of an EM algorithm applied to a mixture density in which the mixture components are Gaussian distributions *(41)*.

1. Randomly assign each point a cluster number between 1 and k.
2. For each of the k clusters, calculate the mean of all points, which are assigned the same cluster number.
3. For each data point, determine the distance of the point to each of the k means and assign the point to the cluster number of the closest of these means.
4. Repeat steps 2 and 3 until no data point is assigned to a different cluster than in the previous iteration.

CAST models the data points as vertices in a graph in which each pair of vertices has an edge between them representing the similarity of their expression vectors. CAST then searches for groups of vertices in the graph (cliques) in which every vertex in the group has a sufficiently high similarity to the other vertices in the group. Like k means, CAST has a threshold parameter, which effectively dictates the number of clusters, into which the algorithm will group points.

1. Choose a nonclustered point and put it in its own cluster.
2. If the average distance from any nonclustered point to the points within the cluster is greater than some threshold, then add the nonclustered point to the cluster.
3. If the average distance from any point in the cluster to the other points in the cluster is less than the threshold, then remove the point from the cluster.
4. Repeat steps 2 and 3 until the cluster stabilizes.
5. Mark all points in the cluster as clustered, and if any nonclustered points remain, return to step 1.

Once expression data have been clustered, there are many opportunities for further analysis and obtaining meaningful biological results. Clustered data are easily visualized. A researcher can verify whether similarly regulated genes or genes in the same cellular pathway cluster together. Clusters that contain groups of genes with unknown function as well as functionally characterized genes aid in the annotation of the unknown genes.

However, clustering has its limits. Although clustering can capture evident trends in expression profiles, many genes exhibit expression profiles that do not fit the pattern of any cluster and hence are given misleading classifications. Also, most clustering techniques require the user to dictate some threshold parameter that ultimately determines the number of clusters into which expression profiles are grouped. Often, there is no good technique for determining how many clusters to mandate because most genes are involved in mutiple cellular pathways.

One of the most promising areas for cluster analysis of expression data arises in its use in combination with sequence analysis methods. For instance, the DNA regions upstream of genes with expression profiles that cluster together can be searched or locally aligned to identify common transcription factor binding sites under the assumption that genes with similar expression patterns are likely to be regulated similarly *(37)*. Until these expression analysis approaches are further honed, however, their power lies not in their accuracy for individual gene expression so much as their ability to provide functional indications for large groups of genes and a more global picture of cellular pathways (Fig. 2).

TRANSCRIPTIONAL WHOLE GENOME ANALYSIS

As discussed in this chapter, microarrays can be used to assay the expression levels of all genes in an organism for which the genomic sequence is known. However, microarrays are used increasingly not only to determine the expression of known genes, but also to locate new genes *(42,43)*, study RNA decay *(44)*, and identify transcription elements that are not necessarily translated *(42,43,45)*. Once the genomic sequence of an organism has been established, oligonucleotide probes are designed to assay the expression across the entire genome, not just those sequences ultimately translated.

For instance, the first step for scientists after newly sequencing a microbial genome is to annotate the genes by computational sequence analysis. Gene prediction programs are fairly accurate, particularly for identifying coding sequences of genes, but they tend to be less reliable for distinguishing transcribed but untranslated regions (UTRs) immediately upstream of a start codon and downstream of a stop codon. An oligonucleotide microarray, which assays the expression level across the entire genome, can provide evidence for these UTR regions elucidating transcription initiation and termination sites.

Figure 8 shows the expression level as measured by 24 microarray experiments of the *pfkA* gene, which codes for a phosphofructokinase enzyme in *E. coli*. The figure shows the signal intensity of probes assaying the coding sequence as well as probes assaying the region immediately downstream of the stop codon. The downstream expression indicates the 3' UTR and the location of transcription termination. Understanding transcribed but untranslated regions surrounding genes is particularly useful in identifying and comprehending the role of the regulatory elements of genes.

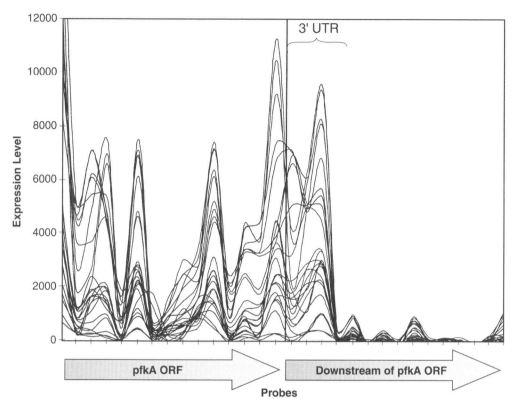

Fig. 8. The expression level of the *pfkA* gene in *E. coli* as well as the expression level of the downstream transcribed, but untranslated, region as measured in 24 experiments.

Genes that are transcribed into a single mRNA are called polycistronic, or multigene, operons. When gene expression indicates evidence of a transcript extending from a gene's coding sequence all the way to a neighboring gene's coding sequence, then the two genes are likely candidates for a polycistronic mRNA. In the case of operon genes, the intergenic region between the genes is also transcribed, and this may be identified in a microarray experiment. Figure 9 shows a picture of a microarray hybridization image with the intensity of the probes assaying the *manXYZ* operon in *E. coli (46)*. The operon is composed of three genes (*manX*, *manY*, and *manZ*), and this particular array contains contiguous probes representing adjacent genes for easy visualization of similar expression in neighboring genes. Identifying which genes are part of polycistronic mRNAs is especially useful in understanding the function of the genes because cotranscribed genes are clearly coregulated at the transcription level, and the functions of genes within the same operon are generally tightly coupled.

In addition to transcribed regions surrounding known genes, microarrays can be used to identify new transcripts, such as small-RNA molecules. Small RNAs are short transcripts that are not translated, but rather have a functional role frequently regulating the expression of other genes *(47)*. These molecules are often expressed under specific conditions, such as late stationary phase growth of microbial cells. Figure 10 shows the

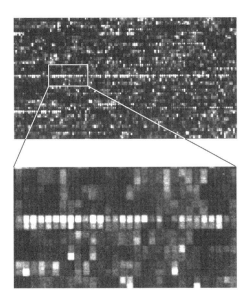

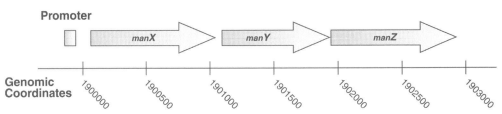

Fig. 9. An oligonucleotide array assaying the expression of the genes *manX*, *manY*, and *manZ*, which compose an operon involved in a mannosphosphotransferase system. The oligonucleotide probes on the array that assay these genes are located contiguously on the array, just as the genes are adjacently located in the *E. coli* genome.

expression of the *csrB* RNA in *E. coli* as measured in three separate microarray experiments *(43)*.

Only a handful of small RNAs have been studied experimentally, and these molecules have proven particularly challenging to identify based on primary sequence analysis because so much of their functional role is defined by higher-level structure. Evidence from microarray experiments of transcripts produced from sequence outside annotated regions of known genes is a good indication of the presence of regions coding for small-RNA molecules or new, previously unannotated open reading frames *(42)*. Additional primary sequence analysis of genomic regions corresponding to these novel transcripts can elucidate the functional role the transcript is likely to play and whether the transcript merits further experimental corroboration.

This is an example of the great enabling power of microarrays. A single microarray experiment can provide a snapshot of the entire transcript expression of a microbial organism under a given growth condition. By assaying expression of known genes as well as providing predictive guides for identifying new genes, microarrays facilitate the analysis of rapidly growing genomic data to aid our understanding of the cellular machinery of microbial organisms (Fig. 3).

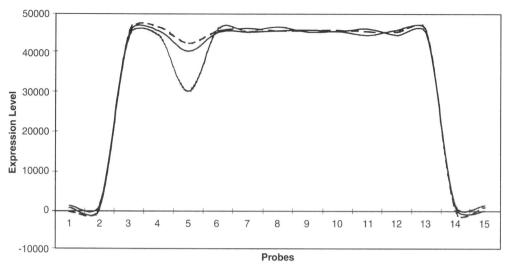

Fig. 10. The expression of the intergenic region between genes b2793 and b2792 in *E. coli* as measured by three oligonucleotide microarray experiments. The high expression level of probes 3 through 13 correspond to the 360-bp untranslated small RNA (sRNA) *csrB*.

CONCLUSIONS

Microarrays have the potential to challenge traditional technologies used in basic research, target identification, drug development, and diagnostics. The simultaneous collection of large data sets combined with the computational advances in translating these data into meaningful information will push this technology not only in the drug development arena, but also into the diagnostic market (Fig. 1). The concurrent expression analysis of the host response during a microbial infection with the expression changes of the infectious agent could discover new essential metabolic pathways in the human and the bacterial cell and uncover novel drug targets.

Unless there is some corresponding increase in the capacity for downstream processing, these targets will be held in the database of pharmaceutical companies. Using microarrays in subsequent drug screening and clinical validation processes will enable these companies to follow up faster on the large number of drug targets discovered.

The next step in these value-building chains is the use of *in silico* absorption, distribution, metabolism, and excretion/toxicity methods in combination with high-throughput technology for in vitro toxicology. This combination will accelerate the preclinical phase by enabling researchers to characterize drug properties more accurately. Building a database of identified physiological markers and characterizing their correlation with a known drug response could ultimately result in a more efficient prediction of drug effects and will lead to better drug dosage information with fewer undesirable or adverse side effects.

REFERENCES

1. Harrington CA, Rosenow C, Retief J. Monitoring gene expression using DNA microarrays. Curr Opin Microbiol 2000; 3:285–291.

2. Lipshutz RJ, et al. High density synthetic oligonucleotide arrays. Nat Genet 1999; 21(1 Suppl): 20–24.
3. Gershon D. Microarray technology: an array of opportunities. Nature 2002; 416:885–891.
4. Lockhart DJ, et al. Expression monitoring by hybridization to high-density oligonucleotide arrays. Nat Biotechnol 1996; 14:1675–1680.
5. Rosenow C, et al. Prokaryotic RNA preparation methods, useful for high density array analysis: comparison of two approaches. Nucleic Acids Res 2001; 29:e112.
6. Richmond CS, et al. Genome-wide expression profiling in *Escherichia coli* K-12. Nucleic Acids Res 1999; 27:3821–3835.
7. Blattner FR, et al. The complete genome sequence of *Escherichia coli* K-12. Science 1997; 277: 1453–1474.
8. Wodicka L, et al. Genome-wide expression monitoring in *Saccharomyces cerevisiae*. Nat Biotechnol 1997; 15:1359–1367.
9. Tao H, et al. Functional genomics: expression analysis of *Escherichia coli* growing on minimal and rich media. J Bacteriol 1999; 181:6425–6440.
10. Wilson M, et al. Exploring drug-induced alterations in gene expression in *Mycobacterium tuberculosis* by microarray hybridization. Proc Natl Acad Sci USA 1999; 96:12,833–12,838.
11. Diehn M, Relman DA. Comparing functional genomic datasets: lessons from DNA microarray analyses of host-pathogen interactions. Curr Opin Microbiol 2001; 4:95–101.
12. Debouck C, Goodfellow PN. DNA microarrays in drug discovery and development. Nat Genet 1999; 21(1 Suppl):48–50.
13. Cohen P, et al. Monitoring cellular responses to *Listeria monocytogenes* with oligonucleotide arrays. J Biol Chem 2000; 275:11,181–11,190.
14. Mandell LA, et al. The battle against emerging antibiotic resistance: should fluoroquinolones be used to treat children? Clin Infect Dis 2002; 35:721–727.
15. Hooper DC. Fluoroquinolone resistance among Gram-positive cocci. Lancet Infect Dis 2002; 2:530–538.
16. Ball CA, et al. Standards for microarray data. Science 2002; 298:539.
17. Brazma A, et al. Minimum Information About a Microarray Experiment (MIAME)—toward standards for microarray data. Nat Genet 2001; 29:365–371.
18. Sherlock G, et al. The Stanford Microarray Database. Nucleic Acids Res 2001; 29:152–155.
19. National Cancer Center (NCI). Center for Cancer Research. Development of Bio-Informatics to manage access, and analyze cDNA μArray data generated by the NCI/CCR μArray Center. 12/ 2003. http://nciarray.nci.nih.gov/
20. Cutler DJ, et al. High-throughput variation detection and genotyping using microarrays. Genome Res 2001; 11:1913–1925.
21. Kato-Maeda M, et al. Comparing genomes within the species *Mycobacterium tuberculosis*. Genome Res 2001; 11:547–554.
22. Hayashi T, et al. Complete genome sequence of enterohemorrhagic *Escherichia coli* O157:H7 and genomic comparison with a laboratory strain K-12. DNA Res 2001; 8:11–22.
23. Wong RM, et al. Sample sequencing of a *Salmonella typhimurium* LT2 lambda library: comparison to the *Escherichia coli* K12 genome. FEMS Microbiol Lett 1999; 173:411–423.
24. Himmelreich R, et al. Comparative analysis of the genomes of the bacteria *Mycoplasma pneumoniae* and *Mycoplasma genitalium*. Nucleic Acids Res 1997; 25:701–712.
25. Tettelin H, et al. Complete genome sequence of *Neisseria meningitidis* serogroup B strain MC58. Science 2000; 287:1809–1815.
26. Parkhill J, et al. Complete DNA sequence of a serogroup A strain of *Neisseria meningitidis* Z2491. Nature 2000; 404:502–506.
27. Ochman H, Lawrence JG, Groisman EA. Lateral gene transfer and the nature of bacterial innovation. Nature 2000; 405:299–304.

28. Hiramatsu K, et al. The emergence and evolution of methicillin-resistant *Staphylococcus aureus.* Trends Microbiol 2001; 9:486–493.

29. Kuroda M, et al. Whole genome sequencing of meticillin-resistant *Staphylococcus aureus.* Lancet 2001; 357:1225–1240.

30. Ochman H, Jones IB. Evolutionary dynamics of full genome content in *Escherichia coli.* EMBO J 2000; 19:6637–6643.

31. Hakenbeck R, et al. Mosaic genes and mosaic chromosomes: intra- and interspecies genomic variation of *Streptococcus pneumoniae.* Infect Immun 2001; 69:2477–2486.

32. Tettelin H, et al. Complete genome sequence and comparative genomic analysis of an emerging human pathogen, serotype V Streptococcus agalactiae. Proc Natl Acad Sci USA 2002; 99:12,391–12,396.

33. Li C, Wong WH. Model-based analysis of oligonucleotide arrays: expression index computation and outlier detection. Proc Natl Acad Sci USA 2001; 98:31–36.

34. Rocke DM, Durbin B. A model for measurement error for gene expression arrays. J Comput Biol 2001; 8:557–569.

35. Ideker T, et al. Testing for differentially-expressed genes by maximum-likelihood analysis of microarray data. J Comput Biol 2000; 7:805–817.

36. Eisen MB, et al. Cluster analysis and display of genome-wide expression patterns. Proc Natl Acad Sci USA 1998; 95:14,863–14,868.

37. Tavazoie S, et al. Systematic determination of genetic network architecture. Nat Genet 1999; 22:281–285.

38. Ben-Dor Shamir AR, Yakhini Z. Clustering gene expression patterns. J Comput Biol 1999; 6:281–297.

39. Tamayo P, et al. Interpreting patterns of gene expression with self-organizing maps: methods and application to hematopoietic differentiation. Proc Natl Acad Sci USA 1999; 96:2907–2912.

40. Yeung KY, et al. Model-based clustering and data transformations for gene expression data. Bioinformatics 2001; 17:977–987.

41. Bishop CM. *Neural Networks for Pattern Recognition.* Oxford, UK: Oxford University Press, 1995.

42. Tjaden B, et al. Transcriptome analysis of *Escherichia coli* using high-density oligonucleotide probe arrays. Nucleic Acid Res 2002; 30:1–7.

43. Wassarman KM, et al. Identification of novel small RNAs using comparative genomics and microarrays. Genes Dev 2001; 15:1637–1651.

44. Selinger DW, et al. Global RNA half-life analysis in *Escherichia coli* reveals positional patterns of transcript degradation. Genome Res 2003; 13:216–223.

45. Kapranov P, et al. Large-scale transcriptional activity in chromosomes 21 and 22. Science 2002; 296:916–919.

46. Plumbridge J. Control of the expression of the manXYZ operon in *Escherichia coli*: Mlc is a negative regulator of the mannose PTS. Mol Microbiol 1998; 27:369–380.

47. Wassarman KM. Small RNAs in Bacteria. Diverse regulators of gene expression in response to environmental changes. Cell 2002; 109:141–144.

23

Microbial Population Genomics and Ecology

A New Frontier

Edward F. DeLong

INTRODUCTION: HISTORICAL UNDERPINNINGS

Microbial Diversity and Ecology

Microbial life has been integral to the history and function of life on earth for over 3.5 billion years. The cycles of the elements on earth, including the carbon, sulfur, and nitrogen cycles, are balanced by and function because of microbial activities. Ironically, it is exactly when ecosystems are healthy and geochemical cycles are functioning homeostatically that the presence of microbes is least noticed, and then only indirectly. This is partly why the central importance of microbes to the evolution and contemporary environment of the planet is often overlooked. Microbes still are most often considered from an anthropocentric perspective, with attention focused on the relatively few species that cause human disease and the potential of the microbes to provide useful products and processes. Considering their central role in earth's geological, climatic, biogeochemical, and biological evolution, it is remarkable that the general importance of microbes has only recently been appreciated by more than just a few specialists.

Extant archaeal and bacterial biodiversity can be viewed from a variety of different perspectives. Taxonomic definitions provide the backdrop for a detailed cataloging of microbial life and are traditionally based on both genotypic and phenotypic descriptors. Alternatively, microbes may often be placed into broad functional groups or guilds (for instance, sulfate-reducing bacteria, nitrifying bacteria, methanotrophs, oxyphototrophs) that make reference to a specific functional property or ecological role of a microbe in its natural habitat. Both perspectives provide important information for different purposes, yet each type of biological definition also has inherent deficiencies. As discussed here, contemporary approaches, including molecular phylogenetic and genomic techniques, have called into question the adequacy of traditional taxonomic and functional descriptors for accurately describing the true nature of the natural microbial world.

Taxonomic assessments of archaea and bacteria are largely rooted in operational definitions because the biological species concept *(1)* does not truly hold for prokaryotes. Unlike metazoa, there are few conspicuous morphological features for systematically defining, identifying, and interrelating different microbial species. In addition, the microbial fossil record is neither sufficiently extensive nor informative to provide much insight into ancestral microbial life. Therefore, prokaryotic taxonomic schemes have largely relied

From: *Microbial Genomes*
Edited by: C. M. Fraser, T. D. Read, and K. E. Nelson © Humana Press Inc., Totowa, NJ

almost exclusively on phenoptypic features and a few aggregate (and often nonhomologous) properties such as Gram stain, flagellar arrangement, guanine and cytosine (G+C) content in deoxyribonucleic acid (DNA). In phenotypically based taxonomic schemes, there are few features that can be generally applied to interrelate all taxa. DNA–DNA hybridization has also found wide use for species determinations, but can be generally applied only for interrelating very closely related species or strains. Other methodologies, including multilocus sequence typing and DNA microarrays (see Chapter 22), are now being seriously considered as tools in modern microbial taxonomy as well *(2)*.

More ecologically oriented descriptions of the microbial world parse microorganisms into different functional groups that often have no bearing on their evolutionary or taxonomic relationships. This perspective can be empirically very useful because the microbial world harbors so much biogeochemically relevant functional diversity not found in eukaryal macrobiota. From a microbial ecologist's point of view, function is what matters, and general guilds describe the activities and roles of microbes in a useful way. However, broad functional groups often clump together unrelated organisms with similar functions that actually have quite disparate evolutionary origins and mechanistic properties. In addition, the simplistic view of categorizing a microbe into a single functional group overlooks the complexity of the organism and its multiple interactions within the community and environment in which it lives.

Until only very recently, the evolutionary relationships and true ecological diversity of microbial life were essentially unknown and undescribed. There have been several major stumbling blocks to characterizing indigenous microbial life. First, there has been great reliance on isolation and pure culture techniques for characterizing microbial species. Although extremely powerful, the fact remains that this approach has so far successfully recovered only a very small fraction of extant microbial life. In addition, microbial genetic diversity is vastly undersampled by cultivation-based methods, even for those microbes that are readily cultivated.

Its microscopic nature presents formidable observational challenges, and the microbial world is still mostly observed only via indirect methods. Techniques for observing microbial ecosystems generally disrupt and perturb them. In analogy to the vagaries of Heisenberg's uncertainty principle, in studying the microbial world we more often than not change it. The lack of methods to identify and describe naturally occurring microbes has been a long-standing problem, and methods development in this context remains an intense area of focus. One extremely productive solution to these inherent difficulties has been the adoption of molecular biological techniques to the study of naturally occurring microbial diversity.

Cultivation-Independent Phylogenetic Surveys in the Environment

The ability to infer evolutionary relationships among all lifeforms changed dramatically when, as Zuckerandl and Pauling advocated *(3)*, informational molecules (semantides) began to be exploited as documents of evolutionary history. Evolutionary relationships can now be deduced from the sequence differences observed between orthologous macromolecules. For the first time, universal comparisons of homologous macromolecular features from virtually all (known) cellular lifeforms have become a practical reality.

In the early 1980s, developments in comparative molecular phylogenetics were instrumental in removing the roadblocks preventing accurate description of natural microbial diversity *(4,5)*. These cultivation-independent surveys extract phylogenetically informative gene sequences directly from nucleic acids of naturally occurring microbial assemblages. DNA isolated from mixed microbial populations is fragmented and recovered in recombinant clones that are then screened, sorted, and sequenced. Molecular sequence comparisons provide phylogenetic identification of the original population constituents. Small-subunit ribosomal ribonucleic acid (rRNA) genes have been the most commonly used phylogenetic markers to date because of their ubiquity and conserved nature.

Molecular phylogenetic surveys of natural microbial populations also provide markers useful for tagging specific groups or species for identification purposes. Small-subunit rRNAs, targets for these nucleic acid–based hybridization probes, have proven extraordinarily useful for such molecular tagging approaches. Relatively high levels of intracellular rRNA provide an abundant target for phylogenetic identification of individual cells using fluor-labeled oligonucleotide probes *(6,7)*. Individual microbial cells can now be stained with color-coded probes and identified via fluorescent *in situ* hybridization techniques. Nested suites of probes, specific for different taxonomic levels (e.g., domains, genera, species) can now be designed that allow hierarchical taxonomic dissection of complex, naturally occurring microbial assemblages *(7)*. Indeed, it has even become possible to link identity to function using such approaches. Phylogenetic identification (via fluorescent *in situ* hybridization) can now be combined with the determination of stable isotope composition of indivual cells by mass spectometry *(8,9)*.

The methodologies developed for cultivation-independent phylogenetic surveys laid the foundation for future efforts in environmental microbial genomics. The general scheme for cloning large DNA fragments from natural populations as originally proposed by Pace and colleagues in 1985 *(4,5)*, is shown in Fig. 1. Although newer enabling technologies are available today, variations on the scheme shown in Fig. 1 form the basis for contemporary studies of microbial population genomics. Initial schemes used bacteriophage-λ as the vector to archive natural population DNA. Early applications that surveyed the phylogenetic diversity of marine bacterioplankton used shotgun libraries and demonstrated the efficacy of the general approach *(10)*.

In the late 1980s, the coupled use of the polymerase chain reaction (PCR) with thermostable DNA polymerases *(11)* had a major practical impact on developments in the field of microbial ecology. Because it is so technically straightforward, PCR became the major tool for surveying naturally occurring microbial biodiversity using Pace's scheme (Fig. 1). The very first application of PCR to mixed microbial communities in ocean waters led to the discovery of ubiquitous and abundant new microbial lineages, as well as apparently high genetic microheterogeneity in closely related phylogenetic types *(12)*. At this point, most microbial ecologists abandoned the technically more difficult shotgun library approach, favoring instead PCR amplification of single genetic loci (like rRNA genes) from mixed microbial communities. This accelerated studies of naturally occurring phylogenetic diversity, but did not contribute much toward the functional description or characterization of naturally occurring microbes.

Cultivation-independent approaches have invigorated the field of microbial ecology over the past two decades. Phylogenetic comparison of rRNA genes retrieved directly from the environment has fast become the standard for surveying natural microbial diversity *(7,13)*. The approach has led to the discovery of many novel microbial taxa,

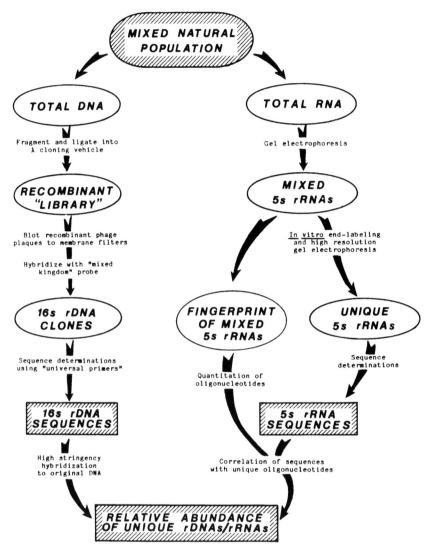

Fig. 1. Phylogenetic gene surveys in natural microbial populations according to the original scheme as proposed by Pace and collaborators (in 1985), involving cultivation-independent environmental surveys of phylogenetically informative genes. (Reprinted with permission from the American Society of Microbiology, ref. *4*.)

ranging from new species, to new phyla, even new "kingdoms." These newly recognized microbes are not minor players in the environment; they often represent the major taxa present in both terrestrial and marine ecosystems. These studies suggest that the pheno-typic properties of many of the most abundant microbes inhabiting earth remain to be determined.

In his treatise on bacterial evolution in 1987, Woese identified 12 major divisions in the domain Bacteria, inferred from rRNA sequence data, that were known at the time *(14)*. These 12 divisions still represent most of the taxa that have been readily cultivated and characterized using cultivation methods. Nevertheless, recent cultivation-independent

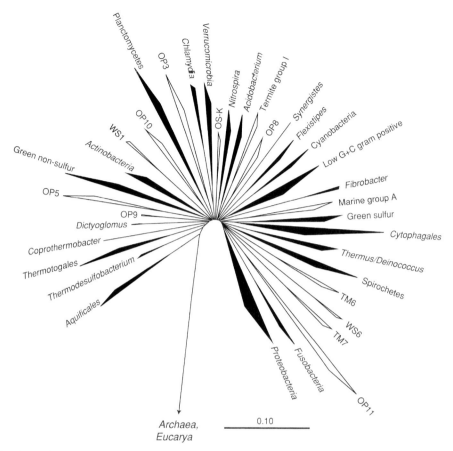

Fig. 2. Phylogenetically identified divisions within the domain Bacteria according to ref. *15.* Bacterial divisions with cultivated representatives are indicated in solid black, and those identified solely by rRNA genes recovered in environmental samples are indicated as open wedges. Scale represents substitutions per site. (Reprinted with permission from the American Society of Microbiology, ref. *15.*)

molecular surveys have revealed that the bacterial domain consists of many more divisions, having few or no cultured representatives (Fig. 2). The current tree of bacteria now contains upwards of 40 phylogenetically well-resolved bacterial divisions *(13,15).* The newly discovered groups within the domain Bacteria, most with no cultivated representatives, indicate that the microbial species of past culture collections have provided only a skewed and incomplete picture of extant microbial life on earth. This picture may be changing for the better as molecular phylogenetic surveys begin to inform and direct novel cultivation efforts *(16,17).* These efforts may provide at least some isolates from major ecological and phylogenetic groups that currently have no cultivated representatives.

Cultivation-independent surveys have also revealed new lineages of Archaea, now known to be widespread and abundant in many diverse habitats. The discovery of widespread diversity of archaea in nonextreme habitats is one of the particularly striking findings of culture-independent surveys. Previously, there was no reason to believe

ARCHAEA

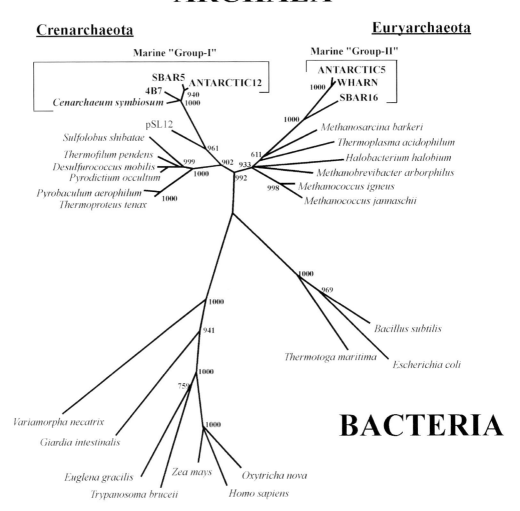

Fig. 3. Diversity of the most abundant cultivated and uncultivated marine planktonic Archaea, after ref. *21*. Two common marine archaeal groups (marine group I and marine group II) are indicated on the tree. (Reprinted with permission from the American Society of Microbiology, ref. *21*.)

that Archaea contributed significantly to the ecology of aerobic marine or terrestrial habitats. Yet, it is now apparent that at least two major groups (Fig. 3) of Archaea are common and very abundant components of marine plankton *(18–22)*. New groups keep cropping up in unexpected places, both terrestrial and marine *(18,19,23,24)*. It is now clear that Archaea are commonly present in the marine environment at cell densities of 1×10^5/mL at depths ranging from 100 m to the oceans' abyssal plains 5000 m deep *(20,*

22). At these cell densities, archaea represent about 20% or more of all microbial cells in the oceans, a habitat in which 10 years ago they were thought not to even exist. A more recent example of unanticipated archaeal diversity was the discovery and cultivation of the archaeon *Nanoarchaeum (25),* which lives in apparently obligate association with another archaeon belonging to the genus *Igneococcus.* This newly discovered archaeon has diverged so far from known groups that its rRNA sequence is not recognized by so-called universal oligodeoxynucleotide primers.

The recognition of novel, environmentally abudant microbial lineages was not the only discovery precipitated by cultivation-independent surveys. Cultivation-independent rRNA surveys have also shown a tremendous amount of microdiversity or microheterogeneity among very closely related microbes within and between populations *(12, 26–28).* This phenomena of genetic microheterogeneity (observed as highly similar, but distinct, rRNA sequences in a single population) appears independent of the specific taxonomic group, environment, or methodology used. How this rRNA microvariation is reflected in higher-order genome structure and organization, how this variation molds and impacts microbial population dynamics, and which mechanisms generate and maintain this variation remain to be explored fully.

GENOMICS MEETS MICROBIAL ECOLOGY

There are a number of conceptual hurdles to overcome when combining disciplines as disparate as genomics and microbial ecology. Genomics is technologically driven and has exploded as a discipline in as little as half a decade; its strength lies in the massive amounts of data that can be handled without preliminary culling. Microbial ecology, on the other hand, is at least a century old, yet it is still a severely technologically limited discipline. Genomic's technology push and microbial ecology's question-rich (but technique-limited) pull actually complement one another quite well. The difficulties sometimes arise when interfacing the predominantly laboratory-oriented, model systems–based approaches of genomics with the field-based, ecosystems approaches of microbial ecology.

Although the interfaces between the two disciplines are not always fully aligned, it is clear that each discipline has much to offer the other. Microbial ecologists benefit tremendously from the high-throughput, massively parallel approaches developed for genomic studies. Microbial genomics, in turn, will benefit greatly from direct study of genome evolutionary dynamics in the appropriate environmental context (via collaborative efforts with microbial ecologists).

Contemporary Genomic Methodologies

As discussed elsewhere in this volume (see Chapter 1), whole genome shotgun sequencing became possible when automated DNA-sequencing technologies achieved single pass 500-bp reads. This phememal throughput allowed shotgun sequencing of the first completed whole microbial genome sequence (of *Haemophilus influenzae*) in only 24,304 sequence reads *(29,30).* Since that time, raw throughput has increased many fold, and faster, cheaper, and more accessible DNA-sequencing technologies are just around the corner. In tandem with accelerated sequence production, software for sequence assembly and automated annotation is improving also, but probably not necessarily in pace with sequencing technologies.

The development of new vectors that can stably maintain and propagate large pieces of DNA was another important milestone in the genome revolution. In particular, bacterial artificial chromosome (BAC) vectors developed in the early 1990s for use in the human genome project have been especially useful. DNA replication in BACs is controlled by the F1 origin of replication of the *Escherichia coli* F plasmid *(31,32)*. By virtue of their low copy number, BACs can be used to propagate stably very large DNA fragments that are otherwise unstable *(32)*.

A large percentage (about 75%) of prokaryotic genomes sequenced so far originate from clinically important bacteria. Of the total prokaryotic sequences, about 90% are from the domain Bacteria *(33)*. Archaeal genome sequences are much less well represented in the databases. As discussed in detail throughout this volume, lessons learned from complete prokaryote genome sequences include

1. The discovery that a large fraction of newly identified open reading frames encode genes of unknown function
2. The recognition of the significance and pervasiveness of lateral gene transfer in genome (macro) evolution
3. The identification of specific genomic signatures of pathogenesis
4. A recognition of the diverse genomic consequences of becoming obligately symbiotic or parasitic

These are important lessons indeed, but the fact remains that an estimated more than 99.9% of the natural microbial world remains uncultivated, unknown, and understudied. Accessing this unseen microbial world requires strategies different from those used in the past. Microbial population genomics represents one of those strategies.

Extending Genomic Methodologies to Natural Microbial Populations

A number of microbial whole genome sequences now available were derived from microbes that have never been cultured in pure form. These include pathogens like *Mycobacterium leprae (34)*, *Rickettsia prowazekii (35)*, and *Plasmodium falciparum* (see Chapter 20) *(36)*, which can be purified free from other contaminating cells, but must be cultured parasitically in the presence of host animal cells. In addition, the complete genome sequences of two strains of the aphid endosymbiont *Buchnera aphidicola* have been reported *(37,38)*. In this case, the DNA content of *B. aphidicola* endosymbionts is extremely high (100 genome copies per cell), so it was possible to extract symbionts from bacteriocytes (symbiont-containing insect organelles) derived from 2000 aphids and purify the liberated symbionts through a 5-μm filter before performing direct whole genome shotgun sequencing *(38)*.

The applications discussed above demonstrate that the modern genomic procedures, such as whole genome shotgun cloning and sequencing, are entirely applicable to microbes that have never been isolated in pure culture. In most examples to date, this has included physical purification of the cells of interest from other organisms prior to shotgun sequencing. From this step onward, whole genome sequence determination was no different from the approach used for laboratory-cultivated strains.

An obvious next question then is, can similar approaches and technologies be applied to mixed microbial populations found in nature? In theory, the approaches, techniques, and analytical strategies used for genomic analysis of individual strains are entirely applicable to naturally occurring microbes. Questions to ask before embarking on such wild microbial genome safaris might include the following: Is physical purification or separation of specific target cells an absolute prerequisite for obtaining whole genome sequences from naturally occurring microbes in the environment? Do large genome fragment cloning and sequencing techniques represent a viable cultivation-independent strategy for characterizing microbes in natural populations? Given the extensive genetic microheterogeneity found in natural microbial populations, can current technologies provide the necessary resolution to deconvolute authentic whole genome sequences from the shotgun-sequenced DNA of a complex mixed population?

ENVIRONMENTAL MICROBIAL GENOMICS: METHODOLOGIES

Sample Collection

Sampling strategies are always context dependent and dictated by the nature of the microbial community in question and the environment it occupies. A number of questions need to be addressed in the design of sampling strategies.

Will the cells need to be purified away from a soil, sediment, or rock matrix? The interference of contaminating materials in DNA purification and downstream steps requiring enzymatic modifications are major considerations. Do the cells need to be concentrated in some fashion before DNA extraction? For some procedures, a minimum of 1×10^9 cells or more are required, and this may be more or less easy to achieve depending on the microbial community under study. What size DNA fragments are required, and what cloning strategy will be used? The answer to these questions will dictate whether the cells need to be embedded in agarose to maintain high molecular weight, intact chromosomes during DNA preparation. (For inserts 40 kb or smaller, embedding cells in agarose is not an absolute requirement.) What is the complexity, richness and evenness, and specific composition of the population/sample in question? For example, it would take only a small proportion of eukaryotic genomes in a sample, each having 10 to 50 times greater genome size, to effect the representation of prokaryotic genome sequences in any given library dramatically.

All these considerations, and many others, are quite important to consider before embarking on studies in environmental genomics. Prior knowledge of the nature, complexity, and composition of the original sample, both matrix and microbes, will largely influence experimental design and the final outcome of library construction and downstream analyses.

Library Construction Approaches

Shotgun sequencing approaches, extensively discussed in this volume (see Chapter 1), are only just now being seriously considered for complex natural microbial populations. The main reason that shotgun sequencing has not been extensively applied so far for microbial population analyses is that, for any given population, the magnitude and scale (and economics) of sequencing required for adequate coverage is uncertain. Nonetheless, shotgun sequencing strategies will undoubtedly play a large role in the future of microbial population genomics. However, to date, only more directed strategies using lambda, cosmid, or BAC libraries to recover natural microbial population DNA have been reported. Next is a brief discussion of large-insert DNA-cloning methodologies as applied to natural microbial populations.

Fosmids are essentially identical in sequence and other properties to BACs once packaged via lambda extracts and transfected into *E. coli (31)*. The main difference between BACs and fosmids is the method used to introduce them into *E. coli*. BACs are introduced into host cells via electroporation, whereas fosmids are packaged in phage λ-heads and transfected into *E. coli* host cells. These differences in methodology influence the maximum size of recombinant DNA inserts recovered. BACs recovered from natural microbial populations can be as large as 200 kb *(39)*, whereas fosmid sizes range from 32- to 45-kb *(31,40)*.

The general strategy for BAC cloning is shown in Fig. 4 and is essentially identical to original protocols introduced in the early 1990s *(31,32,34)*. In brief, cells are first embedded in agarose to prevent physical shearing during downstream lysis and purification steps. Although very efficient, any other material copurifying with cells will also be trapped in the agarose and may inhibit downstream enzymatic modification (e.g., restriction digestion or ligation). Therefore, cells to be embedded in agarose need to be relatively free of contaminating materials. Cells concentrated from relatively clean, dilute aqueous environments may be more amenable to embedding than those derived from more challenging environments like soils or sediments.

After proteinase K/detergent digestion of the agarose-embedded cells, the liberated DNA is subjected to partial digestion by restriction endonucleases (*Bam*H1 and *Hind*III are commonly used). Partial digestion is adjusted to generate a maximum amount of DNA in the appropriate size range, around 100–300 kb. Often, there is sheared DNA in the original preparation, so partial digestion conditions need to be monitored carefully above and below the size regions of interest (see, for instance, Fig. 5A). After purification (via Pulse Field Gel Electrophoresis [PFGE]), the DNA fragments are recovered from the agarose, ligated into the BAC vector, and electroporated into *E. coli*.

Improvements in screening and optimization of electroporation conditions have elevated the efficiency of the entire process. Importantly, these general procedures need to be tuned to optimize them for any given sample examined. For standard BAC libraries, it is particularly important to ensure that there is sufficient material, in general a minimum of 1×10^{10} cells/mL in the original agarose plug.

Although smaller in size on average (ca. 40 kb), the fosmid-cloning approach has several potential advantages compared to standard BAC cloning procedures for construction of high-quality libraries from environmental samples. In general, less DNA, with a smaller average size, is required. This lessens the stringent requirement for standard BAC cloning for isolation of largely intact chromosomes from many (1×10^{10}) cells (as discussed in this section). The original fosmid-cloning procedure *(34)* is outlined in Fig. 5A. In the original protocol, the vector arms are prepared, and *Sau*3A partial digests of the DNA are prepared, generally with a size selection step. After ligation, lambda extract packaging, and transfection, the libraries are arrayed in microtiter dishes, and standard plasmid purification and screening procedures are used for evaluation.

Newer approaches have greatly improved the efficiency of fosmid library preparation tremendously. These new technques incorporate random physical sheering of DNA inserts, followed by end repair and blunt-end ligation into the circularized fosmid (Fig. 5B), as opposed to partial restriction endonuclease digestion and "sticky-end" cloning as with standard BAC vectors (Figs. 4 and 5A). In general, DNA prepared by standard purification procedures produces DNA of the required size (40 kb), eliminating the

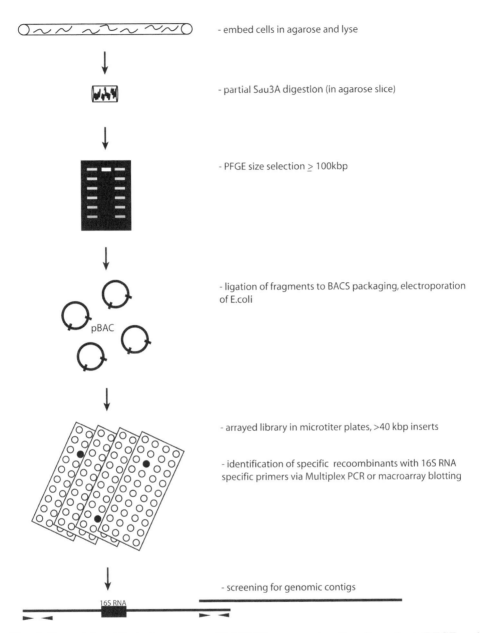

- embed cells in agarose and lyse

- partial Sau3A digestion (in agarose slice)

- PFGE size selection ≥ 100kbp

- ligation of fragments to BACS packaging, electroporation of E.coli

pBAC

- arrayed library in microtiter plates, >40 kbp inserts

- identification of specific recoombinants with 16S RNA specific primers via Multiplex PCR or macroarray blotting

- screening for genomic contigs

16S RNA

Fig. 4. Bacterial artifical chromosome (BAC) library construction flow chart (PFGE, pulse field gel electrophoresis).

agarose embedding of cells, partial restriction endonuclease digestion, and PFGE steps (Fig. 5B). In addition, more rigorous DNA purification protocols (for instance, CsCl equilibrium density gradient centrifugation) can be used quite successfully with the blunt-end fosmid-cloning approach (E. F. DeLong, unpublished results).

These advantages allow the preparation of high-qualtity fosmid libraries from very small amounts of material. Such blunt-end fosmid libraries may even be more representative

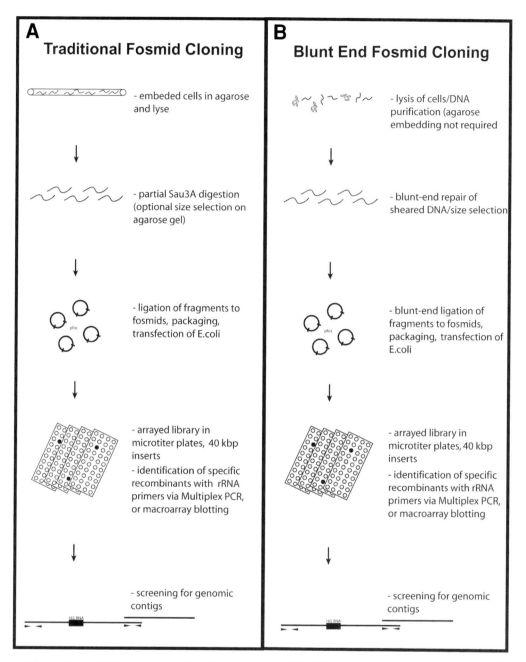

Fig. 5. Fosmid library construction flow charts. (**A**) Traditional fosmid cloning using partially digested DNA and "sticky end" cloning methods. (**B**) Alternative blunt-end fosmid-cloning strategy.

than standard BAC libraries (H. Shizuya, personal communication, 2002). This is because regions with few Hind III or BamHI sites are expected to be less well represented in BAC or fosmids prepared using partial digests (Figs. 4 and 5A), which are not required in the blunt-end fosmid-cloning approach (Fig. 5B). With this streamlined procedure, it is now possible to prepare high-quality fosmid libraries successfully using as little as 1 µg or less of natural population DNA (E. F. DeLong, unpublished results).

A significant concern in library construction centers is the possibility that chimeric BACs might be formed from DNA originating from different organisms. Although a potential problem, the methods employed during library construction, particularly size selection of DNA fragments before cloning, help reduce this risk. In addition, protocols that use lambda packaging to introduce fosmids into *E. coli* incorporate an additional size selection step during packaging, which further reduces the risk of chimeric BAC recovery.

Several methods can also be used to rule out the possibility that any given BAC is chimeric. BAC population screening helps tremendously: Homologous but nonidentical BACs can generally be identified in libraries, which represent closely related but nonidentical strain variants in any given microbial population (see, for instance, refs. *42* and *43*). Near-identical genomic structures in these BAC variants, including gene content, organization, and synteny, in essence prove that the genomic structures in question are not artifacts (e.g., not chimeric). In addition, there are high-resolution optical mapping methods becoming available *(44)* that might be used to test and verify the integrity of individual clones. Similarly, chromosome-painting techniques *(45)* might also be used to verify the integrity and presence of paired-end BAC sequences in individual clones.

Library Screening and Analysis

One of the advantages of BACs and fosmids is that standard screening approaches developed for conventional plasmids are all applicable. High-density colony blots (macroarrays) that allow screening of ten thousand clones on a 20 cm × 20 cm blot are one typical approach *(40)*. Multiplex screening by PCR is another rapid and very useful way to screen for specific recombinant clones. Both approaches have been successfully used to identify clones of interest in environmental BAC and fosmid libraries *(39,40)*. For screening of rRNA genes, other rapid multiplexing approaches that exploit automated capillary electrophoresis include length heterogeneity polymorphism *(46)* or terminal restriction fragment PCR *(47)* screening.

In addition, functional gene screening can be very useful for gaining information about specific metabolic pathways and processes *(48)*. Another screening method that has proven very useful is random end sequencing, which can be used for library quality assessment, gene surveys, contig identification, and comparative community analyses (see, for example, http://www.tigr.org/tdb/MBMO/BAC_end_ann_info.shtml).

Although end sequencing of single-copy BACs and fosmids was once challenging, improvements in sequencing technologies have now removed technical barriers to high-throughput BAC end sequencing. This is probably the first-pass method of choice for gaining an in-depth qualitative and quantitative assessment of any given environmental library.

A significant point to consider in library assessments is the fact that, even in low copy number BAC vectors *(49)*, there may be differential recovery of different genomic sequences. These artifacts can arise from the presence of highly repetitive, tandem sequence motifs *(49)* or because of gene expression of recombinant BAC DNA in *E. coli* (especially if it is of bacterial origin) *(50)*, which may sometimes prove lethal.

Assessing the extent and fidelity of genome recovery from complex assemblages is important in many contexts, but remains difficult to quantify accurately.

Screening for phylogenetic representation within the library (via rRNA genes) is one useful approach for assessing the diversity of genomes recovered in libraries *(51,52)* relative to the representation in the starting DNA. Statistical methods to estimate total genome recovery in populations may also prove useful *(53)*, but may require a fairly extensive sampling regime.

The type of coverage desired in recombinant libraries also largely depends on the purpose of the investigation. For genome-oriented population biology studies, sample normalization is not necessarily desirable because quantitative information about genome representation is a major goal. Bioprospectors, on the other hand, may wish to maximize the diversity of recovered types and therefore seek methods to amplify low-abundance genome types in the libraries.

MICROBIAL POPULATION GENOMICS
AND ECOLOGY: REAL-WORLD APPLICATIONS

There are a variety of motivators that prompt researchers to embark on studies in microbial population genomics [also referred to as *environmental genomics (40,42)* or *metagenomics (51)*]. Curiosity-driven ecological or population genetic studies, bioprospecting, and process-oriented biogeochemical or biogeological inquiries all stand to benefit from the general approach. The field is nascent and evolving symbiotically with technical developments in genomics.

The scheme in Fig. 6 illustrates some potential applications that flow from application of genomics to the study of natural microbial assemblages. Gene discovery, metabolic pathway characterization and exploitation, biochemical studies, and correlation of genotype with ecological distribution represent some of the activities that flow from the scheme illustrated in Fig. 6. Next, a few case studies in environmental genomics illustrate the potential of various approaches and their synergy with other fields and disciplines. Combined, some of these examples have followed the full path depicted in Fig. 6 and have led to some unexpected surprises *(39,52,54)*.

Directed Characterization of Naturally Occurring Microbes

As discussed previously, one approach for cloning large genomic fragments from environmentally derived DNA is to use high-fidelity, low copy number BACs *(31,32)*. The first application of BAC vectors to recover DNA from a mixed microbial population was reported in 1996 *(40)*. In this study, a BAC library was constructed from microbes collected at a depth of 200 m in the Pacific Ocean to characterize uncultivated marine archaea. A 40-kbp DNA fragment from a planktonic archaeon was identified on a BAC clone and partially sequenced, providing the first glimpse into the gene arrangement, content, and protein encoding genes of marine archaea *(40)*. Large genome fragments derived from bacterioplankton of the order Planctomycetales were also later identified in the same library *(55)*.

A similar approach was later used systematically to retrieve genome fragments from the nonthermophilic uncultivated archaeal sponge symbiont, *Cenarchaeum symbiosum*

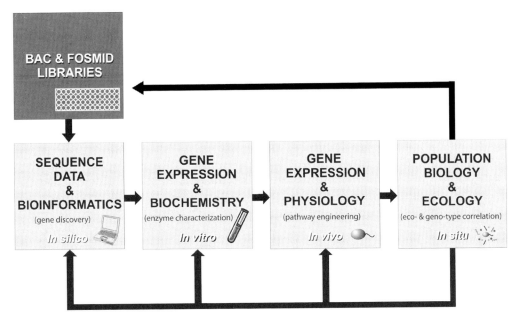

Fig. 6. Generalized flow path of integrated studies in microbial population genomics.

(42,56,57). A genome fragment harboring the rRNA operon was also shown to encode the gene for a type B archaeal DNA polymerase that shared highest sequence similarity with orthologs from hyperthermophilic crenarchaea. The *C. symbiosum* DNA polymerase was expressed, biochemically characterized *(56),* and shown to catalyze biochemical activities identical to those of characterized thermophilic archaeal polymerases. However, the *C. symbiosum* polymerase was thermostable only to 40°C, unlike related archaeal polymerases, which are stable to at least 75°C. This demonstrated the utility of identifying, expressing, and characterizing specific gene products from uncultivated microbes by gene walking from phylogenetic markers *(56).* In a follow-up study, the nature and extent of genomic heterogeneity in strains of *C. symbiosum* were explored in considerable detail using *C. symbiosum* BAC libraries *(42).*

To assess the nature and function of naturally occurring planktonic microbes, many of which had resisted cultivation, Béjà et al. *(52)* constructed a BAC library from surface water microbes of Monterey Bay. These BAC libraries had an average insert size of about 80 kbp, with some clones exceeding 150 kbp. These results represent some of the most enouraging so far and demonstrate the feasibility of generating very large insert BAC libraries from mixed microbial population DNA.

Analysis of this surface seawater BAC library showed that the phylogenetic representation mirrored that of PCR-amplified rRNA clone libraries *(52).* A 60-kb genome fragment from an uncultivated marine archaeon was located in the library, fully sequenced, and annotated. The screening and analysis of this BAC library suggested that significant genomic information, providing insight into the diversity and biological properties of naturally occurring microbial assemblages, could be derived from the approach.

Bioprospecting

There has been a good deal of interest in recovering native microbial DNA from soil, with most studies focusing on bioprospecting for drugs, enzymes, and other natural products. The rationale is simply that, because more than 99% of the microbes in these environments have not been cultivated, the majority of natural products (antibiotics, enzymes, etc.) remain to be discovered and exploited. This type of approach has been in use for over a decade in the biotechnology industry *(58)*. Directed bioprospecting in small (5- to 8-kbp) insert environmental DNA libraries has been fruitful on several fronts, for instance, for isolating novel lipolytic 4-hydroxybutyrate dehydrogenase or chitinase genes *(59–62)*. A two-prong approach of combining environmental discovery with directed in vitro evolution to identify and optimize an industrially important biocatalyst (in this case, α-amylase) was reported *(63)*.

In another report *(51)*, microbial DNA was extracted directly from soil and cloned into BAC libraries. Two different so-called metagenome libraries were recovered and analyzed, one with an average DNA insert size of 27 kb and the other averaging 44.5 kb. The phylogenetic composition of rRNA genes in the library was consistent with other phylogenetic surveys in soil. Although the average insert size of these libraries did not much exceed that of conventional lambda or cosmid library approaches, this report was one of the first to retrieve large DNA fragments from soil microbial populations *(51)*. The authors also showed that a small fraction of the BAC clones expressed identifiable phenotypes, including DNAse, lipase, and amylase activities *(51)*, consistent with results of a previous report *(50)*. Another study focused on soil microbes recovered from a BAC library with an average insert size of 37 kbp and a range of 5 to 120 kbp *(64)*. Similarly, Brady et al. reported the recovery of the biosynthetic gene cluster for the broad-spectrum antibiotic violacein from a soil DNA cosmid library *(65)*.

Microbial Population Genomics

Environmentally oriented microbial genomics is also catalyzing in-depth studies of microbial population genetics and placing them in an ecological context. Current studies have focused on cultivated strains with habitat-variable phenotypes (e.g., ecotypes) *(26)*, as well as large DNA fragments cloned directly from the environment *(42,43)*. An example of blending comparative genomics with microbial population biology is from recent comparisons of two closely related *Prochlorococcus* species *(66)*. *Prochlorococcus* is a chlorophyll b–containing marine cyanobacterial group that accounts for as much as 50% of the photosynthetic biomass in the open ocean. High- and low-light *Prochlorococcus* types differ in their chlorophyll b/chlorophyll a ratios, in their irradiance optima for photosynthesis, and in their relative distributions in the water column. Yet, their small-subunit rRNA sequences differ by only 3%.

The complete genome sequence of two strains, a strain adapted to high light (MED4) and a strain adapted to low light (MIT9313), were determined *(66)*. Comparative analyses revealed the genomic origins of some of the physiological and ecological variability in the closely related *Prochlorococcus* ecotypes. The type adapted to low light has a significantly larger genome (2.4 Mbp) than the high-light type (1.7 Mbp) and has more genes associated with the photosynthetic apparatus, including phycoerythrin biosynthetic genes. The high-light type, on the other hand, has more genes in the high-light inducible protein category, as well as more genes for ultraviolet damage repair *(66)*. Differences in the complement of nitrogen assimilation genes between these closely related *Prochlorococus* strains may also influence their distribution in the water column *(67)*.

Population genetics represents another outstanding area in which detailed characterization of naturally occurring microbial genomes (and their microvariability) is likely to have high impact. In the past, a good deal of information about microbial population genetics was inferred from two main types of data sets: multilocus enzyme electrophoresis or multilocus sequence typing, largely focused on pathogens *(68)*. Such studies involved mainly comparisons of cultivated pathogenic strains and not so much genuine within- or between-population comparisons. In general, the extent cultivation introduces sampling biases in such studies in unknown. As has been acknowledged by microbial population geneticists, "sampling bias, ecological substructuring, and the emergence of transient adaptive clones" need to be considered carefully *(68–70)*. Now, using genomics, it will be possible to approach microbial population biology on its own terms— in authentic populations, with approaches that rule out cultivation biases.

Some of the major questions, for instance, those focusing on the relative contributions of recombination vs point mutation in microbial diversification, should be more directly addressable via microbial population genomics. For example, one study demonstrated tremendous genomic variation within a single population of marine archaea, which are virtually indistinguishable by rRNA sequence *(43)*. Even archaea with identical rRNA sequences were shown to be significantly different in flanking protein-encoding genes, even though they coexist in the same population. The implication was that enormous allelic variation, the direct result of point mutations and genetic drift, exists in free-living, sympatric microbial species.

Microbial population genomics is now poised to clarify better the relative importance of lateral transfer, recombination, and genetic drift to the diversification of microbial species within and between populations. It may also allow better estimations of central population genetic parameters *(71)*, which will inform theorists and provide generalized models with robust and much-needed data. Such studies should spur the development of new theory and greatly improve the understanding of microbial population biology as it occurs in different organismal, ecological, and population contexts.

Metabolic Reconstruction of Naturally Occurring Microbial Activities

In a follow-up study on a Monterey Bay surface water BAC library (discussed in the section on directed characterization), the sequence of a 130-kb BAC clone derived from a yet-uncultivated SAR86 type was determined. Sequence downstream of the SAR86 rRNA operon revealed the unanticipated presence of a rhodopsin-like protein (Fig. 7; *39)*. This class of membrane protein had never before been observed in the domain Bacteria or in the ocean. When expressed in the presence of its chromophore retinal, the novel membrane protein (dubbed proteorhodopsin) proved to be a light-driven proton pump *(39)*. Proteorhodopsin was shown to be similar in biochemical function and photocycle properties to classical bacteriorhodopsins, originally discovered in extremely halophilic archaea. These data suggested that the SAR86 bacterioplankton represented a new, abundant type of phototroph distributed in oceanic surface waters worldwide. The postulated light-driven energy-generating mechanism for SAR86 is shown in Fig. 7, and it

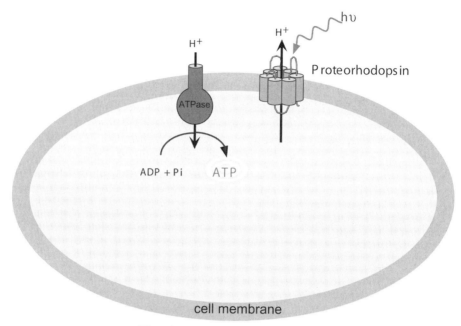

Fig. 7. Postulated energy-generating mechanism in uncultivated SAR86 planktonic marine bacteria. ADP, adenosine 5'-diphosphate; ATP, adenosine triphosphate; Pi, inorganic phosphate. (After reference *39* and E. F. DeLong and O. Béja, unpublished).

is currrently presumed (not shown) that these microbes are photoheterotrophic, deriving considerable energy from respiration of organic carbon sources as well *(39)*. Later, the presence of native proteorhodopsin in ocean waters and the discovery of multiple proteorhodopsin genetic variants spectrally "tuned" as a function of depth were also demonstrated in the field *(54)*.

In aggregate, this work showed that genomic studies applied to natural microbial populations are more than just an exercise in bioinformatics. Hypothesis and reagents are generated as part of the process and can lead to the identification and verification of novel properties and processes occurring in natural ecosystems. Genome-based studies also provide the tools to study newly discovered processes in greater depth and detail. These points are also illustrated more generally in the flow chart of Fig. 6, emphasizing the feedback and interplay between sequence-based, biophysical, biochemical, and ecological studies.

The same surface water Monterey Bay BAC library was later screened for evidence of bacteriochlorophyll-containing aerobic anoxygenic phototrophic (AAP) bacteria *(48)*. Bacteria with bacteriochlorophyll-based photosystems had recently been reported as unexpectedly abundant in seawater *(72,73)*. A number of surface water BAC clones were shown to contain about 40 kbp superoperons that encoded the photosynthetic reaction center, carotenoid, and bacteriochlorophyll biosynthetic AAP bacteria genes. Analyses of the genomic structure of the AAP bacterial photosynthetic superoperons showed that some of the most prevalent planktonic phototrophs were not the types predicted from known cultivated strains. Again, the importance of cultivation-independent approaches to understand the natural world is evident in this example. Genome-based discoveries are bound to be plentiful in the context of naturally occurring microbial populations.

Comparative Ecosystem Genomics

Comparative ecosystem genomics represents a new and virtually unexplored frontier in microbial biology that will inevitably emerge from studies in microbial population genomics. Soon, it will become possible to inventory a reasonably representative sampling of the phylogenetic and functional repertoire of genes and genomes that comprise entire microbial communities and ecosystems. The interesting comparisons will emerge when patterns of gene content can be correlated to within- and between-ecosystem differences. It is likely that significant microbe–microbe interrelationsips will become evident, and that specific indicators of organismal interactions will result from careful study of genomic patterns. The ecological significance of genomic variation and its functional consequences are also likely to become much clearer on closer examination of the relationships between microscale genomic variation and environmental differences.

From a different perspective, higher-order trophic relationships and ecosystem characteristics are also likely to emerge from careful inspection of habitat-specific genomic patterns. As a thought experiment, consider, for example, the simple concept shown in Fig. 8. As a point of departure, consider some ongoing analyses of BAC libraries retrieved from several depths in the Monterey Bay water column that includes random BAC end sequencing (Fig. 8 and http://www.tigr.org/tdb/MBMO/).

It is already known that microbial communities are stratified in the ocean's water column *(74)*. As data genomic data accumulate, it should become possible to gain a more detailed view of the nature of the genomes that make up these different communities. Questions that could then be considered include the following: Does the functional gene representation found at each depth reflect the current understanding of the ecology of these systems? Do the gene and genome distribution patterns provide information about processes as yet unsuspected within each habitat? What overlaps or core gene suites (indicated by the cross hatches in the Venn diagram in Fig. 8) are shared between these diverse habitats and communities? Are there nonhomologous, but functionally analogous, core biochemical and metabolic pathways that are shared by each ecosystem as well? Alternatively, what are the uniquely identifying genomic characters (white areas in circles of Fig. 8) that define each depth-stratified microbial ecosystem, and what do they tell us about each ecosystem from a functional standpoint?

The possibilities for comparative (microbial) ecosystem genomics are enormous. As new data and technologies emerge, this nascent field promises to provide entirely new ways of perceiving the world.

CONCLUSION

Leveraging technologies and general economic considerations continue to evolve in favorable directions for enabling studies in microbial population genomics and ecology. Improvements in technical areas, including automated DNA sequencing, bioinformatics, proteomics, microarray technology, and automated environmental sensors all are likely to play important roles in future integrated studies of microbial population genomics and ecology. One of the more challenging issues in such a data-dense, reductionist activity

Sample Origin	Total Cells	Average Insert Size	Estimated Coverage	Cumulative Genomes
Seawater 0m	2.5×10^{11}	80 kb	458 MB	~152
Seawater 80m	1.5×10^{11}	74 kb	1,184 MB	~390
Seawater 750m	5×10^{10}	60 kb	96 MB	~32

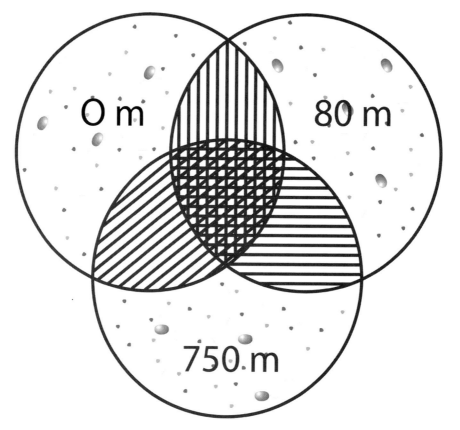

Fig. 8. Comparative community genomics. The table shows current BAC libraries being examined from different depths in the water column in Monterey Bay (also see http://www.tigr. org/tdb/MBMO/). The Venn diagram outlines the potentially shared as well as unique subsets of each genomic community occupying the different depths.

includes coordination of disparate observations and data sets at comparable spatial and temporal resolution. Likely developments in tandem are the maturation of genomic and ecological theory, which coincide with the large amount of newly available data and help direct the future acquisition of more.

Zucherandl and Pauling *(3)* posited that individual macromolecules are documents of evolutionary history. A logical extension of this view is that naturally occurring genomes are documents of environmental, evolutionary, and ecological history and dynamics. Environmental genomic approaches are cultivation independent and so provide equivalent access to the collective genomes of abundant, yet-uncultivated microbial species.

Genomic analyses of uncultured microbes can provide significant insight into the biological properties of individuals within microbial populations. Applied microbial population genomics can now facilitate predictions about phenotype from genotype, broadening the knowledge of natural environmental processes. Access to this raw and unadulterated diversity extends the ability to exploit natural microbial products and processes. Genomic patterns that give rise to emergent properties within and between populations of interacting genomes, and their relationship to ecosystems, are likely to become increasingly apparent.

Studies of genomics in natural microbial populations are only now beginning. Opportunities abound. It soon will become realistic to think about studying the dynamics of genome populations, which are likely to fluctuate and evolve via mechanisms that include genetic drift, recombination, dispersal, competition, migration, and succession. The combined application of classical population genetic theory coupled with a new-found ability to study the specifics of genome evolution and exchange within and between microbial populations is one area in which studies of microbial ecology and genomics will merge in the future. The natural microbial world provides ideal case studies for understanding the processes and dynamics of genome evolution. Conversely, genomics provides the necessary data to elucidate the complexity of form and process in genome evolution underlying microbial population structure and dynamics. Microbial ecology and genomics are now bound to inform each other in a synergism that should result in new theory and a deeper understanding of the natural microbial world around us.

REFERENCES

1. Ward DM. A natural species concept for prokaryotes. Curr Opin Microbiol 1998; 1:271–277.
2. Stackebrandt E, Frederiksen W, Garrity GM, et al. Report of the ad hoc committee for the re-evaluation of the species definition in bacteriology. Int J Syst Evol Microbiol 2002; 52:1043–1047.
3. Zuckerkandl E, Pauling L. Molecules as documents of evolutionary history. Theor Biol 1965; 8:357–366.
4. Pace NR, Stahl DA, Olsen GJ, Lane DJ. Analyzing natural microbial populations by rRNA sequences. ASM News 1985; 51:4–12.
5. Olsen GJ, Lane DJ, Giovannoni SJ, Pace NR, Stahl DA. Microbial ecology and evolution: a ribosomal RNA approach. Annu Rev Microbiol 1986; 40:337–365.
6. DeLong EF, Wickham GS, Pace NR. Phylogenetic stains: ribosomal RNA-based probes for the identification of single cells. Science 1989; 243:1360–1363.
7. Amann RI, Ludwig W, Schleifer K-H. Phylogenetic identification and *in situ* detection of individual microbial cells without cultivation. Microbiol Rev 1995; 59:143–169.
8. Orphan VJ, House CH, Hinrichs KU, McKeegan KD, DeLong EF. Methane-consuming archaea revealed by directly coupled isotopic and phylogenetic analysis. Science 2001; 293:484–487.
9. Orphan VJ, House CH, Hinrichs KU, McKeegan KD, DeLong EF. Multiple archaeal groups mediate methane oxidation in anoxic cold seep sediments. Proc Natl Acad Sci USA 2002; 99: 7663–7668.
10. Schmidt TM, DeLong EF, Pace NR. Analysis of a marine picoplankton community by 16S rRNA gene cloning and sequencing. J Bacteriol 1991; 173:4371–4378.
11. Saiki RK, Gelfand DH, Stoffel S, et al. Primer-directed enzymatic amplification of DNA with a thermostable DNA polymerase. Science 1988; 239:487–491.
12. Giovannoni SJ, Britschgi TB, Moyer CL, Field KG. Genetic diversity in Sargasso Sea bacterioplankton. Nature 1990; 345:60–63.

13. Pace NR. A molecular view of microbial diversity and the biosphere. Science 1997; 276:734–740.

14. Woese CR. Bacterial evolution. Microbiol Rev 1987; 51:221–271.

15. Hugenholtz P, Goebel BM, Pace NR. Impact of culture-independent studies on the emerging phylogenetic view of bacterial diversity. J Bacteriol 1998; 180:4765–4774.

16. Rappe MS, Connon SA, Vergin KL, Giovannoni SJ. Cultivation of the ubiquitous SAR11 marine bacterioplankton clade. Nature 2002; 418:630–663.

17. Connon SA, Giovannoni SJ. High-throughput methods for culturing microorganisms in very-low-nutrient media yield diverse new marine isolates. Appl Environ Microbiol 2002; 68:3878–3885.

18. DeLong EF. Everything in moderation: archaea as "non-extremophiles." Curr Opin Genet Dev 1998; 8:649–654.

19. DeLong E. Archael means and extremes. Science 1998; 280:542–543.

20. DeLong EF, Taylor LT, Marsh TL, Preston CM. Visualization and enumeration of marine planktonic archaea and bacteria by using polyribonucleotide probes and fluorescent *in situ* hybridization. Appl Environ Microbiol 1999; 65:5554–5563.

21. DeLong EF, King LL, Massana R, et al. Dibiphytanyl ether lipids in nonthermophilic crenarchaeotes. Appl Environ Microbiol 1998; 64:1133–1138.

22. Karner MB, DeLong EF, Karl DM. Archaeal dominance in the mesopelagic zone of the Pacific Ocean. Nature 2001; 409:507–510.

23. Lopez-Garcia P, Lopez-Lopez A, Moreira D, Rodriguez-Valera F. Diversity of free-living prokaryotes from a deep-sea site at the Antarctic Polar Front. FEMS Microbiol Ecol 2001; 36:193–202.

24. Lopez-Garcia P, Moreira D, Lopez-Lopez A, Rodriguez-Valera F. A novel haloarchaeal-related lineage is widely distributed in deep oceanic regions. Environ Microbiol 2001; 3:72–78.

25. Huber H, Hohn MJ, Rachel R, Fuchs T, Wimmer VC, Stetter KO. A new phylum of Archaea represented by a nanosized hyperthermophilic symbiont. Nature 2002; 417:63–67.

26. Moore LR, Rocap G, Chisholm SW. Physiology and molecular phylogeny of coexisting Prochlorococcus ecotypes. Nature 1998; 393:464–467.

27. Rocap G, Distel DL, Waterbury JB, Chisholm SW. Resolution of Prochlorococcus and Synechococcus ecotypes by using 16S-23S ribosomal DNA internal transcribed spacer sequences. Appl Environ Microbiol 2002; 68:1180–1191.

28. DeLong E, Pace NR. Environmental diversity of bacteria and archaea. Systematic Biol. 2001; 50:1–9.

29. Fleischmann RD, Adams MD, White O, et al. Whole-genome random sequencing and assembly of *Haemophilus influenzae* Rd. Science 1995; 269:496–512.

30. Nelson KE, Paulsen IT, Fraser CM. Microbial genome sequencing: a window into evolution and physiology. ASM News 2001; 67:310–317.

31. Kim U-J, Shizuya H, Dejong P, Birren B, Simon M. Stable propagation of cosmid sized human DNA inserts in an F-factor based vector. Nucleic Acids Res 1992; 20:1083–1185.

32. Shizuya H, Birren B, Kim UJ, et al. Cloning and stable maintenance of 300-kilobase-pair fragments of human DNA in *Escherichia coli* using an F-factor-based vector. Proc Natl Acad Sci USA 1992; 89:8794–8797.

33. Doolittle RF. Microbial genomes opened up. Nature 2002; 392:339–342.

34. Cole ST, Eiglmeier K, Parkhill J, et al. Massive gene decay in the leprosy bacillus. Nature 2001; 409:1007–1011.

35. Andersson SG, Zomorodipour A, Andersson JO, et al. The genome sequence of *Rickettsia prowazekii* and the origin of mitochondria. Nature 1998; 396:133–140.

36. Gardner MJ, Tettelin H, Carucci DJ, et al. The malaria genome sequencing project: complete sequence of *Plasmodium falciparum* chromosome 2. Parassitologia 1999; 41:69–75.

37. Tamas I, Klasson L, Canback B, et al. Fifty million years of genomic stasis in endosymbiotic bacteria. Science 2002; 296:2376–2379.

38. Shigenobu S, Watanabe H, Hattori M, Sakaki Y, Ishikawa H. Genome sequence of the endocellular bacterial symbiont of aphids *Buchnera* sp APS. Nature 2000; 407:81–86.

39. Béjà O, Aravind L, Koonin EV, et al. Bacterial rhodopsin: evidence for a new type of phototrophy in the sea. Science 2000; 289:1902–1906.

40. Stein JL, Marsh TL, Wu KY, Shizuya H, DeLong EF. Characterization of uncultivated prokaryotes: isolation and analysis of a 40-kilobase-pair genome fragment from a planktonic marine archaeon. J Bacteriol 1996; 178:591–599.

41. Shizuya H, Kouros-Mehr H. The development and applications of the bacterial artificial chromosome cloning system. Keio J Med 2001; 50:26–30.

42. Schleper C, DeLong EF, Preston CM, Feldman RA, Wu KY, Swanson RV. Genomic analysis reveals chromosomal variation in natural populations of the uncultured psychrophilic archaeon *Cenarchaeum symbiosum*. J Bacteriol 1998; 180:5003–5009.

43. Béjà O, Koonin EV, Aravind L, et al. Comparative genomic analysis of archaeal genotypic variants in a single population and in two different oceanic provinces. Appl Environ Microbiol 2002; 68:335–345.

44. Cai W, Jing J, Irvin B, et al. High-resolution restriction maps of bacterial artificial chromosomes constructed by optical mapping. Proc Natl Acad Sci USA 1998; 95:3390–3395.

45. Lanoil BD, Carlson CA, Giovannoni SJ. Bacterial chromosomal painting for *in situ* monitoring of cultured marine bacteria. Environ Microbiol 2000; 2:654–665.

46. Suzuki M, Rappe MS, Giovannoni SJ. Kinetic bias in estimates of coastal picoplankton community structure obtained by measurements of small-subunit rRNA gene PCR amplicon length heterogeneity. Appl Environ Microbiol 1998; 64:4522–4529.

47. Marsh TL, Saxman P, Cole J, Tiedje J. Terminal restriction fragment length polymorphism analysis program, a Web-based research tool for microbial community analysis. Appl Environ Microbiol 2000; 66:3616–3620.

48. Béjà O, Suzuki MT, Heidelberg JF, et al. Unsuspected diversity among marine aerobic anoxygenic phototrophs. Nature 2002; 415:630–633.

49. Song J, Dong F, Lilly JW, Stupar RM, Jiang J. Instability of bacterial artificial chromosome (BAC) clones containing tandemly repeated DNA sequences. Genome 2001; 44:463–469.

50. Rondon MR, Raffel SJ, Goodman RM, Handelsman J. Toward functional genomics in bacteria: analysis of gene expression in *Escherichia coli* from a bacterial artificial chromosome library of *Bacillus cereus*. Proc Natl Acad Sci USA 1999; 96:6451–6455.

51. Rondon MR, August PR, Bettermann AD, et al. Cloning the soil metagenome: a strategy for accessing the genetic and functional diversity of uncultured microorganisms. Appl Environ Microbiol 2000; 66:2541–2547.

52. Béjà O, Suzuki MT, Koonin EV, et al. Construction and analysis of bacterial artificial chromosome libraries from a marine microbial assemblage. Environ Microbiol 2000; 2:516–529.

53. Hughes JB, Hellmann JJ, Ricketts TH, Bohannan BJ. Counting the uncountable: statistical approaches to estimating microbial diversity. Appl Environ Microbiol 2001; 67:4399–4406.

54. Béjà O, Spudich EN, Spudich JL, Leclerc M, DeLong EF. Proteorhodopsin phototrophy in the ocean. Nature 2001; 411:786–789.

55. Vergin KL, Urbach E, Stein JL, DeLong EF, Lanoil BD, Giovannoni SJ. Screening of a fosmid library of marine environmental genomic DNA fragments reveals four clones related to members of the order Planctomycetales. Appl Environ Microbiol 1998; 64:3075–3078.

56. Schleper C, Swanson RV, Mathur EJ, DeLong EF. Characterization of a DNA polymerase from the uncultivated psychrophilic archaeon *Cenarchaeum symbiosum*. J Bacteriol 1997; 179:7803–7811.

57. Preston CM, Wu KY, Molinski TF, DeLong EF. A psychrophilic crenarchaeon inhabits a marine sponge: *Cenarchaeum symbiosum* gen nov, sp nov. Proc Natl Acad Sci USA 1996; 93:6241–6246.

58. Short JM. Recombinant approaches for accessing biodiversity. Nat Biotechnol 1997; 15:1322–1323.

59. Majernik A, Gottschalk G, Daniel R. Screening of environmental DNA libraries for the presence of genes conferring Na(+)(Li(+))/H(+) antiporter activity on *Escherichia coli*: characterization of the recovered genes and the corresponding gene products. J Bacteriol 2001; 183:6645–6653.

60. Henne A, Daniel R, Schmitz RA, Gottschalk G. Construction of environmental DNA libraries in *Escherichia coli* and screening for the presence of genes conferring utilization of 4-hydroxy-butyrate. Appl Environ Microbiol 1999; 65:3901–3907.

61. Henne A, Schmitz RA, Bomeke M, Gottschalk G, Daniel R. Screening of environmental DNA libraries for the presence of genes conferring lipolytic activity on *Escherichia coli*. Appl Environ Microbiol 2000; 66:3113–3116.

62. Cottrell MT, Moore JA, Kirchman DL. Chitinases from uncultured marine microorganisms. Appl Environ Microbiol 1999; 65:2553–2557.

63. Richardson TH, Tan X, Frey G, et al. A novel, high performance enzyme for starch liquefaction: discovery and optimization of a low pH thermostable alpha-amylase. J Biol Chem, 2002.

64. MacNeil IA, Tiong CL, Minor C, et al. Expression and isolation of antimicrobial small molecules from soil DNA libraries. J Mol Microbiol Biotechnol 2001; 3:301–308.

65. Brady SF, Chao CJ, Handelsman J, Clardy J. Cloning and heterologous expression of a natural product biosynthetic gene cluster from eDNA. Org Lett 2001; 3:1981–1984.

66. Hess WG, Rocap G, Ting CS, et al. The photosynthetic apparatus of *Prochlorococcus*: insights through comparative genomics. Photosynthesis Res 2001; 70:53–71.

67. Ting CS, Rocap G, King J, Chisholm SW. Cyanobacterial photosynthesis in the oceans: the origins and significance of divergent light-harvesting strategies. Trends Microbiol 2002; 10:134–142.

68. Spratt BG, Hanage WP, Feil EJ. The relative contributions of recombination and point mutation to the diversification of bacterial clones. Curr Opin Microbiol 2001; 4:602–606.

69. Smith JM, Feil EJ, Smith NH. Population structure and evolutionary dynamics of pathogenic bacteria. Bioessays 2000; 22:1115–1122.

70. Feil EJ, Spratt BG. Recombination and the population structures of bacterial pathogens. Annu Rev Microbiol 2001; 55:561–590.

71. Curtis TP, Sloan WT, Scannell JW. Estimating prokaryotic diversity and its limits. Proc Natl Acad Sci USA 2002; 99:10,494–10,499.

72. Kolber ZS, Plumley FG, Lang AS, et al. Contribution of aerobic photoheterotrophic bacteria to the carbon cycle in the ocean. Science 2001; 292:2492–2495.

73. Kolber ZS, Van Dover CL, Niederman RA, Falkowski PG. Bacterial photosynthesis in surface waters of the open ocean. Nature 2000; 407:177–179.

74. Field KG, Gordon D, Wright T, et al. Diversity and depth-specific distribution of SAR11 cluster rRNA genes from marine planktonic bacteria. Appl Environ Microbiol 1997; 63:63–70.

24

Application of Genomics to Biocatalysis and Biodegradation

Lawrence P. Wackett

"The world we inhabit today teeters between becoming either the lovely garden or the barren desert that our contrary impulses strive to bring about. Our future is now closely tied to human creativity."

Mihaly Csikszentmihalyi

INTRODUCTION

Prokaryotes are thought to possess the most diverse metabolism of any major group of organisms on earth *(1)*. This stems from their great phylogenetic diversity and their expansion into many exotic niches on planet earth. Prokaryotes are the main recyclers of carbon in the biosphere. Their success is exemplified by their sheer numbers (5×10^{30}) and a biomass greater than that of all the green plants combined *(2)*. Biodegradation and biocatalysis are manifestations of this naturally diverse microbial metabolism.

Biodegradation, the degradation of organic matter by microorganisms, has been observed by humans for thousands of years, for example, the natural decay of wood in the forest by wood rot fungi. Only very recently in human history have the biochemical reactions, or metabolism, underlying wood rot begun to be revealed *(3,4)*. Metabolism directed toward the breakdown of compounds to obtain energy is called *catabolism*. In some instances, the growing knowledge of microbial catabolism has led to biological solutions for cleaning up hazardous wastes, a process known as *bioremediation*.

Biocatalysis is related to biodegradation, but most commonly the term is used to denote microbial catabolic or biosynthetic reactions for making chemical compounds that can be sold commercially. For example, antibiotics are produced by a mixture of fermentation, enzymatic, and synthetic technologies. The use of biocatalysis in industry is expanding, in part because of the discovery of new biodegradative reactions. In one example, the enzyme nitrile hydratase is being used to transform acrylonitrile to acrylamide (Fig. 1). This has formed the basis of a large-scale biotechnological process developed by Nitto Chemical Company in Japan based on the clean, high-yield conversion that the enzyme affords *(5)*. Nitrile hydratase is likely produced by bacteria to biodegrade, and thus detoxify, the many natural product nitriles, and cyanide released from them, biosynthesized by plants to ward off insect pathogens *(6)* (Fig. 1). Some

From: *Microbial Genomes*
Edited by: C. M. Fraser, T. D. Read, and K. E. Nelson © Humana Press Inc., Totowa, NJ

CATABOLISM   BIODEGRADATION   BIOCATALYSIS

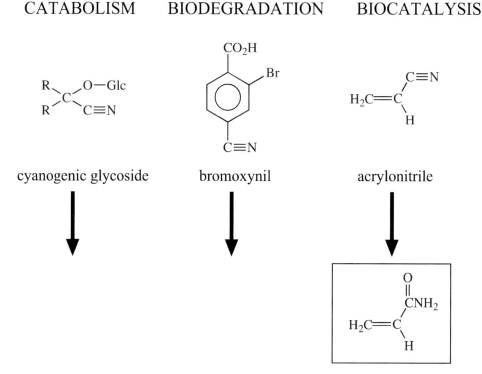

Fig. 1. Microbial enzyme-catalyzed hydration of the cyano, or nitrile, functional group in different compounds may be denoted catabolism, biodegradation, or biocatalysis depending on the significance of the reaction to the microbe or humans. Acrylamide is a major commodity chemical produced by biocatalysis.

forms of bacterial nitrile hydratase have also been observed to biodegrade synthetic herbicides such as bromoxynil *(7)* (Fig. 1). In this context, naturally evolved enzymes may have uses with natural products, for biodegradation of pollutants, and for biosynthesis of industrial chemicals.

BIODEGRADATION

Biodegradation in the Prokaryotic World

Bacteria in natural soils and waters often have sufficiently broad catabolic metabolism such that they can survive in environments in which there is fierce competition for carbon, nitrogen, phosphorus, and other nutrients. As a consequence, soil microorganisms likely have more extensive biodegradative metabolism than enteric and pathogenic bacteria. An *Escherichia coli* strain may survive poorly in soil and water, but thrive in an animal's digestive tract. In the last environment, it will receive a fairly uniform and restricted diet. It has been observed that *E. coli* can catabolize a range of aromatic acids, a finding that is not surprising considering that this class of compounds is derived from aromatic amino acids and plant natural products found in the typical *E. coli* intestinal environment *(8,9)*.

Are there important classes of well-studied prokaryotes that are not prominent in known biodegradative metabolism? Consider the Archaea, a major division of life. This group has been studied heavily in the last 10 years, and several whole genome-sequencing projects have now been completed. These organisms include extremophiles that thrive in high-salt environments or at high temperatures. A major metabolic group within the Archaea is known as methanogens, prokaryotes that transform carbon dioxide or acetate to methane, deriving energy in the process and contributing to the global pool of atmospheric methane gas.

Pure cultures of methanogens have been studied for their ability to reduce chlorinated aliphatic compounds, but these reactions are generally considered fortuitous *(10)*. Moreover, chlorinated methanes have long been known to be toxic to methanogens, perhaps by competing with physiological substrates for cobalamin (vitamin B_{12}) reactive centers or by generating reactive carbene intermediates following their reduction. In general, methanogens are not considered to have a prominent role in the catabolism of diverse organic structures. Similarly, halophilc Archaea are not considered to have broad catabolic abilities. Although some have been reported to degrade aliphatic hydrocarbons, they are generally thought to be quite restricted in the carbon substrates that they utilize *(11)*. Moreover, the biodegradation of hydrocarbons, particularly the aromatic fractions in petroleum spills, is observed to decrease when the salt concentration becomes elevated, and there is a switch over from eubacterial to archaebacterial populations.

Nonetheless, catabolic capabilities are fairly broadly distributed in the taxonomic tree of life (Table 1). For example, the genera of bacteria listed in Table 1 are considered prominent for their catabolic metabolism, which has been applied for biodegradation and biocatalysis by humans. These bacteria are largely members of the low guanine and cytosine (G+C) Gram-positives, high G+C Gram-positives, and Proteobacteria.

Yet, although many bacteria are likely important in biodegradation, some might be more important than others. As early as 1926, den Dooren de Jong *(12)* reported on the ability of a *Pseudomonas* species to catabolize hundreds of organic compounds. The compounds include alkanes and aromatic ring compounds, and this type of bacterial substrate preference is still considered "exotic" by some people. But, alkanes and aromatic ring compounds are ubiquitous and have been present globally in significant quantities for millions of years *(13)*; microbes could hardly resist such thermodynamically rich compounds.

Genomics of Bacteria Important in Biodegradation

Genomics is just beginning to have an impact on biodegradation research in major ways (Table 2). The first few years of genome sequencing were almost exclusively focused on pathogenic bacteria. This was because the first sequencing projects were funded by the U.S. National Institutes of Health; the intent was to have a direct impact on human health. However, there is now much broader support for sequencing the genomes of phylogentically diverse prokaryotes. Approximately 500 bacterial genome-sequencing projects are now completed or under way in the public domain; it is likely that several hundred complete bacterial genomes will be available shortly. This includes a significant number of phylogenetically diverse soil bacteria (Table 2).

Genomics will increasingly provide insights into metabolic activities of bacteria in the soil. In one example, the *Bacillus subtilis* genome-sequencing article reported that

Table 1
Bacterial Genera Represented in the University of Minnesota Biocatalysis/Biodegradation Database (UM-BBD) Based on Their Catabolic Metabolism[a]

High G+C Gram +	Low G+C Gram +	Proteobacteria				Cytophagales green sulfur	Green nonsulfur
		α	β	γ	δ/ε		
Arthrobacter	Bacillus	Agrobacterium	Achromobacter	Acinetobacter	Desulfovibrio	**Flavobacterium**	Dehalococcoides
Brevibacterium	**Clostridium**	Ancylobacter	Alcaligenes	Aeromonas			
Clavibacter	Desulfitobacterium	Brevundimonas	Azoarcus	Azotobacter			
Corynebacterium	Eubacterium	Chelatobacter	**Burkholderia**	Enterobacter			
Dehalobacter	Staphylococcus	Hyphomicrobium	Comomonas	**Escherichia**			
Nocardia		Methylobacterium	Hydrogenophyga	Klebsiella			
Rhodococcus		Paracoccus	Ralstonia	Methylobacter			
Streptomyces		Rhodobacter	Thavera	Methylococcus			
Terrabacter		**Sphingomonas**	Thiobacillus	Moraxella			
				Pseudomonas			

[a]Bold genera are most heavily represented.

Table 2
Prokaryotic Genome Projects in the Public
Domain Relevant to Biocatalysis and Biodegradation[a]

Organism	Genome size (kb)	Application in biocatalysis or biodegradation
Completed genomes		
Acinebacter sp ADP1, ATCC 33305	3583	Model for alkane/benzoate catabolism
Clostridium acetobutylicum ATCC 824D	4100	Acetone/butanol fermentation
Corynebacterium glutamicum	3309	Glutamic acid fermentation
Corynebacterium glutamicum ATCC-13032	3309	Amino acid fermentation
Deinococcus radiodurans R1	3284	Biodegradation in radioactive environments
Streptomyces avermitilis MA-4680	8700	Antibiotic production
Streptomyces coelicolor A3(2)	8667	Antibiotic production
Streptomyces diversa	N/A	Antibiotic production
Zymomonas mobilis ZM4	2052	Ethanol/sorbitol fermentation
Zymomonas mobilis	1833	Ethanol/sorbitol fermentation
Genomes in progress		
Burkholderia (Pseudomonas) cepacia J2315	7600	Pesticide biodegradation
Corynebacterium efficiens YS-314T	3140	Glutamic acid fermentation
Corynebacterium glutamicum ATCC 13032	3309	Glutamic acid fermentation
Corynebacterium thermo-aminogenes FERM9246	N/A	Amino acid fermentation
Dehalococcoides ethenogenes	1500	Reductive dechlorination of solvents
Desulfitobacterium hafniense	4600	Reductive dechlorination
Geobacter metallireducens	6800	Toxic metal reduction/immobilization
Geobacter sulfurreducens	2500	Metal reduction
Nitrosomonas europaea ATCC 25978	2980	Nitrogen cycling; solvent oxidation
Pseudomonas fluorescens Pf0-01	3500	Diverse biodegradative capabilities
Pseudomonas fluorescens SBW25	6600	Diverse biodegradative capabilities
Pseudomonas putida KT2440	6100	Diverse biodegradative capabilities
Pseudomonas putida PRS1	6100	Diverse biodegradative capabilities
Ralstonia metallidurans (eutropha) CH34	3000	Heavy metal resistance
Rhodococcus sp I24	5487	Indene biotransformation
Rhodococcus sp RHA1	N/A	Polychlorinated biphenyl biodegradation
Rhodopseudomonas palustris CGA009	5460	Phototrophic aromatic degrader
Sphingomonas aromaticivorans F199	3800	Diverse aromatic biodegradation
Streptomyces ambofaciens	8000	Antibiotic production

[a]Data from Genomes Online Database (GOLD) database, June 12, 2002.
N/A, not available.

the bacterium had genes thought to encode for the catabolism of certain plant natural products *(14)*. This was considered somewhat surprising given that those genes were previously described in a taxonomically very different Gram-negative soil organism. However, in studies using ribosomal ribonucleic acid (rRNA) indicators, a gram of soil has been proposed to contain as many as 10,000 distinct species of bacteria *(15)*. Consider further that some plants make flavonoid compounds at a concentration as high as 27% of the leaf mass, and this leaf matter provides the major new carbon input in soils under the plant *(6)*. Thus, divergent taxa would be expected to contain genes encoding the catabolism of widely available compounds in temperate soils where the particular plant is found, and these genes might not be found in hot springs, arctic environments, or deserts. Thus, gene clusters can begin to be linked not only with single organisms, but also with biologically complex environments, thus making a start in the derivation of a global genomic composition. Although such an undertaking will never be close to complete, it will help define a broader context for genomics, moving the information into the sphere of ecological research.

Genomics of Catabolic Plasmids

Many soil bacteria harbor extrachromosomal deoxyribonucleic acid (DNA) elements called plasmids. The definition of what constitutes a plasmid has blurred with the increased discovery of numerous small DNA elements in bacteria because of the advent of widespread genome sequencing. Historically, plasmids were first described in *E. coli* in 1952 *(16)*, and since then, plasmids have been shown to carry genes functional in antibiotic resistance, pathogenesis, or the catabolism of diverse chemical compounds. Many plasmids also carry genes that promote their transfer to other bacteria by conjugation. Plasmids with a broad host range can transfer and replicate in diverse bacterial genera. In this manner, plasmids are considered highly important in the transfer of catabolic genes in the environment. Plasmids encoding the catabolism of octane, toluene, camphor, naphthalene, nicotine, *p*-toluenesulfonic acid, and 2,4-dichlorophenoxyacetate are known *(17–21)*.

Plasmid DNA can constitute a significant fraction of the total DNA of some soil bacteria, such as *Pseudomonas* sp. In one example, a *Pseudomonas* species isolated for its ability to catabolize the herbicide atrazine may contain as many as five distinct catabolic plasmids that together comprise approximately 1 Mb of DNA *(22)*.

Genomic approaches will certainly add much to the understanding of plasmid structure and evolution. However, only several hundred plasmids have been sequenced, and most of those are small vector plasmids used in molecular biology or antibiotic resistance plasmids in pathogenic bacteria. Several catabolic plasmids that have been or are being sequenced are shown in Table 3. The catabolic plasmid pNL1, from *Sphingomonas aromaticivorans* strain F199, has been completely sequenced and contains genes encoding enzymes for the metabolism of biphenyl, naphthalene, *m*-xylene, and *p*-cresol *(23)*. The complete genome of *S. aromaticivorans* strain F199 is also being sequenced.

The catabolic plasmid pADP-1 from *Pseudomonas* sp ADP was more recently sequenced *(24)* (Fig. 2). Annotation revealed that the gene regions encoding plasmid pADP-1 replication, transfer, and maintenance functions were virtually identical to analogous genes on pR751, an IncPβ plasmid from *Enterobacter aerogenes (25)*. The *atz* genes encoding the entire atrazine catabolic pathway reside on pADP-1; however, they are not all

Table 3
Complete DNA Sequences of Catabolic Plasmids

Organism	Plasmid	Size (kb)	Functions encoded
Completed			
Sphingomonas aromaticivorans F199	pNL-1	186	Aromatic catabolism
Pseudomonas sp ADP	pADP-1	107	Atrazine catabolism
Pseudomonas putida TOL	pWWO	117	Toluene/xylene catabolism
In progress			
Pseudomonas sp ND6	pND6-1	102	Naphthalene catabolism
Arthrobacter nicotinovorans	pAO1	160	Nicotine catabolism

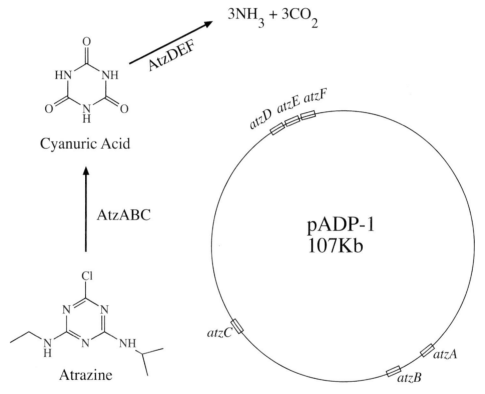

Fig. 2. Metabolic pathway for the catabolism of atrazine by *Pseudomonas* sp ADP and plasmid pADP-1, which contains the genes encoding the enzymes of atrazine metabolism.

contiguously organized into an operonic structure (Fig. 2). In fact, enzymes catalyzing the first three reactions in atrazine catabolism are encoded by genes that are singly displaced and flanked by insertion sequence elements. These genes appear to be constitutively expressed. One of them, *atzC*, has a G+C content of 44%, much less than that of *atzA* (58%) and *atzB* (61%). In total, these observations are consistent with the idea that these *atz* genes have been recently acquired by a plasmid backbone with a broad host range to generate a new plasmid that encodes a pathway allowing *Pseudomonas* sp ADP to grow on atrazine as its sole nitrogen source.

In contrast, the genes encoding the enzymes catalyzing the second set of three reactions in the atrazine pathway are clustered and are proposed to be coordinately regulated *(24)*. It is suggested that the atrazine catabolic genes are composed of an "upper" and a "lower" pathway, and that the former enzymes may have recently evolved to acquire enzymatic activity directed against synthetic *s*-triazine herbicides. An observation consistent with this is the recent discovery of a deaminase gene *triA*, homologous to the atrazine chlorohydrolase gene, *atzA*. It is remarkable that TriA shows 98% sequence identity to AtzA, yet TriA is a deaminase with negligible dechlorinase activity with atrazine. This suggests that a few amino acid changes could convert a TriA enzyme to an AtzA enzyme, which has been shown experimentally via DNA shuffling and screening recombinant proteins with altered catalytic activities *(26)*.

Genomics of a Bacterium "Engineered" for Bioremediation: Deinococcus radiodurans

For bioremediation to be used most broadly, it may be desirable to engineer biodegradation traits into microorganisms with unique characteristics that make them amenable to use at specific sites. An emerging example of this is the metabolic engineering of *Deinococcus radiodurans* (Fig. 3) for transforming mixed organic waste containing radioisotopes. This situation exists at numerous US Department of Energy (DOE) sites and at comparable places in other countries where radioisotopes have been concentrated for use in civilian nuclear reactors and military nuclear warheads.

With the end of the Cold War in the early 1990s, the DOE initiated a major program to safely maintain and decontaminate 3×10^6 M^3 of mixed waste buried at 7000 sites *(27)*. In some places, radioactive wastes are leaking into soil and groundwater *(28)*. It is estimated that 40 million m^3 of soil and 4 trillion L of groundwater have become contaminated, and cleanup costs over the next 10 years are estimated at over $60 billion. In an effort to reduce the cost of remediating these sites, the DOE is supporting efforts to develop *in situ* bioremediation approaches.

A significant number of DOE sites contain ionizing radiation at levels lethal to the microorganisms listed in Table 2, and this provides the impetus for using a radiation-resistant bacterium for bioremediation. *Deinococcus radiodurans* possesses the required radiation tolerance. It was originally isolated from a can of irradiated meat and has subsequently been shown capable of surviving acute exposures to radiation in excess of 15,000 gy *(29)*. *Deinococcus radiodurans* is also highly resistant to other DNA-damaging conditions, such as desiccation, ultraviolet radiation, and oxidizing reagents. These traits are thought to be a complex phenotype, arising from a spectrum of enzyme activities and perhaps its genome structure. This mitigates against a strategy of engineering radiation resistance into sensitive bacteria and focuses attention on engineering biodegradation functions into *D. radiodurans*. This realization provided impetus for the genome sequencing and annotation of *D. radiodurans*. The genomic data rapidly provided insight into the range of metabolism native to this organism and a roadmap for metabolic engineering to construct complementary metabolic pathways that feed intermediates into indigenous pathways.

The *D. radiodurans* R1 genome consists of four replicons of 2649 kb, 412 kb, 177 kb, and 45 kb *(30)*. Clearly, the larger two replicons contain essential genes, and all four appear to be stably maintained. Annotation suggested that 91% of the genome con-

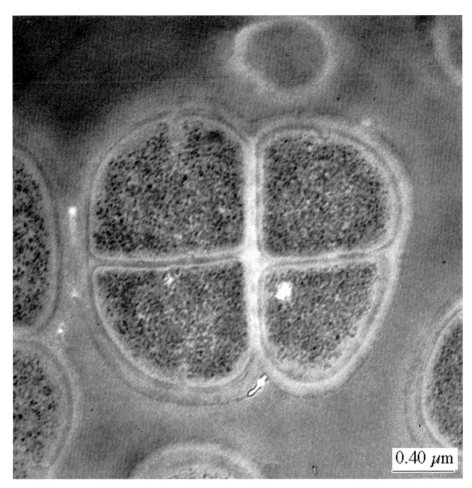

0.40 μm

Fig. 3. Electron micrograph of a thin section of *Deinococcus radiodurans* R1 showing the characteristic tetrad cell arrangement.

sisted of coding regions, providing 3187 open reading frames. Initially, 69% of the open reading frames matched sequences in the databases, and further analysis of these provided insights into the organism's metabolic pathways.

Deinoccus radiodurans appears to possess complete sets of genes encoding glycolytic, gluconeogenic, pentose phosphate shunt, tricarboxylic acid cycle, and glyoxylate shunt pathways. The glyoxylate shunt is not present in most prokaryotes, but in *D. radiodurans* appears to be strongly expressed *(31)*. The genome annotation also revealed that *D. radiodurans* is largely devoid of enzymes involved in the catabolism of organic pollutants such as aromatic hydrocarbons. Thus, the types of organic pollutants present at DOE sites would not be expected to be metabolized by native *D. radiodurans*. Direct laboratory experiments have confirmed the lack of biodegradative phenotypes in wild-type *D. radiodurans* R1 (S. McFarlan, unpublished data). This provided the baseline for metabolic engineering of *D. radiodurans* to transform organic and inorganic toxicants.

It was advantageous that *D. radiodurans* is readily transformable with foreign DNA and that suitable vectors for replication in the organism are available *(29)*. Moreover,

Daly and coworkers have devised an amplifiable DNA cassette system *(29)*. Foreign DNA is flanked by an antibiotic resistance marker and tandem repeat sequences, which will recombine into the *D. radiodurans* genome. By stepping up the antibiotic concentration in the medium during sequential growth cycles, the cassette will amplify in the genome and thus increase gene dosage. In this way, as many as 200 copies of a recombinant gene cassette have been amplified. Considering that a single copy of the amplified *mer* operon (mercury resistance operon) is approximately 20 kb, this constitutes a doubling of the genomic load carried by *D. radiodurans* cells.

Using these methods, the *mer* operon and toluene catabolism genes have been cloned, amplified, and expressed in *D. radiodurans (32,33)*. The *mer* operon encodes a soluble mercuric reductase, *merA*, and transporters for mercuric ion *(34)*. The Hg^{2+} mercuric ions are highly toxic to bacteria, but a number of bacteria deploy a system to withstand Hg^{2+} by transporting it into the cell and then reducing it to metallic mercury, Hg^0. Metallic mercury is much less toxic and sufficiently volatile that it will readily escape from the cell. Native *D. radiodurans* is not resistant to mercury, and the genome project did not identify genes involved in mercury detoxification. The complete mercury operon from *E. coli,* along with an ampicillin resistance gene, was cloned into *D. radiodurans* as described here.

As expected, increasing ampicillin concentration in the growth medium led to increased resistance to mercuric ion and a corresponding increase in the copy number of *mer* genes per cell. Moreover, recombinant *D. radiodurans* containing both the *mer* operon and genes involved in toluene oxidation oxidized toluene in the presence of mercuric ion concentrations that would be toxic to the native strain. Significant quantities of mercury are present at DOE sites, and thus *mer* genes may be a crucial component of any bacterial agent used for bioremediation at these sites. Native *D. radiodurans* can reduce certain metals, including uranium and technetium, which are important radionuclides found at DOE sites *(35)*. Reduction of metals can render them less mobile in soils and thus diminish their chances for contaminating nearby sites.

Biological metal reductions require the presence of an oxidizable substrate, and it would be most desirable if the substrate were a cocontaminant at a waste site. Many sites contain fuel hydrocarbons in which toluene is prevalent. In this context, the *tod ABC1C2* genes encoding toluene dioxygenase were cloned into *D. radiodurans (32)*. The recombinant strain was shown to oxidize toluene, chlorobenzene, and trichloroethylene, even in the presence of ionizing radiation. However, the initiating reaction in hydrocarbon oxidation does not generate electrons that could be used for bioreduction reactions. To this end, a follow-up cloning experiment was conducted in which the *todABC1C2DE* and *xylFJQK* gene cassettes were put into *D. radiodurans* (H. Brim et al., unpublished data). The genes act in concert to oxidize toluene to generate pyruvate and acetate. The latter compound is a good carbon source for *D. radiodurans*, consistent with the genome annotation research, which demonstrated that the glyoxylate shunt is a highly expressed metabolic pathway.

BIOCATALYSIS

Prokaryotes in Biocatalysis

A historic triumph in the use of microbial biocatalysis was the fermentation of cornstarch to produce acetone using *Clostridium acetobutylicum* in 1917. The development of the process is largely credited to Chaim Weizmann, and it made major contributions to the ability of the British to fight successfully in World War I *(36)*. Acetone was needed to make explosives, and supplies of the solvent from the German chemical industry were halted by the war. The process reached beyond England and attained an impressive scale; 22 fermentors with a capacity of 30,000 gallons each were operating in Canada in 1918 *(37)*. After the war, however, world markets opened up, and a petroleum-based process to make acetone again became predominant. Today, the genome of *C. acetobutylicum* ATCC 824D has been completed (Table 2), and a better understanding of the metabolism underlying the acetone/butanol fermentation may be forthcoming. This could lead to a more economically competitive biocatalytic process for solvent production.

The classical *Clostridium* acetone fermentation presaged current developments in bio-catalysis in several ways. The application of biotechnology for chemical manufacture is based on biomass as a feedstock instead of petroleum. Many sources of biomass are transformed to glucose, either biologically or chemically, as the starting material for microbial fermentations to produce desired chemicals. In some cases, enzymes may be used in place of microbes. An example of a currently important biotransformation is the conversion of cornstarch to glucose by the enzyme α-amylase and its subsequent enzymatic transformation to fructose by glucose isomerase for high-fructose syrup, a widely used food sweetener *(1)*.

The chemical industry is anticipating a major transformation to a future in which biomass is used as a feedstock instead of petroleum. An important feedstock will be cellulose, the major biopolymer on earth. Similar to starch, cellulose can be processed to yield glucose by the action of enzymes, in this case, cellulases. A number of prokaryotes and fungi excrete extracellular cellulases to provide glucose for growth. For example, the fungus *Trichoderma viride* is among the most proficient secretors of extracellular cellulases known *(38)*. Cellulase from *Trichoderma reesei* is being used to produce ethanol through fermentation in industrial processes. Thermophilic bacteria such as *Clostridium thermocellum* have also been studied for their cellulase, which is packaged into a cellulosome, a large complex consisting of at least 15 different proteins *(39)*.

Hydrolysis of biopolymers releases glucose and other sugars that serve as the starting material for fermentation processes to produce important industrial compounds. Most bacteria are capable of catabolizing glucose. Thus, different microbes can be exploited for their unique biochemistry and used in glucose-fed fermentations to produce desired chemical end products. For specialty chemical production, many different microorganisms are currently used, and genomics will be important for the most efficient application of these bacteria. *Streptomyces coelicolor*, important in antibiotic production, has been sequenced to date, as have several *Corynebacteria* species, some of which are currently used in amino acid fermentation (Table 2). The genomes of many more such strains have been sequenced in the industrial domain, a trend that will no doubt increase. Genomics will contribute to shortening the time required to go from the discovery of a bacterial strain to its use in fermentative production of chemicals.

Genomics of Bacteria Important in Biocatalysis

Against the backdrop of prokaryotic genome sequencing in the public domain, there is a vast underbelly of proprietary prokaryotic genomics. Much of that is being done with bacterial strains catalyzing industrially relevant biotransformation reactions. Industries are using genomics to gain insight into the overall metabolism of important bacterial strains that produce antibiotics, amino acids, biopesticides, vitamins, organic acids, and alcohols. When those products are already being made microbiologically, genomics offers the hope of increasing the yield per unit volume, thus enhancing an already profitable industrial process. Fortunately for those denied access to this privately held information, there is an emerging compendium of public genomic data relevant to microbes important in biocatalysis. Some of those microorganisms are shown in Table 2.

There are several classes of biocatalytic reactions that genomics research will impact. For example, *Corynebacterium* strains are prominent in fermentations to produce amino acids such as lysine. Lysine is used in large volume in cattle feed because it is often a limiting factor in the diet of farm animals, and its supplementation increases weight gain substantially. Currently, genomics projects are under way for *Corynebacterium glutamicum* strains for which data will be publicly available (Table 2).

Rhodococcus strains are important in a number of industrial biotransformation reactions and in biodegradation. *Rhodococcus* strains catabolize a diverse set of substrates: small gaseous compounds, the fuel additive methyl *tert*-butyl ether, terpenes, and the herbicide atrazine. In addition, interest in rhodococci derives from their use in biotechnological processes. In fact, one of the largest volume biocatalytic industrial processes is the conversion of acrylonitrile to acrylamide by a *Rhodococcus* species *(5)* (Fig. 1). Industry produces 400 million pounds of acrylamide, and the conventional chemical synthesis, based on a copper catalyst, is plagued with the high costs of catalyst removal and undesirable purity of the product. Use of a *Rhodococcus* strain containing the enzyme nitrile hydratase circumvents these problems and is suitable for very high-yield production of acrylamide. Nitto Chemical Company in Japan has developed a commercial process based entirely on this microbiological catalyst.

Rhodococcus strains have also been used for biocatalytic desulfurization of fossil fuels *(40)*. Fossil fuels contain variable levels of organic sulfur, and burning that fuel leads to the formation of sulfur dioxide and acid rain. Chemical catalysts to remove fuel sulfur often foul and are fairly expensive, paving the way for a biotechnological process that would require a very large-scale process to be applicable to the current problem. Several *Rhodococcus* strains have been identified that oxidatively remove the sulfur from sulfur heterocyclic rings in fossil fuels. A current challenge is to scale up this process to treat the very large volumes of fuel required to meet the needs of the petroleum industries.

One of the *Rhodococcus* sp currently undergoing genome sequencing, strain I24, has been investigated for its potential to produce a fine chemical, enantiomerically pure 1-amino-2-hydroxyindan, which comprises a key structural fragment of Indinavir, a new drug treatment for human immunodeficiency virus. *Rhodococcus* sp strain I24 is one of a number of strains that oxidize indene to a *cis*-1,2-dihydrodiol that can be transformed by chemical synthesis to 1-amino-2-hydroxyindan *(41)*. A major issue is stereochemical purity of the indandiol and the production of other oxidized side products formed during the oxidation of indan. The production of chiral intermediates for the manufacture of chiral pharmaceutical compounds is a growing arena in biotechnology *(1)*. As more is learned about the enzyme determinants controlling stereospecificity, genomics will be increasingly useful in these cases.

A major class of biocatalytic microorganisms for pharmaceutical production are members of the genus *Streptomyces*. *Streptomyces* is the leading bacterial genus for producing natural product pharmaceutical compounds: antibiotics, antitumor agents, and immunosuppresants. The classes of chemicals include polyketides, chalcones, and nonribosomal peptides. For this reason, the genome-sequencing project for *S. coelicolor* A3(2) was anxiously awaited and has been completed (Table 2) *(42)*. The large 8.7-Mb linear chromosome of *S. coelicolor* A3(2) contains genes that encode the biosynthesis of actinorhodin, geosmin, and coelichelin. This genome encodes the largest number of genes in any bacterium yet identified (7825) and should provide the basis for further natural product prospecting for many years.

BIOCATALYSIS AND BIODEGRADATION INFORMATICS

Genome sequence data is most useful when complemented by readily available information on prokaryotic metabolism. Typically, more than 75% of the genome in prokaryotes encode proteins. Thus, most gene annotation involves mapping DNA sequences onto proteins likely to catalyze a single biochemical reaction or a set of related reactions. Microbial reactions are increasingly being catalogued by databases that are publicly available on the World Wide Web.

Metabolism databases may be specialized, focusing on specific bacterial strains, or broadly trained on classes of metabolism, such as common intermediary metabolism or biocatalysis/biodegradation. An example of a specialized metabolic database is EcoCyc, which depicts the metabolism of *E. coli* (Table 4). A more broadly focused database is KEGG, the Kyoto Encyclopedia of Genes and Genomes *(43)*. KEGG depicts metabolic pathways, largely of intermediary metabolism, that are found in many different microorganisms; these are organized according to type, for example, amino acid metabolism, nucleotide metabolism, or carbohydrate metabolism (see Chapter 6).

There is an ongoing effort to compile information on biodegradative and biocatalytic microorganisms, genes, enzymes, and substrates on the University of Minnesota Biocatalysis/Biodegradation Database *(44)* (UM-BBD). A complementary database (Table 4) focused on microorganismal data has been developed at Michigan State University and is known as the Biodegradative Strain Database (BSD). The BSD and UM-BBD have forged reciprocal links so users may comprehensively access biodegradation information by searching first for microorganisms, enzymes, or specific substrates.

The UM-BBD developers seek to represent the breadth of microbial biocatalytic reactions, whether they are currently exploitable by industry or whether they depict the wealth of microbial biochemistry that exists in nature. The breadth of microbial metabolism found in nature is suggested by the wide range of natural product compounds that exist and that provide potential substrates for microbial growth in soils. Currently, over 18 million chemical compounds are known, and more are synthesized every day. Many of those compounds may ultimately prove to be biodegradable by some microorganism found in at least one location on earth. This suggests that millions of different enzyme reactions, as defined by one reaction per substrate, are occurring somewhere around the globe. This represents the frontiers of metabolic biochemistry after a century of concentrated research to elucidate the outlines of microbial intermediary metabolism as depicted in most biochemistry textbooks. In this context, a substantial fraction of undefined genes derived from microbial genome-sequencing projects may turn out to be involved in the catabolism of natural product or synthetic chemical compounds *(1)*.

Table 4
Representative Web Databases Depicting Microbial
Metabolism, With a Focus on Biocatalytic and Biodegradation Metabolism

Database	URL	Domain covered
EcoCyc	http://ecocyc.org/	Metabolism of *Escherichia coli*
KEGG	http://www.genome.ad.jp/kegg/	Clickable metabolic maps; broad metabolism
BSD	http://bsd.cme.msu.edu/bsd/	Prokaryotes important in biodegradation
UM-BBD	http://umbbd.ahc.umn.edu/	Microbial metabolism outside intermediary metabolism
UM-BBD Functional Groups	http://umbbd.ahc.umn.edu/search/FuncGrps.html	Enzyme-catalyzed transformations of specific chemical functional groups
UM-BBD Periodic Table	http://umbbd.ahc.umn.edu/periodic/	Microbial transformation of the chemical elements

To aid in genome annotation efforts, the functional groups section of the UM-BBD (Table 4) lists the known classes of organic functional groups that undergo enzymatic transformation by at least one microorganism or a consortium of microorganisms. A new project shows the breadth of microbial metabolism involved in transforming different chemical elements. In addition to the two dozen most common elements in biological systems (C, H, O, N, P, S, K, Na, Mg, Ca, Se, Fe, Mn, V, Mo, Co, Ni, Zn, B, Cl, Br, F, I, As), microbes interact with, and sometimes change, the oxidation states or chemical speciation of a wide range of other elements. These reactions may be important in bioremediation, for example, in the reduction of uranium metal to minimize its mobility in soils (Fig. 4) *(45)*. Moreover, microbial transformation of minerals is thought to be important in shaping the mineral deposits on the earth's surface and thus may be important in global cycling of those elements.

CONCLUSIONS

Microbial science covers a vast territory. Less than 1% of the bacteria in the world have ever been obtained in pure culture. The rate at which new species of bacteria and catabolic plasmids evolve in nature may be extraordinarily rapid. Against this constantly changing landscape, microbial genome sequencing continues with a much broader phylogenetic range of prokaryotes. These studies will no doubt continue to reveal the genomic and biocatalytic richness of microbial life, contributing to the understanding of life and providing new biological reactions to use for industrial purposes. The continuing discovery of biocatalytic reactions, and the enzymes responsible for them, will enhance the ability to annotate genomes by giving a broader base of reactions against which to map sequences. Thus, the fields of biodegradation, biocatalysis, and genomics will inevitably advance together.

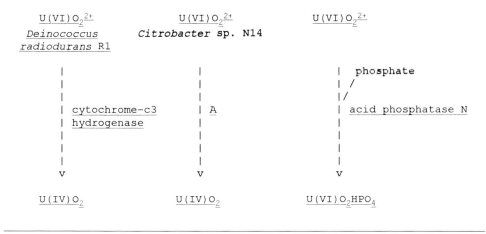

[Compounds and Reactions] [BBD Main Menu]

Fig. 4. Map of uranium metabolism by bacteria as found on the University of Minnesota Biocatalysis/Biodegradation Database (UM-BBD).

ACKNOWLEDGMENTS

I thank Steven Toeniskoetter for assistance in preparing the manuscript. The electron micrograph of *D. radiodurans* was kindly provided by Professor Michael Daly.

REFERENCES

1. Wackett LP, Hershberger CD. Biocatalysis and Biodegradation: Microbial Transformation of Organic Compounds. Washington, DC: American Society for Microbiology Press, 2001.
2. Whitman WB, Coleman DC, Wiebe WJ. Prokaryotes: the unseen majority. Proc Natl Acad Sci USA 1998; 95:6578–6583.
3. D'Souza TM, Merritt CS, Reddy CA. Lignin-modifying enzymes of the white rot basidiomycete *Ganoderma lucidum*. Appl Environ Microbiol 1999; 65:5307–5313.
4. Kuan IC, Tien M. Stimulation of Mn peroxidase activity: a possible role for oxalate in lignin biodegradation. Proc Natl Acad Sci USA 1993; 90:1242–1246.
5. Nagasawa T, Yamada H. Bioconversion of nitriles to amides and acids. In: Abramowicz DA (ed). Biocatalysis. New York: Van Nostrand Reinhold, 1990, pp. 277–318.
6. Harbourne J. Ecological Biochemistry, 3rd ed. New York: Academic Press, 1988.
7. Vokounova M, Vacek O, Kunc F. Degradation of the herbicide bromoxynil in *Pseudomonas putida*. Folia Microbiol 1992; 37:122–127.
8. Blattner FR, Plunkett G, Bloch CA, et al. The complete genome sequence of *Escherichia coli* K-12. Science 1997; 277:1453–1462.
9. Diaz E, Ferrandez A, Prieto MA, Garcia JL. Biodegradation of aromatic compounds by *Escherichia coli*. Microbiol Mol Biol Rev 2001; 65:523–569.
10. Krone UE, Thauer RK, Hogenkamp HP, Steinbach K. Reductive formation of carbon monoxide from CCl4 and FREONs 11, 12, and 13 catalyzed by corrinoids. Biochemistry 1991; 30: 2713–2719.
11. Ohren A, Gurevich P, Azachi M, Henis Y. Microbial degradation of pollutants at high salt concentrations. Biodegradation 1992; 3:387–398.
12. den Dooren de Jong LE. Bijdrage tot de kennis van het mineralisatieproces. Rotterdam, The Netherlands: Nijgh and van Ditmar, 1926.

13. Blumer M. Polycyclic aromatic compounds in nature. Sci Am 1976; 234:35–45.

14. Kunst F, Ogasawara N, Moszer I, et al. The complete genome sequence of the Gram-positive bacterium *Bacillus subtilis*. Nature 1997; 390:249–256.

15. Torsvik V, Ovreas L. Microbial diversity and function in soil: from genes to ecosystems. Curr Opin Microbiol 2002; 5:240–245.

16. Lederberg J. Cell genetics and hereditary symbiosis. Physiol Rev 1952; 32:403–430.

17. Austen RA, Dunn NW. Isolation of mutants with altered metabolic control of the NAH plasmid-encoded catechol meta-cleavage pathway. *Aust J Biol Sci* 1977; 30:583–592.

18. Chaudhry GR, Huang GH. Isolation and characterization of a new plasmid from a *Flavobacterium* sp which carries the genes for degradation of 2,4-dichlorophenoxy-acetate. J Bacteriol 1988; 170:3897–3902.

19. Rheinwald JG, Chakrabarty AM, Gunsalus IC. A transmissible plasmid controlling camphor oxidation in *Pseudomonas putida*. *Proc Natl Acad Sci USA* 1973; 70:885–889.

20. Tralau T, Cook AM, Ruff J. Map of the IncP1β plasmid pTSA encoding the widespread genes (*tsa*) for *p*-toluenesulfonate degradation in *Comamonas testosteroni* T2. Appl Environ Microbiol 2001; 67:1508–1516.

21. Williams PA, Murray K. Metabolism of benzoate and the methylbenzoates by *Pseudomonas putida* (arvilla) mt-2: evidence for the existence of a TOL plasmid. J Bacteriol 1974; 1:416–423.

22. Wackett LP, Sadowsky MJ, Martinez B, Shapir N. Biodegradation of atrazine and related triazine compounds: from enzymes to field studies. Appl Microbiol Biotechnol 2002; 58:39–45.

23. Romine MF, Stillwell LC, Wong KK, et al. Complete sequence of 184-kilobase catabolic plasmid from *Sphingomonas aromaticivorans* F199. J Bacteriol 1999; 181:1585–1602.

24. Martinez B, Tomkins J, Wackett LP, Wing R, Sadowsky MJ. Complete nucleotide sequence and organization of the atrazine catabolic plasmid pADP-1 from *Pseudomonas* sp ADP. J Bacteriol 2001; 183:5684–5697.

25. Thorsted PB, Marcarteney DP, Akhtar P, et al. Complete sequence of the IncPβ plasmid R751: implications for the evolution and organisation of the IncP backbone. J Mol Biol 1998; 282: 969–990.

26. Raillard S-A, Krebber A, Chen Y, et al. Novel enzyme activities and functional plasticity revealed by recombining homologous enzymes. Chem Biol 2001; 125:1–9.

27. Riley RG, Zachara JM, Wobber FJ. Chemical contaminants on DOE lands, DOE/ER-0547T. Washington, DC: US Department of Energy, Office of Energy Research, Subsurface Science Program, 1992.

28. Johnson J. Hanford on fast forward. Chem Eng News, 2002; June 10:24–33.

29. Daly MJ. Engineering radiation-resistant bacteria for environmental biotechnology. Curr Opin Biotechnol 2000; 11:280–285.

30. White O, Eisen JA, Heidelberg JF, et al. Genome sequence of the radioresistant bacterium *Deinococcus radiodurans* R1. Science 1999; 286:1571–1577.

31. Karlin S, Mrazek J. Predicted highly expressed and putative alien genes of *Deinococcus radiodurans* and implications for resistance to ionizing radiation damage. Proc Natl Acad Sci USA 2001; 98:5240–5245.

32. Lange CC, Wackett LP, Minton K, Daly M. Engineering a recombinant *Deinococcus radiodurans* for organopollutant degradation in radioactive mixed waste environments. Nature Biotech 1998; 16:929–933.

33. Brim H, McFarlan SC, Fredrickson JK, et al. Engineering *Deinococcus radiodurans* for metal remediation in radioactive mixed waste environments. Nature Biotechnol 2000; 15:85–90.

34. Silver S, Phung LT. Bacterial heavy metal resistance: new surprises. Annu Rev Microbiol 1996; 50 :753–789.

35. Fredrickson JK, Kostandarithes HM, Li SW, Plymale AE, Daly MJ. Reduction of Fe(III), Cr(VI), U(VI), and Tc(VII) by *Deinococcus radiodurans* R1. Appl Environ Microbiol 2000; 66:2006–2011.

36. Dixon B. Power Unseen: How Microbes Rule the World. Oxford, UK: Oxford University Press, 1984.

37. Kluyver AJ. Microbiology and industry. In: Kamp AF, La Riviere JWM, Verhoeven W (eds). A. J. Kluyver: His Life and Work. Amsterdam: North-Holland, 1957, pp. 165–185.

38. Glazer AN, Nakaido H. Microbial Biotechnology: Fundamentals of Applied Microbiology. New York: Freeman, 1995.

39. Lamed R, Setter E, Bayer EA. Characterization of a cellulose-binding, cellulase-containing complex in *Clostridium thermocellum*. J Bacteriol 1983; 156:828–836.

40. Gray KA, Pogrebinsky OS, Mrachko GT, Xi L, Monticello DJ, Squires CH. Molecular mechanisms of biocatalytic desulfurization of fossil fuels. Nature Biotechnol 1996; 14:1705–1709.

41. Treadway SL, Yanagimachi KS, Lankenau E, Lessard PA, Stephanopoulos G, Sinskey AJ. Isolation and characterization of indene bioconversion genes from *Rhodococcus* strain I24. Appl Microbiol Biotechnol 1999; 51:786–793.

42. Bentley SD, Chater KF, Cerdeno-Tarraga AM, et al. Complete genome sequence of the model actinomycete *Streptomyces coelicolor* A3(2). Nature 2002; 417:141–147.

43. Goto S, Okuno Y, Hattori M, Nishioka T, Kanehisa M. LIGAND: database of chemical compounds and reactions in biological pathways. Nucleic Acids Res 2002; 30:402–404.

44. Ellis LBM, Hershberger CD, Wackett LP. The University of Minnesota Biocatalysis/Biodegradation Database: microorganisms, genomics and prediction. Nucleic Acids Res 2000; 28:377–379.

45. Lovley DR, Widman PK, Woodward JC, Phillips EJ. Reduction of uranium by cytochrome c3 of *Desulfovibrio vulgaris*. Appl Environ Microbiol 1993; 59:3572–3576.

25

Enzyme Discovery and Microbial Genomics

Robert M. Kelly and Keith R. Shockley

INTRODUCTION

More than two decades ago, the enzyme industry was focused on a small number of bio-catalysts derived from a narrow phylogenetic range of microorganisms. These enzymes were used for a limited set of applications, mostly in large-scale processes for starch processing and as cleaning agents in detergents *(1–3)*. The difficulties associated with discovering new enzymes, producing them in large scale, and integrating them, some-times imperfectly, into existing bioprocesses constrained the development of this tech-nological sector. Even if a microorganism with physiological features that hinted at the existence of an important biocatalyst could be found, isolating a particular enzyme from among the thousands of similar biomolecules was especially challenging. If the microorganism (wild type or mutant) had no phenotype that generated copious amounts of the enzyme of interest in large-scale fermentation, the likelihood that this biocata-lyst would find its way into a technologically significant use was low. Simply put, before the dawn of industrial biotechnology in the 1980s, commercializing a valuable enzyme embedded in the genotype of an uncharacterized microbial source was the domain of a limited number of industrial practitioners.

But, that was then, and this is now. Beginning with the first successful commercializa-tion of enzymes produced in recombinant organisms (often involving amplification of gene expression in the native host; *1*), the tools of molecular biology have been exploited to revise completely the nature of enzyme discovery *(4)*. Once there were few enzymes considered for many potential applications; now, there is a virtually unlimited number that can be considered based on information encoded in sequenced microbial genomes. The challenge today is to find a technological fit for particular biocatalysts. Furthermore, recombinant approaches have been used to create engineered versions of specific enzymes that can be customized to the application through site-directed mutagenesis, deoxyribo-nucleic acid (DNA) shuffling and directed evolution *(5,6)*. At this point, those focused on biocatalyst technology are seeking answers to questions concerning the diversity of nature's treasure chest of enzymes and whether biological routes can replace or create processes for important chemicals and biochemicals.

Many industrial enzymes have traditionally come from microorganisms isolated from natural environments or from specific niches by which a biological process mimics the industrial use *(4)*. However, it is clear that conventional methods for enzyme discovery

From: *Microbial Genomes*
Edited by: C. M. Fraser, T. D. Read, and K. E. Nelson © Humana Press Inc., Totowa, NJ

have revealed only a fraction of nature's biocatalytic inventory. Comparative analyses based on small-subunit (16S or 18S) ribosomal ribonucleic acid (rRNA) sequences demonstrate that most of life on earth is microbial, yet it is estimated that more than 99% of microorganisms observed in nature have not been cultivated by standard techniques *(7,8)*. Furthermore, fewer than 2% of all microorganisms have been identified, and fewer than that have been closely examined *(9)*. The question arises as to whether the current sampling of microbial enzymes that has been studied represents the larger set. With available microbial genome sequence data, this issue can begin to be investigated.

It is not clear how much microbial biodiversity has been captured in the microbial genomes sequenced to date. Certainly, significant genetic diversity is present in sequenced microbial genomes, even among closely related species or strains within the same species *(10)*. Until now, pathogenic organisms have been the subjects of most genomic comparison studies *(11–15)*, but nonpathogenic organisms have been examined as well. For instance, almost one-third of the open reading frames (ORFs) in *Bacillus halodurans* did not have a clear match to those in another organism in its genus, *Bacillus subtilis (10)*. Also, the genome of the hyperthermophilic archaeon *Pyrococcus furiosus (16)* is about 10% larger than the genome of another organism in this genus, *Pyrococcus horikoshii (17)*. Most of the difference can be attributed to additional amino acid biosynthetic pathways and routes for the uptake of carbohydrates, such as cellobiose, maltose, trehalose, laminarin, and chitin *(16,18)*. The very limited perspective gained from interspecies genome sequence comparisons suggests that, at this point, only a glimpse of nature's biocatalytic repertoire has been seen.

Clearly, the gene-by-gene, protein-by-protein approach to enzyme discovery is being replaced by the use of new methodologies and information arising from advances in the molecular biology and genomics sciences *(9,19,20)*. When complete microbial genome sequences appeared in the mid-1990s, the first look into the complete enzymatic inventory within microbial species was possible. More than 140 microbial genomes have been completed, and at least 300 projects are in progress *(21)*. For the first time, it is now possible to compare indirect evidence for the presence of specific biocatalysts in an organism against its own genetic blueprint.

Much uncertainty still exists, given that half or more of the ORFs in microbial genomes were initially not coupled to specific function. For instance, even in the much-studied *Escherichia coli*, with 4288 annotated protein-coding genes, 38% initially had no attributed function *(22)*. Also, of the 2977 proteins predicted to be encoded by the extremely thermoacidophilic crenarchaeon *Sulfolobus solfataricus* P2, about one-third have no detectable homologs in other sequenced genomes *(23)*. Over 50% of the ORFs discovered in the *P. horikoshii* genome have functions that have not been identified through database similarity searches *(24)*. The correct annotation of ORFs in microbial genomes remains an ongoing effort that currently employs a diverse set of tools, including in vivo, in vitro, and *in silico* approaches. As has been the case to date, the annotation process has probably raised as many new questions as it has provided answers. Nonetheless, the technologically relevant enzyme inventory for particular microorganisms is becoming more clear.

Armed with the prospect of using the polymerase chain reaction to amplify a gene of interest from genomic DNA for subsequent cloning and overexpression in a suitable host, each microbial genome can potentially provide thousands of biocatalysts that can be examined for technological importance. Selection of candidates for further characterization can be facilitated through the use of a variety of bioinformatics tools (Table 1). If several candidates are identified, the list of those to be produced and characterized biochemically can be shortened considerably by examining structural and catalytic traits *in silico*, preferably in conjunction with amino acid sequence-based classification schemes, such as those developed for glycosyl hydrolases *(25)*.

MINING HYPERTHERMOPHILE GENOMES FOR USEFUL BIOCATALYSTS

For several reasons, initial efforts for sequencing microbial genomes focused on microorganisms that typically inhabit biologically hostile environments *(26)*. Hyperthermophiles, extremophilic microorganisms that belong to the domains Archaea and Bacteria, have optimal growth temperatures of 80°C or higher. Stemming from their evolutionary placement, small genome size, and potential importance as a source of stable biocatalysts, a number of genome sequences for these microorganisms have been reported (Table 2).

Although a myriad of recombinant techniques now exist for improving enzyme characteristics (see below), these are best applied to a biocatalyst with natural properties close to the final optimal form. One particularly sought-after trait in industrial biocatalysts is stability to heat, an intrinsic feature of enzymes from hyperthermophiles. The interest in thermostable biocatalysts arises from the fact that most industrial processes still utilize enzymes from mesophilic microorganisms, even though many reactions are conducted at elevated temperatures *(4,27–29)*. At higher temperatures, lower enzyme concentrations and higher conversions may result from decreased viscosity and larger diffusion coefficients of organic compounds *(27,30,31)*. Elevated temperatures also can lead to higher substrate bioavailability, along with reduced risks of biological contamination *(26,27,31)*. As an added benefit, enzymes from hyperthermophilic organisms are often relatively resistant to chemical denaturants, such as detergents, chaotropic agents, and organic solvents, making them useful as industrial biocatalysts *(26,31–33)*. Thus, if stability is important, an already hyperthermophilic enzyme can be modified through recombinant approaches to improve catalytic traits. Given the number of hyperthermophile genome sequences completed and available, there is a good possibility of finding a thermostable enzyme with biocatalytic properties that either match specific needs or can be modified to meet specific requirements.

EXAMINING THE BIOCATALYST INVENTORY IN HYPERTHERMOPHILE GENOMES

Most often, ORFs in genomic sequence data are annotated through full sequence alignment (e.g., Basic Local Alignment Search Tool [BLAST]; *34*) to those found in databases such as GenBank (see Chapter 3). Similar approaches can be used to determine the inventory of specific enzymes among selected organisms, assisted by specialized databases (e.g., ref. *35*) and handbooks (e.g., ref. *36*). For example, Table 3 shows the inferred protease inventory from publicly available genome sequences for hyperthermophiles, based on all known (isolated and characterized biochemically) or putative (inferred from bioinformatics tools, such as those shown in Table 1) proteases in *P. furiosus (37)*. The protease homologs in Table 1 were defined based on more than 30% amino acid sequence identity over more than 50% of the sequence; this criterion is

Table 1
Useful Bioinformatic Tools for Microbial Systems[a]

Search Tool	Web address	Description
Protein sequence databases		
BLAST	http://www.ncbi.nlm.nih.gov/BLAST/	Basic Local Alignment Search Tool; allows the rapid comparison of sequences (34); a fast way to compare sequences in a database, identify gene or protein sequences, and identify regions of similarity between a sequence of interest and those in a database; Variations on BLAST, such as PSI-BLAST or Gapped-BLAST, allow higher sensitivity
PROSITE	http://ca.expasy.org/prosite/	Helps elucidate the function of an unknown protein sequence (translated from cDNA or genomic sequence) by comparison with known families of proteins (53)
Pfam	http://pfam.wustl.edu/hmmsearch.shtml	A database of multiple protein domain alignments capable of assessing and identifying proteins with multiple domains, detecting end-to-end similarity among protein sequences (91)
Blocks	http://blocks.fhcrc.org	Detects local regions of similarity in proteins (92)
eMOTIF	http://motif.stanford.edu/emotif	Determines and searches for protein motifs (93)
SMART	http://smart.embl-heidelberg.de	Simple Modular Architecture Research Tool; allows the analysis of domain architecture (94)
PRINTS	http://www.bioinf.man.ac.uk/dbbrowswer/PRINTS/	Compendium of conserved protein motif sets (95,96)
CDD	http://www.ncbi.nlm.nih.gov/Structure/cdd/cdd.shtml	Conserved Domain Database; composed of multiple sequence alignments for conserved regions of proteins (97)
TOPITS	http://www.embl-heidelberg.de/predictprotein/predictprotein.html	Prediction-based threading program; useful for relating sequence motifs to protein structure/function (53)

Genome comparisons

Name	URL	Description
STRING	http://www.bork.embl-heidelberg.de/STRING/	Search tool for recurring instances of neighboring genes able to locate and display the genes that repeatedly occur in clusters on published genome sequences (98); genes that occur in repeated clusters across a wide range of genomic sequences often indicate a functional association
COG	http://www.ncbi.nlm.nih.gov/COG/	Clusters of Orthologous Groups of proteins, determined by comparing protein sequences in complete genomes; orthologs typically have the same function (99,100)
PEDANT	http://pedant.gsf.de/	High-throughput processing of genomic data to assign functional and structural categories to proteins using a wide range of bioinformatic approaches (101)
AlignACE	http://arep.med.harvard.edu/mrnadata/mrnasoft.html	Aligns Nucleic Acid Conserved Elements; predicts functional interactions based on comparative genomics (102,103)
Genome Information Broker	http://gib.genes.nig.ac.jp	Allows the retrieval and visualization of regions of any sequenced microbial genome that are of interest along with biological annotation (104)

Metabolic databases

Name	URL	Description
EcoCyc	http://ecocyc.org	Annotation on all known metabolic and signal transduction pathways for *E. coli*, including a description of the genome and biochemical machinery of the organism (65)
ENZYME	http://www.expasy.org/enzyme/	Provides nomenclature, catalytic activities, and cofactors associated with an enzyme of interest (105)

(continued)

Table 1 (Continued)

Search Tool	Web address	Description
Metabolic databases		
LIGAND	http://www.genome.ad.jp/ligand/	Composed of three main areas; designed to provide information that links biological and chemical aspects of life; COMPOUND gives information about metabolites and associated chemical compounds; REACTION is for the collection of metabolic reactions; ENZYME gives all known enzymatic reactions pertaining to the protein of interest (106)
KEGG	http://www.genome.ad.jp/kegg	Kyoto Encyclopedia of Genes and Genomes; provides functional information derived from genome sequence data (64)
WIT2	http://wit.mcs.anl.gov/WIT2/	What Is There database; contains information on metabolic pathways and is based on sequence comparisons and biochemical and phenotypic data (107)
Other useful tools		
SignalP	http://www.cbs.dtu.dk/services/SignalP/	Identification of signal peptides and cleavage sites based on neural networks for prokaryotic and eukaryotic systems (108)
TMpred	http://www.ch.embnet.org/software/TMPRED_form.html	Predicts membrane-spanning regions of proteins and orientation; based on the statistical algorithms of TMbase (109)
BRENDA	http://www.brenda.uni-koeln.de/	Collection of enzyme functional data (110)
ClustalW	http://www.ebi.ac.uk/clustalw/	Multiple sequence alignment program for DNA or proteins that allows sequences to be aligned to view similarities or differences between molecules (111)

PSORT	http://psort.nibb.ac.jp/form.html	Prediction of protein-sorting signals, or protein localization sites in cells, from amino acid sequence data (*112*)
TMHMM	http://www.cbs.dtu.dk/services/TMHMM/	Can predict the occurrence of transmembrane helices; based on hidden Markov models (*113*)
DAS	http://www.sbc.su.se/~miklos/DAS/	Dense alignment surface method to predict trans membrane regions from amino acid sequence data for any integral membrane protein (*114*)
TIGRFAMs	http://www.tigr.org/TIGRFAMs/	Functional identification of proteins; based on hidden Markov models (*115*)

*See ref. *116* for a more extensive listing of bioinformatic databases. See also ref. *117*.

Table 2
Available Genome Sequences for Hyperthermophilic Microorganisms

Name	Year	T_{opt} (°C)	Genome size (Mbp)	ORFs	Genes with unknown function[a]	Unique genes[a]	%G+C	Isolation site	Reference
Archaea									
Aeropyrum pernix	1999	95	1.67	2694	523 (19%)	1538 (57%)	56	Kodakara	118
Archaeoglobus fulgidus	1997	83	2.18	2436	1315 (54%)	641 (25%)	49	Vulcano	44
Methanococcus jannaschii	1996	85	1.66	1729	1076 (62%)	525 (30%)	31	East Pacific Rise	45
Pyrococcus abyssi	2001	98	1.77	1765	NR	NR	45	North Fiji Basin	119
Pyrococcus horikoshii	1998	98	1.74	2061	859 (42%)	453 (22%)	42	Okinawa Trough	17
Pyrococcus furiosus	2002	98	1.91	2208	NR	NR	40	Vulcano	43
Sulfolobus solfataricus	2001	80	2.99	3032	577 (22%)	743 (25%)	NR	Pisciarelli Solfatara	23
Pyrobaculum aerophilum	2002	100	2.22	2587	NR	302 (12%)	51	Maronti Beach	120
Bacteria									
Aquifex aeolicus	1998	95	1.55	1512	663 (43%)	407 (27%)	43	Not reported	121
Thermotoga maritima	1999	80	1.86	1877	863 (43%)	373 (20%)	46	Vulcano	122

Descriptions:
- Aeropyrum pernix: Strictly aerobic crenarchaeon
- Archaeoglobus fulgidus: Strictly anaerobic Archaeoglobales; sulfur metabolizing
- Methanococcus jannaschii: Anaerobic, autotrophic, methanogenic Methanococcales
- Pyrococcus abyssi: Anaerobic Thermococcales
- Pyrococcus horikoshii: Anaerobic, obligately heterotrophic Thermococcales
- Pyrococcus furiosus: Anaerobic Thermococcales; grows well on sugars and peptides
- Sulfolobus solfataricus: Aerobic Solfolobales; grows best at low pH
- Pyrobaculum aerophilum: Facultatively aerobic, nitrate-reducing crenarchaeon
- Aquifex aeolicus: Microaerophilic Aquificaceae; obligate chemolithoautotroph
- Thermotoga maritima: Anaerobic Thermotogales; metabolizes simple and complex carbohydrates

%G+C, percentage guanine and cytosine.
[a]Refers to the time of genome sequence publication.
NR, not reported.

Table 3
Proteases in *Pyrococcus furiosus*

	Locus	S[a]	Nuc.	a.a.	Ph	Pa	Mj	Af	Pae	Ap	Ss	Tm	Aa
ATP-dependent proteases													
Proteasome, subunit beta (PsmB-1)	PF1404	N	621	206	x	x	x	x	x	x	x		
Proteasome, subunit beta (PsmB-2)	PF0159	N	599	199	x	x	x	x	x	x	x		
ATP-dependent regulatory subunit (PAN)	PF0115	N	1199	399	x	x	x	x	x	x	x		
ATP-dependent LA (Lon)	PF0467	N	3140	1046	x	x	x	x					
Proteasome, subunit alpha (PsmA)	PF1571	N	798	265	x	x	x	x	x	x	x		
ATP-independent proteases													
Subtilisin-like protease	PF0688	N	593	197					x	x[b]			
Intracellular protease I (PfpI)	PF1719	N	582	193	x	x	x	x	x	x	x		x
Periplasmic serine protease, putative	PF0240	Y	842	280	x	x	x	x	x	x		x	x
Metalloprotease	PF0392	N	1253	417	x	x		x		x			
Alkaline serine protease	PF1670	Y	1992	663				x	x	x			
Metalloprotease	PF0167	Y	1133	377	x	x	x	x	x	x			x
Putative bacteriocin/protease	PF1191	N	785	261									
Pyrolysin	PF0287	Y	4238	1412					x[b]				
Hydrogenase maturation protease (hyc I)	PF0617	N	485	161	x	x	x	x					
Hypothetical protein	PF0760	N	1022	340	x						x	x	
Protease IV	PF1583	Y	990	329	x	x	x	x					x
Putative protease	PF1905	Y	1332	443		x		x[b]					
Metalloprotease	PF0457	N	629	209	x	x	x	x	x	x	x		x
Peptidases													
Acetylornithine deacetylase (ArgE)/peptidase	PF1185	N	1061	353	x	x	x	x	x	x			
HtpX heat shock protein	PF1135	N	875	291	x	x	x	x		x			x
Proline dipeptidase–related protein	PF0702	N	521	173	x	x						x	
Protein similar to endo-1,4-β-glucanase (ytoP)	PF1861	N	1040	346	x	x	x	x	x	x		x	
D-Aminopeptidase	PF1924	N	1098	365	x	x							
Signal sequence peptidase I, SEC 11	PF0313	N	264	87	x	x	x	x	x	x			

(continued)

Table 3 (Continued)

	Locus	S[a]	Nuc.	a.a.	Ph	Pa	Mj	Af	Pae	Ap	Ss	Tm	Aa
Peptidases													
Hypothetical protein	PF0669	N	932	310	x	x							x[b]
O-Sialoglycoprotein endopeptidase (gcp-2)	PF0473	N	680	226	x	x		x	x	x	x		
O-Sialoglycoprotein endopeptidase (gcp-1)	PF0172	N	974	324	x	x	x	x	x	x	x	x	x
Succinyl-diaminopimelate desuccinylase/peptidase	PF2048	N	1343	447	x	x				x			x[b]
Pyroglutamyl-peptidase I	PF1299	N	654	217	x	x					x		
XAA-Pro dipeptidase (proline dipeptidase)	PF1343	N	1047	348	x	x	x	x	x	x	x	x	x
Protein similar to acylaminoacyl peptidase (acylaminoacid–releasing enzyme homolog)	PF0318	N	1862	620	x	x			x	x			
Prolyl endopeptidase	PF0825	N	1863	620	x	x							
Endoglucanase (CelM)/aminopeptidase	PF1547	N	1046	348	x	x	x	x	x	x	x[b]	x	
Methionine aminopeptidase (MAP) (Pep M)	PF0541	N	887	295	x	x	x	x	x	x	x	x	x
Putative proline dipeptidase	PF0747	N	1076	358	x	x	x						
Heat shock protein X	PF1597	Y	800	266	x	x							
Carboxypeptidase I	PF0456	N	1499	499	x	x	x	x	x	x	x		
Endoglucanase/peptidase	PF0369	N	998	332	x	x	x	x	x	x		x	
Putative aminopeptidase	PF2059	Y	1704	567									
Putative aminopeptidase	PF2063	Y	1755	584									
Putative aminopeptidase	PF2065	Y	1776	591									
Membrane dipeptidase	PF0874	HN	1140	379	x	x				x			

Proteins in database were considered present if they contained >30% identity over >50% length of protein in database. *P. furiosus* (strain DSM3638) was compared against Ph, *Pyrococcus horikoshii* OT3; Pa, *Pyrococcus abyssi* GE5; Mj, *Methanococcus jannaschii* DSM 2661; Af, *Archaeoglobus fulgidus* VC-16; Pae, *Pyrobaculum aerophilum* IM2; Ap, *Aeropyrum pernix* K1; Ss, *Sulfolobus solfataricus* P2; Tm, *Thermotoga maritima* MSB8; Aa, *Aquifex aeolicus* VF5.

[a]S—indicates the presence of a signal peptide as identified from signal P (Y, yes; N, no) *(108)*.

[b]Predicted amino acid length of protein encoded by ORF differs significantly from predicted *P. furiosus* protein length.

arbitrary and can be relaxed or made more stringent. Table 3 illustrates the biodiversity of proteases within a given organism and between organisms. As might be expected, the three pyrococci examined share many similar protease-encoding genes, although they also clearly exhibit differences. In some cases, there are no homologs to the *P. furiosus* proteases among the hyperthermophiles listed; in other cases, homologs appear to exist, but can differ significantly in molecular mass. To illustrate, a putative protease (with a signal peptide; PF1905), three aminopeptidases (with signal peptides; PF2059, PF2063, and PF2065), an intracellular putative bacteriocin/protease (PF1191), and a transmembrane protease pyrolysin (PF0287) appear to be unique to *P. furiosus*; an intracellular *o*-sialoglycoprotein endopeptidase (PF0172), a proline dipeptidase (PF1343), and a methionine dipeptidase (PF0541) are ubiquitous in the hyperthermophile genome sequences examined (Table 3).

Some of the first enzymes studied from hyperthermophiles were glycosyl hydrolases *(38,39)*. These attracted attention because of their industrial significance in the starch-processing industries and their physiological importance for heterotrophic hyperthermophiles growing on glycan-based media *(40,41)*. Hyperthermophile genome sequences revealed significant differences among hyperthermophiles in their available enzyme inventory for carbohydrate hydrolysis *(42)*. Although the genome sequence for the hyperthermophilic archaeon *P. furiosus (43)* shows the presence of a range of glucan-degrading enzymes (Table 4), the genome sequence of *Archaeoglobus fulgidus (44)*, a hyperthermophilic archaeon, is apparently devoid of such enzymes. Furthermore, even among three pyrococci, glycosyl hydrolase content varies, especially with respect to enzymes capable of degrading laminarin *(40)* and chitin *(44a)*. In fact, *P. furiosus* appears to contain a chitin utilization pathway that includes a chitin deacetylase and glucoaminidase not evident from initial genome annotations *(44a)*. Also, the *Methanococcus jannaschii* genome *(45)* revealed the presence of several glycosidases, one of which (MJ1610) is a family 15 glucoamylase. Given the small sampling of hyperthermophilic genomes, it is difficult to assess the diversity of enzymes such as glycosidases, although many surprises should be expected as these organisms are probed further.

At one level, genome sequence annotation provides a glimpse into the actual and putative enzymatic inventory of a specific microorganism. However, genome sequence annotation itself presents significant challenges for biocatalyst identification, and the results can be misleading at times *(20,46)*. For instance, two putative enzymes that share a similar substrate-binding domain may have dissimilar catalytic domains, although a sequence alignment tool may classify both as related and reflect this in the annotation (see Chapter 3). Furthermore, when simple homology searches based on full sequence length are the basis for annotation, these may not detect less-obvious relationships found in enzyme superfamilies and will not recognize nonorthologous genes carrying the same function *(4,47,48)*. Another problem occurs when incorrect functions are assigned to specific ORFs in databases, and this assignment propagates in subsequent sequences that are reported. Analysis based on simple amino acid sequence homology alone is not powerful enough to confirm the absence of an enzyme from a genome sequence or rule out the possibility that an enzyme has multiple functions. Incomplete understanding of cellular metabolism also creates problems. For instance, some of the genes encoding enzymes that are used in microbial tryptophan biosynthesis pathways were missing from the genome of *P. horikoshii (24)*. Although this organism requires tryptophan for

Table 4
Glycosidase Inventory from *Pyrococcus furiosus* Based on Genomic Sequence Data

Locus	Annotation	Reference	Activity	S[a]	Ph	Pa	Mj	Af	Pae	Ap	Ss	Tm	Aa
	Cellulases												
PF0073	β-Glucosidase	123,124	Cel1A		x								
PF0442	β-Glucosidase	125	Cel1B										
PF0854	Endo-1,4-β-glucanase	126	Cel12	Yes					x			**x**	
	Mannosidases												
PF1208	β-Mannosidase	127	Man1		x	x							
	Laminarinases												
PF0076	Endo-β-1,3-glucanase	123,128	Lam16	Yes								**x**	
	Chitinases												
PF1234	Chitinase	44a,129	Chi18A	Yes									
PF1233	Chitinase	44a,129	Chi18B										
	Amylases/pullulanases												
PF0477	α-Amylase	130,131	Amy13	Yes			**x**					x	
PF0272	α-Amylase	132	Amy57A		x	x	**x**		x				x
PF1935	Amylopullulanase	133	Amy57B			x			x		x		
PF0478	α-Amylase[b]		Amy13								x		x
PF1939	Neopullulanase		Pul13						x		x	x	
	Galactosidases												
PF0444	α-Galactosidase		Gal57		x								
PF0356	β-Galactosidase[b]		Gal1			x					**x**		
PF0363	β-Galactosidase precursor[b]		Gal35		x	x							x

[a]S – indicates the presence of a signal peptide as identified from SignalP (108).
[b]Putative proteins.
Characterized proteins are noted in bold.

cell viability and growth, it is not clear whether it is auxotrophic for this amino acid or whether a complete synthetic pathway exists, but contains unidentified elements.

When BLAST searches alone are not sufficient to elucidate gene function, other bioinformatic approaches may complement the analyses. For example, enzymatic resolution of racemic mixtures of 2-aryl propionic esters, such as those used for nonsteroidal anti-inflammatory drugs *(49)*, has been reported using esterases and lipases from mesophilic sources. BLAST searches, in conjunction with protein structure–based motif analysis (Threading One-dimensional Predictions Into Three-dimensional Structures [TOPITS]; *50)*, identified a carboxylesterase in the genome of *S. solfataricus* P1 (SsoEST1) *(51)*. This enzyme proved to be more effective for the resolution of naproxen methyl esters than other mesophilic candidates, despite the fact that temperatures more than 50°C below this enzyme's optimum were used *(52)*. While BLAST searches alone turned up other thermostable esterase/lipase posibilities, the combination of several bioinformatics tools led to the most promising candidate.

Searching databases, such as PROSITE *(53)*, for short sequence patterns or motifs to identify functional domains in predicted proteins can eliminate some of the problems associated with matching entire sequences *(54)*. For instance, a database termed IDENTIFY *(55)*, may be able to identify a protein superfamily even when BLAST results are not suggestive. A total of 833 ORFs in the yeast genome had not been assigned a function when the genome was first published, but 172 of the unknown proteins were subsequently assigned putative functions based on the IDENTIFY algorithm *(55)*.

Other approaches have also been used. Threading or *ab initio* folding methods allow the prediction of tertiary structure from sequence information and can be further screened by descriptors of protein active sites called *fuzzy functional forms (47,54)*. This approach helped identify the function of two proteins in the glutaredon/thioredoxin disulfide oxidoreductase family in the yeast genome with functions that could not be predicted by BLAST searches or local sequence alignment algorithms *(54)*. Another approach, based on combining methods of prediction and experimental data, examined correlated evolution, correlated messenger RNA expression patterns, and patterns of domain fusion among the 6217 proteins of the yeast *Saccharomyces cerevisiae* to assign functions to more than half of the 2557 previously uncharacterized yeast proteins *(56)*.

Intergenomic comparisons among microbial genomes can also be used to find promising biocatalysts. For example, conservation of gene clusters across a wide range of organisms can help determine candidates for homologous function or indicate the presence of an essential role. The malaria parasite *Plasmodium falciparum* uses an essential isoprenoid biosynthesis pathway commonly found in plants but not in mammals, which led to the discovery of a herbicide that targets this pathway as a specific antimalarial agent *(48,57)*. Sometimes, lateral gene transfer has complicated single-gene studies of evolutionary relatedness. However, full genome sequence information can be used to identify horizontally transferred genes more rapidly by looking for regional differences, such as base composition in bacterial chromosomes *(48)*.

FUNCTIONAL GENOMICS AND ENZYME DISCOVERY

While genomic information pertaining to identifiable ORFs can provide a useful blueprint of an organism's repertoire of genes, the challenge remains to discover how an organism uses its genetic information to accomplish biological tasks. From this kind of insight, purposeful uses of specific enzymes can be projected. Genetic regulation has been described as the process by which a cell decides whether a gene is active or inactive *(58, 59)*. Active genes are those that are being transcribed, that have transcripts that are being translated, and that have enzymatic products that are actively performing their function. Functional genomics includes both transcriptional analyses and proteomics approaches, making use of gene expression data, systematic mass mutagenesis, and protein interaction maps to elucidate functions of genes *(47,60)*. In the Bacteria and the Eukarya (and presumably the Archaea), the majority of gene regulation is most often controlled at the level of transcription initiation *(48,61,62)*. Therefore, an understanding of the mechanisms that regulate the initiation of gene transcription is a good way to discover enzymes of interest and is essential to the knowledge of the underlying biological processes.

As mentioned, genome sequence comparison is a principal first step toward elucidating the metabolic role of a given protein encoded by genomic sequence information, but results from bioinformatic predictions need to be confirmed through both transcriptional and biochemical analyses. For example, *Thermotoga maritima*, a hyperthermophilic bacterium capable of growth on a spectrum of α- and β-linked glycosides *(63)*, produces several glycosyl hydrolases, including an endoglucanase (Cel5A) and a mannanase (Man5). When compared against proteins found in the GenBank database through BLAST searching *(42)*, Man5 was most similar (46% identity at the amino acid sequence level) to a β-mannanase (ManF) from *Bacillus stearothermophilus*, while Cel5A showed highest similarity (38% amino acid sequence identity) to a family 5 endoglucanase (CelD) from *Clostridium celluloyticum*. Northern blotting and complementary DNA (cDNA) microarray experiments demonstrated that *man5* was induced when *T. maritima* was grown on carob galactomannan, konjac glucomannan, and to a lesser extent, on carboxymethyl cellulose. Surprisingly, *cel5A* was induced only on konjac glucomannan.

To investigate this unexpected result further, the activities of recombinant forms of Man5 and Cel5A were tested against a variety of polysaccharide substrates. Man5 was active only on mannose-based polysaccharides; Cel5A was active against glucans, xylans, and mannans. Remarkably, the activity of Cel5A was comparable to that of Man5 against galactomannan and higher than Man5 against glucomannan, a finding much different from what was predicted by genome sequence comparison alone. Further investigation of the two enzymes revealed that Man5 contained a signal peptide; Cel5A did not. Taken together, these results indicate that the primary physiological role of Cel5A (and a related enzyme Cel5B) is to break down glucomannan oligosaccharides that are transported into the cell following extracellular hydrolysis by exported glycosidases, such as Man5, even though the genes encoding Cel5A and Man5 are not proximal in the genome (see Fig. 1). The functional assignment for Cel5A is based on biochemical properties of the enzyme in conjunction with gene regulation patterns for the native organism growing on various substrates.

Biological function of an enzyme encoded on a sequenced genome can sometimes be determined using database comparisons of sequence data with previously characterized proteins or bioinformatic tools. These include the Kyoto Encyclopedia of Genes and Genomes (or KEGG) *(64)* and EcoCyc *(65)*, both of which present genomic information in a comprehensive form based on biological pathways and molecular assemblies. However, it is also useful to have functional data directly from expression analysis studies

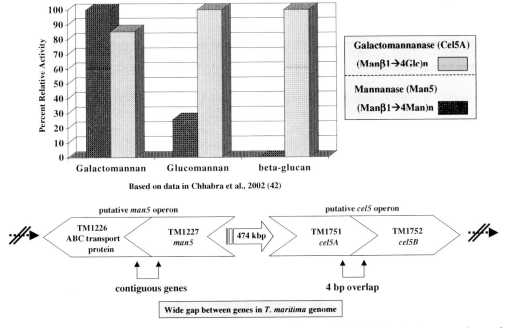

Fig. 1. Man5 and Cel5A in *Thermotoga maritima*. Shown is the biochemical comparison of the activity of each enzyme against galactomannan, glucomannan, and β-glucan as well as the genomic position of the genes encoding the enzymes. (Based on data in ref. *42*.)

through the identification and quantification of differentially expressed transcripts between two or more biologically varying samples.

Numerous advances have been made that allow the identification and quantification of differentially expressed genes, especially cDNA microarrays *(66–73)*, which were first used to monitor the expression of 45 different *Arabidopsis* genes simultaneously *(74)*. Since the debut of this technology, DNA microarrays have been used to study relative wide-scale gene expression patterns in many different kinds of organisms, including bacteria *(66–73)*, archaea *(75)*, yeast *(76–78)*, plants *(79,80)*, fruit flies *(81)*, mice *(82)*, and human beings *(83–85)*. Microarrays provide a mechanism to identify differentially expressed genes in microorganisms growing on a specific substrate, the modification of which is of technological importance. For example, using targeted cDNA microarrays, the genes encoding Cel5A and Man5 in *T. maritima* were indeed found to be co-regulated during growth on mannan-based compounds *(86)*. As microarray technology improves and becomes less expensive, use of environmental arrays to follow gene expression patterns in microbial consortia can be envisioned as a means of biocatalyst discovery.

BIOCATALYST DESIGN BY GENOME SCANNING, FUNCTIONAL SCREENING, AND IMPROVEMENT

High-throughput screening techniques are improving every year *(87,88)*, such that recombinant approaches for biocatalyst improvement can produce enzymes for specific applications with optimal properties. Directed molecular evolution and related methods can be used to generate useful variants of existing enzymes, such that the resulting biocatalyst works better than the natural one (see Chapter 24). This approach consists of multiple rounds of mutagenesis and screening, followed by amplification of selected variants *(4,5,89)*. As genome annotation becomes more sophisticated, so will selection processes for biomolecules that can serve as strategic starting points for evolution techniques.

One recent example of directed molecular evolution is the creation of an ampicillin resistance activity from a functionally unrelated DNA fragment in *P. furiosus (89)*. The resulting mutant enzyme conferred resistance to other drugs that target bacterial cell wall synthesis as well, although the mechanism of action is unclear. *P. furiosus* is a hyperthermophilic archaeon that does not contain peptidoglycan in its cell wall. As such, it is not susceptible to common antibiotics that are directed against the synthesis of cell walls, including β-lactam antibiotics like ampicillin. Nevertheless, an expression library of *P. furiosus* DNA fragments was screened for a gene that created ampicillin resistance activity (amp^R) in *E. coli*, which corresponded to a 1.2-kb fragment (including an ORF encoding a 226-amino acid protein) *(89)*. This DNA fragment was subjected to 50 rounds of directed evolution, in which mutations and DNA recombinations were randomly introduced. The resulting DNA fragments contained two genetic regions that had coevolved during the course of the experiment; one region was essential for the ampicillin resistance activity, and the other was able to enhance the activity. This experiment illustrates the potential utility of choosing genes from genome sequences that can be evolved to function in applications unrelated to their natural roles.

MICROBIAL GENOMICS: FUTURE DIRECTIONS FOR ENZYME DISCOVERY

It is still too early to tell how effective genomics-based enzyme discovery approaches will be. Beyond identifying homologs to enzymes of interest through bioinformatics tools, there is great interest in novel physiological systems that give rise to unique enzymes or pathways of technological importance. These can be inferred from differential expression experiments in which whole genome or targeted microarrays are used to follow genetic response to environmental and nutritional changes. For example, *P. furiosus*, like other hyperthermophiles, lacks a phosphotransferase system for carbohydrate uptake, instead relying on adenosine-triphosphate binding cassette transporters (see Fig. 2). Glycosidase–transporter couplings can be used to track cellular response to various carbohydrates, thereby providing hints to yet to be annotated genes encoding enzymes capable of hydrolyzing substrates of interest. Thus, new enzymes involved in various stages of polysaccharide hydrolysis can be identified from differential expression analysis on specific carbohydrates. Similar approaches for other types of enzymes can be used if sufficient insight into the organism's physiological patterns is available.

Much has changed since the early days of enzyme discovery. The arrival of microbial genomics will no doubt stimulate creative approaches to finding biocatalysts that until this point have been hidden within certain microbial genotypes. By combining newly developed methods for high-throughput screening, directed evolution, and biocatalyst production with bioinformatics tools, microbial genomes can be fully utilized for significant technological advances related to important biotransformations.

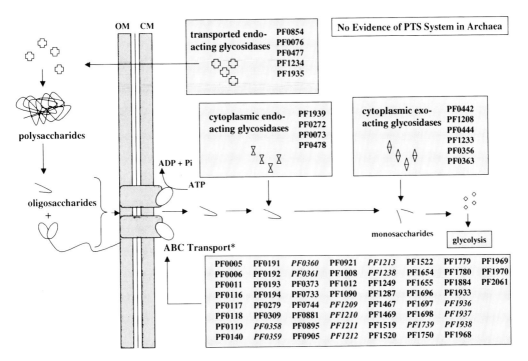

Fig. 2. Sugar metabolism in *Pyrococcus furiosus* represented by all known intracellular and extracellular glycosidases and (ABC) transporters identified from genome sequence information and the on-line CAZY database (http://afmb.cnrs-mrs.fr/~cazy/CAZY/), which classifies glycosidases according to the family-based scheme of ref. *25*. *Known and putative ABC transport proteins from the *P. furious* genome are given, including those in the annotation listed as belonging to the sugar transport or carbohydrate uptake transporter family and the dipeptide/oligopeptide transport (Opp-) family *(90)*. Transporters that are characterized or are located close to known or putative glycosidases on the *P. furiousus* genome are shown in italics. OM, outer membrane; CM, cell membrane; PTS, phosphotransferase system; Pi, free phosphate.

ACKNOWLEDGMENTS

This work was supported in part by grants from the Biotechnology Program, National Science Foundation, and the Energy Biosciences Program, US Department of Energy.

REFERENCES

1. Dordick JS. The general uses of biocatalysts. In: Dordick JS (ed). Biocatalysts for Industry. New York: Plenum Press, 1991, pp. 1–19.

2. Neidleman SL. Historical perspective on the industrial uses of biocatalysts. In: Dordick JS (ed). Biocatalysts for Industry. New York: Plenum Press, 1991, pp. 21–33.

3. Uhlig H. Industrial Enzymes and Their Applications. Linsmaier-Bednar (EM (trans). New York: Wiley, 1998.

4. Marrs B, Delagrave S, Murphy D. Novel approaches for discovering industrial enzymes. Curr Opin Microbiol 1999; 2:241–245.

5. Arnold FH, Volkov AA. Directed evolution of biocatalysts. Curr Opin Chem Biol 1999; 3: 54–59.

6. Crameri A, Raillard SA, Bermudez E, Stemmer WP. DNA shuffling of a family of genes from diverse species accelerates directed evolution. Nature 1998; 391:288–291.

7. Aaman RI, Ludwig W, Schleifer K-H. Phylogenetic identification and *in situ* detection of individual microbial cells without cultivation. Microbiol Rev 1995; 59:143–169.

8. Hugenholtz P, Goebel BM, Pace NR. Impact of culture-independent studies on the emerging phylogenetic view of bacterial diversity. J Bacteriol 1998; 180:4765–4774.

9. Bull AT, Ward AC, Goodfellow M. Search and discovery strategies for biotechnology: the paradigm shift. Microbiol Mol Biol Rev 2000; 64:573–606.

10. Boucher Y, Nesbo CL, Doolittle WF. Microbial genomes: dealing with diversity. Curr Opin Microbiol 2001; 4:285–289.

11. Stephens RS, Kalman S, Lammel C, et al. Genome sequence of an obligate intracellular pathogen of humans: *Chlamydia trachomatis*. Science 1998; 282:754–759.

12. Alm RA, Ling LS, Moir DT, et al. Genomic-sequence comparison of two unrelated isolates of the human gastric pathogen *Helicobacter pylori*. Nature 1999; 397:176–180.

13. Alm RA, Trust TJ. Analysis of the genetic diversity of *Helicobacter pylori*: the tale of two genomes. J Mol Med 1999; 77:834–846.

14. Kalman S, Mitchell W, Marathe R, et al. Comparative genomes of *Chlamydia pneumoniae* and *C. trachomatis*. Nat Genet 1999; 21:385–389.

15. Read TD, Brunham RC, Shen C, et al. Genome sequences of *Chlamydia trachomatis* MoPn and *Chlamydia pneumoniae* AR39. Nucleic Acids Res 2000; 28:1397–1406.

16. Robb FT, Maeder DL, Brown JR, et al. Genomic sequence of hyperthermophile, *Pyrococcus furiosus*: implications for physiology and enzymology. Methods Enzymol 2001; 330:134–157.

17. Kawarabayasi Y, Sawada M, Horikawa H, et al. Complete sequence and gene organization of the genome of a hyper-thermophilic archaebacterium, *Pyrococcus horikoshii* OT3 (supplement). DNA Res 1998; 5:147–155.

18. Lecompte O, Ripp R, Puzos-Barbe V, et al. Genome evolution at the genus level: comparison of three complete genomes of hyperthermophilic archaea. Genome Res 2001; 11:981–993.

19. Dean PM, Zanders ED, Bailey DS. Industrial-scale, genomics-based drug design and discovery. Trends Biotechnol 2001; 19:288–292.

20. Tang CM, Moxon ER. The impact of microbial genomics on antimicrobial drug development. Annu Rev Genomics Hum Genet 2001; 2:259–269.

21. The Institute for Genomic Research. Completed genomes available at: http://www.tigr.org/tigr-scripts/CMR2/CMRGenomes.spl; genomes in progress available at: http://www.tigr.org/tdb/mdb/mdbinprogress.html. Accessed November 24, 2003.

22. Blattner FR, Plunkett G, Bloch CA, et al. The complete genome sequence of *Escherichia coli* K-12. Science 1997; 277:1453–1462.

23. She Q, Singh RK, Confalonieri F, et al. The complete genome of the crenarchaeon *Sulfolobus solfataricus* P2. Proc Natl Acad Sci USA 2001; 98:7835–7840.

24. Kawarabayasi Y. Genome of *Pyrococcus horikoshii* OT3. Methods Enzymol 2001; 330:124–134.

25. Henrissat B, Teeri TT, Warren RA. A scheme for designating enzymes that hydrolyse the polysaccharides in the cell walls of plants. FEBS Lett 1998; 425:352–354.

26. Adams MW, Perler FB, Kelly RM. Extremozymes: expanding the limits of biocatalysis. Biotechnology 1995; 13:662–668.

27. Zamost BL, Nielsen HK, Starnes RL. Thermostable enzymes for industrial applications. J Indust Microbiol 1991; 8:71–82.

28. Stetter KO. Hyperthermophiles: isolation, classification, and properties. In: Horikoshi K, Grant WD (eds). Extremophiles: Microbial Life in Extreme Environments. New York: Wiley-Liss, 1998, pp. 1–24.

29. Demirjian DC, Moris-Varas F, Cassidy CS. Enzymes from extremophiles. Curr Opin Chem Biol 2001; 5:144–151.

30. Kalisz MH. Microbial proteinases. Adv Biochem Eng Biotechnol 1988; 36:17–55.

31. Niehaus F, Bertoldo C, Kahler M, Antranikian G. Extremophiles as a source of novel enzymes for industrial application. Appl Microbiol Biotechnol 1999; 51:711–729.

32. von der Osten C, Branner S, Hastrup S, et al. Protein engineering of subtilisins to improve stability in detergent formulations. J Biotechnol 1993; 28:55–68.

33. Cowan DA. Protein stability at high temperatures. Essays Biochem 1995; 29:193–207.

34. Altschul SF, Gish W, Miller W, Myers EW, Lipman DJ. Basic local alignment search tool. J Mol Biol 1990; 215:403–410.

35. Barrett AJ, Rawlings ND, O'Brien EA. The MEROPS database as a protease information system. J Struct Biol 2001; 134:95–102.

36. Barrett AJ, Rawlings ND, Woessner JF. Handbook of Proteolytic Enzymes. London: Academic, 1998.

37. Ward DE, Shockley KR, Chang LS, et al. Proteolysis in hyperthermophilic microorganisms. Archaea 2002; 1:63–74.

38. Costantino HR, Brown SH, Kelly RM. Purification and characterization of an alpha-glucosidase from a hyperthermophilic archaebacterium, *Pyrococcus furiosus*, exhibiting a temperature optimum of 105 to 115 degrees C. J Bacteriol 1990; 172:3654–3660.

39. Brown SH. Saccharidases from high-temperature bacteria: physiological and enzymological studies. PhD thesis, Johns Hopkins University, Baltimore, MD, 1992.

40. Bauer MW, Driskill LE, Kelly RM. Glycosyl hydrolases from hyperthermophilic microorganisms. Curr Opin Biotechnol 1998; 9:141–145.

41. Driskill LE, Kusy K, Bauer MW, Kelly RM. Relationship between glycosyl hydrolase inventory and growth physiology of the hyperthermophile *Pyrococcus furiosus* on carbohydrate-based media. Appl Environ Microbiol 1999; 65:893–897.

42. Chhabra SR, Shockley KR, Ward DE, Kelly RM. Regulation of endo-acting glycosyl hydrolases in the hyperthermophilic bacterium *Thermotoga maritima* grown on glucan- and mannan-based polysaccharides. Appl Environ Microbiol 2002; 68:545–554.

43. Weiss RB. Direct Submission: *Pyrococcus furiosus* genomic sequence. Salt Lake City, UT: Human Genetics, University of Utah, 2002.

44. Klenk HP, Clayton RA, Tomb JF, et al. The complete genome sequence of the hyperthermophilic, sulphate-reducing archaeon *Archaeoglobus fulgidus*. Nature 1997; 390:364–370.

44a. Gao J, Bauer MW, Shockley KR, Pysz MA, Kelly RM. Growth of hyperthermophilic archaeon *Pyrococcus furiosus* on chitin involves two family 18 chitinases. Appl Environ Microbiol 2003; 69:3119–3128.

45. Bult CJ, White O, Olsen GJ, et al. Complete genome sequence of the methanogenic archaeon, *Methanococcus jannaschii*. Science 1996; 273:1058–1073.

46. Pennisi E. Keeping genome databases clean and up to date. Science 1999; 286:447–450.

47. Pallen MJ. Microbial genomes. Mol Microbiol 1999; 32:907–912.

48. Sassetti C, Rubin EJ. Genomic analyses of microbial virulence. Curr Opin Microbiol 2002; 5:27–32.

49. Sehgal D. Effect of reaction environment on biocatalysis and enantioselectivity of hyperthermophilic esterases. PhD thesis, North Carolina State University, Raleigh, NC, 2002.

50. Rost B. TOPITS: threading one-dimensional predictions into three-dimensional structures. Proc Int Conf Intell Syst Mol Biol 1995; 3:314–321.

51. Sehgal AC, Callen W, Mathur EJ, Short JM, Kelly RM. Carboxylesterase from *Sulfolobus solfataricus* P1. Methods Enzymol 2001; 330:461–471.

52. Sehgal AC, Kelly RM. Enantiomeric resolution of 2-aryl propionic esters with hyperthermophilic and mesophilic esterases: contrasting thermodynamic mechanisms. J Am Chem Soc 2002; 124:8190–8191.

53. Hofmann K, Bucher P, Falquet L, Bairoch A. The PROSITE database, its status in 1999. Nucleic Acids Res 1999; 27:215–219.

54. Fetrow JS, Skolnick J. Method for prediction of protein function from sequence using the sequence-to-structure-to-function paradigm with application to glutaredoxins/thioredoxins and T1 ribonucleases. J Mol Biol 1998; 281:949–968.

55. Nevill-Manning CG, Wu TD, Brutlag DL. Highly specific protein sequence motifs for genome analysis. Proc Natl Acad Sci USA 1998; 95:5865–5871.

56. Marcotte EM, Pellegrini M, Thompson MJ, Yeates TO, Eisenberg D. A combined algorithm for genome-wide prediction of protein function. Nature 1999; 402:83–86.

57. Jomaa H, Wiesner J, Sanderbrand S, et al. Inhibitors of the nonmevalonate pathway of isoprenoid biosynthesis as antimalarial drugs. Science 1999; 285:1573–1576.

58. Kornberg RD. Eukaryotic transcriptional control. Trends Biochem Sci 2000; 24:M46–M49.

59. Klug WS, Cummings MR. Genetics. Upper Saddle River, NJ: Prentice-Hall, 2000.

60. Kotra LP, Vakulenko S, Mobashery S. From genes to sequences to antibiotics: prospects for future developments from microbial genomics. Microbes Infect 2000; 2:651–658.

61. Neidhardt FC, Ingraham JL, Schaechter M. Physiology of the Bacterial Cell: A Molecular Approach. Sunderland, MA: Sinauer Associates, 1990.

62. Thomm M. Archaeal transcription factors and their role in transcription initiation. FEMS Microbiol Rev 1996; 18:159–171.

63. Huber R, Langworthy TA, Konig H, et al. *Thermotoga maritima* sp nov represent a new genus of unique extremely thermophilic eubacteria growing up to 90°C. Arch Microbiol 1986; 144:324–333.

64. Kanehisa M, Goto S, Kawashima S, Nakaya A. The KEGG databases at GenomeNet. Nucleic Acids Res 2002; 30:42–46.

65. Karp PD, Riley M, Saier M, et al. The EcoCyc Database. Nucleic Acids Res 2002; 30:56–58.

66. Richmond CS, Glasner JD, Mau R, Jin H, Blattner FR. Genome-wide expression profiling in *Escherichia coli* K-12. Nucleic Acids Res 1999; 27:3821–3835.

67. Tao H, Bausch C, Richmond C, Blattner FR, Conway T. Functional genomics: expression analysis of *Escherichia coli* growing on minimal and rich media. J Bacteriol 1999; 181:6425–6440.

68. Helmann JD, Wu MF, Kobel PA, et al. Global transcriptional response of *Bacillus subtilis* to heat shock. J Bacteriol 2001; 183:7318–7328.

69. Merrell DS, Butler SM, Qadri F, et al. Host-induced epidemic spread of the cholera bacterium. Nature 2002; 417:642–645.

70. Oh MK, Liao JC. DNA microarray detection of metabolic responses to protein overproduction in *Escherichia coli*. Metab Eng 2000; 2:201–209.

71. Oh MK, Liao JC. Gene expression profiling by DNA microarrays and metabolic fluxes in *Escherichia coli*. Biotechnol Prog 2000; 16:278–286.

72. Loos A, Glanemann C, Willis LB, et al. Development and validation of *Corynebacterium* DNA microarrays. Appl Environ Microbiol 2001; 67:2310–2318.

73. Oh MK, Rohlin L, Kao KC, Liao JC. Global expression profiling of acetate-grown *Escherichia coli*. J Biol Chem 2002; 277:13,175–13,183.

74. Schena M, Shalon D, Davis RW, Brown PO. Quantitative monitoring of gene expression patterns with a complementary DNA microarray. Science 1995; 270:467–470.

75. Schut GJ, Zhou JZ, Adams MWW. DNA microarray analysis of the hyperthermophilic archaeon *Pyrococcus furiosus*: evidence for a new type of sulfur-reducing enzyme complex. J Bacteriol 2001; 183:7027–7036.

76. ter Linde JJ, Liang H, Davis RW, Steensma HY, van Dijken JP, Pronk JT. Genome-wide transcriptional analysis of aerobic and anaerobic chemostat cultures of *Saccharomyces cerevisiae*. J Bacteriol 1999; 181:7409–7413.

77. Lashkari DA, DeRisi JL, McCusker JH, et al. Yeast microarrays for genome wide parallel genetic and gene expression analysis. Proc Natl Acad Sci USA 1997; 94:13,057–13,062.

78. Hughes TR, Marton MJ, Jones AR, et al. Functional discovery via a compendium of expression profiles. Cell 2000; 102:109–126.

79. Desprez T, Amselem J, Caboche M, Hofte H. Differential gene expression in *Arabidopsis* monitored using cDNA arrays. Plant J 1998; 14:643–652.

80. Kawasaki S, Borchert C, Deyholos M, et al. Gene expression profiles during the initial phase of salt stress in rice. Plant Cell 2001; 13:889–905.

81. Jin W, Riley RM, Wolfinger RD, White KP, Passador-Gurgel G, Gibson G. The contributions of sex, genotype and age to transcriptional variance in *Drosophila melanogaster*. Nat Genet 2001; 29:389–395.

82. Kaminski N, Allard JD, Pittet JF, et al. Global analysis of gene expression in pulmonary fibrosis reveals distinct programs regulating lung inflammation and fibrosis. Proc Natl Acad Sci USA 2000; 97:1778–1783.

83. Schena M, Shalon D, Heller R, Chai A, Brown PO, Davis RW. Parallel human genome analysis: microarray-based expression monitoring of 1000 genes. Proc Natl Acad Sci USA 1996; 93:10,614–10,619.

84. Alizadeh AA, Eisen MB, Davis RE, et al. Distinct types of diffuse large B-cell lymphoma identified by gene expression profiling. Nature 2000; 403:503–511.

85. Coller HA, Grandori C, Tamayo P, et al. Expression analysis with oligonucleotide microarrays reveals that MYC regulates genes involved in growth, cell cycle, signaling, and adhesion. Proc Natl Acad Sci USA 2000; 97:3260–3265.

86. Chhabra SR, Shockley KR, Conners SB, Scott K, Wolfinger RD, Kelly RM. Carbohydrate-induced differential gene expression patterns in the hyperthermophilic bacterium *Thermotoga maritima*. J Biol Chem 2003; 278:7740–7752.

87. Demirjian DC, Shah PC, Moris-Varas F. Screening for novel enzymes. Top Curr Chem 1999; 200:1–29.

88. Dautin N, Karimova G, Ullmann A, Ladant D. Sensitive genetic screen for protease activity based on a cyclic AMP signaling cascade in *Escherichia coli*. J Bacteriol 2000; 182:7060–7066.

89. Yano T, Kagamiyama H. Directed evolution of ampicillin-resistant activity from a functionally unrelated DNA fragment: a laboratory model of molecular evolution. Proc Natl Acad Sci USA 2001; 98:903–907.

90. Koning SM, Albers SV, Konings WN, Driessen AJM. Sugar transport in (hyper)thermophilic archaea. Res Microbiol 2002; 153:61–67.

91. Bateman A, Birney E, Cerruti L, et al. The Pfam protein families database. Nucleic Acids Res 2002; 30:276–280.

92. Henikoff S, Henikoff JG. Automated assembly of protein blocks for database searching. Nucleic Acids Res 1991; 19:6565–6572.

93. Huang JY, Brutlag DL. The EMOTIF database. Nucleic Acids Res 2001; 29:202–204.

94. Schultz J, Copley RR, Doerks T, Ponting CP, Bork P. SMART: a web-based tool for the study of genetically mobile domains. Nucleic Acids Res 2000; 28:231–234.

95. Attwood TK, Beck ME. PRINTS—a protein motif fingerprint database. Protein Eng 1994; 7: 841–848.

96. Attwood TK, Beck ME, Bleasby AJ, Parry-Smith DJ. PRINTS—a database of protein motif fingerprints. Nucleic Acids Res 1994; 22:3590–3596.

97. Marchler-Bauer A, Panchenko AR, Shoemaker BA, Thiessen PA, Geer LY, Bryant SH. CDD: a database of conserved domain alignments with links to domain three-dimensional structure. Nucleic Acids Res 2002; 30:281–283.

98. Snel B, Lehmann G, Bork P, Huynen MA. STRING: a Web-server to retrieve and display the repeatedly occurring neighbourhood of a gene. Nucleic Acids Res 2000; 28:3442–3444.

99. Tatusov RL, Koonin EV, Lipman DJ. A genomic perspective on protein families. Science 1997; 278:631–637.

100. Tatusov RL, Natale DA, Garkavtsev IV, et al. The COG database: new developments in phylogenetic classification of proteins from complete genomes. Nucleic Acids Res 2001; 29:22–28.

101. Frishman D, Albermann K, Hani J, et al. Functional and structural genomics using PEDANT. Bioinformatics 2001; 17:44–57.

102. Roth FP, Hughes JD, Estep PW, Church GM. Finding DNA regulatory motifs within unaligned noncoding sequences clustered by whole-genome mRNA quantitation. Nat Biotechnol 1998; 16:939–945.

103. McGuire AM, Hughes JD, Church GM. Conservation of DNA regulatory motifs and discovery of new motifs in microbial genomes. Genome Res 2000; 10:744–757.

104. Fumoto M, Miyazaki S, Sugawara H. Genome Information Broker (GIB): data retrieval and comparative analysis system for completed microbial genomes and more. Nucleic Acids Res 2002; 30:66–68.

105. Bairoch A. The ENZYME database in 2000. Nucleic Acids Res 2000; 28:304–305.

106. Goto S, Okuno Y, Hattori M, Nishioka T, Kanehisa M. LIGAND: database of chemical compounds and reactions in biological pathways. Nucleic Acids Res 2002; 30:402–404.

107. Overbeek R, Larsen N, Pusch GD, et al. WIT: integrated system for high-throughput genome sequence analysis and metabolic reconstruction. Nucleic Acids Res 2000; 28:123–125.

108. Nielsen H, Engelbrecht J, Brunak S, von Heijne G. A neural network method for identification of prokaryotic and eukaryotic signal peptides and prediction of their cleavage sites. Int J Neural Syst 1997; 8:581–599.

109. Hoffmann K, Stoffel W. TMbase—a database of membrane spanning proteins segments. Biol Chem Hoppe-Seyler 1993; 347:166.

110. Schomburg I, Chang A, Schomburg D. BRENDA, enzyme data and metabolic information. Nucleic Acids Res 2002; 30:47–49.

111. Thompson JD, Higgins DG, Gibson TJ. CLUSTAL W: improving the sensitivity of progressive multiple sequence alignment through sequence weighting, position-specific gap penalties and weight matrix choice. Nucleic Acids Res 1994; 22:4673–4680.

112. Nakai K, Horton P. PSORT: a program for detecting sorting signals in proteins and predicting their subcellular localization. Trends Biochem Sci 1999; 24:34–36.

113. Krogh A, Larsson B, von Heijne G, Sonnhammer EL. Predicting transmembrane protein topology with a hidden Markov model: application to complete genomes. J Mol Biol 2001; 305: 567–580.

114. Cserzo M, Wallin E, Simon I, von Heijne G, Elofsson A. Prediction of transmembrane α-helices in prokaryotic membrane proteins: the dense alignment surface method. Protein Eng 1997; 10: 673–676.

115. Haft DH, Loftus BJ, Richardson DL, et al. TIGRFAMs: a protein family resource for the functional identification of proteins. Nucleic Acids Res 2001; 29:41–43.

116. Baxevanis AD. The Molecular Biology Database Collection: 2002 update. Nucleic Acids Res 2002; 30:1–12.

117. Nelson KE, Paulsen IT, Heidelberg JF, Fraser CM. Status of genome projects for nonpathogenic bacteria and archaea. Nat Biotechnol 2000; 18:1049–1054.

118. Kawarabayasi Y, Hino Y, Horikawa H, et al. Complete genome sequence of an aerobic hyperthermophilic crenarchaeon, *Aeropyrum pernix* K1. DNA Res 1999; 6:83–101, 145–152.

119. Direct submission: *Pyrococcus abyssi* Genomic Sequence. National Center for Biotechnology Information, National Institutes of Health, Bethesda, MD, 2001.

120. Fitz-Gibbon ST, Ladner H, Kim UJ, Stetter KO, Simon MI, Miller JH. Genome sequence of the hyperthermophilic crenarchaeon *Pyrobaculum aerophilum*. Proc Natl Acad Sci USA 2002; 99:984–989.

121. Deckert G, Warren PV, Gaasterland T, et al. The complete genome of the hyperthermophilic bacterium *Aquifex aeolicus*. Nature 1998; 392:353–358.

122. Nelson KE, Clayton RA, Gill SR, et al. Evidence for lateral gene transfer between archaea and bacteria from genome sequence of *Thermotoga maritima*. Nature 1999; 399:323–329.

123. Voorhorst WGB, Rik IL, Luesink EJ, Devos WM. Characterization of the celB gene coding for β-glucosidase from the hyperthermophilic archaeon *Pyrococcus furiosus* and its expression and site-directed mutation in *Escherichia coli*. J Bacteriol 1995; 177:7105–7111.

124. Kengen SWM, Luesink EJ, Stams AJM, Zehnder AJB. Purification and characterization of an extremely thermostable β-glucosidase from the hyperthermophilic archaeon *Pyrococcus furiosus*. Eur J Biochem 1993; 213:305–312.

125. Verhees CH. Direct submission: *Pyrococcus furiosus* Genomic Sequence. Laboratory of Microbiology, Wageningen University and Research Center, Wageningen, The Netherlands, 1999.

126. Bauer MW, Driskill LE, Callen W, Snead MA, Mathur EJ, Kelly RM. An endoglucanase, eglA, from the hyperthermophilic archaeon *Pyrococcus furiosus* hydrolyzes β-1,4 bonds in mixed-linkage (1->3),(1->4)-β-D-glucans and cellulose. J Bacteriol 1999; 181:284–290.

127. Bauer MW, Bylina EJ, Swanson RV, Kelly RM. Comparison of a β-glucosidase and a β-mannosidase from the hyperthermophilic archaeon *Pyrococcus furiosus*. Purification, characterization, gene cloning, and sequence analysis. J Biol Chem 1996; 271:23,749–23,755.

128. Gueguen Y, Voorhorst WGB, van der Oost J, deVos WM. Molecular and biochemical characterization of an endo-β-1,3-glucanase of the hyperthermophilic archaeon *Pyrococcus furiosus*. J Biol Chem 1997; 272:31,258–31,264.

129. Tanaka T, Fujiwara S, Nishikori S, Fukui T, Takagi M, Imanaka T. A unique chitinase with dual active sites and triple substrate binding sites from the hyperthermophilic archaeon *Pyrococcus kodakaraensis* KOD1. Appl Environ Microbiol 1999; 65:5338–5344.

130. Jorgensen S, Vorgias CE, Antranikian G. Cloning, sequencing, characterization, and expression of an extracellular α-amylase from the hyperthermophilic archaeon *Pyrococcus furiosus* in *Escherichia coli* and *Bacillus subtilis*. J Biol Chem 1997; 272:16,335–16,342.

131. Savchenko A, Vieille C, Kang S, Zeikus JG. *Pyrococcus furiosus* α-amylase is stabilized by calcium and zinc. Biochemistry 2002; 41:6193–6201.

132. Laderman K, Davis B, Krutzsch H, et al. The purification and characterization of an extremely thermostable α-amylase from the hyperthermophilic archaebacterium *Pyrococcus furiosus*. J Biol Chem 1993; 268:24,394–24,401.

133. Dong GQ, Vieille C, Zeikus JG. Cloning, sequencing, and expression of the gene encoding amylopullulanase from *Pyrococcus furiosus* and biochemical characterization of the recombinant enzyme. Appl Environ Microbiol 1997; 63:3577–3584.

26

Integration of Genomics
in the Drug Discovery Process

Jacques Ravel

INTRODUCTION

Over the last 60 years, the discovery of antibiotics has made an extraordinary contribution to medicine, especially in the treatment of bacterial infectious diseases. These compounds were in large part natural products isolated from soil microorganisms, especially from members of the genus *Streptomyces*, which have provided two-thirds of the naturally derived antibiotics in current use *(1)*. Despite the large number of antibiotics available today, bacterial infectious diseases remain one of the leading causes of death worldwide *(2)*.

At the end of the 20th century, there were outbreaks and epidemics of new infectious diseases as well as the reemergence of old ones thought to be under control *(3)*. Some microorganisms previously considered innocuous have recently evolved strains of powerful pathogens, such as *Escherichia coli* and the foodborne pathogen *E. coli* OH:157, outbreaks of which have been associated with many deaths in the United States *(2,3)*. This trend represents a serious threat to public health and is compounded by the increase in antibiotic-resistant clinical isolates.

The use (and misuse) of antibiotics provides a powerful selection pressure for pathogenic microorganisms to develop or acquire resistance to overcome the action of many antibiotics. Antibacterials are now the second most commonly prescribed category of drugs *(4)*. The increased use of antibiotics, combined with other human social and economic factors, has compromised the efficacy of some of the most powerful antibiotics in the clinician's tool kit. The recent isolation in the US of vancomycin-resistant enterococci and staphylococci *(5)*, penicillin-resistant pneumococci *(6)*, and multidrug-resistant *Mycobacterium tuberculosis* is extremely alarming *(7)*.

Ironically, the success of antibiotics as therapeutics has slowed the production and discovery of new chemical classes. Rather, the recent successes in the field have resulted from improving existing antibiotics by direct chemical modification (e.g., second- and third-generation cephalosporins) or the use of secondary inhibitors such as clavulanic acid, a β-lactamase inhibitor used in combination with β-lactam antibiotics. Until the recent approval by the US Food and Drug Administration of a new oxazolidinone antibiotic linezolid *(8)*, no new class of antibiotics had been developed in 30 years. Disturbingly, clinical isolates of *Staphylococcus aureus* that are resistant to linezolid (and many other antibiotics, including vancomycin) have already been reported *(5)*.

From: *Microbial Genomes*
Edited by: C. M. Fraser, T. D. Read, and K. E. Nelson © Humana Press Inc., Totowa, NJ

The dark picture outlined above highlights the need for novel antibacterial compounds. The pharmaceutical industry has now awakened to the need for new therapies, and there is growing interest in the development of novel approaches for the discovery of the next generations of antibacterial drugs. Since the completion of the first microbial genome sequence of *Haemophilus influenzae* in 1995 *(9)*, more than 170 microbial genomes have been sequenced, and at least another 330 are in progress (http://wit.integrated genomics.com/GOLD/). This global effort has focused primarily on pathogens, which encompass the majority of all genome projects, and has generated a large amount of raw material for *in silico* analysis. In addition, multiple strains of the same organism or multiple species of the same genus are being sequenced or have been completely sequenced, opening the possibility to use comparative genomics tools for discovering novel drug targets. A major challenge in the postgenomic era is to exploit and decipher this new wealth of information fully to fulfill the urgent need for novel antimicrobial agents.

Rapid and revolutionary developments in genome sciences and bioinformatics are now poised to have a major impact on drug discovery. Genomics is now an integral part of the drug discovery pipeline for the identification of novel drug targets, but it is also applied to the discovery of novel microbial natural products. Some of the novel aspects and technologies are reviewed in this chapter.

TARGETS FOR DRUG DISCOVERY/DESIGN

The advent of genomics research has revitalized the hope for the development of new classes of antibacterial agents and other categories of drugs. It is now possible to use bioinformatics tools to analyze genome sequence data and identify novel putative targets with potential for therapeutic intervention. Prioritizing targets using bioinformatics technologies is now commonplace in the industry, but represents just the first step in a very long process. The subsequent steps, including target validation, assay development, and small molecule library screening, still represent the most time- and labor-intensive parts of the drug discovery process. It can take years between the identification of a target and the selection of a drug candidate (Fig. 1).

Target Identification Using In Silico Analysis

What Defines a "Good" Target Gene

Comparative genomics tools have made it possible for the bioinformatician to identify open reading frames (ORFs) that represent potential targets for drug discovery. There is a general agreement in the literature as to the criteria for a good antimicrobial drug target: essentially, selectivity, spectrum, practicality, and functionality.

ESSENTIALITY

The target must be essential for the growth, replication, viability, or survival of the microorganism in the laboratory because primary assays for antimicrobial activity are typically performed under such conditions. However, other criteria need to be considered as different growth conditions may compensate for the loss of certain gene functions. The target can also be essential if it is required for survival of the bacterium in the susceptible host because the gene product might be needed to establish and maintain infection *(10)*. This implies that virulence-associated genes could be considered

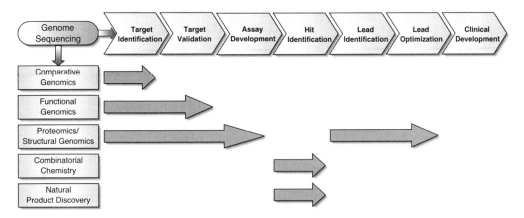

Fig. 1. Genome sciences are an integral part of the drug discovery pipeline. Many aspects, from genome sequencing to genome analysis and structural genomics, are used at different stages. Time-scale from target identification to clinical development is about 10 years.

essential, a fact that is still questioned because functional assignment is difficult to make *(11)*.

In addition, functional paralogs have to be considered as they might be essential only when inhibited in concert. For example, the double mutant of two Gram-negative helicases, UvrD and Rep, is not viable, whereas single mutants are viable *(12)*. Also, no ortholog should exist in genomes of microbes sharing the same environment with the targeted pathogens as transmission of the substitute from a nonpathogen to a pathogen isolate could rapidly negate the effect of the drug. This exemplifies the fact that genomic technology is still better served when intimately coupled with knowledge of microbial physiology.

SELECTIVITY

The microbial target should not have any well-conserved homolog in the mammalian host to be treated to reduce cytotoxicity issues. The recent completion of the human and mouse genomes therefore represents a major step for the drug industry *(13,14)*. Ideally, the selected targets should represent a unique feature of the biochemistry of the microorganism.

However, targeting a unique activity does not guarantee nontoxicity to the host. For example, uridine diphosphate (UDP)-*N*-acetylglucosamine enolpyruvyl transferase (MurA), an enzyme responsible for an early step in cell wall biosynthesis, has no homolog in the human genome. However, fosfomycin, a potent inhibitor of this enzyme, is poorly tolerated *(15)*. On the other hand, the well-established drug trimethoprim inhibits dihydrofolate reductase *(16)*, a key enzyme in the synthesis of thymidylate and, therefore, of deoxyribonucleic acid (DNA). Although this enzyme has a human homolog, trimethoprim is still well tolerated without any major toxicity *(15)*.

Selection of potential targets by primary (amino acid) sequence comparison alone has many limitations as similar protein folds (and hence inhibitor-binding sites) may be found in proteins that have only weak primary sequence homology. Therefore, secondary and tertiary structure information should also be used when assessing target selectivity.

SPECTRUM

Targets can be separated into two groups depending on their distribution among microorganisms and the expected spectrum of potential inhibitors: (1) Targets that are present in a large number of pathogens should be useful in the discovery of broad-spectrum agents. In view of the challenges that exist in rapidly identifying the causative agent of an infection in clinical situations, a single treatment with a broad-spectrum agent may be more appropriate, cost-effective, and clinically easy to use *(17)*. (2) Targets specific to one or a subgroup of pathogens can be used to develop agents that may have narrower spectra of activity. Narrow-spectrum agents may be extremely useful in the context of chronic infections or long-term therapy to minimize detrimental effects on the normal gastrointestinal microbial flora. Moreover, the use of narrow-spectrum antimicrobials may also minimize the transfer of drug resistance to other pathogens *(18,19)*.

PRACTICALITY

Once a protein has been selected using bioinformatics tools, a significant amount of work is generally required to determine if it represents a feasible target for drug discovery. Assays must be developed, and for this purpose the protein should be amenable to high-throughput screening. Biochemical assays often rely on soluble proteins, such that some classes of proteins may be difficult to use as targets. For example, membrane proteins tend to be poorly soluble when overexpressed in vitro and present a special challenge in assay development.

Surface-exposed membrane proteins, if essential, represent almost the "perfect" target. Drugs designed to attack such targets do not have to penetrate the cell, thus simplifying chemical design. Such drugs can have increased molecular weight and are easier to optimize. However, as mentioned, testing the efficacy of the drug is difficult because assay development is hampered by poor in vitro expression and solubility. Many antibiotics used in medicine target membrane proteins (e.g., cell wall biosynthesis machinery), but have been discovered through whole-cell assay, with their specific target elucidated afterward. Membrane transporters are numerous in pathogenic bacteria because these microorganisms depend on their host to provide essential nutrients, and they are attractive targets. However, drugs targeting such transporters might also be toxic to humans *(11)*. One exception appears to be the adenosine triphosphate/adenosine 5'-diphosphate translocase of *Chlamydia* and *Rickettsia*, which is more related to plant chloroplast than human mitochondria *(20)*.

FUNCTIONALITY

Information on the function of the target can be critical in generating viable chemical leads from a particular screening assay. For example, functional information can help rationally design a library of chemical compounds to screen for target inhibition. Structural genomics is emerging as a critical part of this process (see the section on structural genomics). In addition, biochemical and physiological information combined with structural data can greatly improve the chances of success.

Genome Comparison and Target Prioritization

As defined in the sections above, potential targets are characterized by a set of parameters that will influence the way genome sequence data is analyzed. These parameters can be applied to whole genome data and the resulting targets prioritized accordingly.

Bioinformatics tools that facilitate this process have been developed in both the private and the public sectors. These tools often take advantage of genome sequence databases that can be queried with user input parameters. These queries are often based on protein sequence similarities using the BLASTP (Basic Local Alignment Search Tool P) algorithm *(21)*.

Other types of comparative analyses make use of conserved functional motifs rather than simple linear alignments and include the use of hidden Markov models *(22,23)*, PSI-BLAST (Position-Specific Iterated BLAST; *21*), PHI-BLAST (Pattern-Hit Initiated BLAST; *24*), or Clusters of Orthologous Groups (COGs; *25*). Approaches taking advantage of the Enzyme Commission (EC) classification have also been reported when proteins with low sequence similarity but similar function (EC number) are identified *(26)*. The HOBACGEN (Homologous Bacterial Genes) database for comparative genomics allows the user to select gene families and select for homology common to the taxa in the database, and it can be useful for identifying orthologous or paralogous genes *(27)*.

The Institute for Genomic Research Comprehensive Microbial Resource (CMR) provides an extensive database of complete microbial genome sequences (http://www.tigr. org/tigr-scripts/CMR2/CMRHomePage.spl). Read et al. *(18)* used the CMR to screen for putative genes for targeted therapy in the genomes of the upper respiratory pathogens *Streptococcus pneumoniae*, *H. influenzae*, and *Neisseria meningitidis*. Comparison of these organisms may also provide insight into common strategies used by microbes to survive in the respiratory tract. Using protein sequence identity (BLASTP) with a threshold of 25% amino acid sequence similarity, they identified a list of 32 proteins that are common to these three pathogens, but absent in the nonpathogenic strains *E. coli* K12 and *Bacillus subtilis*. This analysis revealed genes with unknown functions, as well as known virulence factors conserved in other pathogens, and provides an example of how public on-line databases can be used to perform bioinformatics analysis with user input parameters.

Similarly, using concordance analysis (and the COGs database), Bruccoleri et al. *(28)* identified 89 putative targets for broad-spectrum antibiotics. This set of genes was obtained by identifying all *E. coli* proteins also present in *B. subtilis*, *H. influenzae*, *Helicobacter pylori*, and *M. tuberculosis*. The 89 proteins identified by this analysis include a number of validated antibacterial targets, such as DNA gyrase (*gyrA*), the target of the quinolone class of antibiotics *(29)*, and *murA*, the fosfomycin target *(30)*.

Chalker et al. *(31)* used another comparative genomics approach to identify ORFs that are conserved between two strains of *H. pylori* (26695 and J99) and 13 other highly diverged eubacteria. Using a series of screening parameters, such as BLASTP scores, hydropathy profiles (to eliminate putative membrane proteins), and putative functional assignments, the authors identified 73 ORFs considered to satisfy the definition of a "good" target. Among this set, 45 ORFs spanning several functional categories were chosen to test for essentiality, and 33 of these were shown to be essential in vitro using a rapid vector-free allelic replacement mutagenesis technique. Interestingly, 12 genes found essential were expected to be nonessential by analogy with orthologous genes in other bacteria. The authors suggested that the products of these highly diverged essential genes may be suitable for investigation as potential targets for novel and highly specific anti–*H. pylori* drugs. A combination of stepwise prioritization of ORFs using simple biological criteria directly relevant to antibacterial drug discovery may provide an efficient way to select for potential target genes and to identify those that are critical for normal microbial cell function. This analysis resulted in a rather small set of genes to be tested for essentiality, an advantage in terms of cost-effectiveness of the downstream process.

Given the extensive gaps in the knowledge of microbial physiology, bioinformatics analyses typically yield a large set of ORFs with no functional assignments. These genes may still fulfill the requirements for a good target, even in the absence of functional information. Most drug development programs require some knowledge or prediction of the function of a protein before it can be pursued as a potential target. Thus, a large set of potential targets may be eliminated outright from further study given that, on average, one-third of the ORFs in a sequenced genome have no assigned functions.

However, genomic technologies have been developed that attempt to extract functional information from these unknown proteins. Transcriptional profiling by DNA microarray studies can provide information on the timing of expression of an unknown gene. By varying environmental growth conditions (e.g., growth media, growth in a host, or exposure to antibiotics), valuable information can be gathered that may lead to a functional assignment because genes that share expression patterns often share functions.

Not all cellular processes are controlled at the level of gene expression, however, and protein profiling (proteomics) provides a complementary tool for assessing gene function. In particular, very sensitive and versatile mass spectrometry techniques now make it possible to analyze proteins on a genomewide scale, and these techniques can be used to assess protein levels under different growth conditions *(32–34)*. In addition to being useful in identifying putative functions of unknown genes, these new transcriptome and proteome analysis technologies are also being used to identify and validate novel targets *(15)*.

Testing for Essentiality

To be considered for further study, potential targets identified through bioinformatics analysis are put through a series of biological screens to test for essentiality. As defined in the section on target identification, a protein must be essential for viability of the microorganism in order to be considered a valid target. Genes may be identified as essential under in vitro test conditions (e.g., laboratory growth conditions) or as essential in vivo during infection in animal models.

In vitro essentiality has long been used to identify putative antimicrobial targets. The use of temperature-sensitive mutants of *Salmonella typhimurium* allowed the identification of sets of essential genes required for growth in rich laboratory media *(35)*. These approaches are based on the generation of libraries of random mutants, followed by genetic mapping of the mutations, and are laborious and time consuming.

Sequence-directed approaches provide an attractive alternative for identifying essential genes when a subset of potential targets has already been identified by bioinformatics analysis.

Plasmid insertion mutagenesis makes use of a suicide plasmid (unable to replicate in the host) containing a selectable marker (e.g., antibiotic resistance). A portion of the targeted gene is cloned in the plasmid, which is then transformed into a nonpermissive host. Gene disruption occurs by homologous recombination between the plasmid-borne target gene sequence and the native chromosomal locus. The identification of *S. aureus* genes essential for growth in vitro was achieved using such a strategy *(36)*. High-throughput gene disruption has been achieved in *S. pneumoniae*; of 347 candidate ORFs tested, 113 were found to be essential *(37)*. However, these insertional inactivation techniques suffer some disadvantages, such as polar effects caused by the insertion of the plasmid into an operon, truncated proteins that retain activity, and mutation reversion via plasmid excision.

An alternative method, termed allelic replacement mutagenesis, offers the advantage of achieving stable, complete gene deletion without detrimental polar effects. Allelic replacement constructs consist of a selectable marker cassette flanked on either side by sequences homologous to the regions upstream and downstream of the target locus. The construct can be introduced into wild-type cells directly or by means of a suicide plasmid, for which a double-crossover event occurs by homologous recombination between the two flanking regions of homology. The target locus is deleted and replaced by the selectable marker cassette. If numerous constructs are required, multiple overlapping polymerase chain reaction (PCR) provides a system amenable to automation. PCR primers can be chosen so that the flanking genes and potential promoters remain intact in the deletion mutant, minimizing polar effects. Target genes are scored as essential when three or more independent, controlled transformations fail to produce mutants.

Allelic replacement mutagenesis was used by Chalker et al. to identify genes essential for *H. pylori* growth *(31)*. In a similar study, Wilding et al. *(38)* identified *S. pneumoniae* 3-hydroxy-3-methylglutaryl-coenzyme A synthase as essential for survival in vitro. This enzyme is part of the mevalonate pathway for isopentenyl diphosphate biosynthesis in Gram-positive cocci. Allelic replacement of the 3-hydroxy-3-methylglutaryl-coenzyme A synthase gene with an erythromycin resistance cassette rendered the organism auxotrophic for mevalonate and severely attenuated in a murine respiratory tract infection model *(38)*. Thus, by using genome prioritization and in vitro and in vivo target validation methods, the authors demonstrated that the enzymes involved in the biosynthesis of isopentyl diphosphate may represent good drug targets in Gram-positive cocci.

Structural Genomics

Structural genomics is a newcomer to the world of "omics." As highlighted in the previous sections, a major challenge in the postgenomic era is to exploit the wealth of information provided by genome-sequencing projects. Structural genomics ultimately aims to provide three-dimensional structure information for every tractable protein *(39)*. The large number of completed microbial genomes, because of their relative simplicity, represents the perfect template to implement this goal. Structural information can prove to be an invaluable guide to the scientist because it can be applied to assign biological roles to unknown proteins and in other ways to help the drug discovery process. Gilliland et al. *(40)* undertook the functional assignment for hypothetical *H. influenzae* gene products through structural genomics, and this effort is actually identifying a number of possible new targets for drug development.

As of October 1, 2002, there were over 18,000 structures in the Protein Data Bank (PDB), the public repository for three-dimensional protein structures *(41)*. However, only about 5000 of these structures represent unique wild-type proteins; the remainder are mutants, duplicates, or enzyme–ligand complexes.

Solving the structure of a particular protein has long been thought to be a slow and time-consuming undertaking, making it inadequate for high-throughput automation.

However, recent development in the fields of genome sequencing, protein expression *(42)*, selenomethionine-labeling techniques, multiwavelength anomalous diffraction phasing, and synchrotron crystallography have greatly facilitated the process of protein structure determination *(43)*. Although X-ray crystallography has clear advantages in defining structures to a finer resolution, nuclear magnetic resonance (NMR) has the ability to analyze proteins in their native state and does not require the often rate-limiting step of crystal formation *(44,45)*.

Overall, the quality and the resolution of protein structures are improving rapidly. Although there are still many challenges ahead, scientists from diverse backgrounds are working together to make high-throughput structure determination a reality. Advances in computer technology and bioinformatics are contributing greatly in providing the tools necessary to manage the large amounts of data generated by three-dimensional structure analysis *(46,47)*.

Structural Genomics Technical Difficulties

A large quantity of soluble proteins is required for three-dimensional structure determination, and protein production is key to any structural genomics project *(48)*. Considerable effort has been focused on automation of the protein expression/purification process. It is expected that, in the next few years, a small research group will be capable of expressing and purifying tens of thousands of proteins per year *(48)*. The development of procedures for the production of a large number of functional proteins for structural analysis will also yield benefits in other fields, such as protein microarray, and in the development of screening assays for drug discovery. Optimization of protein expression and purification in *E. coli* (choice of vector, purification tag, growth media) is aided by the development of high-throughput techniques and the miniaturization of the processes *(42)*.

Despite all these efforts, it is estimated that one-third to one-half of the known prokaryotic proteins cannot be expressed in a soluble form *(48)*. The problems associated with protein expression and purification present a major challenge to structural genomics efforts and target screening programs alike as both require large amounts of soluble protein. One approach *(49)* is to exploit natural protein variants by screening for the most soluble ortholog of a given protein in parallel expression and purification experiments in an *E. coli* expression system (for a recent review, see ref. *50*). The authors found that, in 80% of the proteins studied, at least one and often multiple soluble variants were identified, increasing the chances of obtaining a crystal for X-ray diffraction studies. In addition, multiple subclones of any given gene can be constructed to express individual structural domains, potentially improving solubility and crystal diffraction quality. Although experimental analysis, such as limited proteolysis coupled with mass spectrometry, provided a better indication of domain boundaries than sequence comparisons (e.g., *51*), bioinformatics tools are being developed that may accurately predict such structural information *(52,53)*.

As discussed in section on Practicality, integral membrane proteins such as transporters and cell envelope biosynthesis components provide some of the most promising antimicrobial drug targets. However, because of their poor solubility, they are often excluded from the pool of putative drug targets. Membrane proteins represent one of the greatest challenges for X-ray crystallography and NMR analysis. New technologies such as lipid cubic phase–mediated crystallization are showing promising results in

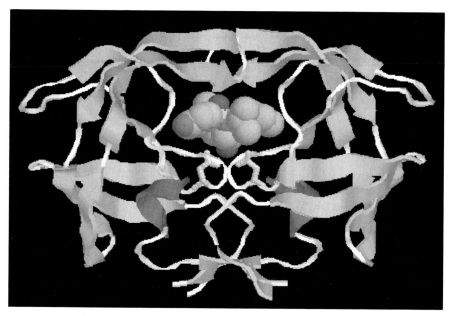

Fig. 2. Structure-based drug design. Amprenavir (space filled), a human immunodeficiency virus protease (ribbon) inhibitor bound in the active site.

this regard *(54,55)*. As a consequence, the number of membrane protein domains in the Protein Data Bank *(41)* should increase rapidly in the future.

Integration of Structural Genomics in the Drug Discovery Process

The pharmaceutical industry now has a track record in using protein crystal structures to design drugs, such as enzyme inhibitors, using rational structure-based approaches. However, this track record has been limited by the difficulty in obtaining high-quality protein structure information. In addition to target discovery, structural genomics will have a major impact on at least two aspects of the drug discovery process: (1) lead identification (the development and screening of chemical libraries of small molecules active against a particular target) and (2) lead optimization (the chemical refinement of a lead molecule to improve selectivity and potency) (Fig. 1). In both cases, access to the three-dimensional structure of the biological target (or a related protein) will allow scientists to generate and improve lead molecules in a more rational and efficient manner *(39)*. Structure-based drug design has generated successfully marketed compounds derived from structure–function studies. Such drugs include the neuramidase inhibitor zanamivir (Relenza), the human immunodeficiency virus protease inhibitors amprenavir (Agenerase) and nefelvir (Viracept) *(43)*, and the Bcr-Abl tyrosine kinase inhibitor imatinib (Gleevec) *(56)* (Fig. 2).

Functional information about the putative targets is extremely valuable in therapeutic design and can be critical to assay development, screening, and validation. It is believed that identifying function first is the key to success in drug discovery in the postgenomic era *(57)*. Because the function of a protein is more a consequence of its three-dimensional shape than its amino acid sequence, the development of structural genomic technologies should play a key role in many drug discovery programs. Function is often inferred by sequence similarities at the DNA or protein level. Such inferences can lead to misannotation of genes and ORFs and misinterpretation of function because it is known that proteins with similarity in amino acid sequence can have different three-dimensional structures and hence different functions *(58)*. Enlarging the pool of three-dimensional structures of proteins will most likely provide structural templates for the majority of known proteins.

When known folds are not obvious from sequence analysis, biochemical and functional information may be gleaned from elucidation and global architecture analysis (e.g., presence of homologous protein folds and level of oligomerization) *(49)*. The active site of *E. coli* FabH, a β-ketoacyl carrier protein synthase, was characterized by the presence of a common fold found in other condensing enzymes *(59)*. Local features such as clusters of polar residues or hydrophobic patches, net charge, and bound small molecule ligand and metal ions may identify functional sites. Surface characteristics such as shape and electrostatic properties can be used to deduce functional traits *(49)*.

As a proof of principle, a combination of these approaches has been used to demonstrate the function of unknown proteins from genome-sequencing projects *(60–63)*. Structural genomics was used to elucidate the role of LuxS, a conserved protein required for signaling by a two-component quorum-sensing pathway and involved in the expression of virulence genes in several Gram-negative and Gram-positive pathogens *(64)*. LuxS acts as a homodimer and cleaves the ribose ring of *S*-ribosylhomocysteine during the biosynthesis of the essential autoinducer molecule AI-2.

This approach is not always successful because a protein fold may be novel, or a particular type of fold may have multiple functions, such as the triose phosphate isomerase barrel fold and the Rossman fold. However, as more protein structures are deposited in databases, the ability to build predictive models that assign function from structure will improve. Web-based initiatives such as Gen3D *(65)* use the CATH structure classification *(66)* superfamily protein (class, architecture, topology, and homologous) to perform protein structure assignments at the whole genome level.

Another goal of structural genomics is to provide high-resolution templates of a representative set of protein domains *(49)*. These templates can then be used to predict protein folds, structures, and ultimately function. Sequence–structure homology models are in development, and programs such as FUGUE are achieving high success in fold prediction *(67,68)*. These predictive models can be applied to target discovery for drug design. Identification of protein folds specific to a pathogenic organism may be of significant value. Three-dimensional structure can vary even among proteins with strong amino acid sequence similarity. Thus, structure analysis may identify unique folds and potential drug-binding sites that would be missed by simple amino acid sequence comparisons. Similarly, stretches of protein that have only limited amino acid similarity may nonetheless adopt a similar three-dimensional fold and thus present similar binding sites for small molecules.

Therefore, further development in structural genomics technologies can be expected to provide new opportunities for drug discovery and design. NMR and crystallographic methods are now routinely applied to probe for ligand-binding interactions, providing alternatives to activity-based screening for identifying small molecules that bind to proteins of unknown function. This type of approach can also be used to obtain functional information by screening libraries of biological compounds (such as cofactors and inhibitors) for binding to a particular uncharacterized protein target.

Structural genomics, although still in its infancy, is developing at a fast pace and is becoming an integral part of the drug discovery process. Knowledge of the structure of the full proteome will help address such issues as drug metabolism and toxicity. Initiatives such as the Ligand Depot database, which will contain information on more than 250,000 small molecules, in combination with other protein structure databases will become increasingly important as more protein structures are solved *(69)*.

NOVEL NATURAL PRODUCT DISCOVERY AND GENOMICS

Not only can genomics be applied to target discovery, but it also can help in the search for novel natural products as lead compounds in the drug development process. The application of genomic developments to drug discovery is rapidly changing the industry. Over the past 60 years, the pharmaceutical industry has exploited the biosynthetic abilities of microorganisms to produce antimicrobials.

Historically, antibiotics were defined as low molecular weight organic natural products, secondary metabolites, made by microorganisms active against other microorganisms at very low concentration *(70)*. Progress in chemistry has somewhat changed that definition, but microorganisms are still the principal source of antimicrobials used in medicine. Many pharmaceutical companies are phasing out their search for microbial secondary metabolites in favor of chemical libraries generated through combinatorial chemistry, in spite of the fact that this technology has yet to deliver a new drug to the market.

Secondary metabolites have evolved in nature in response to the needs and challenges of the natural local environment. Nature is continually carrying out its own version of combinatorial chemistry *(70)*. It should also be kept in mind that microbial diversity is enormous, and that only a small proportion of microorganisms have been examined for the production of secondary metabolites. Nature's chemical biodiversity has barely been scratched, with only an estimated 1 to 10% of the microorganisms able to be cultivated, depending on the environment. Many initiatives have set out to tap into the wealth, and this section highlights a few that take advantage of the progress made in the field of microbial genomics. These novel approaches do not limit themselves to the discovery of antibacterials, but to secondary metabolites in general with activities that are anticancer, antiinflammatory, immunosuppressant, and antiparasitic, and there are those that are herbicides and insecticides.

Genome Scanning

A high-throughput method, termed *genome scanning*, was developed to discover secondary metabolic loci independent of their expression *(71)*. This approach takes advantage of the fact that the genes required for secondary metabolite biosynthesis are typically clustered together in a bacterial genome (Fig. 3). Using this method, all of the natural product gene clusters in a microorganism can be cloned, sequenced, and analyzed without sequencing the entire bacterial genome.

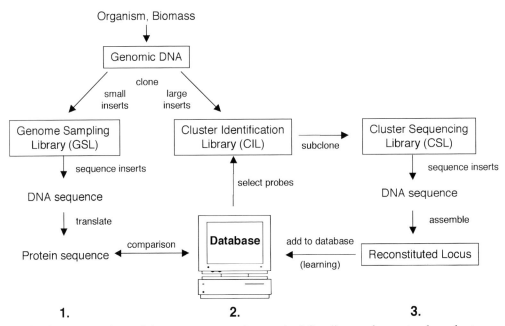

Fig. 3. An overview of the genome-scanning method for discovering natural product gene clusters in microbial genomes *(71)*. Genome scanning provides an efficient way to discover natural product gene clusters without sequencing entire genomes. This approach takes advantage of the fact that the genes required for the biosynthesis of a natural product are typically clustered together in the bacterial genome. (1) High molecular weight genomic DNA is randomly fragmented; small fragments are used to prepare a genome sampling library (GSL) in a plasmid vector; large fragments are used to prepare a cluster identification library (CIL) in a cosmid or BAC vector. A limited number of gene sequences are generated from the GSL clones using a universal primer located in the plasmid vector, and the sequences are compared to a database of natural product biosynthetic genes to identify clones derived from genes involved in natural product biosynthesis. (2) Genes identified by bioinformatics analysis are then used as probes to identify CIL clones containing the corresponding genes as well as their neighboring genes that may constitute a biosynthetic locus. (3) The selected CIL clones are randomly fragmented and used to prepare a second plasmid library that provides further templates for sequencing. Sequencing and assembly of the selected CIL clones results in a complete natural product gene cluster, which is then annotated and entered into the database. The efficiency of the method increases as new natural product gene clusters are entered into the database because the probability that a particular gene cluster will be identified by analysis of a given number of GSL clones is improved if the database contains a large number of gene clusters representing a broad range of natural product classes. The Decipher™ database (Ecopia BioSciences, Inc., Saint-Laurent, Quebec, Canada) was initially populated with gene clusters representing a diverse range of natural product classes collected from the public databases, such as GenBank, and subsequently enriched with new gene clusters discovered by genome scanning.

To date, the genome-scanning method has been used to discover more than 400 natural product gene clusters encoding a structurally diverse range of bioactive molecules, including the antibiotics everninomicin (an oligosaccharide), ramoplanin (a glycosylated lipo-peptide), and rosaramicin (a polyketide), as well as the antitumor antibiotics calicheamicin (an enediyne) and anthramycin (a benzodiazepinone), to name only a few *(72–76)*. Analysis of more than 50 actinomycete strains representing at least 10 genera showed that it is typical to find a dozen or more natural product gene clusters in a single organism, even ones that were previously analyzed by fermentation broth screening and reported to produce only a single natural product (C. Farnet, personal communication, February, 2003).

Thus, although the biosynthetic talents of the actinomycetes have been appreciated for many years, genome scanning indicates that their ability to produce natural products has been vastly underestimated. These findings bode well for the discovery of new natural products from this source and highlight the increasing role that genomics will play in discovering new bioactive molecules from nature.

Exploring the Environmental Metagenome for Novel Natural Products

Cultured microorganisms have produced some of the most biologically active and medically useful compounds. However, rediscovery of already characterized molecules is now limiting the use of culturable organisms (77). As mentioned in section on novel natural product discovery, these organisms represent a minute proportion of the microbial diversity in the environment, and if this diversity is an indicator of the chemical ability of the uncultured microorganisms, many more novel compounds remain to be discovered (see Chapter 23).

To access this untapped source of metabolites, the environmental DNA can be cloned into BAC (bacterial artificial chromosome) vectors, which can stably maintain large fragments of DNA (>100 kb), thereby accessing the collective genomes of the environmental community or metagenome (78). The metagenome represents the genetic and functional diversity of culturable and nonculturable microorganisms of a given environmental sample. These metagenomic BAC libraries are then screened for novel phenotypic expression in a surrogate host such as *E. coli*, *Bacillus* sp, or *Streptomyces* sp. Varying the heterologous host is expected to increase the chances of producing novel activity.

This cultivation-independent genomic approach has been applied successfully to antibacterial agent discovery. Typically, secondary metabolite biosynthetic genes are clustered together with the genes necessary for self-resistance and can be fairly large (>100 kb). The use of BACs maximizes the chances of expression of the entire pathway in the host from a cloned cluster. Once a novel activity has been detected, the genetic material is immediately accessible for analysis and manipulation.

Two new families of natural products derived from long-chain *N*-acyltyrosine have been isolated by screening an environmental DNA library for clones with the ability to biosynthesize biologically active compounds (79). Sequence analysis of the clones and transposon knockout of individual genes showed that a cluster of 13 ORFs was responsible for the activity. Using *Streptomyces lividans* as the heterologous host, a novel family of compounds, the terragines, was discovered from a soil metagenomic library (80).

These reports demonstrate that culture-independent approaches to access environmental chemical diversity have the potential to reveal novel biosynthetic activities. As a proof of principle, violacein, an already known broad-spectrum antibiotic, was produced by an *E. coli* soil metagenomic library, and its biosynthetic cluster was characterized (81). The recent discovery of the turbomycins using metagenomics also showed that this approach would be useful to access novel activity by creating chimeric pathways between the environmental DNA and the genetic material of the host, thus enhancing the biosynthetic ability of the host.

Courtois et al. *(82)*, have successfully screened by PCR a small (5000 clones) soil metagenomic library for polyketide biosynthetic genes, some of which were responsible for the production of novel molecules when heterologously expressed in *S. lividans*. These data reinforce the idea that exploiting previously unknown or uncultivated microorganisms for the discovery of novel natural products has potential value and, most important, suggest a strategy for developing this technology into a realistic and effective drug discovery tool.

Metagenomics for identification of novel natural products has so far been applied only to the soil environment, which in the past has been a rich source of novel compounds from culturable microorganisms *(77)*. However, in light of the growing emphasis on screening of marine microbes as a source of natural products and because some of the products originally found in marine invertebrates are likely actually produced by symbiotic microbes associated with these invertebrates *(83)*, the marine environment should provide a fertile area for the metagenomic approach to finding novel natural products. In addition, organisms from the marine environment can provide novel structures not seen in compounds from terrestrial sources *(83)*.

In many cases, it has not been possible to culture the symbiotic microorganisms producing natural product isolated from marine invertebrate hosts *(84)* (Fig. 4). Metagenomics can provide access to the genetic pathway responsible for the biosynthesis of these secondary metabolites. Still limited to the biotechnology sector, metagenomics may ultimately provide a source of new compounds for pharmaceutical industry pipelines.

Genome Shuffling for the Engineering of Novel Natural Products

DNA shuffling allows fast evolution of genes and subgenomic DNA fragments *(85, 86)*. Genome shuffling addresses the directed evolution of whole microorganisms by combining the recombinational ability of protoplast fusion and DNA shuffling between multiple parents for phenotypic improvement in bacteria *(87)*. Classical strain improvement directs the evolution of a microorganism by successive rounds of random mutagenesis and screening for improved phenotype, by which the best mutant is selected at each cycle for further improvement. Genome shuffling takes advantage of the genetic diversity created after the first round of mutagenesis by shuffling a population of improved mutants. A new population of mutants with better performance is created, which can then be shuffled again in further rounds.

This technique has been used for the rapid improvement of tylosin production from *Streptomyces fradiae* and demonstrated that two rounds of genome shuffling (1 year) are sufficient to obtain tylosin production levels that were achieved in 20 years of classical strain improvement *(87)*. Similarly, other phenotypes can be improved, such as acid tolerance, as demonstrated by genome shuffling of an industrial strain of *Lactobacillus* *(88)*. Tolerance for pH is encoded by at least 18 loci and more than 60 genes, hence the significance and applicability of such random methods over a more rational engineering approach.

It can be seen how this technology can be developed to engineer novel secondary metabolites, such as antibiotics with improved performance. In addition, novel structures can be evolved using a whole genome combinatorial biosynthesis approach. These new genomic technologies are adding to the tools that can be used in the field of drug discovery.

Fig. 4. Is the next blockbuster drug in the marine environment? Marine sponges such as *Xestospongia muta* at Key Largo, Florida, harvest a plentitude of novel bioactive compounds. Genomics might give access to the biosynthetic source. (Photograph used with permission by Jayme Lohr, laboratory of Russell T. Hill, Center of Marine Biotechnology, University of Maryland Biotechnology Institute; *[80].*)

CONCLUSION

The genomics era is providing a flood of potential targets that will require increased chemical diversity for drug discovery. The rapid growth of combinatorial chemistry technologies will undoubtedly provide some of that diversity, but there is still a bright future for microbial natural products as novel and creative genomics-based technologies are being developed to access the boundless chemical diversity found in nature. Natural product discovery should not be abandoned as many environments have not been screened for novel compounds, and many organisms are yet to be cultured. The war on pathogenic microbes needs to be fought relentlessly on multiple fronts.

Genomics is having a broad impact on drug discovery. Genomics and bioinformatics technologies allow dissection of the genetic basis of infection and pathogenicity and identification of targets that can be screened for new medicines. These new approaches to drug discovery are expanding treatment options and defensive strategies against infectious disease. In the last decade, chemists, biochemists, biologists, microbiologists, structural scientists, computer scientists, medical doctors, pharmacologists, and genomicists have worked together to address the increasingly complex problems that drug research is facing. Such a multidisciplinary approach promises to generate unprecedented advances in the war against pathogenic microbes.

Genomics has established itself as the driving force for the development of creative ways to search for novel drugs, and the pharmaceutical industry is effectively incorporating these new technologies into its drug discovery programs. Although the field is still young, the recent advances indicate that genomics is here to stay and will contribute significantly to the improvement of human health.

ACKNOWLEDGMENTS

I would like to thank Chris Farnet, Ecopia Bioscience, Inc., for providing Fig. 3 and for helpful comments and suggestions on the manuscript. I also acknowledge Brian Dougherty for his critical review and suggestions.

REFERENCES

1. Bentley SD, Chater KF, Cerdeno-Tarraga AM, et al. Complete genome sequence of the model actinomycete *Streptomyces coelicolor* A3(2). Nature 2002; 417:141–147.
2. Cassell GH, Mekalanos J. Development of antimicrobial agents in the era of new and reemerging infectious diseases and increasing antibiotic resistance. JAMA 2001; 285:601–605.
3. Cohen ML. Changing patterns of infectious disease. Nature 2000; 406:762–767.
4. McCaig LF, Hughes JM. Trends in antimicrobial drug prescribing among office-based physicians in the United States. JAMA 1995; 273:214–219.
5. *Staphylococcus aureus* resistant to vancomycin—United States, 2002. MMWR Morb Mortal Wkly Rep 2002; 51:565–567.
6. Geographic variation in penicillin resistance in *Streptococcus pneumoniae*—selected sites, United States, 1997. MMWR Morb Mortal Wkly Rep 1999; 48:656–661.
7. Glynn JR, Whiteley J, Bifani PJ, Kremer K, van Soolingen D. Worldwide occurrence of Beijing/W strains of *Mycobacterium tuberculosis*: a systematic review. Emerg Infect Dis 2002; 8:843–849.
8. Shinabarger DL, Marotti KR, Murray RW, et al. Mechanism of action of oxazolidinones: effects of linezolid and eperezolid on translation reactions. Antimicrob Agents Chemoth 1997; 41: 2132–2136.
9. Fleischmann RD, Adams MD, White O, et al. Whole-genome random sequencing and assembly of *Haemophilus influenzae* Rd. Science 1995; 269:496–512.
10. Buysse JM. The role of genomics in antibacterial target discovery. Curr Med Chem 2001; 8: 1713–1726.
11. Galperin MY, Koonin EV. Searching for drug targets in microbial genomes. Curr Opin Biotechnol 1999; 10:571–578.
12. Petit MA, Ehrlich D. Essential bacterial helicases that counteract the toxicity of recombination proteins. EMBO J 2002; 21:3137–3147.
13. Mural RJ, Adams MD, Myers EW, et al. A comparison of whole-genome shotgun-derived mouse chromosome 16 and the human genome. Science 2002; 296:1661–1671.
14. Venter JC, Adams MD, Myers EW, et al. The sequence of the human genome. Science 2001; 291:1304–1351.
15. Haney SA, Alksne LE, Dunman PM, Murphy E, Projan SJ. Genomics in anti-infective drug discovery getting to endgame. Curr Pharm Des 2002; 8:1099–1118.
16. Schweitzer BI, Dicker AP, Bertino JR. Dihydrofolate reductase as a therapeutic target. FASEB J 1990; 4:2441–2452.
17. Brown JD, Warren PV. Antibiotic discovery: is it all in the genes? Drug Discov Today 1998; 3: 564–566.

18. Read TD, Gill SR, Tettelin H, Dougherty BA. Finding drug targets in microbial genomes. Drug Discov Today 2001; 6:887–892.

19. Chalker A, Lunsford R. Rational identification of new antibacterial drug targets that are essential for viability using a genomics-based approach. Pharmacol Ther 2002; 95:1.

20. Wolf YI, Aravind L, Koonin EV. *Rickettsiae* and *Chlamydiae*: evidence of horizontal gene transfer and gene exchange. Trends Genet 1999; 15:173–175.

21. Altschul SF, Madden TL, Schaffer AA, et al. Gapped BLAST and PSI-BLAST: a new generation of protein database search programs. Nucleic Acids Res 1997; 25:3389–3402.

22. Bateman A, Birney E, Durbin R, Eddy SR, Finn RD, Sonnhammer EL. Pfam 3.1: 1313 multiple alignments and profile HMMs match the majority of proteins. Nucleic Acids Res 1999; 27: 260–262.

23. Haft DH, Loftus BJ, Richardson DL, et al. TIGRFAMs: a protein family resource for the functional identification of proteins. Nucleic Acids Res 2001; 29:41–43.

24. Zhang Z, Schaffer AA, Miller W, et al. Protein sequence similarity searches using patterns as seeds. Nucleic Acids Res 1998; 26:3986–3990.

25. Tatusov RL, Koonin EV, Lipman DJ. A genomic perspective on protein families. Science 1997; 278:631–637.

26. Galperin MY, Walker DR, Koonin EV. Analogous enzymes: independent inventions in enzyme evolution. Genome Res 1998; 8:779–790.

27. Perriere G, Duret L, Gouy M. HOBACGEN: database system for comparative genomics in bacteria. Genome Res 2000; 10:379–385.

28. Bruccoleri RE, Dougherty TJ, Davison DB. Concordance analysis of microbial genomes. Nucleic Acids Res 1998; 26:4482–4486.

29. Domagala JM, Hanna LD, Heifetz CL, et al. New structure-activity relationships of the quinolone antibacterials using the target enzyme. The development and application of a DNA gyrase assay. J Med Chem 1986; 29:394–404.

30. Schonbrunn E, Sack S, Eschenburg S, et al. Crystal structure of UDP-*N*-acetylglucosamine enolpyruvyltransferase, the target of the antibiotic fosfomycin. Structure 1996; 4:1065–1075.

31. Chalker AF, Minehart HW, Hughes NJ, et al. Systematic identification of selective essential genes in *Helicobacter pylori* by genome prioritization and allelic replacement mutagenesis. J Bacteriol 2001; 183:1259–1268.

32. Zhou H, Ranish JA, Watts JD, Aebersold R. Quantitative proteome analysis by solid-phase isotope tagging and mass spectrometry. Nat Biotechnol 2002; 20:512–515.

33. Washburn MP, Wolters D, Yates JR III. Large-scale analysis of the yeast proteome by multidimensional protein identification technology. Nat Biotechnol 2001; 19:242–247.

34. Cagney G, Emili A. *De novo* peptide sequencing and quantitative profiling of complex protein mixtures using mass-coded abundance tagging. Nat Biotechnol 2002; 20:163–170.

35. Schmid MB, Kapur N, Isaacson DR, Lindroos P, Sharpe C. Genetic analysis of temperature-sensitive lethal mutants of *Salmonella typhimurium*. Genetics 1989; 123:625–633.

36. Xia M, Lunsford RD, McDevitt D, Iordanescu S. Rapid method for the identification of essential genes in *Staphylococcus aureus*. Plasmid 1999; 42:144–1449.

37. Thanassi JA, Hartman-Neumann SL, Dougherty TJ, Dougherty BA, Pucci MJ. Identification of 113 conserved essential genes using a high-throughput gene disruption system in *Streptococcus pneumoniae*. Nucleic Acids Res 2002; 30:3152–3162.

38. Wilding EI, Brown JR, Bryant AP, et al. Identification, evolution, and essentiality of the mevalonate pathway for isopentenyl diphosphate biosynthesis in Gram-positive cocci. J Bacteriol 2000; 182:4319–4327.

39. Russell RB, Eggleston DS. New roles for structure in biology and drug discovery. Nat Struct Biol 2000; 7(Suppl):928–930.

40. Gilliland GL, Teplyakov A, Obmolova G, et al. Assisting functional assignment for hypothetical *Haemophilus influenzae* gene products through structural genomics. Curr Drug Targets Infect Disord 2002; 2:339–353.

41. Berman HM, Westbrook J, Feng Z, et al. The Protein Data Bank. Nucleic Acids Res 2000; 28: 235–242.

42. Chambers SP. High-throughput protein expression for the post-genomic era. Drug Discov Today 2002; 7:759–765.

43. Blundell TL, Jhoti H, Abell C. High-throughput crystallography for lead discovery in drug design. Nature Rev Drug Disc 2002; 1:45–54.

44. Renfrey S, Featherstone J. Structural proteomics. Nature Rev Drug Disc 2002; 1:175–176.

45. Yee A, Chang X, Pineda-Lucena A, et al. An NMR approach to structural proteomics. Proc Natl Acad Sci USA 2002; 99:1825–1830.

46. Berman HM, Bhat TN, Bourne PE, et al. The Protein Data Bank and the challenge of structural genomics. Nat Struct Biol 2000; 7(Suppl):957–959.

47. Gerstein M. Integrative database analysis in structural genomics. Nat Struct Biol 2000; 7(Suppl): 960–963.

48. Edwards AM, Arrowsmith CH, Christendat D, et al. Protein production: feeding the crystallographers and NMR spectroscopists. Nat Struct Biol 2000; 7(Suppl):970–972.

49. Buchanan SG, Sauder JM, Harris T. The promise of structural genomics in the discovery of new antimicrobial agents. Curr Pharm Des 2002; 8:1173–1788.

50. Lesley SA. High-throughput proteomics: protein expression and purification in the postgenomic world. Prot Expr Purif 2001; 22:159–164.

51. Pfuetzner RA, Bochkarev A, Frappier L, Edwards AM. Replication protein A. Characterization and crystallization of the DNA binding domain. J Biol Chem 1997; 272:430–434.

52. Udwary D, Merski M, Townsend C. A method for prediction of the locations of linker regions within large multifunctional proteins, and application to a type I polyketide synthase. J Mol Biol 2002; 323:585–598.

53. Elofsson A, Sonnhammer EL. A comparison of sequence and structure protein domain families as a basis for structural genomics. Bioinformatics 1999; 15:480–500.

54. Ostermeier C, Michel H. Crystallization of membrane proteins. Curr Opin Struct Biol 1997; 7: 697–701.

55. Chiu ML, Nollert P, Loewen MC, et al. Crystallization *in cubo*: general applicability to membrane proteins. Acta Crystallogr D Biol Crystallogr 2000; 56:781–784.

56. Capdeville R, Buchdunger E, Zimmermann J, Matter A. Glivec (STI571, imatinib), a rationally developed, targeted anticancer drug. Nature Rev Drug Disc 2002; 1:493–502.

57. Betz SF, Baxter SM, Fetrow JS. Function first: a powerful approach to post-genomic drug discovery. Drug Discov Today 2002; 7:865–871.

58. Baxter SM, Fetrow JS. Sequence- and structure-based protein function prediction from genomic information. Curr Opin Drug Discov Dev 2001; 4:291–295.

59. Davies C, Heath RJ, White SW, Rock CO. The 1.8 A crystal structure and active-site architecture of beta-ketoacyl-acyl carrier protein synthase III (FabH) from *Escherichia coli*. Structure Fold Des 2000; 8:185–195.

60. Colovos C, Cascio D, Yeates TO. The 1.8 A crystal structure of the ycaC gene product from *Escherichia coli* reveals an octameric hydrolase of unknown specificity. Structure 1998; 6:1329–1337.

61. Cort JR, Yee A, Edwards AM, Arrowsmith CH, Kennedy MA. Structure-based functional classification of hypothetical protein MTH538 from *Methanobacterium thermoautotrophicum*. J Mol Biol 2000; 302:189–203.

62. Minasov G, Teplova M, Stewart GC, Koonin EV, Anderson WF, Egli M. Functional implications from crystal structures of the conserved *Bacillus subtilis* protein Maf with and without dUTP. Proc Natl Acad Sci USA 2000; 97:6328–6333.

63. Teplova M, Tereshko V, Sanishvili R, et al. The structure of the *yrdC* gene product from *Escherichia coli* reveals a new fold and suggests a role in RNA binding. Protein Sci 2000; 9: 2557–2566.

64. Lewis HA, Furlong EB, Laubert B, et al. A structural genomics approach to the study of quorum sensing: crystal structures of three LuxS orthologs. Structure 2001; 9:527–537.

65. Buchan DW, Shepherd AJ, Lee D, et al. Gene3D: structural assignment for whole genes and genomes using the CATH domain structure database. Genome Res 2002; 12:503–514.

66. Pearl FM, Lee D, Bray JE, Buchan DW, Shepherd AJ, Orengo CA. The CATH extended protein-family database: providing structural annotations for genome sequences. Protein Sci 2002; 11:233–244.

67. Williams MG, Shirai H, Shi J, et al. Sequence-structure homology recognition by iterative alignment refinement and comparative modeling. Proteins 2001; Suppl 5:92–97.

68. Shi J, Blundell TL, Mizuguchi K. FUGUE: sequence–structure homology recognition using environment-specific substitution tables and structure-dependent gap penalties. J Mol Biol 2001; 310:243–257.

69. Liu Y, Luscombe NM, Alexandrov V, et al. Structural genomics: a new era for pharmaceutical research. Genome Biol 3: REPORTS4004. 2002.

70. Demain AL. Microbial natural products: alive and well in 1998. Nat Biotechnol 1998; 16: 3–4.

71. Farnet CM, Staffa A, Zazopoulos E. High throughput method for discovery of gene clusters. Canadian Appl. CA 2,352,451. Ecopia Biosciences, Inc., Canada. 2001.

72. Farnet CM, Mercure S, Nowacki P, Staffa A, Zazopoulos E. Gene cluster for everinomicin biosynthesis. PCT Int. Appl. WO 0155180. Ecopia Biosciences, Inc., 2001.

73. Farnet CM, Staffa A, Zazopoulos E. Gene cluster for ramoplanin biosynthesis. PCT Int. Appl. WO 0231155. Ecopia Biosciences, Inc., 2002.

74. Farnet CM, Staffa A, Yang X. Gene and proteins for rosaramicin biosynthesis. Canadian Appl. CA 2,391,131. Ecopia Biosciences, Inc., Canada, 2002.

75. Farnet CM, Staffa A. Genes and proteins for the biosynthesis of anthramycin. Canadian Appl. CA 2,386,587. Ecopia Biosciences, Inc, Canada, 2002.

76. Ahlert J, Shepard E, Lomovskaya N, et al. The calicheamicin gene cluster and its iterative type I enediyne PKS. Science 2002; 297:1173–1176.

77. Newman DJ, Cragg GM, Snader KM. The influence of natural products upon drug discovery. Nat Prod Rep 2000; 17:215–234.

78. Rondon MR, August PR, Bettermann AD, et al. Cloning the soil metagenome: a strategy for accessing the genetic and functional diversity of uncultured microorganisms. Appl Environ Microbiol 2000; 66:2541–2547.

79. Brady SF, Chao CJ, Clardy J. New natural product families from an environmental DNA (eDNA) gene cluster. J Am Chem Soc 2002; 124:9968–9969.

80. Wang GY, Graziani E, Waters B, et al. Novel natural products from soil DNA libraries in a streptomycete host. Org Lett 2000; 2:2401–2404.

81. Brady SF, Chao CJ, Handelsman J, Clardy J. Cloning and heterologous expression of a natural product biosynthetic gene cluster from eDNA. Org Lett 2001; 3:1981–1984.

82. Courtois S, Cappellano CM, Ball M, et al. Recombinant environmental libraries provide access to microbial diversity for drug discovery from natural products. Appl Environ Microbiol 2003; 69:49–55.

83. Jensen PR, Fenical W. Marine microorganisms and drug discovery: Current status and future potential. In: Fusetani N (ed). Drugs from the Sea. Basel: Karger, 2000, pp. 6–29.

84. Faulkner DJ, Harper MK, Haygood MG, Salomon CE, Schmidt EW. Symbiotic bacteria in sponges: source of bioactive substances. In: Fusetani N (ed). Drugs from the Sea. Basel: Karger, 2000, pp. 107–119.

85. Ness JE, Welch M, Giver L, et al. DNA shuffling of subgenomic sequences of subtilisin. Nat Biotechnol 1999; 17:893–896.

86. Stemmer WP. DNA shuffling by random fragmentation and reassembly: in vitro recombination for molecular evolution. Proc Natl Acad Sci USA 1994; 91:10,747–10,751.

87. Zhang YX, Perry K, Vinci VA, Powell K, Stemmer WP, del Cardayre SB. Genome shuffling leads to rapid phenotypic improvement in bacteria. Nature 2002; 415:644–646.

88. Patnaik R, Louie S, Gavrilovic V, et al. Genome shuffling of *Lactobacillus* for improved acid tolerance. Nat Biotechnol 2002; 20:707–712.

27

A Genomic Approach to Vaccine Development

Rino Rappuoli, Vega Masignani, Mariagrazia Pizza, Guido Grandi, and John L. Telford

INTRODUCTION

The microbial database maintained at the Institute for Genomic Research lists more than 140 completed bacterial genomes, and more than 300 other microorganisms are being sequenced in various laboratories around the world. This mass of information, inconceivable even a few years ago, together with the accompanying rapid development of sophisticated computational tools has changed the understanding of the prokaryotic world and will influence the approach of microbiological research over the next years. The availability of larger and larger sets of databases and specialized tools allows researchers to infer protein functions quickly based only on sequence analyses instead of following the classical biochemical approaches, which are generally laborious, expensive, and slow.

The principal goal of sequencing pathogenic bacteria is the understanding of the infectious disease process. Along with such understanding comes the ability to develop molecular diagnostic probes and to define new drug targets and preventive measures to treat infections caused by these organisms. In this context, one of the more promising applications of bioinformatics is in the vaccine field. The complete sequence of a bacterial genome can in fact provide an opportunity to tackle vaccine development from a totally different point of view. Within an annotated genome, all the protein antigens are equally visible at once, no matter which of them are more or less abundantly expressed in vitro and in vivo and in which phase of growth. This allows the identification not only of all the antigens selected using the conventional biochemical, serological and microbiological methods, but also the discovery of novel antigens *(1)*.

Sophisticated computer programs are available to predict the function of gene products, to search homologies with known virulence factors produced by other pathogens, and to predict cellular localization of newly identified open reading frames (ORFs) (Table 1). Hence, it has become possible to identify all potentially protective antigens in a bacterial pathogen directly by *in silico* analysis and then test them in models of protective immunity. This approach, which we have termed *reverse vaccinology*, has already been used to identify novel vaccine candidates against serogroup B *Neisseria meningitidis (2)* and is currently being used against a number of other pathogens *(3)*. In this chapter, we describe the basic approach and how it is being applied.

From: *Microbial Genomes*
Edited by: C. M. Fraser, T. D. Read, and K. E. Nelson © Humana Press Inc., Totowa, NJ

FROM GENOME TO ANTIGEN: A NOVEL PARADIGM

"In Silico" *Identification of Vaccine Candidates*

The primary condition for a bacterial protein to be considered an antigen is its cellular localization. Proteins restricted to the cytosolic compartment are unlikely to be immunological targets, whereas surface-associated and secreted structures are more easily accessible to antibodies, the primary immune effector molecules against bacterial pathogens. Figure 1 summarizes the types of proteins that may be considered as vaccine targets. In bacteria, there are a number of systems that control the targeting of newly synthesized proteins to the extracytoplasmic compartment and that therefore contribute to the surface expression of both extracellular enzymes and proteins, such as adhesins and virulence factors, involved in the interaction with the host. Although some of these systems are shared between Gram-negative and Gram-positive bacteria, others are specific for one or the other class of microorganisms.

For both groups, most of the secreted proteins are synthesized as precursors with an amino-terminal "Zip code"—the signal peptide—the recognition of which leads the nascent protein into the general secretory pathway (Sec) *(4)* and that is subsequently cleaved off during translocation through the inner membrane by a specific signal peptidase (SPase). Although the primary structures of different amino-terminal signal peptides show little similarity, it is possible to identify three domains with conserved features, characterized by a positively charged N-region, followed by a hydrophobic core and a neutral but polar C-region that precedes the cleavage site (e.g., Ala-X-Ala) *(5)*. Structural features of the leader peptides can vary slightly depending on the secretion apparatus that the proteins use for their export.

The signals of lipoproteins, for example, are characterized by a well-conserved lipobox that contains an invariable cysteine residue (generally within the amino acid pattern–Leu-X-X-Cys) that is lipid modified by diacylglyceryl transferase prior to precursor cleavage *(6)*. After translocation across the cytoplasmic membrane, lipoproteins remain anchored to the inner or to the outer membrane by their amino-terminal lipid-modified cysteine residue. A particular and nonubiquitous Sec-independent secretory system, initially discovered in plants, is the twin-arginine-translocation apparatus (TAT system), shared by a subgroup of Gram-positive and Gram-negative bacteria *(7)*. Substrates of this pathway are characterized by a particular leader sequence that contains a pair of consecutive arginine residues before the hydrophobic core and within the consensus Arg-Arg-X-Phe-Leu-Lys.

The extensive knowledge of these characteristics has permitted the development of algorithms capable of predicting secreted proteins on the basis of their amino-terminal sequence. Several software products (see Table 1) can be used to do this rapidly and automatically to identify potentially secreted proteins within an entire bacterial genome. Although signal peptide identification reveals the more likely secreted proteins in a bacterial genome, other characteristics of surface-exposed proteins can be used to support the predictions and to identify potentially surface-exposed proteins that lack a classical signal peptide.

The translocation of complex proteins across the outer membrane of Gram-negative bacteria is achieved in several different ways. The simplest case is the so-called autotransporter secretion mechanism *(8)*, which was first described for the IgA1 protease

Table 1
Computer Programs for Gene Investigation

Program	Web address	Suited for
BLAST/PSI-BLAST	http://www.ncbi.nlm.nih.gov/BLAST/	Homology search
FASTA	GCG Package, in house	Homology search
PSORT	http://psort.nibb.ac.jp/	Prediction of signal peptides, transmembrane segments, general localization
SignalP	http://www.cbs.dtu.dk/services/SignalP/	Signal peptide prediction
SPScan	GCG Package, in house	Signal peptide prediction
TMpred	http://www.ch.embnet.org/software/TMPRED_form.html	Prediction of transmembrane proteins and orientation
TopPred2	http://bioweb.pasteur.fr/seqanal/interfaces/toppred.html	Hydrophobic segments and topology of membrane proteins
Motifs	GCG Package, in house	Known protein motifs
FindPatterns	GCG Package, in house	User-defined protein motifs
InterPro	http://www.ebi.ac.uk/interpro/	Integrated resource for the identification of signatures and protein families A
PredictProtein	http://www.embl-heidelberg.de/predictprotein/	Structure prediction
PSIPRED	http://bioinf.cs.ucl.ac.uk/psipred/	Structure prediction

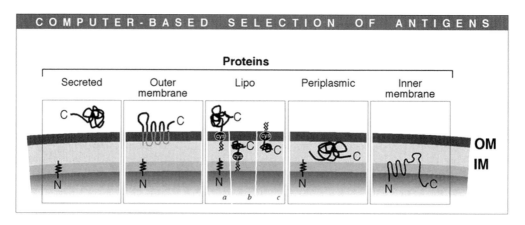

Fig. 1. Types of predicted proteins selected as possible antigens. From left to right: secreted proteins; outer membrane (OM) proteins; lipoproteins; periplasmic proteins; inner membrane (IM) proteins.

of *Neisseria gonorrhoeae (9)*. Proteins belonging to the class of autotransporters are exported across the outer membrane through their carboxy-terminal domain, which is arranged within the outer membrane in a porelike fashion composed of antiparallel, amphipathic β-sheets. These proteins can be recognized by the presence of an almost-invariable terminal Phe or Trp residue, essential for anchoring to the outer membrane, which is preceded by alternating charged or polar and aromatic or hydrophobic amino acid, thus resulting in a well-defined signature of the type (Y, F, W, L, I, V)-X-(F, W).

More complex mechanisms of export across both inner and outer membranes include the recently discovered type III *(10)* and type IV machineries *(11)*. These systems involve a variable number of components, which assemble into a large structure that spans both the inner and the outer membranes, allowing the secretion of specific factors directly to the exterior of the cell or even through the host cell membrane. Prototypes of these systems are the *Yersinia* sp Yop apparatus *(12)* and the VirB system of *Agrobacterium tumefaciens (13)*. Identification of these systems is not straightforward because no particular sequence or structural motifs have so far been associated with proteins exported by the type III and type IV mechanisms, and the level of sequence similarity among the various components is often very low. However, the genes coding for these secretion systems are generally cotranscribed or at least arranged consecutively on the genome and comprise several adenosine triphosphate–binding components, as well as integral membrane, periplasmic, and secreted factors.

Prediction of hydrophobic segments is widely used for both Gram-positive and Gram-negative bacteria to identify proteins spanning the cytoplasmic membrane. Members of this group are often involved in transportation and in signal transduction mechanisms, as in the case of permeases and histidine kinases, respectively.

Given the different organization of the surface structures in Gram-positive bacteria, the prediction of surface-associated proteins involves other criteria in addition to the already mentioned general export pathway. In particular, we should mention the detection of surface proteins by the presence of C-terminal LPXTG cell wall anchor motifs, which are necessary for correct anchoring of the protein to the peptidoglycan structure

(14). This amino acid pattern is located approximately 25–30 residues from the C-terminus, is preceded by a domain rich in Pro-Gly or Ser-Thr domain and is followed by a hydrophobic segment that spans the inner membrane and by a final short, positively charged tail. Other Gram-positive specific surface-associated structures include the group of proteins anchored via hydrophobic interactions or charged domains and those with repeats involved in the binding to lipoteichoic acid. Finally, vaccine candidates in both Gram-positive and Gram-negative bacteria can be identified by homology to known virulence factors or to proteins known to be located on the surface of other microorganisms.

High-Throughput Expression

Application of the above criteria to identify possible vaccine candidates results in selection of a large number of genes, covering as much as 25% of the total number of ORFs in the genome. To produce recombinant proteins corresponding to each of these genes, it is necessary to use relatively simple procedures that permit large numbers of genes to be cloned and expressed. Fortunately, the development of robotics and the polymerase chain reaction (PCR) reaction make this a feasible exercise.

Oligonucleotides for the PCR are designed on the basis of the genome sequences. Each pair of oligonucleotides also contains sequences corresponding to appropriate restriction sites for ligation cloning into expression vectors or recognition sites for recombinases to use with technologies that employ in vitro recombination to construct plasmids. The product of each PCR reaction is then cloned into two separate expression vectors that contain in-frame sequence coding for either a tag consisting of six consecutive histidine residues or glutathione-S-transferase. These tags permit rapid purification of the recombinant protein by simple column chromatography.

Testing the Antigens

The crux of the reverse vaccinology approach is the availability of a rapid assay that gives indication of immunological protection against the pathogen. In the simplest case, the recombinant antigens can be used to immunize mice and the immune serum obtained can be tested by enzyme-linked immunosorbent assay (ELISA) or flow cytometry for its capacity to recognize the native antigen on the surface of the bacteria. This is very rapid, but surface recognition by antibodies does not necessarily correlate with protection. The complement-mediated lysis of Gram-negative bacteria in the presence of specific antibody is only slightly more laborious than simply assessing surface recognition, and in some cases this bactericidal activity correlates well with protection against infection in humans. For Gram-positive bacteria, an opsonophagocytosis that involves antibody- and complement-dependent bacterial killing in the presence of neutrophils isolated from fresh blood can often be used, although this is considerably more complex than the straightforward bactericidal assay. In some cases, however, the only screen for protection is using animal challenge models.

A VACCINE AGAINST MENINGOCOCCUS B

Antigen Identification

Following the success of the glycoconjugate vaccine against *Haemophilus influenzae* introduced into clinical practice in 1988, the same approach has been pursued for

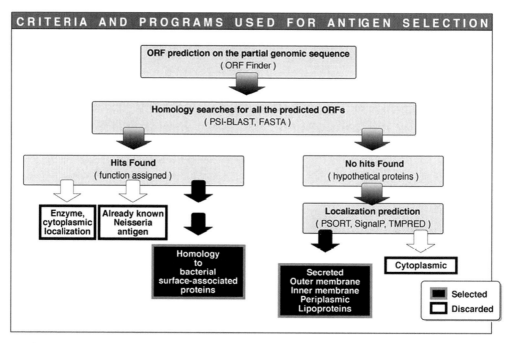

Fig. 2. *In silico* antigen selection strategy showing the decision steps and computer programs used to select potential antigens.

the construction of similar products against *Streptococcus pneumoniae* and *N. mening-itidis (15)*. Although in the case of *S. pneumoniae* and serogroup A, C, Y, and W135 meningococcus, polysaccharide-based vaccines are now available or in the latest steps of development *(16)*, meningococcus B has been a major challenge. The main problem for the design of a conventional capsular-based vaccine against this serogroup is the peculiar structure of the type B polysaccharide, which closely resembles the structure of a component molecule of human brain tissue. The serogroup B polysaccharide is thus poorly immunogenic in humans, and attempts to break tolerance are likely to lead to autoimmune responses *(17)*. For this reason, more recent strategies have been mainly directed toward the detection of protein molecules, which might be able to confer immune protection against the disease rather than polysaccharides *(18)*. However, despite the continuous efforts of a number of research groups over the last four decades, all conventional biochemical and microbiological approaches have failed to provide a universal vaccine against meningococcus B.

Taking advantage of the novel computational techniques and of the complete genomic sequence of a highly virulent isolate of *N. meningitidis* serogroup B *(19)*, we have applied the reverse vaccinology approach to the discovery of vaccine candidates against meningococcus B *(2)*. The general strategy used for antigen selection in *N. meningitidis* is summarized in Fig. 2. The *in silico* screening yielded 570 novel ORFs predicted to be secreted or surface exposed and could therefore represent new potential vaccine candidates. Approximately half of the selected gene products display homologies to proteins of defined function, whereas the others are more or less equally distributed between conserved hypothetical proteins (sharing significant similarities to hypothetical proteins present in several organisms, but without any putative function assigned) and hypothetical proteins (no homologies in databases and very likely species-specific proteins). The majority of putative candidates are represented by integral membrane proteins, followed by periplasmic proteins, lipoproteins, and outer membrane and secreted proteins, which represent less than 15% of the total.

Of the 570 selected ORFs, 350 were successfully cloned and expressed either as 6xHis or GST fusion proteins in *Escherichia coli*. The majority of ORFs that failed to be expressed were predicted to code for proteins with more than two transmembrane-spanning regions. These proteins are notoriously difficult to express in *E. coli* and, when expressed, are frequently toxic to the cell. Each of the expressed proteins was then purified by single-column chromatography on either a nickel chelating resin, or a glutathione conjugate resin, and 344 of them were used to immunize groups of four CD1 mice. The sera obtained were subsequently assayed by immunoblot of bacterial cell lysates to determine if the protein was expressed in *N. meningitidis* and, if so, by ELISA on whole bacterial cells and by indirect immunofluorescence and flow cytometry to determine whether it was detectable on the surface of the bacteria. From these experiments, 91 novel surface-exposed proteins were identified.

To assess the potential of these antigens as vaccine candidates, the immune sera raised against the recombinant proteins were tested for their capacity to promote complement-dependent bacterial killing. High serum titers in the bactericidal assay correlate well with protection against disease in humans, and in clinical trials is currently accepted as a surrogate for protection by most regulatory agencies. These assays identified 29 new antigens that were capable of inducing high serum titers of serum bactericidal activity. Thus, in the space of about 2 years, many more potential vaccine candidates were identified than in the previous four decades of research on *N. meningitidis*. The results of the antigen screening are shown in Fig. 3.

Antigen Variability and Cross Protection

Most known surface-exposed antigens are highly variable in amino acid sequence and hence in antigenicity. In fact, vaccines based on outer membrane vesicle preparations containing the major surface antigens have been shown to confer good protection against infection by the strain from which they were made, but little or no protection against unrelated strains circulating in the population *(20)*. It was therefore important to assess the variability of the antigens identified in the genomic screen and their capacity to induce a broadly protective immune response.

To do this, seven antigens were selected based on their surface exposure and bactericidal titers, and the presence of the corresponding genes was determined by PCR and blot hybridization to deoxyribonucleic acid (DNA) from a panel of *N. meningiditis* isolates that covered the major disease-causing lineages circulating in the world population. Genes coding for all of the seven antigens were present in all meningococcal strains tested, including those of serogroups A, C, Y, Z, and W135. The genes for some of the antigens were also found in *Neisseria lactamica*, *Neisseria cinerea*, or *N. gonorrhoeae*. Sequencing of the genes from these strains demonstrated that five of the antigens were very highly conserved, whereas two genes showed some variability in regions of the proteins. The sera against these proteins has also been shown to confer cross protection against a subset of the strain panel that was suitable for analysis in the bactericidal assay.

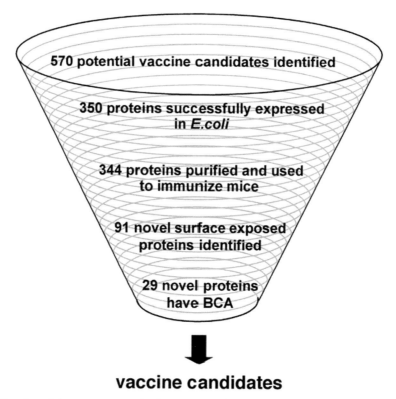

vaccine candidates

Fig. 3. Results of the reverse vaccinology approach to a vaccine against meningococcus B.

Thus, in a remarkably short period of time, a number of novel protein antigens that perform well in preclinical experimentation have been identified. These antigens will now enter clinical trials in humans to determine if they are suitable as a vaccine to protect against this medically important pathogen.

A GRAM-POSITIVE EXPERIENCE

Group B Streptococci

To demonstrate the generality of the reverse vaccinology approach, we decided to apply it to the design of a vaccine against the Gram-positive human pathogen, group B streptococcus (*Streptococcus agalactiae*), which is the major cause of neonatal sepsis in the developed world *(21)*. This bacterium is passed from colonized mothers to babies during birth and can result in devastating bacteraemia and death, usually in the first 24 h after delivery. It is believed that a vaccine given to mothers would induce antibodies against the bacterium; these antibodies could be transferred across the placenta to the baby before birth and could protect it from invasive infection *(22)*. This form of maternal protection has been demonstrated in mouse models *(23)*, and studies in people have shown that babies rarely become infected if the mother already has high-titer antibodies against the bacterium *(24)*.

The best protective antigen to date in animal models is the polysaccharide capsule of the bacterium *(22)*. Unfortunately, there are at least nine different capsular serotypes, and there is little or no cross protection between serotypes. Very little is known about the genomic variation within and between different serotypes, and it is not clear whether sufficiently conserved protein antigens can be found that could confer cross protection against the range of disease-causing strains in circulation. Thus, the problems involved in developing a vaccine for this Gram-positive organism are somewhat different from those confronted in the search for a vaccine against *N. meningiditis*.

The Complete Group B Streptococcus Genome

In collaboration with the Institute for Genomic Research, we have determined and analyzed the complete genome of a serotype V strain of *S. agalactiae (25)*. The genome is predicted to code for 2175 ORFs, of which 650 were predicted to be exposed on the surface of the bacteria. We have successfully expressed about 350 of these ORFs in *E. coli* and used them to immunize mice. Using the sera in ELISA and flow cytometric analysis against intact bacteria, we demonstrated that 55 of these antigens are in fact measurably expressed on the surface of the bacterium, and these antigens are being evaluated in in vitro and in vivo models for their capacity to protect against invasive infection by group B streptococcus. Because little is known about the strain variation in this bacterium, we have assessed gene variation at both the genomic level and at the level of individual genes.

Serotype Variability at the Genome Level

We have used complete genome hybridization to detect the presence or absence of all genes in a panel of 19 strains of group B streptococci representing multiple serotypes. Short sequences corresponding to all the detected ORFs in the sequenced strain were produced by PCR and arrayed on gene chips. The chips were then hybridized with labeled DNA from each of the 19 strains, and the signals were compared with hybridization with the arrayed genome. Lack of hybridization with any strain indicates that that strain either lacks the gene or that the gene has evolved to a very high degree.

A total of 401 ORFs contained in the reference strain did not hybridize with the DNA of at least one other strain, indicating that they are lacking or very highly diverged in those strains. Of the variable genes, 90% were found in 15 clusters containing at least 5 contiguous genes. In some cases, the clusters had clear features of prophages or were flanked by sequences predicted to be transposons. In addition, 10 of these regions had a nucleotide composition atypical of the rest of the genome, indicating that they have been acquired in the reference strain by horizontal transfer of DNA from unknown sources. Of genes present in only some strains, 37 were found scattered randomly in the genome. Clearly, good vaccine candidates must come from those genes found in at least the majority of circulating strains. Interestingly, there was no clear relationship between gene compliment and serotype.

Variability at the Gene Level

To assess the variability between serotypes at the level of the amino acid sequence of the individual proteins, we determined the DNA sequence of 8 genes predicted to code for cytoplasmic housekeeping genes and 11 genes predicted to code for surface-exposed proteins from 11 of the 19 strains. Genes coding for known abundant surface proteins were not included in the analysis as most of these are known to be highly variable, presumably because of pressure of the immune system of the host. Surprisingly, all the genes analyzed were highly conserved, whether the corresponding proteins were predicted to be located in the cytoplasm or on the surface of the bacteria.

On average, the predicted proteins were over 97% identical in amino acid sequence in the panel of strains tested. It is not clear why the surface-exposed proteins are less variable than those of other invasive pathogens such as *N. meningiditis*, but this may reflect the fact that group B streptococci colonize as a commensal the anal and vaginal regions, which are not particularly active immunological sites. Regardless of the reason, this gene conservation bodes well for the possibility of finding cross-serotype antigens capable of preventing invasive disease.

Interestingly, phylogenetic analysis of the 11 strains based on the nucleotide sequence of the genes revealed that, although there was to some extent clustering corresponding to serotype, some strains clustered with strains of other serotypes. Taken together with the complete genome hybridizations, these data indicate that actual genetic lineages are independent of serotype and suggest that serotype switching may be relatively frequent in group B streptococci. This may not be too surprising given that the genome hybridizations indicate that there is substantial mobility of DNA in the genome.

THE FUTURE OF REVERSE VACCINOLOGY

The basic concept of reverse vaccinology has been demonstrated to be effective by the identification of several novel protective antigens against type B meningococcus in a relatively short time *(2)*. The speed at which these antigens have been identified attests to the validity of the approach. Furthermore, preliminary results of the genomic approach to group B streptococci have led to the identification of a large number of novel highly conserved antigens expressed on the bacterial surface *(25)*; from these, it is highly likely that protective antigens as candidates for inclusion in a vaccine will be found. Thus, the approach appears to be applicable generally to a wide range of pathogens, with the only limits being the availability of the genome sequence and suitable in vitro or in vivo models to test capacity of the antigens to induce a protective immune response. One obvious advantage of the approach is that all the antigens that pass the test for protection have been produced as soluble recombinant proteins in *E. coli*, thus ensuring the relatively straightforward development of a suitable manufacturing process for large-scale production.

The approach, however, may be further refined using other novel technologies based on the knowledge of the complete genome sequence. DNA microarrays representing all the ORFs in a genome can be hybridized with ribonucleic acid extracted from bacteria grown in vitro or isolated from infected animals or even patients with infection. The data generated will permit the identification of genes expressed sufficiently in the bacteria to permit an effective immune response. In addition, identification of genes that are expressed in vivo in models of the human disease can further help refine the selection procedure. We have validated this strategy using microarray hybridizations with ribonucleic acid extracted from bacteria grown adhering to epithelial cells to identify novel protective antigens against meningococcus B that were missed during the direct genomic approach *(26)*.

Proteomic approaches may also help refine the antigen selection. An analysis of surface-associated proteins of *Chlamydia trachomatis* by two-dimensional gel electrophoresis and mass spectroscopy has permitted the identification of a number of potentially protective antigens against this infection *(27)*. These experiments can now be performed extremely rapidly by taking advantage of novel technologies that complement well more traditional proteomic strategies (reviewed in ref. *28*).

Finally, although reverse vaccinology has only been applied to bacterial pathogens, there is no reason in principle why it cannot be applied to viruses or even eukaryotic parasites. The only limitations are the size of the genomes and the availability of a suitable model with which to test the recombinant antigens. It is probably reasonable to state that the availability of complete genome sequence information has revolutionized modern vaccine research.

REFERENCES

1. Rappuoli R. Reverse vaccinology. Curr Opin Microbiol 2000; 3:445–450.
2. Pizza M, Scarlato V, Masignani V, et al. Identification of vaccine candidates against serogroup B meningococcus by whole-genome sequencing. Science 2000; 287:1816–1820.
3. Wizemann TM, Heinrichs JH, Adamou JE, et al. Use of a whole genome approach to identify vaccine molecules affording protection against *Streptococcus pneumoniae* infection. Infect Immun 2001; 69:1593–1598.
4. Economou A. Following the leader: bacterial protein export through the Sec pathway. Trends Microbiol 1999; 7:315–320.
5. von Heijne G. The signal peptide. J Membr Biol 1990; 115:195–201.
6. von Heijne G. The structure of signal peptides from bacterial lipoproteins. Protein Eng 1989; 2:531–534.
7. Berks BC, Sargent F, De Leeuw E, et al. A novel protein transport system involved in the biogenesis of bacterial electron transfer chains. Biochim Biophys Acta 2000; 1459:325–330.
8. Henderson IR, Navarro-Garcia F, Nataro JP. The great escape: structure and function of the autotransporter proteins. Trends Microbiol 1998; 6:370–378.
9. Pohlner J, Halter R, Beyreuther K, Meyer TF. Gene structure and extracellular secretion of *Neisseria gonorrhoeae* IgA protease. Nature 1987; 325:458–462.
10. Cornelis GR, Van Gijsegem F. Assembly and function of type III secretory systems. Annu Rev Microbiol 2000; 54:735–774.
11. Christie PJ. Type IV secretion: intercellular transfer of macromolecules by systems ancestrally related to conjugation machines. Mol Microbiol 2001; 40:294–305.
12. Cornelis GR, Wolf-Watz H. The *Yersinia* Yop virulon: a bacterial system for subverting eukaryotic cells. Mol Microbiol 1997; 23:861–867.
13. Zupan JR, Ward D, Zambryski P. Assembly of the VirB transport complex for DNA transfer from *Agrobacterium tumefaciens* to plant cells. Curr Opin Microbiol 1998; 1:649–655.
14. Navarre WW, Schneewind O. Surface proteins of Gram-positive bacteria and mechanisms of their targeting to the cell wall envelope. Microbiol Mol Biol Rev 1999; 63:174–229.
15. Lindberg AA. Glycoprotein conjugate vaccines. Vaccine 1999; 17:S28–S36.
16. Zollinger WD. New and improved vaccines against meningococcal disease. In: Levine MM, Woodrow GC, Cobon GS (eds). New Generation Vaccines. New York: Decker, 1997, pp. 468–488.
17. Hayrinen J, Jennings H, Raff HV, et al. Antibodies to polysialic acid and its *N*-propyl derivative: binding properties and interaction with human embryonal brain glycopeptides. J Infect Dis 1995; 171:1481–1490.
18. Jodar L, Feavers IM, Salisbury D, Granoff DM. Development of vaccines against meningococcal disease. Lancet 2002; 359:1499–1508.
19. Tettelin H, Saunders NJ, Heidelberg J, et al. Complete genome sequence of *Neisseria meningitidis* serogroup B strain MC58. Science 2000; 287:1809–1815.

20. Rosenstein NE, Fischer M, Tappero JW. Meningococcal vaccines. Infect Dis Clin North Am 2001; 15:155–169.
21. Schuchat A. Group B streptococcus. Lancet 1999; 353:51–56.
22. Baker CJ, Kasper DL. Group B streptococcal vaccines. Rev Infect Dis 1985; 7:458–467.
23. Paoletti LC, Wessels MR, Rodewald AK, Shroff AA, Jennings HJ, Kasper DL. Neonatal mouse protection against infection with multiple group B streptococcal (GBS) serotypes by maternal immunization with a tetravalent GBS polysaccharide-tetanus toxoid conjugate vaccine. Infect Immun 1994; 62:3236–3243.
24. Lin FY, Philips JB 3rd, Azimi PH, et al. Level of maternal antibody required to protect neonates against early-onset disease caused by group B streptococcus type Ia: a multicenter, seroepidemiology study. J Infect Dis 2001; 184:1022–1028.
25. Tettelin H, Masignani V, Cieslewicz MJ, et al. Complete genome sequence and comparative genomic analysis of an emerging human pathogen, serotype V *Streptococcus agalactiae*. Proc Natl Acad Sci USA 2002; 99:12,391–12,396.
26. Grifantini R, Bartolini E, Muzzi A, et al. Previously unrecognized vaccine candidates against group B meningococcus identified by DNA microarrays. Nat Biotechnol 2002; 20:914–921.
27. Montigiani S, Falugi F, Scarselli M, et al. Genomic approach for analysis of surface proteins in *Chlamydia pneumoniae*. Infect Immun 2002; 70:368–379.
28. Grandi G. Antibacterial vaccine design using genomics and proteomics. Trends Biotechnol 2001; 19:181–188.

Microbial Proteomics

Svend Birkelund, Brian B. Vandahl,
Allan C. Shaw, and Gunna Christiansen

INTRODUCTION

Bacterial Proteomes as Complements of Genomes

Since the publication of the first genome in 1995 *(1)*, more than 140 microbial genomes have been sequenced, and over 300 more sequence identifications are in progress at the National Center for Biotechnology Information *(2)*. The focus in the postgenomics era is on functional genomics, for which proteomics is playing an important role. Genomes provide important knowledge of the biological potential of an organism, but the genome is static and gives no information on the expression of a particular gene, of posttranslational modifications, or how a protein is regulated during specific biological situations.

The living cell is a dynamic and complex environment with a composition and an organization that cannot be predicted from the genome sequence. For each bacterial species, approximately 30% of the predicted proteins have no homology to entries in public databases or have homology to proteins with unknown function. Transcript analysis, which is feasible on a large scale by the application of microarray technology *(3)*, reveals which genes are transcribed into ribonucleic acid (see Chapter 22). There is, however, no strict relationship between messenger ribonucleic acid (mRNA) and protein levels *(4,5)*, and no information is given on posttranslational modification, processing, and transportation. Direct investigation of the total content of proteins in a cell is the task of *proteomics*, defined as the complete set of proteins (posttranslationally modified and processed) in a well-defined biological environment under specific circumstances, such as growth conditions and time of investigation *(6)*.

TECHNIQUES

Proteomics can be approached in a number of different ways, but in principle it involves two separate steps. The first step is separation of the proteins in a sample. The second step is identification of the proteins. The most common separation tool is two-dimensional polyacrylamide gel electrophoresis (2-D PAGE). The method of choice for large-scale identification is mass spectrometry (MS), but other well-known identification tools, such as N-terminal sequencing, immunoblotting, overexpression, spot colocalization, and gene knockouts, can be used.

From: *Microbial Genomes*
Edited by: C. M. Fraser, T. D. Read, and K. E. Nelson © Humana Press Inc., Totowa, NJ

Two-Dimensional Polyacrylamide Gel Electrophoresis Separation of Proteins

Because of the high resolving power of 2-D PAGE, this technique currently provides the best solution for global visualization of proteins from microorganisms. In the first dimension, isoelectric focusing is performed, during which proteins are separated in a pH gradient according to their isoelectric point (p*I*). The proteins focus at the pH value at which their net charge is zero. In the second dimension, proteins are separated according to their molecular mass by SDS-PAGE (sodium dodecyl sulfate–PAGE). The resulting gel image is a pattern of spots in which each represents a protein, for which p*I* and relative molecular weight *(M*$_r$*)* can be read as in a coordinate system.

Sample Preparation

A crucial step in 2-D PAGE is sample preparation. There is no single method of sample preparation that can be universally applied because different reagents are superior with respect to different samples. After cell lysis and removal of interfering substances (e.g., deoxyribonucleic acid, phenols) or inhibition of proteases, proteins must be denatured and solubilized. The most common lysis and solubilization solution is that based on the work of O'Farrell *(7)*, using a mixture of 2% Nonidet P40 (NP-40), 9*M* urea, 1% dithiothreitol (DTT), and 0.8% carrier ampholytes.

Chaotropes such as urea, which act by changing the parameters of the solvent, are used in most 2-D PAGE procedures. However, denaturation by urea increases the hydrophobic interactions between proteins, a problem that can partly be solved by addition of thiourea. Thiourea is a stronger denaturant, but cannot be applied in sufficient amounts alone because of limited solubility. In urea–thiourea-based lysis buffers, a detergent must be added to break lipid interactions. A nonionic or zwitterionic detergent such as 3-[(3-cholamidopropyl)dimethylammonio]-1-propane sulfonate (CHAPS) is often used, but good results have also been reported using sulfobetains *(8)*. The anionic detergent SDS, which is superior at disrupting noncovalent protein interactions, interferes with the isoelectric focusing, but SDS can be used to presolubilize samples prior to dilution with lysis buffer. The lysis buffer displaces the SDS from the proteins and replaces it with a nonionic or zwitterionic detergent, thereby maintaining the proteins in a soluble state *(9)*. For reduction of disulfide bridges DTT (or dithioerythritol) is preferred because mercaptoethanol ionizes at basic pH and can ruin the pH gradient.

Major problems in 2-D PAGE are concerned with limited entry into the gel of high molecular weight proteins, highly hydrophobic proteins, and basic proteins *(10,11)*. However, use of new solubilizing agents, new immobilized pH gradient (IPG) drystrips, and new focusing apparatus have helped solve these problems. In addition, advantage can be achieved by prefractionation of samples *(12–14)*.

Protein Separation

The protein mixture is loaded onto an acrylamide gel strip in which a pH gradient is established. On application of a high voltage over the strip, the proteins will focus at the pH at which they carry no net charge. The pH gradient is established during the focusing using either carrier ampholytes in a slab gel *(15)* or a precast polyacrylamide gel with an IPG *(16)*. A major advantage of the IPG drystrips is improved reproducibility, which has made interlaboratory comparison possible *(17)*. Furthermore, the osmotic drift against the cathode is less profound in the IPGs, which gives better resolution for basic proteins, the focusing time can be prolonged, and therefore greater amounts of sample can be loaded onto the strip *(18–20)*. Samples can be applied to IPGs either by cuploading or by rehydration. Rehydration of dried IPGs under application of a low voltage (10–50 V) has markedly improved the recovery of especially high molecular weight proteins *(21)*.

In the second dimension, the isoelectric focused proteins are separated according to their relative molecular weight by SDS-PAGE. Prior to the second dimension, the IPG strips are equilibrated in SDS buffer in two consecutive steps. First, the disulfide bridges are reduced in DTT, and then they are alkylated by iodacetamide, which prevents reoxidation.

Mass Spectrometry

Mass spectrometry has become the method of choice for identification of proteins in proteome projects. In MS, the mass of molecules converted to gas phase ions can be measured with an accuracy better than 50 ppm *(22)*. The two most widespread techniques for ionization are matrix-assisted laser desorption ionization (MALDI) *(23)* and electrospray ionization *(24)*.

In the MALDI-TOF (time of flight) MS approach *(25)*, peptides originating from enzymatic digestion with trypsin of a protein spot excised from a gel are first purified using reverse-phase beads or microcolumns. Purified peptides are then deposited on a metal target together with a matrix substance, usually α-cyano-4-hydroxy cinnamic acid. By firing a laser at the target, the matrix molecules absorb energy and decompose/evaporate. As a result, the peptides are brought into the gas phase in an ionized form *(26)*. MALDI is usually coupled to a TOF device for measurement of the masses. The ionized peptides are accelerated by the application of an electrical field, and the TOF until they reach a detector is used to calculate their mass/charge ratio.

The mass accuracy of 50 ppm *(22)* corresponds to less than 0.1 Da (isotopic resolution) for most peptides. The masses of the collection of peptides, the peptide mass fingerprint, can be used to search a protein database that is theoretically digested with trypsin *(27)*. Positive protein identification is based on the percentage of matching peptides in the spectra, the intensity of peaks from matched peptides, and comparison of observed and calculated molecular weight and p*I* of the protein. A diagram of the identification strategy is shown in Fig. 1.

When doing MALDI-TOF MS, peptide sequence tags (PSTs) *(27)* can be obtained by postsource decay (PSD) analysis *(25)*. Postsource decay utilizes the fact that peptides undergo metastable decay after ionization, meaning that peptide fragments of the same velocity will have different mass and therefore possess different kinetic energy. The differences in kinetic energy can be resolved by reflecting the fragments in a magnetic field. High-energy fragments will penetrate further into the magnetic field than low energy fragments and will therefore be delayed. The spectra resulting from fractionation of a single peptide can be used to deduce the amino acid sequence of the peptide because fragmentation predominantly occurs at the peptide bonds. PSTs can be matched against protein databases, and therefore the protein from which it originates can be identified *(28)*. PSTs of about 10 amino acids, which will give a unique identification, can usually be obtained from several peptides originating from one protein.

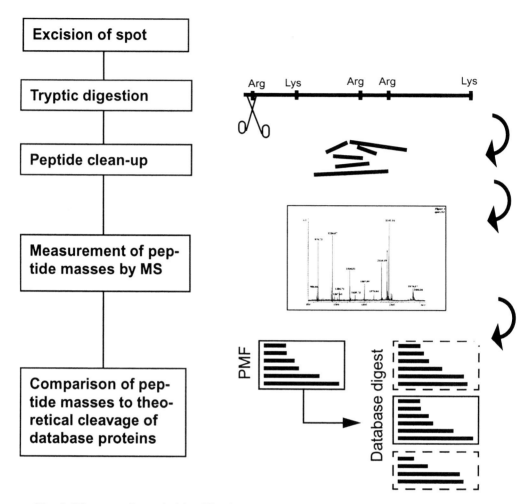

Fig. 1. Diagram of protein identification with mass spectrometry (MS). The excised spot is digested with trypsin; the peptides are extracted, and salt is removed; the sample is analyzed by MS, giving a peptide mass Fingerprint (PMF). The protein is identified by searching the PMF against a database of theoretical PMFs.

PSTs can also be obtained by MS/MS using a triple-quadrupole MS *(29)*. In a quadrupole, the ions pass lengthwise between four metal rods to which potentials can be applied *(30)*. By varying the voltage applied to the rods, ions of certain mass/charge (m/z) ratios can be selected. In the first quadrupole, a specific peptide is selected; in the second quadrupole, which contains an inert gas, the peptide is fragmented by collision-induced dissociation. In the third quadrupole, the collision-induced dissocation spectrum is recorded *(30)*.

In MS/MS, the ionization of peptides is often accomplished by electrospray ionization, in which the peptides are sprayed into the spectrometer *(31)*. The ionization is achieved on evaporation of charged droplets. An alternative means of measurement of masses is the ion trap *(32)*, which selects ions with certain mass/charge ratios by keeping them in sinusoidal motion between two electrodes.

ALTERNATIVES TO GEL SEPARATION

Cutting-edge MS is now at a level at which proteins can be identified quantitatively from a mixture of all proteins present in a cell loaded directly into the mass spectrometer as eluted from a packed capillary column *(4)*. Capillary electrophoresis coupled with MS was reviewed by Shen and Smith *(33)*. The method can be applied in a quantitative way based on reagents called isotope-coded affinity tags and MS/MS *(34)*. Proteins from different samples are tagged with either a nonisotopic label or an isotopic label, $[^{13}C]$ or $[^{15}N]$. The resulting mass differences of the peptides can be resolved in the mass spectra, and the peak intensities can be used to calculate the relative amount of protein in the two samples. As many peptides are generated from each protein, the measurement is statistically improved. In another approach, bacteria are grown under different conditions with either $[^{14}N]$ or $[^{15}N]$ as the sole nitrogen source *(35)*. Quantitative MS used to study protein interaction and protein activity pattern was reviewed by Tyers and Mann *(36)*.

APPLIED PROTEOMICS

Bacteria are interesting to study by proteomics because they have relatively small genomes, many of which have been sequenced. Moreover, it is possible to obtain sample material at different points in time, under different growth conditions, or of fractionated microbial components. The following review of bacterial proteomics describes examples of the information that can be gained through proteome analysis.

Microbial Proteomics

The complete genome sequences provide the basis for experimental identification of expressed proteins at the cellular level, but only a few attempts have been made to identify all expressed and potentially modified proteins. The *Escherichia coli* proteome constructed using of narrow-range IPG strips showed that it was possible to resolve 4950 protein spots corresponding to 70% of the theoretical proteome. However, only 313 spots were identified by MS, corresponding to 222 different proteins; it was not determined whether specific classes of proteins are missing *(37)*.

In an attempt to determine the complete proteome, Ueberle et al. *(38)* analyzed the wall-less bacterium *Mycoplasma pneumoniae*, comparing the predicted 688 open reading frames (ORFs) to identified gene products. There was a good recovery of protein identification as 450 protein spots were assigned to 224 genes, but the article clearly illustrates the limitations in the 2-D separation technology because there were an underrepresentation of proteins with basic p*I*, proteins with molecular weight above 100 kDa, and membrane proteins with multiple transmembranic segments. The authors also analyzed subfractions of the microorganism: cytosolic, heparin-binding, heat stable, and ribosomal proteins *(38)*. Isoelectric focusing is not good for separating ribosomal proteins because they generally have a basic p*I*, and even using nonlinear pH gradients stretched between pH 9.0 and pH 12.0, it was possible only to identify 12 of the 52 predicted ribosomal proteins. In comparison, 48 could be identified by MS from one-dimensional SDS-PAGE.

The Mycoplasmas are the smallest known free-living organisms, and the genome of *Mycoplasma genitalium* is only 570 kb *(39)*. From genome analysis, 480 ORFs were predicted, and the proteome covered 112 proteins from 427 protein spots *(40)*. The *M. genitalium* proteome is interesting with respect to the attempt of listing a minimal set of proteins required for a living cell. Proteins with different functions (cell envelope, metabolism, biosynthesis, transport, replication, transcription, and translation) were found in addition to 17 hypothetical proteins. Two important results from this study are that the identified proteins have Codon Adaptation Index values above the average and that there is a clear underrepresentation of basic proteins and high molecular weight proteins. The protein content of the bacteria in the exponential growth phase was compared to the protein content in the stationary phase. In general, a 42% reduction in protein synthesis was observed, and the relative abundance of ribosomal proteins was reduced as much as eightfold. At present, it is impossible with the 2-D PAGE separation technology to obtain full coverage of microbial proteomes, but important information that cannot be predicted from other studies can be obtained.

PROTEOMICS OF FRACTIONS OF MICROORGANISMS

With proteomics, it is possible to identify proteins in different components of a microorganism. Parts that can be studied are the cytoplasm (soluble protein after physical disruption of the bacteria and high-speed centrifugation to remove membranes and ribosomes), ribosomes, organelles as pili, flagella, the type III secretion apparatus, and the outer membrane complex of Gram-negative bacteria, which can be purified by sarkosyl extraction. Finally, secreted proteins can be analyzed from the supernatant of the culture.

Buttner et al. *(41)* made an extensive analysis of *Bacillus subtilis* cytosolic proteins by 2-D PAGE and identification of the proteins by MS. They disrupted the bacteria using a French press and subsequently removed cell debris by centrifugation. They identified 346 proteins; nearly all were cytoplasmic. The method was thus valid for *B. subtilis*.

Nearly the opposite approach was used to identify the membrane subproteome of *Pseudomonas aeruginosa (42)*. Only a small amount of the major cytosolic protein GroEL was detected, indicating that the method can remove cytosolic protein. The authors observed a major difference in the number of proteins seen depending on whether zwitterionic detergents CHAPS or amido-sulfobetaine 14 was used for solubilizing the proteins in the first dimension, with the latter clearly solubilizing more proteins.

The outer membrane of Gram-negative bacteria can be extracted with the detergent sarkosyl, then named the outer membrane complex (OMC). The chlamydial outer membrane complex OMC *(43)* has been extensively studied. *Chlamydia* are important human pathogens. They are obligate intracellular Gram-negative bacteria with a unique biphasic developmental cycle in which chlamydiae alternate between being extracellular, infectious elementary bodies (EBs) of about 0.3 μm in diameter and intracellular, noninfectious, replicating reticulate bodies (RBs) of about 1 μm in diameter. EBs attach to the host cell and induce their own uptake into a specialized vacuole termed the *chlamydial inclusion*. Internalization of EBs is followed by reorganization into RBs, which multiply by binary fission. Toward the end of the intracellular stage, RBs are transformed into EBs, and ultimately a new generation of infectious EB is released on disruption of the host cell *(44)*.

In one-dimensional gels, the OMC of *Chlamydia trachomatis* shows three major bands (Omp2, Omp3, and MOMP) and *Chlamydia pneumoniae* has additional bands of approximately 100 kDa. These bands contain several proteins called polymorphic outer membrane proteins (Pmp) *(45)*. In the *C. pneumoniae* genome, 21 *pmp* genes were found *(46)*.

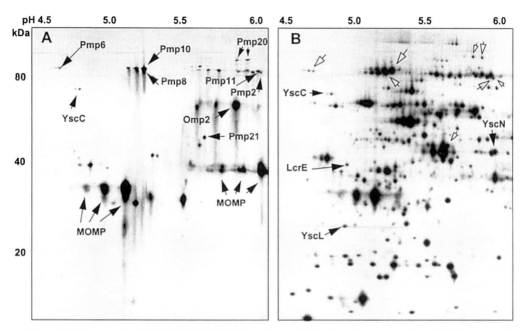

Fig. 2. Silver-stained IPG4-7 2-D PAGE gels loaded with *C. pneumoniae* OMC prepared from (**A**) 1.5 mg purified EBs and (**B**) 250 µg of *C. pneumoniae* EBs. The labeled proteins were identified by MS of trypsin-digested protein spots from EB gels. Open arrows (B) indicate positions of Pmp proteins.

Grimwood et al. *(47)* used antibodies generated to each of the Pmps and showed by immunoblotting that most of the genes were expressed. Vandahl et al. *(48)* separated the *C. pneumoniae* OMC proteins by 2D-PAGE and found that all 7 of the identified Pmps within the investigated pH range (of a total of 10 Pmps identified in the proteome of the whole microorganism) were present in the OMC (Fig. 2), indicating that expressed Pmps are associated with the OMC *(49)*.

Pmps have the structure of autotransporter proteins *(50)*, with a C-terminal part forming a β-barrel in the outer membrane and an N-terminal passenger domain with the structure of parallel β-helices *(51)*. Interestingly, the proteome showed three Pmps (Pmp6, Pmp20, and Pmp21) were cleaved, and the cleavage site, identified by N-terminal sequencing, was found in the region between the autotransporter domain and the passenger domain *(48)*.

Several pathogenic Gram-negative bacteria have a secretion apparatus that injects proteins synthesized in the bacterial cytoplasm into the eukaryotic host cell (type III secretion) *(52)*. The genes encoding the type III secretion apparatus have similarities between species, but there are no common traits to identify the secreted effector proteins. *Chlamydia* has the genes for type III secretion, and Vandahl et al. *(52)* identified YscC, YscN, YscL, and LcrE from the *C. pneumoniae* EBs. YscN and YscL are located in the cytoplasmic membrane, YscC is located in the outer membrane, and LcrE (also known as CopN) controls the release of effector proteins when chlamydiae are in contact with the host cell cytoplasm/vacuole membrane *(53)*. In agreement with this model, only YscC was found in the OMC (Fig. 2), indicating that the chlamydial type III secretion apparatus is assembled as in other bacteria *(49)*. LcrE is not an integrated outer membrane protein, and in *C. trachomatis*, it was shown to be secreted by the type III apparatus and localized in the inclusion membrane *(54)*.

PROTEOMICS USED TO STUDY PROTEIN DYNAMICS

Interferon-γ (IFN-γ) is a potent immunoregulator involved in the control of chlamydial infections and may be implicated in the development of chronic infections. The trachoma serovars A, B, and C of *C. trachomatis* are very sensitive to treatment with IFN-γ. Immunofluorescence microscopy of HeLa cells infected with *C. trachomatis* A and treated with IFN-γ showed atypically large RBs and stained weakly with anti-MOMP antibodies *(55–57)*, in contrast to what was seen for *C. trachomatis* D and L2.

To visualize differences in proteome profiles caused by IFN-γ treatment, HeLa cells infected with *C. trachomatis* A, D, or L2 were cultivated in the presence or absence of IFN-γ and [^{35}S]methonie-labeled 22–24 h postinfection *(58)*. Host cell protein synthesis was stopped by adding cycloheximide to the cell culture. To reveal up- or downregulated proteins, autoradiographs of 2-D PAGE separations were compared.

IFN-γ Downregulated Proteins in Chlamydia trachomatis

2-D PAGE analysis of *C. trachomatis* A revealed that several proteins besides MOMP were downregulated because of IFN-γ treatment compared to steady levels of GroEL. These included ClpC protease and fructose–bisphosphate aldolase class I (Fba) *(56,57)*. The observed downregulated proteins corresponded well with previous observations *(55, 59,60)*, but seemed to be more general and not restricted to the important immunogens. Such IFN-γ–dependent downregulations were not observed on 2D gels of *C. trachomatis* D and L2 *(56,57)*.

Induction of Chlamydia trachomatis Tryptophan Synthase by IFN-γ

A tryptophan operon (*trp* operon) containing tryptophan synthase A (*trp*A) and B (*trp*B) subunit and the tryptophan repressor (*trp*R) is present in the *C. trachomatis* D genome *(61)*, and 2-D PAGE revealed that *C. trachomatis* A, D, and L2 respond to IFN-γ by a strong induction of TrpA and TrpB *(57)*. In contrast to what was seen for TrpB, TrpA of *C. trachomatis* A migrated with a remarkably lower molecular weight than that of *C. trachomatis* D and L2 *(56,57)* because of a 1-bp deletion, which resulted in a frameshift and a premature stop codon truncating 70 amino acids (~7.7 kDa). *Chlamydia trachomatis* C had the same truncation as observed for *C. trachomatis* A, but *C. trachomatis* B had a chromosomal deletion of the genomic region containing the *trp* operon *(62)*. The truncation or absence of TrpBA in the serovars A–C may render these serovars more sensitive to IFN-γ–mediated tryptophan degradation, increasing the ability of these serovars to infect their host cells persistently and cause chronic infections *(57)*.

SECRETED PROTEINS

During their entire intracellular life, chlamydiae are localized within the chlamydial inclusion surrounded by the inclusion membrane, a modified phagosomal membrane. Contact with the host cell may be mediated by secreted proteins, which may modify the host cell during the intracellular development of the inclusion. Furthermore, secreted

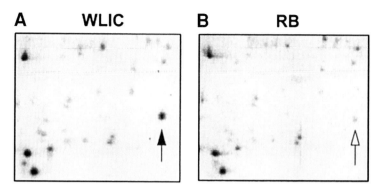

Fig. 3. Autoradiographs of 2-D PAGE separated proteins from whole lysate of *C. pneumoniae* infected cells (WLIC) and lysates of *C. pneumoniae* RBs. The protein syntheses of host cells were stopped by addition of cycloheximide. The infected cells were labeled with [^{35}S] methionine for 36–38 h, and the RBs were purified at 38 h. Arrows point to the N-terminal fragment of CPAF (CPN1016 N-term).

proteins are natural candidates for major histocompatibility complex (MHC) class I antigen presentation and thus possible targets for development of a vaccine.

A proteomics-based strategy was developed for comparing [^{35}S]-labeled chlamydial proteins from whole lysates of infected cells (WLIC) with [^{35}S]-labeled chlamydial proteins from purified RBs and EBs. Secreted proteins are present in WLIC, but absent from purified chlamydiae (Fig. 3). Candidate secreted proteins were identified using MS. The chlamydial protease- or proteasomelike activity factor (CPAF), is known to be secreted into the host cell cytoplasm *(63)*. CPAF is secreted by *C. trachomatis* L2 *(63)* and by *C. pneumoniae (64)* into the host cell cytoplasm, where it degrades the host cell transcription factors RFX5 and USF-1 *(65)*. RFX5 and USF-1 are required for MHC class I and II antigen presentation.

The results of Zhong et al. *(63)* were verified and extended in studies by Shaw et al. *(66)*. The N-terminal and C- terminal fragments of *C. trachomatis* and *C. pneumoniae* CPAF were identified on 2D gels of WLIC, but were absent in RBs and EBs. It was shown that *C. trachomatis* A, D, and L2 and *C. pneumoniae* secrete CPAF from the middle of the developmental cycle and toward the end of the cycle. The CPAF sequence is very conserved among *C. trachomatis, C. pneumoniae*, and *Chlamydia psittaci*, and it is thus likely that all *Chlamydia* strains secrete CPAF during the developmental cycle. Based on pulse-chase studies in combination with 2-D PAGE, the two fragments of CPAF showed high stability in the host cell cytoplasm. The very slow turnover of the CPAF fragments reflects that CPAF may have evolved an amino acid sequence not readily processed in the host cell proteasome. *Chlamydia* thus secrete a protein that inhibits MHC class I and II antigen presentation without itself being a target for MHC class I antigen presentation. The development of a rapid proteomics-based method to screen for secreted proteins facilitates identification of other proteins secreted from *Chlamydia*.

TWO-DIMENSIONAL GEL ELECTROPHORESIS DATABASES

A well-ordered database system mirrored to many sites exists for DNA sequences (European Bioinformatics Institute/National Center for Biotechnology Information) and protein data (SwissProt). A 2D gel database must contain both gel images and text data linked to a protein spot entry. The Swiss 2-D PAGE is a two-dimensional gel electrophoresis database established in 1993. It contains 24 reference maps among others from *Saccharomyces cerevisiae, E. coli*, and *Dictyostelium discoideum* in addition to human and mouse tissue.

Visualization and differentiation of spots is facilitated using the make two-dimensional-gel database (2D-DB) software from ExPASy *(67)*. When "clicking on a spot" in one of the 2-D maps, the information on the protein (isoelectric point, molecular weight, accession number, and link to SwissProt) and all the 2-D maps in which the protein is identified are displayed. The 2-D maps containing a protein can also be found by searching the 2-D database with the protein's accession number. Furthermore, in Swiss Prot protein entries, there are links to the 2D databases of identified proteins. The make 2D-DB software is at present used by seven sites in addition to ExPASy Swiss 2-D PAGE. For an overview of all 2-D PAGE databases and services, see http://www.expasy.org.

The new version Make2ddb is greatly extended in functionality. It is a relational database in which more advanced search queries can be designed. Because of the complex relations between 2-D gel images, spot entries, and the ongoing addition of data to each 2-D image, the system is distributed, and each research group runs its own database site. With Make2ddb, an inquiry can be sent to all 2-D PAGE databases. Use of Make2ddb also makes it possible for the research groups to let ExPASy host their database, securing long-time availability of the data.

The Max Planck Institute for Infection Biology maintains an internet database containing information on proteome analysis of *Mycobacterium, Mycoplasma, Chlamydia, Francisella, Helicobacter*, and *Borrelia (68)*. The institute has developed a data entry software, TopSpot, and run a central database site with both their own and other group's 2-D PAGE data *(68)*. This is a relational database with many search possibilities. The proteomes from two virulent strains of *Mycobacterium tuberculosis* (H37Rv and Erdman) and two nonvirulent strains of *Mycobacterium bovis* BCG (Chicago and Copenhagen) have been compared, and a total of more than 800 spots were identified *(69)*. Sixteen proteins that differed between the virulent strains and 25 proteins that differed between H37Rv and the nonvirulent strains were identified. Differences in the protein profiles for virulent strains compared to nonvirulent strains can possibly account for differences in virulence *(70)*, and it is hoped that these pathogenicity candidates may prove useful targets for medical intervention of tuberculosis.

From *Helicobacter pylori,* 126 different proteins are identified *(71)*. The identifications cover the most abundant proteins, but all protein classes are represented, including transcription and translation factors. *Helicobacter pylori* is capable of growing under extreme acidic conditions, which is important with respect to stomach infections. Upregulated virulence factor proteins were identified during acidic growth. Proteins such as HtrA that were not earlier described as being involved in acidic tolerance as well as the well-characterized "acidic" proteins (VacA) were identified. Strong antigens are vaccine candidates *(70)*. Additional potential virulence factors were identified by immunoblotting with patient sera, and both earlier-described and novel antigens were found. It is important to note that all *H. pylori* proteins that had earlier been described as protective in animal vaccine studies were found to be highly abundant proteins *(71)*.

The *Borrelia garinii* proteome was screened for antigens by immunoblotting using sera from patients suffering from Lyme borreliosis *(72)*. The reactions of 217 protein species with 20 sera from different disease states were monitored. Sixty-five antigens were found, and 20 of these were identified. Two of the identified antigens were novel, the glycolytic enzyme glyceraldehyde-3-phosphate dehydrogenase and the ATP-binding cassette transporter, oligopeptide permease.

Both database systems can integrate identification methods and MS data. As of April 2003, there were 44 Internet 2-D PAGE databases (http://www.expasy.org/ch2d/2d-index.html), most of which focus on diseases and microorganisms.

PROTEOMICS AS A COMPLEMENT FOR GENOMICS

Even though it is not yet possible to identify the complete set of proteins in a proteome, it is clear that there are several ways by which proteomics can complement genomics. Identification of the exact start of proteins by Edman degradation of excised spots *(73)*, identification of unannotated proteins *(38,74)*, finding cleaved proteins, and identification of secreted proteins *(48,66)* are some examples for which proteomics has complemented genomics. Major advantages of the 2-D PAGE system are in studies of protein turnover, for which pulse-labeling and pulse-chase of bacteria can identify synthesis and processing of specific proteins *(66)*. Also, the comparison of protein synthesis under various growth conditions in pulse-labeling and comparison of the 2-D PAGE images are powerful technologies that can identify induced and downregulated proteins *(56, 66)*. Exhaustiveness in proteomics is desirable, but until methods are developed for fast, high-throughput and complete proteomics, the 2-D PAGE system in combination with MS remains an excellent choice for the global study of protein expression and protein processing. In comparison to capillary electrophoresis coupled detection by MS, the 2-D PAGE system in combination with MS provides a picture for the human eye to see and judge, an option that should not be underestimated.

REFERENCES

1. Fleischmann RD, Adams MD, White O, et al. Whole-genome random sequencing and assembly of *Haemophilus influenzae* Rd. Science 1995; 269:496–512.
2. National Center for Biotechnology Information. Prominent Organisms Taxonomy/List. November 13, 2003. http://www.ncbi.nlm.nih.gov:80/PMGifs/Genomes/org.html.
3. Lockhart DJ, Winzeler EA. Genomics, gene expression and DNA arrays. Nature 2000; 405:827–836.
4. Gygi SP, Corthals GL, Zhang Y, Rochon Y, Aebersold R. Evaluation of two-dimensional gel electrophoresis-based proteome analysis technology. Proc Natl Acad Sci USA 2000; 97:9390–9395.
5. Futcher B, Latter GI, Monardo P, McLaughlin CS, Garrels JI. A sampling of the yeast proteome. Mol Cell Biol 1999; 19:7357–7368.
6. Wilkins MR, Pasquali C, Appel RD, et al. From proteins to proteomes: large scale protein identification by two-dimensional electrophoresis and amino acid analysis. Biotechnology (NY) 1996; 14:61–65.
7. O'Farrell PH. High resolution two-dimensional electrophoresis of proteins. J Biol Chem 1975; 250:4007–4021.
8. Rabilloud T, Adessi C, Giraudel A, Lunardi J. Improvement of the solubilization of proteins in two-dimensional electrophoresis with immobilized pH gradients. Electrophoresis 1997; 18:307–316.

9. Dunn MJ, Bradd SJ. Separation and analysis of membrane proteins by SDS-polyacrylamide gel electrophoresis. Methods Mol Biol 1993; 19:203–210.

10. Santoni V, Molloy M, Rabilloud T. Membrane proteins and proteomics: un amour impossible? Electrophoresis 2000; 21:1054–1070.

11. Adessi C, Miege C, Albrieux C, Rabilloud T. Two-dimensional electrophoresis of membrane proteins: a current challenge for immobilized pH gradients. Electrophoresis 1997; 18:127–135.

12. Herbert B. Advances in protein solubilisation for two-dimensional electrophoresis. Electrophoresis 1999; 20:660–663.

13. Rabilloud T. Use of thiourea to increase the solubility of membrane proteins in two-dimensional electrophoresis. Electrophoresis 1998; 19:758–760.

14. Gorg A, Obermaier C, Boguth G, et al. The current state of two-dimensional electrophoresis with immobilized pH gradients. Electrophoresis 2000; 21:1037–1053.

15. Righetti PG, Gianazza E. New developments in isoelectric focusing. J Chromatogr 1980; 184: 415–456.

16. Bjellqvist B, Ek K, Righetti PG, et al. Isoelectric focusing in immobilized pH gradients: principle, methodology and some applications. J Biochem Biophys Methods 1982; 6:317–339.

17. Corbett JM, Dunn MJ, Posch A, Gorg A. Positional reproducibility of protein spots in two-dimensional polyacrylamide gel electrophoresis using immobilised pH gradient isoelectric focusing in the first dimension: an interlaboratory comparison. Electrophoresis 1994; 15:1205–1211.

18. Sanchez JC, Rouge V, Pisteur M, et al. Improved and simplified in-gel sample application using reswelling of dry immobilized pH gradients. Electrophoresis 1997; 18:324–327.

19. Gorg A, Obermaier C, Boguth G, Weiss W. Recent developments in two-dimensional gel electrophoresis with immobilized pH gradients: wide pH gradients up to pH 12, longer separation distances and simplified procedures. Electrophoresis 1999; 20:712–717.

20. Gorg A, Obermaier C, Boguth G, et al. The current state of two-dimensional electrophoresis with immobilized pH gradients. Electrophoresis 2000; 21:1037–1053.

21. Zuo X, Speicher DW. Quantitative evaluation of protein recoveries in two-dimensional electrophoresis with immobilized pH gradients. Electrophoresis 2000; 21:3035–3047.

22. Jensen ON, Podtelejnikov AV, Mann M. Identification of the components of simple protein mixtures by high-accuracy peptide mass mapping and database searching. Anal Chem 1997; 69: 4741–4750.

23. Karas M, Hillenkamp F. Laser desorption ionization of proteins with molecular masses exceeding 10,000 daltons. Anal Chem 1988; 60:2299–2301.

24. Fenn JB, Mann M, Meng CK, Wong SF, Whitehouse CM. Electrospray ionization for mass spectrometry of large biomolecules. Science 1989; 246:64–71.

25. Gevaert K, Vandekerckhove J. Protein identification methods in proteomics. Electrophoresis 2000; 21:1145–1154.

26. Zenobi R, Knochenmuss R. Ion formation in MALDI mass spectrometry. Mass Spectrom Rev 1998; 17:337–366.

27. Mann M, Hojrup P, Roepstorff P. Use of mass spectrometric molecular weight information to identify proteins in sequence databases. Biol Mass Spectrom 1993; 22:338–345.

28. Wilkins MR, Gasteiger E, Sanchez JC, Appel RD, Hochstrasser DF. Protein identification with sequence tags. Curr Biol 1996; 6:1543–1544.

29. Andersen JS, Mann M. Functional genomics by mass spectrometry. FEBS Lett 2000; 480:25–31.

30. Yost RA, Boyd RK. Tandem mass spectrometry: quadrupole and hybrid instruments. Methods Enzymol 1990; 193:154–200.

31. Fenn JB, Mann M, Meng CK, Wong SF, Whitehouse CM. Electrospray ionization for mass spectrometry of large biomolecules. Science 1989; 246:64–71.

32. Cooks RG, Glish GL, Kaiser RE, McLuckey SA. Ion trap mass spectrometry. Chem Eng News 1991; 69:26–41.

33. Shen Y, Smith RD. Proteomics based on high-efficiency capillary separations. Electrophoresis 2002; 23:3106–3124.

34. Gygi SP, Rist B, Gerber SA, Turecek F, Gelb MH, Aebersold R. Quantitative analysis of complex protein mixtures using isotope-coded affinity tags. Nat Biotechnol 1999; 17:994–999.

35. Oda Y, Huang K, Cross FR, Cowburn D, Chait BT. Accurate quantitation of protein expression and site-specific phosphorylation. Proc Natl Acad Sci USA 1999; 96:6591–6596.

36. Tyers M, Mann M. From genomics to proteomics. Nature 2003; 422:193–197.

37. Tonella L, Hoogland C, Binz PA, Appel RD, Hochstrasser DF, Sanchez JC. New perspectives in the *Escherichia coli* proteome investigation. Proteomics 2001; 1:409–423.

38. Ueberle B, Frank R, Herrmann R. The proteome of the bacterium *Mycoplasma pneumoniae*: comparing predicted open reading frames to identified gene products. Proteomics 2002; 2: 754–764.

39. Fraser CM, Gocayne JD, White O, et al. The minimal gene complement of *Mycoplasma genitalium*. Science 1995; 270:397–403.

40. Wasinger VC, Pollack JD, Humphery-Smith I. The proteome of *Mycoplasma genitalium*. Chaps-soluble component. Eur J Biochem 2000; 267:1571–1582.

41. Buttne K, Bernhardt J, Scharf C, et al. A comprehensive two-dimensional map of cytosolic proteins of *Bacillus subtilis*. Electrophoresis 2001; 22:2908–2935.

42. Nouwens AS, Cordwell SJ, Larsen MR, et al. Complementing genomics with proteomics: the membrane subproteome of *Pseudomonas aeruginosa* PAO1. Electrophoresis 2000; 21:3797–3809.

43. Caldwell HD, Kromhout J, Schachter J. Purification and partial characterization of the major outer membrane protein of *Chlamydia trachomatis*. Infect Immun 1981; 31:1161–1176.

44. Birkelund S. The molecular biology and diagnostics of *Chlamydia trachomatis*. Dan Med Bull 1992; 39:304–320.

45. Knudsen K, Madsen AS, Mygind P, Christiansen G, Birkelund S. Identification of two novel genes encoding 97- to 99-kilodalton outer membrane proteins of *Chlamydia pneumoniae*. Infect Immun 1999; 67:375–383.

46. Kalman S, Mitchell W, Marathe R, et al. Comparative genomes of *Chlamydia pneumoniae* and *C. trachomatis*. Nat Genet 1999; 21:385–389.

47. Grimwood J, Olinger L, Stephens RS. Expression of *Chlamydia pneumoniae* polymorphic membrane protein family genes. Infect Immun 2001; 69:2383–2389.

48. Vandahl BB, Pedersen AS, Gevaert K, et al. The expression, processing and localization of polymorphic membrane proteins in *Chlamydia pneumoniae* strain CWL029. BMC Microbiol 2002; 2:36.

49. Vandahl BB, Christiansen G, Birkelund S. 2D-page analysis of the *Chlamydia pneumoniae* outer membrane complex. In: Schachter J, Christiansen G, Clarke IN, et al. (eds). Proceedings of the 10th Symposium on Human Chlamydial Infections. June 16–21, International Chlamydia Symposium, San Francisco, CA, 2002, pp. 547–550.

50. Henderson IR, Lam AC. Polymorphic proteins of *Chlamydia* spp—autotransporters beyond the Proteobacteria. Trends Microbiol 2001; 9:573–578.

51. Birkelund S, Christiansen G, Vandahl BB, Pedersen AS. Are the Pmp proteins parallel β-helices? In: Schachter J, Christiansen G, Clarke IN, et al. (eds). Proceedings of the 10th Symposium on Human Chlamydial Infections. June 16–21, International Chlamydia Symposium, San Francisco, CA, 2002, pp. 551–554.

52. Vandahl BB, Birkelund S, Demol H, et al. Proteome analysis of the *Chlamydia pneumoniae* elementary body. Electrophoresis 2001; 22:1204–1223.

53. Rockey DD, Lenart J, Stephens RS. Genome sequencing and our understanding of chlamydiae. Infect Immun 2000; 68:5473–5479.

54. Fields KA, Hackstadt T. Evidence for the secretion of *Chlamydia trachomatis* CopN by a type III secretion mechanism. Mol Microbiol 2000; 38:1048–1060.

55. Beatty WL, Byrne GI, Morrison RP. Morphologic and antigenic characterization of interferon gamma-mediated persistent *Chlamydia trachomatis* infetion in vitro. Proc Natl Acad Sci USA 1993; 90:3998–4002.

56. Shaw AC, Christiansen G, Birkelund S. Effects of interferon gamma on *Chlamydia trachomatis* serovar A and L2 protein expression investigated by two-dimensional gel electrophoresis. Electrophoresis 1999; 20:775–780.

57. Shaw AC, Christiansen G, Roepstorff P, Birkelund S. Genetic differences in the *Chlamydia trachomatis* tryptophan synthase a-subunit can explain variations in serovar pathogenesis. Microbes Infect 2000; 2:581–592.

58. Shaw AC, Gevaert K, Demol H, et al. Comparative analysis of *Chlamydia trachomatis* serovar A, D and L2. Proteomics 2002; 2:164–186.

59. Beatty WL, Belanger TA, Desai AA, Morrison RP, Byrne GI. Tryptophan depletion as mechanism for gamma-interferin-mediated chlamydial persistence. Infect Immun 1994; 62:3705–3711.

60. Beatty WL, Morrison RP, Byrne GI. Reactivation of persistent *Chlamydia trachomatis* infection in cell culture. Infect Immun 1995; 63:199–205.

61. Stephens RS, Kalman S, Lammel C, et al. Genome sequence of an obligate intracellular pathogen of humans: *Chlamydia trachomatis*. Science 1998; 282:754–759.

62. Stephens RS (Ed.) Chlamydia. Intracellular Biology, Pathogenesis, and Immunity, Washington, DC: ASM Press, 1999, pp. 9–27.

63. Zhong G, Fan P, Ji H, Dong F, Huang Y. Identification of a chlamydial protease-like activity factor responsible for the degradation of host transcription factors. J Exp Med 2001; 193:935–942.

64. Fan P, Dong F, Huang Y, Zhong G. *Chlamydia pneumoniae* secretion of a protease-like activity factor for degrading host cell transcription factors is required for major histocompatibility complex antigen expression. Infect Immun 2002; 70:345–349.

65. Zhong G, Fan T, Liu L. *Chlamydia* inhibits interferon-γ inducible major histocombability complex class II expression by degradation of upstream stimulatory factor 1. J Exp Med 1999; 189: 1931–1938.

66. Shaw AC, Vandahl BB, Larsen MR, et al. Characterization of a secreted *Chlamydia* protease. Cell Microbiol 2002; 4:411–424.

67. Hoogland C, Baujard V, Sanchez JC, Hochstrasser DF, Appel RD. Make2ddb: a simple package to set up a two-dimensional electrophoresis database for the World Wide Web. Electrophoresis 1997; 18:2755–2758.

68. Eifert T, Büttner S. Proteome 2D-PAGE database, November 26, 2003. http://www.mpiib-berlin.mpg.de/2D-PAGE/.

69. Jungblut PR, Schaible UE, Mollenkopf HJ, et al. Comparative proteome analysis of *Mycobacterium tuberculosis* and *Mycobacterium bovis* BCG strains: towards functional genomics of microbial pathogens. Mol Microbiol 1999; 33:1103–1117.

70. Cash P. Proteomics in medical microbiology. Electrophoresis 2000; 21:1187–1201.

71. Jungblut PR, Bumann D, Haas G, et al. Comparative proteome analysis of *Helicobacter pylori*. Mol Microbiol 2000; 36:710–725.

72. Jungblut PR, Grabher G, Stoffler G. Comprehensive detection of immunorelevant *Borrelia garinii* antigens by two-dimensional electrophoresis. Electrophoresis 1999; 20:3611–3622.

73. Wasinger VC, Humphery-Smith I. Small genes/gene-products in *Escherichia coli* K-12. FEMS Microbiol Lett 1998; 169:375–382.

74. Shaw AC, Larsen MR, Roepstorff P, Christiansen G, Birkelund S. Identification and characterization of a novel *Chlamydia trachomatis* reticulate body protein. FEMS Microbiol Lett 2002; 212:193–202.

Index